Advances in Intelligent Systems and Computing

Volume 441

Series editor

Janusz Kacprzyk, Polish Academy of Sciences, Warsaw, Poland
e-mail: kacprzyk@ibspan.waw.pl

About this Series

The series "Advances in Intelligent Systems and Computing" contains publications on theory, applications, and design methods of Intelligent Systems and Intelligent Computing. Virtually all disciplines such as engineering, natural sciences, computer and information science, ICT, economics, business, e-commerce, environment, healthcare, life science are covered. The list of topics spans all the areas of modern intelligent systems and computing.

The publications within "Advances in Intelligent Systems and Computing" are primarily textbooks and proceedings of important conferences, symposia and congresses. They cover significant recent developments in the field, both of a foundational and applicable character. An important characteristic feature of the series is the short publication time and world-wide distribution. This permits a rapid and broad dissemination of research results.

Advisory Board

Chairman

Nikhil R. Pal, Indian Statistical Institute, Kolkata, India
e-mail: nikhil@isical.ac.in

Members

Rafael Bello, Universidad Central "Marta Abreu" de Las Villas, Santa Clara, Cuba
e-mail: rbellop@uclv.edu.cu

Emilio S. Corchado, University of Salamanca, Salamanca, Spain
e-mail: escorchado@usal.es

Hani Hagras, University of Essex, Colchester, UK
e-mail: hani@essex.ac.uk

László T. Kóczy, Széchenyi István University, Győr, Hungary
e-mail: koczy@sze.hu

Vladik Kreinovich, University of Texas at El Paso, El Paso, USA
e-mail: vladik@utep.edu

Chin-Teng Lin, National Chiao Tung University, Hsinchu, Taiwan
e-mail: ctlin@mail.nctu.edu.tw

Jie Lu, University of Technology, Sydney, Australia
e-mail: Jie.Lu@uts.edu.au

Patricia Melin, Tijuana Institute of Technology, Tijuana, Mexico
e-mail: epmelin@hafsamx.org

Nadia Nedjah, State University of Rio de Janeiro, Rio de Janeiro, Brazil
e-mail: nadia@eng.uerj.br

Ngoc Thanh Nguyen, Wroclaw University of Technology, Wroclaw, Poland
e-mail: Ngoc-Thanh.Nguyen@pwr.edu.pl

Jun Wang, The Chinese University of Hong Kong, Shatin, Hong Kong
e-mail: jwang@mae.cuhk.edu.hk

More information about this series at http://www.springer.com/series/11156

George A. Anastassiou · Oktay Duman

Editors

Intelligent Mathematics II: Applied Mathematics and Approximation Theory

Springer

Editors
George A. Anastassiou
Department of Mathematical Sciences
University of Memphis
Memphis, TN
USA

Oktay Duman
Department of Mathematics
TOBB University of Economics
  and Technology
Ankara
Turkey

ISSN 2194-5357   ISSN 2194-5365   (electronic)
Advances in Intelligent Systems and Computing
ISBN 978-3-319-30320-8   ISBN 978-3-319-30322-2   (eBook)
DOI 10.1007/978-3-319-30322-2

Library of Congress Control Number: 2016932751

Printed on acid-free paper

This Springer imprint is published by Springer Nature
The registered company is Springer International Publishing AG Switzerland

Dedicated to World Peace!

AMAT 2015 Conference, TOBB University of Economics and Technology, Ankara, Turkey, May 28–31, 2015

George A. Anastassiou and Oktay Duman Ankara, Turkey, May 29, 2015

Preface

Applied Mathematics and Approximation Theory (AMAT) is an international conference that brings together researchers from all areas of Applied Mathematics and Approximation Theory held every 3 or 4 years. This special volume is devoted to the proceedings having to do more with applied mathematics contents presented in the 3rd AMAT Conference held at TOBB Economics and Technology University, during May 28–31, 2015 in Ankara, Turkey.

We are particularly indebted to the Organizing Committee and the Scientific Committee for their great efforts. We also appreciate the plenary speakers: George A. Anastassiou (University of Memphis, USA), Martin Bohner (Missouri University of Science and Technology, USA), Alexander Goncharov (Bilkent University, Turkey), Varga Kalantarov (Koç University, Turkey), Gitta Kutyniok (Technische Universität Berlin, Germany), Choonkil Park (Hanyang University, South Korea), Mircea Sofonea (University of Perpignan, France), Tamaz Vashakmadze (Tbilisi State University, Georgia).

We would like also to thank the anonymous reviewers who helped us select the best articles for inclusion in this proceedings volume, and also the authors for their valuable contributions.

Finally, we are grateful to "TOBB University of Economics and Technology" for hosting this conference and providing all of its facilities, and also to "Central Bank of Turkey" for financial support.

November 2015
George A. Anastassiou
Oktay Duman

Contents

Contributors

Taraneh Abarin Department of Mathematics and Statistics, Memorial University, St John's, Canada

Melih Ağraz Middle East Technical University, Ankara, Turkey

Naima Aïssa USTHB, Laboratoire AMNEDP, Algiers, Algeria

Hande Günay Akdemir Department of Mathematics, Giresun University, Giresun, Turkey

Ahmet Altundag Istanbul Sabahattin Zaim University, Istanbul, Turkey

George A. Anastassiou Department of Mathematical Sciences, University of Memphis, Memphis, TN, USA

Mekki Aouachria University of Hadj lakhdar, Batna, Algeria

Sharefa Asiri Computer, Electrical and Mathematical Sciences and Engineering Division, King Abdullah University of Science and Technology (KAUST), Thuwal, Kingdom of Saudi Arabia

G.T. Bekova L.N. Gumilyov Eurasian National University, Astana, Kazakhstan

R. Benacer University of Hadj Lakhder Batna, Batna, Algeria

Hilal Benkhelil University of Hadj lakhdar, Batna, Algeria

Farouk Benoumelaz Department of Mathematics, University of Hadj Lakhdar, Batna, Algeria

Ernest Dankwa Department of Mathematics and Statistics, Memorial University, St John's, Canada

El Amir Djeffal Department of Mathematics, University of Hadj Lakhdar, Batna, Algeria

Lakhdar Djeffal Department of Mathematics, University of Hadj Lakhdar, Batna, Algeria

Oktay Duman Department of Mathematics, TOBB Economics and Technology University, Ankara, Turkey

Maryam Esmaeili Alzahra University, Tehran, Iran

Boutheina Gasmi University of Hadj Lakhder Batna, Batna, Algeria

Başak Gever TOBB University of Economics and Technology, Ankara, Turkey

Zulfiye Hanalioglu Karabuk University, Karabuk, Turkey

Iuliana F. Iatan Department of Mathematics and Computer Science, Technical University of Civil Engineering, Bucharest, Romania

Mahmood Jokar School of Mathematics, Iran University of Science and Technology, Tehran, Iran

Vandna Jowaheer University of Mauritius, Moka, Mauritius

Merve Kester Department of Mathematical Sciences, University of Memphis, Memphis, TN, USA

Naushad Mamode Khan University of Mauritius, Moka, Mauritius

Tahir Khaniyev TOBB University of Economics and Technology, Ankara, Turkey; Institute of Control Systems, Azerbaijan National Academy of Sciences, Baku, Azerbaijan

Waleed S. Khedr Erasmus-Mundus MEDASTAR Project, Universidad de Oviedo, Oviedo, Asturias, Spain

Taous-Meriem Laleg-Kirati Computer, Electrical and Mathematical Sciences and Engineering Division, King Abdullah University of Science and Technology (KAUST), Thuwal, Kingdom of Saudi Arabia

Jia Li Department of Mathematics and Statistics, Memorial University, St John's, Canada

Mohamed Maiza Laboratoire de Mathématiques Appliquées-Ecole Militaire Polytechnique, Algiers, Algeria

Zh.R. Myrzakulova L.N. Gumilyov Eurasian National University, Astana, Kazakhstan

Dominikus Noll Institut de Mathématiques de Toulouse, Université Paul Sabatier, Toulouse, France

Gulgassyl Nugmanova L.N. Gumilyov Eurasian National University, Astana, Kazakhstan

Vilda Purutçuoğlu Middle East Technical University, Ankara, Turkey

Amira Rachah Institut de Mathématiques de Toulouse, Université Paul Sabatier, Toulouse, France

Mohammed Said Radjef Laboratoire de Modélisation et Optimisation des Systémes, Université de Béjaia, Béjaia, Algeria

Jalil Rashidinia School of Mathematics, Iran University of Science and Technology, Tehran, Iran

Lakhdar Sais Centre de Recherche en Informatique de Lens CNRS-UMR8188, Université d'Artois, Lens, France

G.N. Shaikhova L.N. Gumilyov Eurasian National University, Astana, Kazakhstan

Kanat Shakenov Al-Farabi Kazakh National University, Almaty, Kazakhstan

Mircea Sofonea University of Perpignan Via Domitia, Perpignan, France

Yuvraj Sunecher University of Technology, Port Louis, Mauritius

H. Tsamda USTHB, Laboratoire AMNEDP, Algiers, Algeria

Jianzhong Wang Department of Mathematics and Statistics, Sam Houston State University, Huntsville, TX, USA

K.R. Yesmahanova L.N. Gumilyov Eurasian National University, Astana, Kazakhstan

Bivariate Left Fractional Polynomial Monotone Approximation

George A. Anastassiou

Abstract Let $f \in C^{r,p}\left([0,1]^2\right)$, $r, p \in \mathbb{N}$, and let L^* be a linear left fractional mixed partial differential operator such that $L^*(f) \geq 0$, for all (x, y) in a critical region of $[0, 1]^2$ that depends on L^*. Then there exists a sequence of two-dimensional polynomials $Q_{\overline{m_1},\overline{m_2}}(x, y)$ with $L^*\left(Q_{\overline{m_1},\overline{m_2}}(x, y)\right) \geq 0$ there, where $\overline{m_1}, \overline{m_2} \in \mathbb{N}$ such that $\overline{m_1} > r$, $\overline{m_2} > p$, so that f is approximated left fractionally simultaneously and uniformly by $Q_{\overline{m_1},\overline{m_2}}$ on $[0, 1]^2$. This restricted left fractional approximation is accomplished quantitatively by the use of a suitable integer partial derivatives two-dimensional first modulus of continuity.

1 Introduction

The topic of monotone approximation started in [5] has become a major trend in approximation theory. A typical problem in this subject is: given a positive integer k, approximate a given function whose kth derivative is ≥ 0 by polynomials having this property.

In [2] the authors replaced the kth derivative with a linear differential operator of order k. We mention this motivating result.

Theorem 1 *Let h, k, p be integers, $0 \leq h \leq k \leq p$ and let f be a real function, $f^{(p)}$ continuous in $[-1, 1]$ with modulus of continuity $\omega\left(f^{(p)}, x\right)$ there. Let $a_j(x)$, $j = h, h + 1, ..., k$ be real functions, defined and bounded on $[-1, 1]$ and assume $a_h(x)$ is either $\geq$ some number $\alpha > 0$ or $\leq$ some number $\beta < 0$ throughout $[-1, 1]$. Consider the operator*

$$L = \sum_{j=h}^{k} a_j(x) \left[\frac{d^j}{dx^j}\right]$$

G.A. Anastassiou (✉)
Department of Mathematical Sciences, University of Memphis,
Memphis, TN 38152, USA
e-mail: ganastss@memphis.edu

© Springer International Publishing Switzerland 2016

G.A. Anastassiou and O. Duman (eds.), *Intelligent Mathematics II:
Applied Mathematics and Approximation Theory*, Advances in Intelligent Systems
and Computing 441, DOI 10.1007/978-3-319-30322-2_1

and suppose, throughout $[-1, 1]$,

$$L(f) \geq 0. \tag{1}$$

Then, for every integer $n \geq 1$*, there is a real polynomial* $Q_n(x)$ *of degree* $\leq n$ *such that*

$$L(Q_n) \geq 0 \text{ throughout } [-1, 1]$$

and

$$\max_{-1 \leq x \leq 1} |f(x) - Q_n(x)| \leq Cn^{k-p} \omega \left(f^{(p)}, \frac{1}{n} \right),$$

where C *is independent of* n *or* f.

We need

Definition 2 (*see Stancu* [6]) Let $f \in C([0, 1]^2)$, $[0, 1]^2 = [0, 1] \times [0, 1]$, where $(x_1, y_1), (x_2, y_2) \in [0, 1]^2$ and $\delta_1, \delta_2 \geq 0$. The first modulus of continuity of f is defined as follows:

$$\omega_1(f, \delta_1, \delta_2) = \sup_{\substack{|x_1 - x_2| \leq \delta_1 \\ |y_1 - y_2| \leq \delta_2}} |f(x_1, y_1) - f(x_2, y_2)|.$$

Definition 3 Let f be a real-valued function defined on $[0, 1]^2$ and let m, n be two positive integers. Let $B_{m,n}$ be the Bernstein (polynomial) operator of order (m, n) given by

$$B_{m,n}(f; x, y) \tag{2}$$
$$= \sum_{i=0}^{m} \sum_{j=0}^{n} f\left(\frac{i}{m}, \frac{j}{n}\right) \cdot \binom{m}{i} \cdot \binom{n}{j} \cdot x^i \cdot (1-x)^{m-i} \cdot y^j \cdot (1-y)^{n-j}.$$

For integers $r, s \geq 0$, we denote by $f^{(r,s)}$ the differential operator of order (r, s), given by

$$f^{(r,s)}(x, y) = \frac{\partial^{r+s} f(x, y)}{\partial x^r \partial y^s}.$$

We use

Theorem 4 (Badea and Badea [3]). *It holds that*

$$\left\| f^{(k,l)} - \left(B_{m,n} f\right)^{(k,l)} \right\|_\infty \leq t(k, l) \cdot \omega_1 \left(f^{(k,l)}; \frac{1}{\sqrt{m-k}}, \frac{1}{\sqrt{n-l}} \right)$$
$$+ \max \left\{ \frac{k(k-1)}{m}, \frac{l(l-1)}{n} \right\} \cdot \left\| f^{(k,l)} \right\|_\infty, \tag{3}$$

where $m > k \geq 0$, $n > l \geq 0$ are integers, f is a real-valued function on $[0, 1]^2$ such that $f^{(k,l)}$ is continuous, and t is a positive real-valued function on $\mathbb{Z}_+ = \{0, 1, 2, ...\}$. Here $\|\cdot\|_\infty$ is the supremum norm on $[0, 1]^2$.

Denote $C^{r,p}\left([0, 1]^2\right) := \{f : [0, 1]^2 \to \mathbb{R}; f^{(k,l)}$ is continuous for $0 \leq k \leq r$, $0 \leq l \leq p\}$.

In [1] the author proved the following main motivational result.

Theorem 5 *Let h_1, h_2, v_1, v_2, r, p be integers, $0 \leq h_1 \leq v_1 \leq r$, $0 \leq h_2 \leq v_2 \leq p$ and let $f \in C^{r,p}\left([0, 1]^2\right)$. Let $\alpha_{i,j}(x, y)$, $i = h_1, h_1 + 1, ..., v_1$; $j = h_2, h_2 + 1, ..., v_2$ be real-valued functions, defined and bounded in $[0, 1]^2$ and assume $\alpha_{h_1 h_2}$ is either $\geq \alpha > 0$ or $\leq \beta < 0$ throughout $[0, 1]^2$. Consider the operator*

$$L = \sum_{i=h_1}^{v_1} \sum_{j=h_2}^{v_2} \alpha_{ij}(x, y) \frac{\partial^{i+j}}{\partial x^i \partial y^j} \tag{4}$$

and suppose that throughout $[0, 1]^2$,

$$L(f) \geq 0.$$

Then for integers m, n with $m > r$, $n > p$, there exists a polynomial $Q_{m,n}(x, y)$ of degree (m, n) such that $L\left(Q_{m,n}(x, y)\right) \geq 0$ throughout $[0, 1]^2$ and

$$\left\| f^{(k,l)} - Q_{m,n}^{(k,l)} \right\|_\infty \leq \frac{P_{m,n}(L, f)}{(h_1 - k)!(h_2 - l)!} + M_{m,n}^{k,l}(f), \tag{5}$$

all $(0, 0) \leq (k, l) \leq (h_1, h_2)$. Furthermore we get

$$\left\| f^{(k,l)} - Q_{m,n}^{(k,l)} \right\|_\infty \leq M_{m,n}^{k,l}(f), \tag{6}$$

for all $(h_1 + 1, h_2 + 1) \leq (k, l) \leq (r, p)$. Also (6) is true whenever $0 \leq k \leq h_1, h_2 + 1 \leq l \leq p$ or $h_1 + 1 \leq k \leq r, 0 \leq l \leq h_2$. Here

$$M_{m,n}^{k,l} \equiv M_{m,n}^{k,l}(f) \equiv t(k, l) \cdot \omega_1\left(f^{(k,l)}; \frac{1}{\sqrt{m-k}}, \frac{1}{\sqrt{n-l}}\right)$$
$$+ \max\left\{\frac{k(k-1)}{m}, \frac{l(l-1)}{n}\right\} \cdot \left\| f^{(k,l)} \right\|_\infty \tag{7}$$

and

$$P_{m,n} \equiv P_{m,n}(L, f) \equiv \sum_{i=h_1}^{v_1} \sum_{j=h_2}^{v_2} l_{ij} \cdot M_{m,n}^{i,j}, \tag{8}$$

where t is a positive real-valued function on $\mathbb{Z}_+^2$ and

$$l_{ij} \equiv \sup_{(x,y)\in[0,1]^2} \left| \alpha_{h_1 h_2}^{-1}(x,y) \cdot \alpha_{ij}(x,y) \right| < \infty. \tag{9}$$

In this article we extend Theorem 5 to the fractional level. Indeed here L is replaced by L^*, a linear left Caputo fractional mixed partial differential operator. Now the monotonicity property is only true on a critical region of $[0,1]^2$ that depends on L^* parameters. Simultaneous fractional convergence remains true on all of $[0,1]^2$.

We need

Definition 6 Let $\alpha_1, \alpha_2 > 0$; $\alpha = (\alpha_1, \alpha_2)$, $f \in C\left([0,1]^2\right)$ and let $x = (x_1, x_2)$, $t = (t_1, t_2) \in [0,1]^2$. We define the left mixed Riemann-Liouville fractional two dimensional integral of order α (see also [4]):

$$\left(I_{0+}^{\alpha} f\right)(x) \tag{10}$$

$$:= \frac{1}{\Gamma(\alpha_1)\,\Gamma(\alpha_2)} \int_0^{x_1} \int_0^{x_2} (x_1 - t_1)^{\alpha_1 - 1} (x_2 - t_2)^{\alpha_2 - 1} f(t_1, t_2)\, dt_1 dt_2,$$

with $x_1, x_2 > 0$.

Notice here $I_{0+}^{\alpha}(|f|) < \infty$.

Definition 7 Let $\alpha_1, \alpha_2 > 0$ with $\lceil \alpha_1 \rceil = m_1$, $\lceil \alpha_2 \rceil = m_2$, ($\lceil \cdot \rceil$ ceiling of the number). Let here $f \in C^{m_1, m_2}\left([0,1]^2\right)$. We consider the left (Caputo type) fractional partial derivative:

$$D_{*0}^{(\alpha_1, \alpha_2)} f(x)$$

$$:= \frac{1}{\Gamma(m_1 - \alpha_1)\,\Gamma(m_2 - \alpha_2)} \tag{11}$$

$$\cdot \int_0^{x_1} \int_0^{x_2} (x_1 - t_1)^{m_1 - \alpha_1 - 1} (x_2 - t_2)^{m_2 - \alpha_2 - 1} \frac{\partial^{m_1 + m_2} f(t_1, t_2)}{\partial t_1^{m_1} \partial t_2^{m_2}} dt_1 dt_2,$$

$\forall \, x = (x_1, x_2) \in [0,1]^2$, where Γ is the gamma function

$$\Gamma(\nu) = \int_0^{\infty} e^{-t} t^{\nu - 1} dt, \quad \nu > 0. \tag{12}$$

We set

$$D_{*0}^{(0,0)} f(x) := f(x), \quad \forall \, x \in [0,1]^2; \tag{13}$$

$$D_{*0}^{(m_1, m_2)} f(x) := \frac{\partial^{m_1 + m_2} f(x)}{\partial x_1^{m_1} \partial x_2^{m_2}}, \quad \forall \, x \subset [0,1]^2. \tag{14}$$

Definition 8 We also set

$$D_{*0}^{(0,\alpha_2)} f(x) := \frac{1}{\Gamma(m_2-\alpha_2)} \int_0^{x_2} (x_2-t_2)^{m_2-\alpha_2-1} \frac{\partial^{m_2} f(x_1,t_2)}{\partial t_2^{m_2}} dt_2, \tag{15}$$

$$D_{*0}^{(\alpha_1,0)} f(x) := \frac{1}{\Gamma(m_1-\alpha_1)} \int_0^{x_1} (x_1-t_1)^{m_1-\alpha_1-1} \frac{\partial^{m_1} f(t_1,x_2)}{\partial t_1^{m_1}} dt_1, \tag{16}$$

and

$$D_{*0}^{(m,\alpha_2)} f(x) := \frac{1}{\Gamma(m_2-\alpha_2)} \int_0^{x_2} (x_2-t_2)^{m_2-\alpha_2-1} \frac{\partial^{m_1+m_2} f(x_1,t_2)}{\partial x_1^{m_1} \partial t_2^{m_2}} dt_2, \tag{17}$$

$$D_{*0}^{(\alpha_1,m_2)} f(x) := \frac{1}{\Gamma(m_1-\alpha_1)} \int_0^{x_1} (x_1-t_1)^{m_1-\alpha_1-1} \frac{\partial^{m_1+m_2} f(t_1,x_2)}{\partial t_1^{m_1} \partial x_2^{m_2}} dt_1. \tag{18}$$

2 Main Result

We present our main result

Theorem 9 *Let h_1, h_2, v_1, v_2, r, p be integers, $0 \leq h_1 \leq v_1 \leq r$, $0 \leq h_2 \leq v_2 \leq p$ and let $f \in C^{r,p}([0,1]^2)$. Let $\alpha_{ij}(x,y)$, $i = h_1, h_1+1, ..., v_1$; $j = h_2, h_2+1, ..., v_2$ be real valued functions, defined and bounded in $[0,1]^2$ and assume $\alpha_{h_1 h_2}$ is either $\geq \alpha > 0$ or $\leq \beta < 0$ throughout $[0,1]^2$. Let*

$$0 \leq \alpha_{1h_1} \leq h_1 \leq \alpha_{11} \leq h_1+1 < \alpha_{12} \leq h_1+2$$
$$< \alpha_{13} \leq h_1+3 < ... < \alpha_{1v_1} \leq v_1 < ... < \alpha_{1r} \leq r,$$

with $\lceil \alpha_{1h_1} \rceil = h_1$;

$$0 \leq \alpha_{2h_2} \leq h_2 < \alpha_{21} \leq h_2+1 < \alpha_{22} \leq h_2+2$$
$$< \alpha_{23} \leq h_2+3 < ... < \alpha_{2v_2} \leq v_2 < ... < \alpha_{2p} \leq p,$$

with $\lceil \alpha_{2h_2} \rceil = h_2$. Consider the left fractional differential bivariate operator

$$L^* := \sum_{i=h_1}^{v_1} \sum_{j=h_2}^{v_2} \alpha_{ij}(x,y) D_{*0}^{(\alpha_{1i},\alpha_{2j})}. \tag{19}$$

Let integers $\overline{m_1}, \overline{m_2}$ with $\overline{m_1} > r$, $\overline{m_2} > p$. Set

$$l_{ij} := \sup_{(x,y)\in[0,1]^2} \left| \alpha_{h_1 h_2}^{-1}(x,y) \cdot \alpha_{ij}(x,y) \right| < \infty.$$

Also set ($\lceil \alpha_{1i} \rceil = i$, $\lceil \alpha_{2j} \rceil = j$, $\lceil \cdot \rceil$ ceiling of number)

$$
\begin{aligned}
M_{\overline{m_1},\overline{m_2}}^{i,j} := M_{\overline{m_1},\overline{m_2}}^{i,j}(f) := & \frac{1}{\Gamma(i - \alpha_{1i} + 1)\,\Gamma(j - \alpha_{2j} + 1)} \\
& \cdot \left\{ t(i,j)\,\omega_1 \left(f^{(i,j)}; \frac{1}{\sqrt{\overline{m_1} - i}}, \frac{1}{\sqrt{\overline{m_2} - j}} \right) \right. \\
& \left. + \max \left\{ \frac{i(i-1)}{\overline{m_1}}, \frac{j(j-1)}{\overline{m_2}} \right\} \cdot \left\| f^{(i,j)} \right\|_\infty \right\},
\end{aligned}
\tag{20}
$$

$i = h_1, ..., v_1$; $j = h_2, ..., v_2$. *Here t is a positive real-valued function on $\mathbb{Z}_+^2$, $\|\cdot\|_\infty$ is the supremum norm on $[0,1]^2$. Call*

$$P_{\overline{m_1},\overline{m_2}} := P_{\overline{m_1},\overline{m_2}}(f) = \sum_{i=h_1}^{v_1} \sum_{j=h_2}^{v_2} l_{ij} \cdot M_{\overline{m_1},\overline{m_2}}^{i,j}. \tag{21}$$

Then there exists a polynomial $Q_{\overline{m_1},\overline{m_2}}(x,y)$ of degree $(\overline{m_1}, \overline{m_2})$ on $[0,1]^2$ such that

$$
\begin{aligned}
& \left\| D_{*0}^{(\alpha_{1k},\alpha_{2l})}(f) - D_{*0}^{(\alpha_{1k},\alpha_{2l})}\left(Q_{\overline{m_1},\overline{m_2}}\right) \right\|_\infty \\
& \leq \frac{\Gamma(h_1 - k + 1)\,\Gamma(h_2 - l + 1)\,P_{\overline{m_1},\overline{m_2}}}{\Gamma(h_1 - \alpha_{1k} + 1)\,\Gamma(h_2 - \alpha_{2l} + 1)\,(h_1 - k)!\,(h_2 - l)!} \\
& \quad + M_{\overline{m_1},\overline{m_2}}^{k,l},
\end{aligned}
\tag{22}
$$

for $(0,0) \leq (k,l) \leq (h_1, h_2)$. If $(h_1 + 1, h_2 + 1) \leq (k,l) \leq (r,p)$, or $0 \leq k \leq h_1$, $h_2 + 1 \leq l \leq p$, or $h_1 + 1 \leq k \leq r$, $0 \leq l \leq h_2$, we get

$$\left\| D_{*0}^{(\alpha_{1k},\alpha_{2l})}(f) - D_{*0}^{(\alpha_{1k},\alpha_{2l})}\left(Q_{\overline{m_1},\overline{m_2}}\right) \right\|_\infty \leq M_{\overline{m_1},\overline{m_2}}^{k,l}. \tag{23}$$

By assuming $L^*(f(1,1)) \geq 0$, we get $L^*\left(Q_{\overline{m_1},\overline{m_2}}(1,1)\right) \geq 0$. Let $1 > x, y > 0$, with $\alpha_{1h_1} \neq h_1$ and $\alpha_{2h_2} \neq h_2$, such that

$$x \geq \left(\Gamma(h_1 - \alpha_{1h_1} + 1) \right)^{\frac{1}{(h_1 - \alpha_{1h_1})}}, \tag{24}$$

$$y \geq \left(\Gamma(h_2 - \alpha_{2h_2} + 1) \right)^{\frac{1}{(h_2 - \alpha_{2h_2})}},$$

and

$$L^* \left(f\left(x,y\right) \right) \geq 0.$$

Then

$$L^* \left(Q_{\overline{m_1},\overline{m_2}} \left(x,y\right) \right) \geq 0.$$

To prove Theorem 9 it takes some preparation. We need

Definition 10 Let f be a real-valued function defined on $[0,1]^2$ and let $\overline{m_1}, \overline{m_2} \in \mathbb{N}$. Let $B_{\overline{m_1},\overline{m_2}}$ be the Bernstein (polynomial) operator of order $(\overline{m_1}, \overline{m_2})$ given by

$$
\begin{aligned}
& B_{\overline{m_1},\overline{m_2}} \left(f; x_1, x_2 \right) \\
& := \sum_{i_1=0}^{\overline{m_1}} \sum_{i_2=0}^{\overline{m_2}} f\left(\frac{i_1}{\overline{m_1}}, \frac{i_2}{\overline{m_2}} \right) \\
& \cdot \binom{\overline{m_1}}{i_1} \binom{\overline{m_2}}{i_2} x_1^{i_1} \left(1 - x_1 \right)^{\overline{m_1}-i_1} x_2^{i_2} \left(1 - x_2 \right)^{\overline{m_2}-i_2}.
\end{aligned}
\tag{25}
$$

We need the following simultaneous approximation result.

Theorem 11 (Badea and Badea [3]). *It holds that*

$$
\begin{aligned}
& \left\| f^{(k_1,k_2)} - \left(B_{\overline{m_1},\overline{m_2}} f \right)^{(k_1,k_2)} \right\|_\infty \\
& \leq t\left(k_1, k_2 \right) \omega_1 \left(f^{(k_1,k_2)}; \frac{1}{\sqrt{\overline{m_1} - k_1}}, \frac{1}{\sqrt{\overline{m_2} - k_2}} \right) \\
& + \max \left\{ \frac{k_1\left(k_1 - 1 \right)}{\overline{m_1}}, \frac{k_2\left(k_2 - 1 \right)}{\overline{m_2}} \right\} \cdot \left\| f^{(k_1,k_2)} \right\|_\infty,
\end{aligned}
\tag{26}
$$

where $\overline{m_1} > k_1 \geq 0, \overline{m_2} > k_2 \geq 0$ *are integers,* f *is a real-valued function on* $[0,1]^2$, *such that* $f^{(k_1,k_2)}$ *is continuous, and* t *is a positive real-valued function on* $\mathbb{Z}_+^2$. *Here* $\|\cdot\|_\infty$ *is the supremum norm on* $[0,1]^2$.

Remark 12 We assume that $\overline{m_1} > m_1 = \lceil \alpha_1 \rceil, \overline{m_2} > m_2 = \lceil \alpha_2 \rceil$, where $\alpha_1, \alpha_2 > 0$. We consider also

$$
\begin{aligned}
& D_{*0}^{(\alpha_1,\alpha_2)} \left(B_{\overline{m_1},\overline{m_2}} f \right) \left(x_1, x_2 \right) \\
& = \frac{1}{\Gamma\left(m_1 - \alpha_1 \right) \Gamma\left(m_2 - \alpha_2 \right)} \\
& \cdot \int_0^{x_1} \int_0^{x_2} \left(x_1 - t_1 \right)^{m_1 - \alpha_1 - 1} \left(x_2 - t_2 \right)^{m_2 - \alpha_2 - 1} \frac{\partial^{m_1+m_2} \left(B_{\overline{m_1},\overline{m_2}} f \right)(t_1,t_2)}{\partial t_1^{m_1} \partial t_2^{m_2}} dt_1 dt_2,
\end{aligned}
\tag{27}
$$

$\forall \left(x_1, x_2 \right) \in [0,1]^2$.

Proposition 13 *Let $\alpha_1, \alpha_2 > 0$ with $\lceil \alpha_1 \rceil = m_1$, $\lceil \alpha_2 \rceil = m_2$, $f \in C^{m_1,m_2}\left([0,1]^2\right)$, where $\overline{m_1}, \overline{m_2} \in \mathbb{N} : \overline{m_1} > m_1, \overline{m_2} > m_2$. Then*

$$\left\| D_{*0}^{(\alpha_1,\alpha_2)} f - D_{*0}^{(\alpha_1,\alpha_2)}\left(B_{\overline{m_1},\overline{m_2}} f\right) \right\|_\infty$$

$$\leq \frac{1}{\Gamma(m_1 - \alpha_1 + 1)\,\Gamma(m_2 - \alpha_2 + 1)}$$

$$\cdot \left\{ t(m_1, m_2)\,\omega_1\left(f^{(m_1,m_2)}; \frac{1}{\sqrt{\overline{m_1} - m_1}}, \frac{1}{\sqrt{\overline{m_2} - m_2}}\right)\right.$$

$$\left. + \max\left\{ \frac{m_1(m_1-1)}{\overline{m_1}}, \frac{m_2(m_2-1)}{\overline{m_2}}\right\} \cdot \left\| f^{(m_1,m_2)}\right\|_\infty\right\}, \tag{28}$$

Proof We observe the following

$$\left| D_{*0}^{(\alpha_1,\alpha_2)} f(x_1,x_2) - D_{*0}^{(\alpha_1,\alpha_2)}\left(B_{\overline{m_1},\overline{m_2}} f\right)(x_1,x_2)\right|$$

$$= \frac{1}{\Gamma(m_1-\alpha_1)\,\Gamma(m_2-\alpha_2)} \tag{29}$$

$$\cdot \left| \int_0^{x_1}\int_0^{x_2} (x_1-t_1)^{m_1-\alpha_1-1}(x_2-t_2)^{m_2-\alpha_2-1} \right.$$

$$\left. \cdot \left(\frac{\partial^{m_1+m_2} f(t_1,t_2)}{\partial t_1^{m_1}\partial t_2^{m_2}} - \frac{\partial^{m_1+m_2}\left(B_{\overline{m_1},\overline{m_2}} f\right)(t_1,t_2)}{\partial t_1^{m_1}\partial t_2^{m_2}}\right) dt_1 dt_2\right|$$

$$\leq \frac{1}{\Gamma(m_1-\alpha_1)\,\Gamma(m_2-\alpha_2)} \tag{30}$$

$$\int_0^{x_1}\int_0^{x_2} (x_1-t_1)^{m_1-\alpha_1-1}(x_2-t_2)^{m_2-\alpha_2-1}$$

$$\cdot \left| \frac{\partial^{m_1+m_2} f(t_1,t_2)}{\partial t_1^{m_1}\partial t_2^{m_2}} - \frac{\partial^{m_1+m_2}\left(B_{\overline{m_1},\overline{m_2}} f\right)(t_1,t_2)}{\partial t_1^{m_1}\partial t_2^{m_2}}\right| dt_1 dt_2$$

$$\overset{(26)}{\leq} \left\{ t(m_1, m_2)\,\omega_1\left(f^{(m_1,m_2)}; \frac{1}{\sqrt{\overline{m_1} - m_1}}, \frac{1}{\sqrt{\overline{m_2} - m_2}}\right)\right. \tag{31}$$

$$\left. + \max\left\{ \frac{m_1(m_1-1)}{\overline{m_1}}, \frac{m_2(m_2-1)}{\overline{m_2}}\right\} \cdot \left\| f^{(m_1,m_2)}\right\|_\infty\right\}$$

$$\cdot \frac{1}{\Gamma(m_1-\alpha_1)\,\Gamma(m_2-\alpha_2)}$$

$$\cdot \int_0^{x_1}\int_0^{x_2} (x_1-t_1)^{m_1-\alpha_1-1}(x_2-t_2)^{m_2-\alpha_2-1}\, dt_1 dt_2$$

$$= \frac{x_1^{m_1 - \alpha_1} x_2^{m_2 - \alpha_2}}{\Gamma(m_1 - \alpha_1 + 1) \Gamma(m_2 - \alpha_2 + 1)} \tag{32}$$

$$\cdot \left\{ t(m_1, m_2) \omega_1 \left(f^{(m_1, m_2)}; \frac{1}{\sqrt{\overline{m_1} - m_1}}, \frac{1}{\sqrt{\overline{m_2} - m_2}} \right) \right.$$

$$\left. + \max \left\{ \frac{m_1(m_1 - 1)}{\overline{m_1}}, \frac{m_2(m_2 - 1)}{\overline{m_2}} \right\} \left\| f^{(m_1, m_2)} \right\|_\infty \right\},$$

$\forall (x_1, x_2) \in [0, 1]^2$.

Proof (*Proof of Theorem* 9) Here we use a lot Proposition 13.

Case (i) Assume that throughout $[0, 1]^2$, $\alpha_{h_1 h_2} \geq \alpha > 0$.

Call

$$Q_{\overline{m_1}, \overline{m_2}}(x, y) := B_{\overline{m_1}, \overline{m_2}}(f; x, y) + P_{\overline{m_1}, \overline{m_2}} \frac{x^{h_1} y^{h_2}}{h_1! \, h_2!}. \tag{33}$$

Then by (28) we get

$$\left\| D_{*0}^{(\alpha_{1k}, \alpha_{2l})} \left(f + P_{\overline{m_1}, \overline{m_2}} \frac{x^{h_1} y^{h_2}}{h_1! \, h_2!} \right) - D_{*0}^{(\alpha_{1k}, \alpha_{2l})} \left(Q_{\overline{m_1}, \overline{m_2}} \right) \right\|_\infty \leq M_{\overline{m_1}, \overline{m_2}}^{k, l}, \tag{34}$$

all $0 \leq k \leq r, 0 \leq l \leq p$. When $(0, 0) \leq (k, l) \leq (h_1, h_2)$, inequality (34) becomes

$$\left\| D_{*0}^{(\alpha_{1k}, \alpha_{2l})}(f) + P_{\overline{m_1}, \overline{m_2}} \frac{\Gamma(h_1 - k + 1)\Gamma(h_2 - l + 1)x^{h_1 - \alpha_{1k}} y^{h_2 - \alpha_{2l}}}{\Gamma(h_1 - \alpha_{1k} + 1)\Gamma(h_2 - \alpha_{2l} + 1)(h_1 - k)!(h_2 - l)!} \right.$$
$$\left. - D_{*0}^{(\alpha_{1k}, \alpha_{2l})} \left(Q_{\overline{m_1}, \overline{m_2}} \right) \right\|_\infty \leq M_{\overline{m_1}, \overline{m_2}}^{k, l}. \tag{35}$$

We prove (35) as follows: In (34) we need to calculate $((0, 0) \leq (k, l) \leq (h_1, h_2))$

$$D_{*0}^{(\alpha_{1k}, \alpha_{2l})} \left(\frac{x^{h_1} y^{h_2}}{h_1! \, h_2!} \right)$$

$$= \frac{1}{\Gamma(k - \alpha_{1k}) \Gamma(l - \alpha_{2l})} \tag{36}$$

$$\cdot \int_0^x \int_0^y (x - t_1)^{k - \alpha_{1k} - 1} (y - t_2)^{l - \alpha_{2l} - 1} \frac{t_1^{h_1 - k}}{(h_1 - k)!} \frac{t_2^{h_2 - l}}{(h_2 - l)!} dt_1 dt_2$$

$$= \left(\frac{1}{\Gamma(k - \alpha_{1k})} \int_0^x (x - t_1)^{k - \alpha_{1k} - 1} \frac{t_1^{h_1 - k}}{(h_1 - k)!} dt_1 \right) \tag{37}$$

$$\cdot \left(\frac{1}{\Gamma(l - \alpha_{2l})} \int_0^y (y - t_2)^{l - \alpha_{2l} - 1} \frac{t_2^{h_2 - l}}{(h_2 - l)!} dt_2 \right)$$

$$= \frac{1}{(h_1 - k)!\Gamma(k - \alpha_{1k})} \int_0^x (x - t_1)^{k - \alpha_{1k} - 1} (t_1 - 0)^{(h_1 - k + 1) - 1} dt_1$$

$$\cdot \frac{1}{(h_2 - l)!\Gamma(l - \alpha_{2l})} \int_0^y (y - t_2)^{l - \alpha_{2l} - 1} (t_2 - 0)^{(h_2 - l + 1) - 1} dt_2$$

$$= \frac{1}{(h_1 - k)!\Gamma(k - \alpha_{1k})} \frac{\Gamma(k - \alpha_{1k})\,\Gamma(h_1 - k + 1)}{\Gamma(h_1 - \alpha_{1k} + 1)} x^{h_1 - \alpha_{1k}}$$
$$\cdot \frac{1}{(h_2 - l)!\Gamma(l - \alpha_{2l})} \frac{\Gamma(l - \alpha_{2l})\,\Gamma(h_2 - l + 1)}{\Gamma(h_2 - \alpha_{2l} + 1)} y^{h_2 - \alpha_{2l}} \tag{38}$$

$$= \left(\frac{\Gamma(h_1 - k + 1)}{(h_1 - k)!\Gamma(h_1 - \alpha_{1k} + 1)} \right) \tag{39}$$
$$\cdot \left(\frac{\Gamma(h_2 - l + 1)}{(h_2 - l)!\Gamma(h_2 - \alpha_{2l} + 1)} \right) x^{h_1 - \alpha_{1k}} y^{h_2 - \alpha_{2l}}.$$

So when $(0, 0) \le (k, l) \le (h_1, h_2)$ we get

$$D_{*0}^{(\alpha_{1k}, \alpha_{2l})} \left(\frac{x^{h_1}\,y^{h_2}}{h_1!\,h_2!} \right) \tag{40}$$
$$= \frac{\Gamma(h_1 - k + 1)\,\Gamma(h_2 - l + 1)\,x^{h_1 - \alpha_{1k}}\,y^{h_2 - \alpha_{2l}}}{\Gamma(h_1 - \alpha_{1k} + 1)\,\Gamma(h_2 - \alpha_{2l} + 1)\,(h_1 - k)!\,(h_2 - l)!}.$$

Hence we plug in (40) into (34) to get (35). Using (35) and triangle inequality we obtain $((0, 0) \le (k, l) \le (h_1, h_2))$ that

$$\left\| D_{*0}^{(\alpha_{1k}, \alpha_{2l})}(f) - D_{*0}^{(\alpha_{1k}, \alpha_{2l})} \left(Q_{\overline{m_1, m_2}} \right) \right\|_{\infty}$$
$$\le \frac{\Gamma(h_1 - k + 1)\,\Gamma(h_2 - l + 1)\,P_{\overline{m_1, m_2}}}{\Gamma(h_1 - \alpha_{1k} + 1)\,\Gamma(h_2 - \alpha_{2l} + 1)\,(h_1 - k)!\,(h_2 - l)!} \tag{41}$$
$$+ M_{\overline{m_1, m_2}}^{k, l},$$

proving (22). Next if $(h_1 + 1, h_2 + 1) \le (k, l) \le (r, p)$, or $0 \le k \le h_1, h_2 + 1 \le l \le p$, or $h_1 + 1 \le k \le r, 0 \le l \le h_2$, we get by (34) that

$$\left\| D_{*0}^{(\alpha_{1k}, \alpha_{2l})}(f) - D_{*0}^{(\alpha_{1k}, \alpha_{2l})} \left(Q_{\overline{m_1, m_2}} \right) \right\|_{\infty} \le M_{\overline{m_1, m_2}}^{k, l}, \tag{42}$$

proving (23).

Furthermore, if (x, y) in critical region, see (24), we get

$$\alpha_{h_1 h_2}^{-1}(x, y) L^* \left(Q_{\overline{m_1, m_2}}(x, y) \right)$$

$$= \alpha_{h_1 h_2}^{-1}(x, y) L^* \left(f(x, y) \right) \tag{43}$$

$$+ P_{\overline{m_1, m_2}} \frac{x^{h_1 - \alpha_{1h_1}} y^{h_2 - \alpha_{2h_2}}}{\Gamma\left(h_1 - \alpha_{1h_1} + 1\right) \Gamma\left(h_2 - \alpha_{2h_2} + 1\right)}$$

$$+ \sum_{i=h_1}^{v_1} \sum_{j=h_2}^{v_2} \alpha_{h_1 h_2}^{-1}(x, y) \alpha_{ij}(x, y)$$

$$\cdot D_{*0}^{\left(\alpha_{1i}, \alpha_{2j}\right)} \left[Q_{\overline{m_1, m_2}}(x, y) - f(x, y) - P_{\overline{m_1, m_2}} \frac{x^{h_1} y^{h_2}}{h_1! \, h_2!} \right]$$

$$\overset{(34)}{\geq} P_{\overline{m_1, m_2}} \frac{x^{h_1 - \alpha_{1h_1}} y^{h_2 - \alpha_{2h_2}}}{\Gamma\left(h_1 - \alpha_{1h_1} + 1\right) \Gamma\left(h_2 - \alpha_{2h_2} + 1\right)}$$

$$- \sum_{i=h_1}^{v_1} \sum_{j=h_2}^{v_2} l_{ij} M_{\overline{m_1, m_2}}^{i, j}$$

$$= P_{\overline{m_1, m_2}} \frac{x^{h_1 - \alpha_{1h_1}} y^{h_2 - \alpha_{2h_2}}}{\Gamma\left(h_1 - \alpha_{1h_1} + 1\right) \Gamma\left(h_2 - \alpha_{2h_2} + 1\right)} - P_{\overline{m_1, m_2}} \tag{44}$$

$$= P_{\overline{m_1, m_2}} \left[\frac{x^{h_1 - \alpha_{1h_1}} y^{h_2 - \alpha_{2h_2}}}{\Gamma\left(h_1 - \alpha_{1h_1} + 1\right) \Gamma\left(h_2 - \alpha_{2h_2} + 1\right)} - 1 \right]$$

$$= P_{\overline{m_1, m_2}} \left[\frac{x^{h_1 - \alpha_{1h_1}} y^{h_2 - \alpha_{2h_2}} - \Gamma\left(h_1 - \alpha_{1h_1} + 1\right) \Gamma\left(h_2 - \alpha_{2h_2} + 1\right)}{\Gamma\left(h_1 - \alpha_{1h_1} + 1\right) \Gamma\left(h_2 - \alpha_{2h_2} + 1\right)} \right] \tag{45}$$

$$=: (*) .$$

We know $\Gamma(1) = 1$, $\Gamma(2) = 1$, and Γ is convex and positive on $(0, \infty)$. Here $0 \leq h_1 - \alpha_{1h_1} < 1$ and $0 \leq h_2 - \alpha_{2h_2} < 1$, hence $1 \leq h_1 - \alpha_{1h_1} + 1 < 2$, $1 \leq h_2 - \alpha_{2h_2} + 1 < 2$. Thus $0 < \Gamma\left(h_1 - \alpha_{1h_1} + 1\right)$, $\Gamma\left(h_2 - \alpha_{2h_2} + 1\right) \leq 1$, and $1 - \Gamma\left(h_1 - \alpha_{1h_1} + 1\right) \Gamma\left(h_2 - \alpha_{2h_2} + 1\right) \geq 0$. Clearly acting as in (43)–(45), when $L^* (f(1, 1)) \geq 0$, we get $L^* \left(Q_{\overline{m_1, m_2}}(1, 1) \right) \geq 0$.

Based on the above comments about Gamma function we get $(*) \geq 0$. That is $L^* \left(Q_{\overline{m_1, m_2}}(x, y) \right) \geq 0$, over the critical region of (24).

Case (ii) Assume that throughout $[0, 1]^2$, $\alpha_{h_1 h_2} \leq \beta < 0$. Consider

$$Q_{\overline{m_1, m_2}}^-(x, y) := B_{\overline{m_1, m_2}}(f; x, y) - P_{\overline{m_1, m_2}} \frac{x^{h_1} y^{h_2}}{h_1! \, h_2!} .$$

Then by (28) we get

$$\left\| D_{*0}^{(\alpha_{1k},\alpha_{2l})} \left(f - P_{\overline{m_1},\overline{m_2}} \frac{x^{h_1}}{h_1!} \frac{y^{h_2}}{h_2!} \right) - D_{*0}^{(\alpha_{1k},\alpha_{2l})} \left(Q_{\overline{m_1},\overline{m_2}}^{-} (x,y) \right) \right\|_{\infty}$$
$$\leq M_{\overline{m_1},\overline{m_2}}^{k,l}, \tag{46}$$

all $0 \leq k \leq r, 0 \leq l \leq p$. When $(0,0) \leq (k,l) \leq (h_1,h_2)$ (46) becomes

$$\left\| D_{*0}^{(\alpha_{1k},\alpha_{2l})} (f) - P_{\overline{m_1},\overline{m_2}} \frac{\Gamma(h_1-k+1)\Gamma(h_2-l+1)x^{h_1-\alpha_{1k}} y^{h_2-\alpha_{2l}}}{\Gamma(h_1-\alpha_{1k}+1)\Gamma(h_2-\alpha_{2l}+1)(h_1-k)!(h_2-l)!} \right.$$
$$\left. - D_{*0}^{(\alpha_{1k},\alpha_{2l})} \left(Q_{\overline{m_1},\overline{m_2}}^{-} \right) \right\|_{\infty} \leq M_{\overline{m_1},\overline{m_2}}^{k,l}. \tag{47}$$

Using (47) and triangle inequality we obtain for $(0,0) \leq (k,l) \leq (h_1,h_2)$ that

$$\left\| D_{*0}^{(\alpha_{1k},\alpha_{2l})} (f) - D_{*0}^{(\alpha_{1k},\alpha_{2l})} \left(Q_{\overline{m_1},\overline{m_2}}^{-} \right) \right\|_{\infty}$$
$$\leq \frac{\Gamma(h_1-k+1)\Gamma(h_2-l+1) P_{\overline{m_1},\overline{m_2}}}{\Gamma(h_1-\alpha_{1k}+1)\Gamma(h_2-\alpha_{2l}+1)(h_1-k)!(h_2-l)!} \tag{48}$$
$$+ M_{\overline{m_1},\overline{m_2}}^{k,l}.$$

Next if $(h_1+1, h_2+1) \leq (k,l) \leq (r,p)$, or $0 \leq k \leq h_1, h_2+1 \leq l \leq p$, or $h_1 + 1 \leq k \leq r, 0 \leq l \leq h_2$, we get by (46) that

$$\left\| D_{*0}^{(\alpha_{1k},\alpha_{2l})} (f) - D_{*0}^{(\alpha_{1k},\alpha_{2l})} \left(Q_{\overline{m_1},\overline{m_2}}^{-} \right) \right\|_{\infty} \leq M_{\overline{m_1},\overline{m_2}}^{k,l}. \tag{49}$$

We proved again (22) and (23). Furthermore, if (x,y) in critical region, see (24), we obtain

$$\alpha_{h_1 h_2}^{-1} (x,y) L^* \left(Q_{\overline{m_1},\overline{m_2}}^{-} (x,y) \right)$$
$$= \alpha_{h_1 h_2}^{-1} (x,y) L^* (f(x,y)) \tag{50}$$
$$- P_{\overline{m_1},\overline{m_2}} \frac{x^{h_1-\alpha_{1h_1}} y^{h_2-\alpha_{2h_2}}}{\Gamma \left(h_1 - \alpha_{1h_1} + 1 \right) \Gamma \left(h_2 - \alpha_{2h_2} + 1 \right)}$$
$$+ \sum_{i=h_1}^{v_1} \sum_{j=h_2}^{v_2} \alpha_{h_1 h_2}^{-1} (x,y) \alpha_{ij} (x,y)$$
$$\cdot D_{*0}^{(\alpha_{1i},\alpha_{2j})} \left[Q_{\overline{m_1},\overline{m_2}}^{-} (x,y) - f(x,y) + P_{\overline{m_1},\overline{m_2}} \frac{x^{h_1}}{h_1!} \frac{y^{h_2}}{h_2!} \right]$$

$$\overset{(46)}{\leq} - P_{\overline{m_1},\overline{m_2}} \frac{x^{h_1-\alpha_{1h_1}} y^{h_2-\alpha_{2h_2}}}{\Gamma \left(h_1 - \alpha_{1h_1} + 1 \right) \Gamma \left(h_2 - \alpha_{2h_2} + 1 \right)}$$

$$+ \sum_{i=h_1}^{v_1} \sum_{j=h_2}^{v_2} l_{ij} M_{\overline{m_1},\overline{m_2}}^{i,j} \tag{51}$$

$$= P_{\overline{m_1},\overline{m_2}} \left(1 - \frac{x^{h_1-\alpha_{1h_1}} y^{h_2-\alpha_{2h_2}}}{\Gamma\left(h_1 - \alpha_{1h_1} + 1\right) \Gamma\left(h_2 - \alpha_{2h_2} + 1\right)} \right), \tag{52}$$

$$= P_{\overline{m_1},\overline{m_2}} \left(\frac{\Gamma\left(h_1-\alpha_{1h_1}+1\right)\Gamma\left(h_2-\alpha_{2h_2}+1\right) - x^{h_1-\alpha_{1h_1}} y^{h_2-\alpha_{2h_2}}}{\Gamma\left(h_1-\alpha_{1h_1}+1\right)\Gamma\left(h_2-\alpha_{2h_2}+1\right)} \right) \tag{53}$$

$$=: (**).$$

We know $\Gamma(1) = 1$, $\Gamma(2) = 1$, and Γ is convex and positive on $(0, \infty)$. Here $0 \le h_1 - \alpha_{1h_1} < 1$ and $0 \le h_2 - \alpha_{2h_2} < 1$, hence $1 \le h_1 - \alpha_{1h_1} + 1 < 2$, $1 \le h_2 - \alpha_{2h_2} + 1 < 2$. Thus $0 < \Gamma\left(h_1 - \alpha_{1h_1} + 1\right)$, $\Gamma\left(h_2 - \alpha_{2h_2} + 1\right) \le 1$, and $1 - \Gamma\left(h_1 - \alpha_{1h_1} + 1\right) \Gamma\left(h_2 - \alpha_{2h_2} + 1\right) \ge 0$. Clearly acting as in (50)–(53), when $L^*(f(1,1)) \ge 0$, we get $L^*\left(Q_{\overline{m_1},\overline{m_2}}^-(1,1)\right) \ge 0$.

Based on the above comments about Gamma function we get $(**) \le 0$. That is $L^*\left(Q_{\overline{m_1},\overline{m_2}}^-(x,y)\right) \ge 0$, over the critical region of (24).

References

1. Anastassiou, G.A.: Bivariate monotone approximation. Proc. Amer. Math. Soc. **112**(4), 959–963 (1991)
2. Anastassiou, G.A., Shisha, O.: Monotone approximation with linear differential operators. J. Approx. Theory **44**, 391–393 (1985)
3. Badea, I., Badea, C.: On the order of simultaneously approximation of bivariate functions by Bernstein operators. Anal. Numé r. Théor. Approx. **16**, 11–17 (1987)
4. Mamatov, T., Samko, S.: Mixed fractional integration operators in mixed weighted Hölder spaces. Fractional Calc. Appl. Anal. **13**(3), 245–259 (2010)
5. Shisha, O.: Monotone approximation. Pacific J. Math. **15**, 667–671 (1965)
6. Stancu, D.D.: Studii Si Cercetari Stiintifice, **XI**(2), 221–233 (1960)

Bivariate Right Fractional Pseudo-Polynomial Monotone Approximation

George A. Anastassiou

Abstract In this article we deal with the following general two-dimensional problem: Let f be a two variable continuously differentiable real valued function of a given order, let $\overline{L}$ be a linear right fractional mixed partial differential operator and suppose that $\overline{L}(f) \geq 0$ on a critical region. Then for sufficiently large $n, m \in \mathbb{N}$, we can find a sequence of pseudo-polynomials $Q_{n,m}^*$ in two variables with the property $\overline{L}\left(Q_{n,m}^*\right) \geq 0$ on this critical region such that f is approximated with rates right fractionally and simultaneously by $Q_{n,m}^*$ in the uniform norm on the whole domain of f. This restricted approximation is given via inequalities involving the mixed modulus of smoothness $\omega_{s,q}$, $s, q \in \mathbb{N}$, of highest order integer partial derivative of f.

1 Introduction

The topic of monotone approximation started in [10] has become a major trend in approximation theory. A typical problem in this subject is: given a positive integer k, approximate a given function whose kth derivative is ≥ 0 by polynomials having this property.

In [3] the authors replaced the kth derivative with a linear differential operator of order k. We mention this motivating result.

Theorem 1 *Let h, k, p be integers, $0 \leq h \leq k \leq p$ and let f be a real function, $f^{(p)}$ continuous in $[-1, 1]$ with modulus of continuity $\omega_1\left(f^{(p)}, x\right)$ there. Let $a_j(x)$, $j = h, h + 1, ..., k$ be real functions, defined and bounded on $[-1, 1]$ and assume $a_h(x)$ is either $\geq$ some number $\alpha > 0$ or $\leq$ some number $\beta < 0$ throughout $[-1, 1]$. Consider the operator*

$$L = \sum_{j=h}^{k} a_j(x) \left[\frac{d^j}{dx^j}\right] \tag{1}$$

G.A. Anastassiou (✉)
Department of Mathematical Sciences, University of Memphis,
Memphis, TN 38152, USA
e-mail: ganastss@memphis.edu

© Springer International Publishing Switzerland 2016   15
G.A. Anastassiou and O. Duman (eds.), *Intelligent Mathematics II:*
Applied Mathematics and Approximation Theory, Advances in Intelligent Systems
and Computing 441, DOI 10.1007/978-3-319-30322-2_2

and suppose, throughout $[-1, 1]$,

$$L(f) \geq 0. \tag{2}$$

Then, for every integer $n \geq 1$, *there is a real polynomial* $Q_n(x)$ *of degree* $\leq n$ *such that*

$$L(Q_n) \geq 0 \text{ throughout } [-1, 1] \tag{3}$$

and

$$\max_{-1 \leq x \leq 1} |f(x) - Q_n(x)| \leq Cn^{k-p}\omega_1\left(f^{(p)}, \frac{1}{n}\right), \tag{4}$$

where C *is independent of* n *or* f.

Next let $n, m \in \mathbb{Z}_+$, P_θ denote the space of algebraic polynomials of degree $\leq \theta$. Consider the tensor product spaces $P_n \otimes C([-1, 1])$, $C([-1, 1]) \otimes P_m$ and their sum $P_n \otimes C([-1, 1]) + C([-1, 1]) \otimes P_m$, that is

$$P_n \otimes C([-1, 1]) + C([-1, 1]) \otimes P_m \tag{5}$$
$$= \left\{ \sum_{i=0}^{n} x^i A_i(y) + \sum_{j=0}^{m} B_j(x) y^j; \ A_i, B_j \in C([-1, 1]), \ x, y \in [-1, 1] \right\}.$$

This is the space of pseudo-polynomials of degree $\leq (n, m)$, first introduced by A. Marchaud in 1924–1927 (see [7, 8]). Here $f^{(k,l)}$ denotes $\frac{\partial^{k+l} f}{\partial x^k \partial y^l}$, the (k, l)-partial derivative of f.

In this section we consider the space $C^{r,p}([-1, 1]^2) = \{f : [-1, 1]^2 \to \mathbb{R}; f^{(k,l)}$ is continuous for $0 \leq k \leq r, 0 \leq l \leq p\}$. Let $f \in C([-1, 1]^2)$; for $\delta_1, \delta_2 \geq 0$, define the mixed modulus of smoothness of order (s, q), $s, q \in \mathbb{N}$ (see [9], pp. 516–517) by

$$\omega_{s,q}(f; \delta_1, \delta_2) \equiv \sup\left\{ \left| {}_x\Delta^s_{h_1} \circ_y \Delta^q_{h_2} f(x, y) \right| : (x, y), \tag{6} \right.$$
$$\left. (x + sh_1, y + qh_2) \in [-1, 1]^2, |h_i| \leq \delta_i, i = 1, 2 \right\}.$$

Here

$${}_x\Delta^s_{h_1} \circ_y \Delta^q_{h_2} f(x, y)$$
$$\equiv \sum_{\sigma=0}^{s} \sum_{\mu=0}^{q} (-1)^{s+q-\sigma-\mu} \binom{s}{\sigma} \binom{q}{\mu} f(x + \sigma h_1, y + \mu h_2) \tag{7}$$

is a mixed difference of order (s, q).

We mention

Theorem 2 (see Gonska [4]) *Let* $r, p \in \mathbb{Z}_+, s, q \in \mathbb{N}$, *and* $f \in C^{r,p}\left([-1,1]^2\right)$. *Let* $n, m \in \mathbb{N}$ *with* $n \geq \max\{4(r+1), r+s\}$ *and* $m \geq \max\{4(p+1), p+q\}$. *Then there exists a linear operator* $Q_{n,m}$ *from* $C^{r,p}\left([-1,1]^2\right)$ *into the space of pseudopolynomials* $(P_n \otimes C([-1,1]) + C([-1,1]) \otimes P_m)$ *such that*

$$\left|\left(f - Q_{n,m}(f)\right)^{(k,l)}(x,y)\right| \tag{8}$$
$$\leq M_{r,s} \cdot M_{p,q} \left(\Delta_n(x)\right)^{r-k} \cdot \left(\Delta_m(y)\right)^{p-l} \cdot \omega_{s,q}\left(f^{(r,p)}; \Delta_n(x), \Delta_m(y)\right),$$

for all $(0,0) \leq (k,l) \leq (r,p), x, y \in [-1,1]$, *where*

$$\Delta_\theta(z) = \frac{\sqrt{1-z^2}}{\theta} + \frac{1}{\theta^2}, \quad \theta = n, m; \; z = x, y \in [-1,1].$$

The constants $M_{r,s}, M_{p,q}$, *are independent of* $f, (x,y)$ *and* (n,m); *they depend only on* $(r,s), (p,q)$, *respectively.*

See also [5], saying that $Q_{n,m}^{(r,p)}(f)$ is continuous on $[-1,1]^2$.

The need following result which is an easy consequence of the last theorem (see [9, p. 517]).

Corollary 3 *Let* $r, p \in \mathbb{Z}_+, s, q \in \mathbb{N}$, *and* $f \in C^{r,p}\left([-1,1]^2\right)$. *Let* $n, m \in \mathbb{N}$ *with* $n \geq \max\{4(r+1), r+s\}$ *and* $m \geq \max\{4(p+1), p+q\}$. *Then there exists a pseudopolynomial*

$$Q_{n,m} \equiv Q_{n,m}(f) \in (P_n \otimes C([-1,1]) + C([-1,1]) \otimes P_m)$$

such that

$$\left\|f^{(k,l)} - Q_{n,m}^{(k,l)}\right\|_\infty \leq \frac{C}{n^{r-k}m^{p-l}} \cdot \omega_{s,q}\left(f^{(r,p)}; \frac{1}{n}, \frac{1}{m}\right), \tag{9}$$

for all $(0,0) \leq (k,l) \leq (r,p)$. *Here the constant* C *depends only on* r, p, s, q.

Corollary 3 was used in the proof of the main motivational result that follows.

Theorem 4 (see [1]) *Let* h_1, h_2, v_1, v_2, r, p *be integers,* $0 \leq h_1 \leq v_1 \leq r, 0 \leq h_2 \leq v_2 \leq p$ *and let* $f \in C^{r,p}\left([-1,1]^2\right)$, *with* $f^{(r,p)}$ *having a mixed modulus of smoothness* $\omega_{s,q}\left(f^{(r,p)}; x, y\right)$ *there,* $s, q \in \mathbb{N}$. *Let* $\alpha_{i,j}(x,y), i = h_1, h_1 + 1, ..., v_1$; $j = h_2, h_2 + 1, ..., v_2$ *be real-valued functions, defined and bounded in* $[-1,1]^2$ *and suppose* $\alpha_{h_1 h_2}$ *is either* $\geq \alpha > 0$ *or* $\leq \beta < 0$ *throughout* $[-1,1]^2$. *Take the operator*

$$L = \sum_{i=h_1}^{v_1} \sum_{j=h_2}^{v_2} \alpha_{ij}(x,y) \frac{\partial^{i+j}}{\partial x^i \partial y^j} \tag{10}$$

and assume, throughout $[-1, 1]^2$ *that*

$$L(f) \geq 0. \tag{11}$$

Then for any integers n, m *with* $n \geq \max\{4(r+1), r+s\}$, $m \geq \max\{4(p+1), p+q\}$, *there exists a pseudopolynomial*

$$Q_{n,m} \in (P_n \otimes C([-1, 1]) + C([-1, 1]) \otimes P_m)$$

such that $L(Q_{m,n}) \geq 0$ *throughout* $[-1, 1]^2$ *and*

$$\left\| f^{(k,l)} - Q_{n,m}^{(k,l)} \right\|_\infty \leq \frac{C}{n^{r-v_1} m^{p-v_2}} \cdot \omega_{s,q}\left(f^{(r,p)}; \frac{1}{n}, \frac{1}{m}\right), \tag{12}$$

for all $(0, 0) \leq (k, l) \leq (h_1, h_2)$. *Moreover we get*

$$\left\| f^{(k,l)} - Q_{n,m}^{(k,l)} \right\|_\infty \leq \frac{C}{n^{r-k} m^{p-l}} \cdot \omega_{s,q}\left(f^{(r,p)}; \frac{1}{n}, \frac{1}{m}\right), \tag{13}$$

for all $(h_1 + 1, h_2 + 1) \leq (k, l) \leq (r, p)$. *Also* (13) *is valid whenever* $0 \leq k \leq h_1$, $h_2 + 1 \leq l \leq p$ *or* $h_1 + 1 \leq k \leq r$, $0 \leq l \leq h_2$. *Here* C *is a constant independent of* f *and* n, m. *It depends only on* r, p, s, q, L.

We are also motivated by [2].
We need

Definition 5 (*see* [6]) Let $[-1, 1]^2$; $\alpha_1, \alpha_2 > 0$; $\alpha = (\alpha_1, \alpha_2)$, $f \in C([-1, 1]^2)$, $x = (x_1, x_2)$, $t = (t_1, t_2) \in [-1, 1]^2$. We define the right mixed Riemann-Liouville fractional two dimensional integral of order α

$$\left(I_{1-}^\alpha f\right)(x) \tag{14}$$

$$:= \frac{1}{\Gamma(\alpha_1)\Gamma(\alpha_2)} \int_{x_1}^1 \int_{x_2}^1 (t_1 - x_1)^{\alpha_1 - 1}(t_2 - x_2)^{\alpha_2 - 1} f(t_1, t_2) \, dt_1 dt_2,$$

with $x_1, x_2 < 1$. Notice here that $I_{1-}^\alpha(|f|) < \infty$.

Definition 6 Let $\alpha_1, \alpha_2 > 0$ with $\lceil \alpha_1 \rceil = m_1$, $\lceil \alpha_2 \rceil = m_2$, ($\lceil \cdot \rceil$ ceiling of the number). Let here $f \in C^{m_1, m_2}([-1, 1]^2)$. We consider the right Caputo type fractional partial derivative:

$$D_{1-}^{(\alpha_1, \alpha_2)} f(x)$$

$$:= \frac{(-1)^{m_1 + m_2}}{\Gamma(m_1 - \alpha_1)\Gamma(m_2 - \alpha_2)} \tag{15}$$

$$\cdot \int_{x_1}^{1} \int_{x_2}^{1} (t_1 - x_1)^{m_1 - \alpha_1 - 1} (t_2 - x_2)^{m_2 - \alpha_2 - 1} \frac{\partial^{m_1 + m_2} f(t_1, t_2)}{\partial t_1^{m_1} \partial t_2^{m_2}} dt_1 dt_2,$$

$\forall x = (x_1, x_2) \in [-1, 1]^2$, where Γ is the gamma function

$$\Gamma(\nu) = \int_{0}^{\infty} e^{-t} t^{\nu - 1} dt, \quad \nu > 0. \tag{16}$$

We set

$$D_{1-}^{(0,0)} f(x) := f(x), \quad \forall x \in [-1, 1]^2; \tag{17}$$

$$D_{1-}^{(m_1, m_2)} f(x) := (-1)^{m_1 + m_2} \frac{\partial^{m_1 + m_2} f(x)}{\partial x_1^{m_1} \partial x_2^{m_2}}, \quad \forall x \in [-1, 1]^2. \tag{18}$$

Definition 7 We also set

$$D_{1-}^{(0,\alpha_2)} f(x) := \frac{(-1)^{m_2}}{\Gamma(m_2 - \alpha_2)} \int_{x_2}^{1} (t_2 - x_2)^{m_2 - \alpha_2 - 1} \frac{\partial^{m_2} f(x_1, t_2)}{\partial t_2^{m_2}} dt_2, \tag{19}$$

$$D_{1-}^{(\alpha_1, 0)} f(x) := \frac{(-1)^{m_1}}{\Gamma(m_1 - \alpha_1)} \int_{x_1}^{1} (t_1 - x_1)^{m_1 - \alpha_1 - 1} \frac{\partial^{m_1} f(t_1, x_2)}{\partial t_1^{m_1}} dt_1, \tag{20}$$

and

$$D_{1-}^{(m_1, \alpha_2)} f(x) := \frac{(-1)^{m_2}}{\Gamma(m_2 - \alpha_2)} \int_{x_2}^{1} (t_2 - x_2)^{m_2 - \alpha_2 - 1} \frac{\partial^{m_1 + m_2} f(x_1, t_2)}{\partial x_1^{m_1} \partial t_2^{m_2}} dt_2, \tag{21}$$

$$D_{1-}^{(\alpha_1, m_2)} f(x) := \frac{(-1)^{m_1}}{\Gamma(m_1 - \alpha_1)} \int_{x_1}^{1} (t_1 - x_1)^{m_1 - \alpha_1 - 1} \frac{\partial^{m_1 + m_2} f(t_1, x_2)}{\partial t_1^{m_1} \partial x_2^{m_2}} dt_1. \tag{22}$$

In this article we extend Theorem 4 to the fractional level. Indeed here L is replaced by $\overline{L}$, a linear right Caputo fractional mixed partial differential operator. Now the monotonicity property holds true only on the critical square of $[-1, 0]^2$. Simultaneously fractional convergence remains true on all of $[-1, 1]^2$.

2   Main Result

We present

Theorem 8 *Let h_1, h_2, v_1, v_2, r, p be integers, $0 \leq h_1 \leq v_1 \leq r, 0 \leq h_2 \leq v_2 \leq p$ and let $f \in C^{r,p}\left([-1,1]^2\right)$, with $f^{(r,p)}$ having a mixed modulus of smoothness $\omega_{s,q}\left(f^{(r,p)}; x, y\right)$ there, $s, q \in \mathbb{N}$. Let $\alpha_{ij}(x, y), i = h_1, h_1 + 1, ..., v_1; j = h_2, h_2 + 1, ..., v_2$ be real valued functions, defined and bounded in $[-1,1]^2$ and suppose $\alpha_{h_1 h_2}$ is either $\geq \alpha > 0$ or $\leq \beta < 0$ throughout $[-1, 0]^2$. Assume that $h_1 + h_2 = 2\gamma$, $\gamma \in \mathbb{Z}_+$. Here $n, m \in \mathbb{N} : n \geq \max\{4(r+1), r+s\}, m \geq \max\{4(p+1), p+q\}$. Set*

$$l_{ij} := \sup_{(x,y)\in[-1,1]^2} \left|\alpha_{h_1 h_2}^{-1}(x, y)\, \alpha_{ij}(x, y)\right| < \infty, \tag{23}$$

for all $h_1 \leq i \leq v_1, h_2 \leq j \leq v_2$. Let $\alpha_{1i}, \alpha_{2j} > 0$, $\alpha_{1i}, \alpha_{2j} \notin \mathbb{N}$, with $\lceil \alpha_{1i} \rceil = i$, $\lceil \alpha_{2j} \rceil = j$, $i = 0, 1, ..., r$; $j = 0, 1, ..., p$, ($\lceil \cdot \rceil$ ceiling of the number), $\alpha_{10} = 0$, $\alpha_{20} = 0$.

Consider the right fractional bivariate differential operator

$$\overline{L} := \sum_{i=h_1}^{v_1} \sum_{j=h_2}^{v_2} \alpha_{ij}(x, y)\, D_{1-}^{(\alpha_{1i}, \alpha_{2j})}. \tag{24}$$

Assume $\overline{L}f(x, y) \geq 0$, on $[-1, 0]^2$. Then there exists

$$Q_{n,m}^* \equiv Q_{n,m}^*(f) \in (P_n \otimes C([-1,1]) + C([-1,1]) \otimes P_m)$$

such that $\overline{L}Q_{n,m}^*(x, y) \geq 0$, on $[-1, 0]^2$. Furthermore it holds:

1.

$$\left\| D_{1-}^{(\alpha_{1i}, \alpha_{2j})}(f) - D_{1-}^{(\alpha_{1i}, \alpha_{2j})} Q_{n,m}^* \right\|_{\infty, [-1,1]^2}$$

$$\leq \frac{C2^{(i+j)-(\alpha_{1i}+\alpha_{2j})}}{\Gamma(i - \alpha_{1i} + 1)\, \Gamma\left(j - \alpha_{2j} + 1\right) n^{r-i} m^{p-j}} \cdot \omega_{s,q}\left(f^{(r,p)}; \frac{1}{n}, \frac{1}{m}\right), \tag{25}$$

where C is a constant that depends only on r, p, s, q; $(h_1 + 1, h_2 + 1) \leq (i, j) \leq (r, p)$, or $0 \leq i \leq h_1, h_2 + 1 \leq j \leq p$, or $h_1 + 1 \leq i \leq r, 0 \leq j \leq h_2$,

2.

$$\left\| D_{1-}^{(\alpha_{1i}, \alpha_{2j})}(f) - D_{1-}^{(\alpha_{1i}, \alpha_{2j})} Q_{n,m}^* \right\|_{\infty, [-1,1]^2}$$

$$\leq \frac{c_{ij}}{n^{r-v_1} m^{p-v_2}} \cdot \omega_{s,q}\left(f^{(r,p)}; \frac{1}{n}, \frac{1}{m}\right), \tag{26}$$

for $(1, 1) \leq (i, j) \leq (h_1, h_2)$, *where* $c_{ij} = \overset{.}{C} A_{ij}$, *with*

$$
\begin{aligned}
A_{ij} &:= \left\{ \left[\sum_{\tau=h_1}^{v_1} \sum_{\mu=h_2}^{v_2} \frac{l_{\tau\mu} 2^{(\tau+\mu)-(\alpha_{1\tau}+\alpha_{2\mu})}}{\Gamma (\tau - a_{1\tau} + 1) \Gamma (\mu - \alpha_{2\mu} + 1)} \right] \right. \\
&\quad \cdot \left(\sum_{k=0}^{h_1-i} \frac{2^{h_1-\alpha_{1i}-k}}{k! \Gamma (h_1 - \alpha_{1i} - k + 1)} \right) \\
&\quad \left. \cdot \left(\sum_{\lambda=0}^{h_2-j} \frac{2^{h_2-\alpha_{2j}-\lambda}}{\lambda! \Gamma \left(h_2 - \alpha_{2j} - \lambda + 1\right)} \right) + \frac{2^{(i+j)-(\alpha_{1i}+\alpha_{2j})}}{\Gamma (i - \alpha_{1i} + 1) \Gamma \left(j - \alpha_{2j} + 1\right)} \right\},
\end{aligned}
\tag{27}
$$

3.

$$
\left\| f - Q_{n,m}^* \right\|_{\infty, [-1,1]^2} \leq \frac{c_{00}}{n^{r-v_1} m^{p-v_2}} \cdot \omega_{s,q} \left(f^{(r,p)}; \frac{1}{n}, \frac{1}{m} \right),
\tag{28}
$$

where $c_{00} := \overset{.}{C} A_{00}$, *with*

$$
A_{00} := \frac{1}{h_1! h_2!} \left(\sum_{\tau=h_1}^{v_1} \sum_{\mu=h_2}^{v_2} l_{\tau\mu} \frac{2^{(\tau+\mu)-(\alpha_{1\tau}+\alpha_{2\mu})}}{\Gamma (\tau - a_{1\tau} + 1) \Gamma \left(\mu - \alpha_{2\mu} + 1\right)} \right) + 1,
$$

4.

$$
\left\| D_{1-}^{(0,\alpha_{2j})} (f) - D_{1-}^{(0,\alpha_{2j})} Q_{n,m}^* \right\|_{\infty, [-1,1]^2}
$$
$$
\leq \frac{c_{0j}}{n^{r-v_1} m^{p-v_2}} \cdot \omega_{s,q} \left(f^{(r,p)}; \frac{1}{n}, \frac{1}{m} \right),
\tag{29}
$$

where $c_{0j} = \overset{.}{C} A_{0j}$, $j = 1, ..., h_2$, *with*

$$
\begin{aligned}
A_{0j} &:= \left[\frac{1}{h_1!} \left(\sum_{\tau=h_1}^{v_1} \sum_{\mu=h_2}^{v_2} l_{\tau\mu} \frac{2^{(\tau+\mu)-(\alpha_{1\tau}+\alpha_{2\mu})}}{\Gamma (\tau - a_{1\tau} + 1) \Gamma \left(\mu - \alpha_{2\mu} + 1\right)} \right) \right. \\
&\quad \left. \cdot \left(\sum_{\lambda=0}^{h_2-j} \frac{2^{h_2-\alpha_{2j}-\lambda}}{\lambda! \Gamma \left(h_2 - \alpha_{2j} - \lambda + 1\right)} \right) + \frac{2^{j-\alpha_{2j}}}{\Gamma \left(j - \alpha_{2j} + 1\right)} \right],
\end{aligned}
\tag{30}
$$

5.

$$\left\| D_{1-}^{(\alpha_{1i},0)}(f) - D_{1-}^{(\alpha_{1i},0)} Q_{n,m}^* \right\|_{\infty,[-1,1]^2}$$

$$\leq \frac{c_{i0}}{n^{r-v_1} m^{p-v_2}} \cdot \omega_{s,q}\left(f^{(r,p)}; \frac{1}{n}, \frac{1}{m}\right),\qquad(31)$$

where $c_{i0} = C A_{i0}$, $i = i, ..., h_1$, with

$$A_{i0}$$

$$:= \left[\frac{1}{h_2!}\left(\sum_{\tau=h_1}^{v_1}\sum_{\mu=h_2}^{v_2} l_{\tau\mu}\frac{2^{(\tau+\mu)-(\alpha_{1\tau}+\alpha_{2\mu})}}{\Gamma(\tau - a_{1\tau}+1)\,\Gamma(\mu - \alpha_{2\mu}+1)}\right)\qquad(32)$$

$$\cdot\left(\sum_{k=0}^{h_1-i}\frac{2^{h_1-\alpha_{1i}-k}}{k!\Gamma(h_1-\alpha_{1i}-k+1)}\right) + \frac{2^{i-\alpha_{1i}}}{\Gamma(i-\alpha_{1i}+1)}\right].$$

Proof By Corollary 3 there exists

$$Q_{n,m} \equiv Q_{n,m}(f) \in (P_n \otimes C([-1,1]) + C([-1,1]) \otimes P_m)$$

such that

$$\left\| f^{(i,j)} - Q_{n,m}^{(i,j)}\right\|_\infty \leq \frac{C}{n^{r-i} m^{p-j}} \cdot \omega_{s,q}\left(f^{(r,p)}; \frac{1}{n}, \frac{1}{m}\right),\qquad(33)$$

for all $(0,0) \leq (i,j) \leq (r,p)$, while $Q_{n,m} \in C^{r,p}([-1,1])^2$. Here C depends only on r, p, s, q, where $n \geq \max\{4(r+1), r+s\}$ and $m \geq \max\{4(p+1), p+q\}$, with $r, p \in \mathbb{Z}_+$, $s, q \in \mathbb{N}$, $f \in C^{r,p}([-1,1]^2)$. Indeed by [5] we have that $Q_{n,m}^{(r,p)}$ is continuous on $[-1,1]^2$. We observe the following $(i = 1, ..., r;\ j = 1, ..., p)$

$$\left| D_{1-}^{(\alpha_{1i},\alpha_{2j})} f(x_1, x_2) - D_{1-}^{(\alpha_{1i},\alpha_{2j})} Q_{n,m}(x_1, x_2)\right|$$

$$= \frac{1}{\Gamma(i-\alpha_{1i})\,\Gamma(j-\alpha_{2j})}\left|\int_{x_1}^1\int_{x_2}^1 (t_1-x_1)^{i-\alpha_{1i}-1}(t_2-x_2)^{j-\alpha_{2j}-1}\qquad(34)\right.$$

$$\left.\cdot\left(\frac{\partial^{i+j} f(t_1, t_2)}{\partial t_1^i \partial t_2^j} - \frac{\partial^{i+j} Q_{n,m}(t_1, t_2)}{\partial t_1^i \partial t_2^j}\right)dt_1 dt_2\right|$$

$$\leq \frac{1}{\Gamma(i-\alpha_{1i})\,\Gamma(j-\alpha_{2j})}\int_{x_1}^1\int_{x_2}^1 (t_1-x_1)^{i-\alpha_{1i}-1}(t_2-x_2)^{j-\alpha_{2j}-1}\qquad(35)$$

$$\cdot \left| \frac{\partial^{i+j} f (t_1, t_2)}{\partial t_1^i \partial t_2^j} - \frac{\partial^{i+j} Q_{n,m} (t_1, t_2)}{\partial t_1^i \partial t_2^j} \right| dt_1 dt_2$$

$$\overset{(9)}{\leq} \frac{1}{\Gamma (i - \alpha_{1i}) \, \Gamma (j - \alpha_{2j})} \tag{36}$$

$$\cdot \left(\int\limits_{x_1}^1 \int\limits_{x_2}^1 (t_1 - x_1)^{i - \alpha_{1i} - 1} (t_2 - x_2)^{j - \alpha_{2j} - 1} \right)$$

$$\cdot \frac{C}{n^{r-i} m^{p-j}} \cdot \omega_{s,q} \left(f^{(r,p)}; \frac{1}{n}, \frac{1}{m} \right)$$

$$= \frac{1}{\Gamma (i - \alpha_{1i}) \, \Gamma (j - \alpha_{2j})} \frac{(1 - x_1)^{i - \alpha_{1i}}}{i - \alpha_{1i}} \frac{(1 - x_2)^{j - \alpha_{2j}}}{j - \alpha_{2j}} \frac{C}{n^{r-i} m^{p-j}} \tag{37}$$

$$\cdot \omega_{s,q} \left(f^{(r,p)}; \frac{1}{n}, \frac{1}{m} \right)$$

$$= \frac{(1 - x_1)^{i - \alpha_{1i}}}{\Gamma (i - \alpha_{1i} + 1)} \frac{(1 - x_2)^{j - \alpha_{2j}}}{\Gamma (j - \alpha_{2j} + 1)} \frac{C}{n^{r-i} m^{p-j}} \omega_{s,q} \left(f^{(r,p)}; \frac{1}{n}, \frac{1}{m} \right).$$

That is there exists $Q_{n,m}$:

$$\left| D_{1-}^{(\alpha_{1i}, \alpha_{2j})} f (x_1, x_2) - D_{1-}^{(\alpha_{1i}, \alpha_{2j})} Q_{n,m} (x_1, x_2) \right| \tag{38}$$

$$\leq \frac{(1 - x_1)^{i - \alpha_{1i}} (1 - x_2)^{j - \alpha_{2j}}}{\Gamma (i - \alpha_{1i} + 1) \, \Gamma (j - \alpha_{2j} + 1)} \frac{C}{n^{r-i} m^{p-j}} \cdot \omega_{s,q} \left(f^{(r,p)}; \frac{1}{n}, \frac{1}{m} \right),$$

$i = 1, ..., r, \; j = 1, ..., p, \; \forall \, (x_1, x_2) \in [-1, 1]^2$.

We proved there exists $Q_{n,m}$ such that

$$\left\| D_{1-}^{(\alpha_{1i}, \alpha_{2j})} (f) - D_{1-}^{(\alpha_{1i}, \alpha_{2j})} Q_{n,m}^* \right\|_\infty \tag{39}$$

$$\leq \frac{2^{(i+j) - (\alpha_{1i} + \alpha_{2j})} C}{\Gamma (i - \alpha_{1i} + 1) \, \Gamma (j - \alpha_{2j} + 1) \, n^{r-i} m^{p-j}} \omega_{s,q} \left(f^{(r,p)}; \frac{1}{n}, \frac{1}{m} \right),$$

$i = 0, 1, ..., r, \; j = 0, 1, ..., p$. Define

$$\rho_{n,m} \equiv C \omega_{s,q} \left(f^{(r,p)}; \frac{1}{n}, \frac{1}{m} \right) \tag{40}$$

$$\cdot \left[\sum_{i=h_1}^{v_1} \sum_{j=h_2}^{v_2} \left(l_{ij} \frac{2^{(i+j)-(\alpha_{1i}+\alpha_{2j})}}{\Gamma\left(i - a_{1i} + 1\right) \Gamma\left(j - \alpha_{2j} + 1\right)} n^{i-r} m^{j-p} \right) \right].$$

I. Suppose, throughout $[-1, 0]^2$, $\alpha_{h_1 h_2}(x, y) \geq \alpha > 0$. Let $Q^*_{n,m}(x, y)$, $(x, y) \in [-1, 1]^2$, as in (39), so that

$$\left\| D_{1-}^{(\alpha_{1i},\alpha_{2j})} \left(f(x, y) + \rho_{n,m} \frac{x^{h_1} y^{h_2}}{h_1! h_2!} \right) - D_{1-}^{(\alpha_{1i},\alpha_{2j})} Q^*_{n,m}(x, y) \right\|_\infty$$

$$\leq \frac{2^{(i+j)-(\alpha_{1i}+\alpha_{2j})} \dot{C}}{\Gamma\left(i - \alpha_{1i} + 1\right) \Gamma\left(j - \alpha_{2j} + 1\right) n^{r-i} m^{p-j}} \omega_{s,q} \left(f^{(r,p)}; \frac{1}{n}, \frac{1}{m} \right) \quad (41)$$

$$=: T_{ij},$$

$i = 0, 1, ..., r$; $j = 0, 1, ..., p$.

If $(h_1 + 1, h_2 + 1) \leq (i, j) \leq (r, p)$, or $0 < i \leq h_1$, $h_2 + 1 \leq j \leq p$, or $h_1 + 1 \leq i \leq r, 0 < j \leq h_2$ we get from the last

$$\left\| D_{1-}^{(\alpha_{1i},\alpha_{2j})}(f) - D_{1-}^{(\alpha_{1i},\alpha_{2j})} Q^*_{n,m} \right\|_\infty \quad (42)$$

$$\leq \frac{2^{(i+j)-(\alpha_{1i}+\alpha_{2j})} \dot{C}}{\Gamma\left(i - \alpha_{1i} + 1\right) \Gamma\left(j - \alpha_{2j} + 1\right) n^{r-i} m^{p-j}} \omega_{s,q} \left(f^{(r,p)}; \frac{1}{n}, \frac{1}{m} \right),$$

proving (25).

If $(0, 0) < (i, j) \leq (h_1, h_2)$, we get

$$\left\| D_{1-}^{(\alpha_{1i},\alpha_{2j})} f(x, y) + \rho_{n,m} D_{1-}^{\alpha_{1i}} \left(\frac{x^{h_1}}{h_1!} \right) D_{1-}^{\alpha_{2j}} \left(\frac{y^{h_2}}{h_2!} \right) \right.$$
$$\left. - D_{1-}^{(\alpha_{1i},\alpha_{2j})} Q^*_{n,m}(x, y) \right\|_\infty \leq T_{ij}. \quad (43)$$

That is for $i = 1, ..., h_1$; $j = 1, ..., h_2$, we obtain

$$\left\| D_{1-}^{(\alpha_{1i},\alpha_{2j})} f(x, y) + \rho_{n,m} \left((-1)^{h_1} \sum_{k=0}^{h_1-i} \frac{(-1)^k (1-x)^{h_1-\alpha_{1i}-k}}{k! \Gamma\left(h_1 - \alpha_{1i} - k + 1\right)} \right) \right.$$

$$\left. \cdot \left((-1)^{h_2} \sum_{\lambda=0}^{h_2-j} \frac{(-1)^\lambda (1-y)^{h_2-\alpha_{2j}-\lambda}}{\lambda! \Gamma\left(h_2 - \alpha_{2j} - \lambda + 1\right)} \right) - D_{1-}^{(\alpha_{1i},\alpha_{2j})} Q^*_{n,m}(x, y) \right\|_\infty$$

$$\leq T_{ij}. \quad (44)$$

Hence for $(1, 1) \leq (i, j) \leq (h_1, h_2)$, we have

$$\left\| D_{1-}^{(\alpha_{1i}, \alpha_{2j})} f - D_{1-}^{(\alpha_{1i}, \alpha_{2j})} Q_{n,m}^* \right\|_\infty$$
$$\leq \rho_{n,m} \left(\sum_{k=0}^{h_1-i} \frac{2^{h_1-\alpha_{1i}-k}}{k! \Gamma(h_1 - \alpha_{1i} - k + 1)} \right) \tag{45}$$
$$\cdot \left(\sum_{\lambda=0}^{h_2-j} \frac{2^{h_2-\alpha_{2j}-\lambda}}{\lambda! \Gamma(h_2 - \alpha_{2j} - \lambda + 1)} \right) + T_{ij}$$

$$= \dot{C} \omega_{s,q} \left(f^{(r,p)}; \frac{1}{n}, \frac{1}{m} \right) \tag{46}$$
$$\cdot \left[\sum_{\bar{i}=h_1}^{v_1} \sum_{\bar{j}=h_2}^{v_2} l_{\bar{i}\bar{j}} \frac{2^{(\bar{i}+\bar{j})-(\alpha_{1\bar{i}}+\alpha_{2\bar{j}})}}{\Gamma(\bar{i} - \alpha_{1\bar{i}} + 1) \Gamma(\bar{j} - \alpha_{2\bar{j}} + 1)} \frac{1}{n^{r-\bar{i}}} \frac{1}{m^{p-\bar{j}}} \right]$$
$$\left(\sum_{k=0}^{h_1-i} \frac{2^{h_1-\alpha_{1i}-k}}{k! \Gamma(h_1 - \alpha_{1i} - k + 1)} \right) \left(\sum_{\lambda=0}^{h_2-j} \frac{2^{h_2-\alpha_{2j}-\lambda}}{\lambda! \Gamma(h_2 - \alpha_{2j} - \lambda + 1)} \right)$$
$$+ \frac{2^{(i+j)-(\alpha_{1i}+\alpha_{2j})} \dot{C} \omega_{s,q} \left(f^{(r,p)}; \frac{1}{n}, \frac{1}{m} \right)}{\Gamma(i - \alpha_{1i} + 1) \Gamma(j - \alpha_{2j} + 1) n^{r-i} m^{p-j}}$$
$$\leq \dot{C} \omega_{s,q} \left(f^{(r,p)}; \frac{1}{n}, \frac{1}{m} \right) \frac{1}{n^{r-v_1} m^{p-v_2}} A_{ij}, \tag{47}$$

where

$$A_{ij}$$
$$:= \left\{ \left[\sum_{\bar{i}=h_1}^{v_1} \sum_{\bar{j}=h_2}^{v_2} l_{\bar{i}\bar{j}} \frac{2^{(\bar{i}+\bar{j})-(\alpha_{1\bar{i}}+\alpha_{2\bar{j}})}}{\Gamma(\bar{i} - \alpha_{1\bar{i}} + 1) \Gamma(\bar{j} - \alpha_{2\bar{j}} + 1)} \right] \tag{48} \right.$$
$$\cdot \left(\sum_{k=0}^{h_1-i} \frac{2^{h_1-\alpha_{1i}-k}}{k! \Gamma(h_1 - \alpha_{1i} - k + 1)} \right) \left(\sum_{\lambda=0}^{h_2-j} \frac{2^{h_2-\alpha_{2j}-\lambda}}{\lambda! \Gamma(h_2 - \alpha_{2j} - \lambda + 1)} \right)$$
$$+ \left. \frac{2^{(i+j)-(\alpha_{1i}+\alpha_{2i})}}{\Gamma(i - \alpha_{1i} + 1) \Gamma(j - \alpha_{2j} + 1)} \right\}.$$

(Set $c_{ij} := \dot{C} A_{ij}$)

We proved, for $(1, 1) \leq (i, j) \leq (h_1, h_2)$, that

$$\left\| D_{1-}^{(\alpha_{1i}, \alpha_{2j})} f - D_{1-}^{(\alpha_{1i}, \alpha_{2j})} Q_{n,m}^* \right\|_\infty \leq \frac{c_{ij}}{n^{r-v_1} m^{p-v_2}} \omega_{s,q} \left(f^{(r,p)}; \frac{1}{n}, \frac{1}{m} \right). \tag{49}$$

So that (26) is established.

When $i = j = 0$ from (41) we obtain

$$\left\| f(x, y) + \rho_{n,m} \frac{x^{h_1}}{h_1!} \frac{y^{h_2}}{h_2!} - Q_{n,m}^*(x, y) \right\|_\infty \leq \frac{\overset{\cdot}{C}}{n^r m^p} \omega_{s,q}\left(f^{(r,p)}; \frac{1}{n}, \frac{1}{m}\right). \quad (50)$$

Hence

$$\left\| f - Q_{n,m}^* \right\|_\infty \leq \frac{\rho_{n,m}}{h_1! h_2!} + \frac{\overset{\cdot}{C}}{n^r m^p} \omega_{s,q}\left(f^{(r,p)}; \frac{1}{n}, \frac{1}{m}\right) \quad (51)$$

$$= \frac{\overset{\cdot}{C}}{h_1! h_2!} \omega_{s,q}\left(f^{(r,p)}; \frac{1}{n}, \frac{1}{m}\right) \quad (52)$$

$$\cdot \left[\sum_{\bar{i}=h_1}^{v_1} \sum_{\bar{j}=h_2}^{v_2} l_{\bar{i}\bar{j}} \frac{2^{(\bar{i}+\bar{j})-(\alpha_{1\bar{i}}+\alpha_{2\bar{j}})}}{\Gamma\left(\bar{i} - \alpha_{1\bar{i}} + 1\right) \Gamma\left(\bar{j} - \alpha_{2\bar{j}} + 1\right)} \frac{1}{n^{r-\bar{i}}} \frac{1}{m^{p-\bar{j}}} \right]$$

$$+ \frac{\overset{\cdot}{C}}{n^r m^p} \omega_{s,q}\left(f^{(r,p)}; \frac{1}{n}, \frac{1}{m}\right)$$

$$\leq \frac{\overset{\cdot}{C}\omega_{s,q}\left(f^{(r,p)}; \frac{1}{n}, \frac{1}{m}\right)}{n^{r-v_1} m^{p-v_2}} A_{00}, \quad (53)$$

where

$$A_{00} := \left[\frac{1}{h_1! h_2!} \sum_{\bar{i}=h_1}^{v_1} \sum_{\bar{j}=h_2}^{v_2} \frac{l_{\bar{i}\bar{j}} 2^{(\bar{i}+\bar{j})-(\alpha_{1\bar{i}}+\alpha_{2\bar{j}})}}{\Gamma\left(\bar{i} - \alpha_{1\bar{i}} + 1\right) \Gamma\left(\bar{j} - \alpha_{2\bar{j}} + 1\right)} + 1 \right]. \quad (54)$$

(Set $c_{00} = \overset{\cdot}{C} A_{00}$).

Then

$$\left\| f - Q_{n,m}^* \right\|_\infty \leq \frac{c_{00}}{n^{r-v_1} m^{p-v_2}} \omega_{s,q}\left(f^{(r,p)}; \frac{1}{n}, \frac{1}{m}\right). \quad (55)$$

So that (28) is established.

Next case of $i = 0$, $j = 1, ..., h_2$, from (41) we get

$$\left\| D_{1-}^{(0,\alpha_{2j})} f(x,y) + \rho_{n,m} \frac{x^{h_1}}{h_1!} \left((-1)^{h_2} \sum_{\lambda=0}^{h_2-j} \frac{(-1)^{\lambda}(1-y)^{h_2-\alpha_{2j}-\lambda}}{\lambda! \Gamma(h_2-\alpha_{2j}-\lambda+1)} \right) \right. $$
$$\left. - D_{1-}^{(0,\alpha_{2j})} Q_{n,m}^*(x,y) \right\|_{\infty} \le T_{0j}. \tag{56}$$

Then

$$\left\| D_{1-}^{(0,\alpha_{2j})} f - D_{1-}^{(0,\alpha_{2j})} Q_{n,m}^* \right\|_{\infty}$$
$$\le \frac{\rho_{n,m}}{h_1!} \left(\sum_{\lambda=0}^{h_2-j} \frac{2^{h_2-\alpha_{2j}-\lambda}}{\lambda! \Gamma(h_2-\alpha_{2j}-\lambda+1)} \right) + T_{0j} \tag{57}$$

$$= \frac{\overset{.}{C}}{h_1!} \omega_{s,q} \left(f^{(r,p)}; \frac{1}{n}, \frac{1}{m} \right) \tag{58}$$
$$\cdot \left[\sum_{\overline{i}=h_1}^{v_1} \sum_{\overline{j}=h_2}^{v_2} l_{\overline{ij}} \frac{2^{(\overline{i}+\overline{j})-(\alpha_{1\overline{i}}+\alpha_{2\overline{j}})}}{\Gamma(\overline{i}-\alpha_{1\overline{i}}+1)\Gamma(\overline{j}-\alpha_{2\overline{j}}+1)} \frac{1}{n^{r-\overline{i}}} \frac{1}{m^{p-\overline{j}}} \right]$$
$$\cdot \left(\sum_{\lambda=0}^{h_2-j} \frac{2^{h_2-\alpha_{2j}-\lambda}}{\lambda! \Gamma(h_2-\alpha_{2j}-\lambda+1)} \right)$$
$$+ \frac{2^{j-\alpha_{2j}} \overset{.}{C}}{\Gamma(j-\alpha_{2j}+1) n^r m^{p-j}} \omega_{s,q} \left(f^{(r,p)}; \frac{1}{n}, \frac{1}{m} \right)$$
$$\le \frac{\overset{.}{C} \omega_{s,q} \left(f^{(r,p)}; \frac{1}{n}, \frac{1}{m} \right)}{n^{r-v_1} m^{p-v_2}} A_{0j}, \tag{59}$$

where

$$A_{0j}$$
$$:= \left[\frac{1}{h_1!} \left(\sum_{\overline{i}=h_1}^{v_1} \sum_{\overline{j}=h_2}^{v_2} l_{\overline{ij}} \frac{2^{(\overline{i}+\overline{j})-(\alpha_{1\overline{i}}+\alpha_{2\overline{j}})}}{\Gamma(\overline{i}-\alpha_{1\overline{i}}+1)\Gamma(\overline{j}-\alpha_{2\overline{j}}+1)} \right) \right. \tag{60}$$
$$\left. \cdot \left(\sum_{\lambda=0}^{h_2-j} \frac{2^{h_2-\alpha_{2j}-\lambda}}{\lambda! \Gamma(h_2-\alpha_{2j}-\lambda+1)} \right) + \frac{2^{j-\alpha_{2j}}}{\Gamma(j-\alpha_{2j}+1)} \right].$$

(Set $c_{0j} := \overset{.}{C} A_{0j}$)

We proved that (case of $i=0$, $j=1,...,h_2$)

$$\left\| D_{1-}^{(0,\alpha_{2j})} f - D_{1-}^{(0,\alpha_{2j})} Q_{n,m}^* \right\|_{\infty} \le \frac{c_{0j}}{n^{r-v_1} m^{p-v_2}} \omega_{s,q} \left(f^{(r,p)}; \frac{1}{n}, \frac{1}{m} \right). \tag{61}$$

establishing (29).

Similarly we get for $i = 1, ..., h_1, j = 0$, that

$$\left\| D_{1-}^{(\alpha_{1i}, 0)} f - D_{1-}^{(\alpha_{1i}, 0)} Q_{n,m}^* \right\|_\infty \leq \frac{c_{i0}}{n^{r-v_1} m^{p-v_2}} \omega_{s,q} \left(f^{(r,p)}; \frac{1}{n}, \frac{1}{m} \right), \qquad (62)$$

where $c_{i0} := C A_{i0}$, with

$$
\begin{aligned}
A_{i0} \\
&:= \left[\frac{1}{h_2!} \left(\sum_{\bar{i}=h_1}^{v_1} \sum_{\bar{j}=h_2}^{v_2} l_{\bar{i}\bar{j}} \frac{2^{(\bar{i}+\bar{j})-(\alpha_{1\bar{i}}+\alpha_{2\bar{j}})}}{\Gamma\left(\bar{i}-\alpha_{1\bar{i}}+1\right) \Gamma\left(\bar{j}-\alpha_{2\bar{j}}+1\right)} \right) \right. \\
&\quad \left. \cdot \left(\sum_{k=0}^{h_1-i} \frac{2^{h_1-\alpha_{1i}-k}}{k! \Gamma\left(h_1-\alpha_{1i}-k+1\right)} \right) + \frac{2^{i-\alpha_{1i}}}{\Gamma\left(i-\alpha_{1i}+1\right)} \right],
\end{aligned} \qquad (63)
$$

deriving (31). So if $(x, y) \in [-1, 0]^2$, then

$$\alpha_{h_1 h_2}^{-1}(x, y) \, \overline{L} \left(Q_{n,m}^*(x, y) \right) \qquad (64)$$

$$
\begin{aligned}
&= \alpha_{h_1 h_2}^{-1}(x, y) \, \overline{L} \left(f(x, y) \right) + \rho_{n,m} \frac{(1-x)^{h_1-\alpha_{1i}}}{\Gamma\left(h_1-\alpha_{1i}+1\right)} \frac{(1-y)^{h_2-\alpha_{2j}}}{\Gamma\left(h_2-\alpha_{2j}+1\right)} \qquad (65) \\
&\quad + \sum_{i=h_1}^{v_1} \sum_{j=h_2}^{v_2} \alpha_{h_1 h_2}^{-1}(x, y) \, \alpha_{ij}(x, y) \\
&\qquad \cdot \left[D_{1-}^{(\alpha_{1i}, \alpha_{2j})} Q_{n,m}^*(x, y) - D_{1-}^{(\alpha_{1i}, \alpha_{2j})} f(x, y) - \rho_{n,m} D_{1-}^{(\alpha_{1i}, \alpha_{2j})} \left(\frac{x^{h_1}}{h_1!} \frac{y^{h_2}}{h_2!} \right) \right]
\end{aligned}
$$

$$
\begin{aligned}
&\overset{(41)}{\geq} \rho_{n,m} \frac{(1-x)^{h_1-\alpha_{1i}}}{\Gamma\left(h_1-\alpha_{1i}+1\right)} \frac{(1-y)^{h_2-\alpha_{2j}}}{\Gamma\left(h_2-\alpha_{2j}+1\right)} \qquad (66) \\
&\quad - \left[\sum_{i=h_1}^{v_1} \sum_{j=h_2}^{v_2} l_{ij} \frac{2^{(i+j)-(\alpha_{1i}+\alpha_{2j})}}{\Gamma(i-\alpha_{1i}+1)\Gamma(j-\alpha_{2j}+1)} \frac{C}{n^{r-i} m^{p-j}} \omega_{s,q} \left(f^{(r,p)}; \frac{1}{n}, \frac{1}{m} \right) \right]
\end{aligned}
$$

$$= \rho_{n,m} \left[\frac{(1-x)^{h_1-\alpha_{1i}}}{\Gamma\left(h_1-\alpha_{1i}+1\right)} \frac{(1-y)^{h_2-\alpha_{2j}}}{\Gamma\left(h_2-\alpha_{2j}+1\right)} - 1 \right] \tag{67}$$

$$\geq \rho_{n,m} \left[\frac{1}{\Gamma\left(h_1-\alpha_{1i}+1\right) \Gamma\left(h_2-\alpha_{2j}+1\right)} - 1 \right]$$

$$= \rho_{n,m} \left[\frac{1-\Gamma\left(h_1-\alpha_{1i}+1\right) \Gamma\left(h_2-\alpha_{2j}+1\right)}{\Gamma\left(h_1-\alpha_{1i}+1\right) \Gamma\left(h_2-\alpha_{2j}+1\right)} \right] \geq 0. \tag{68}$$

Explanation: we have that $\Gamma(1) = 1$, $\Gamma(2) = 1$, and Γ is convex on $(0, \infty)$ and positive there, here $0 \leq h_1 - \alpha_{1h_1}, h_2 - \alpha_{2h_2} < 1$ and $1 \leq h_1 - \alpha_{1h_1} + 1, h_2 - \alpha_{2h_2} + 1 < 2$. Thus $0 < \Gamma\left(h_1 - \alpha_{1h_1} + 1\right), \Gamma\left(h_2 - \alpha_{2h_2} + 1\right) \leq 1$, and

$$0 \leq \Gamma\left(h_1 - \alpha_{1h_1} + 1\right) \Gamma\left(h_2 - \alpha_{2h_2} + 1\right) \leq 1. \tag{69}$$

And

$$1 - \Gamma\left(h_1 - \alpha_{1h_1} + 1\right) \Gamma\left(h_2 - \alpha_{2h_2} + 1\right) \geq 0. \tag{70}$$

Therefore it holds

$$\overline{L}\left(Q^*_{n,m}(x, y)\right) \geq 0, \ \forall \ (x, y) \in [-1, 0]^2. \tag{71}$$

II. Suppose, throughout $[-1, 0]^2$, $\alpha_{h_1 h_2}(x, y) \leq \beta < 0$. Let $Q^{**}_{n,m}(x, y), (x, y) \in [-1, 1]^2$, as in (39), so that

$$\left\| D_{1-}^{(\alpha_{1i}, \alpha_{2j})} \left(f(x, y) - \rho_{n,m} \frac{x^{h_1} y^{h_2}}{h_1! h_2!} \right) - D_{1-}^{(\alpha_{1i}, \alpha_{2j})} Q^{**}_{n,m}(x, y) \right\|_\infty$$

$$\leq \frac{2^{(i+j)-(\alpha_{1i}+\alpha_{2j})} \dot{C}}{\Gamma\left(i-\alpha_{1i}+1\right) \Gamma\left(j-\alpha_{2j}+1\right) n^{r-i} m^{p-j}} \omega_{s,q} \left(f^{(r,p)}; \frac{1}{n}, \frac{1}{m} \right), \tag{72}$$

$i = 0, 1, ..., r$, $j = 0, 1, ..., p$.

As earlier we produce the same convergence inequalities (25), (26), (28), (29), and (31). So for $(x, y) \in [-1, 0]^2$ we get

$$
\alpha_{h_1 h_2}^{-1}(x, y) \, \overline{L} \left(Q_{n,m}^{**}(x, y) \right)
$$

$$
= \alpha_{h_1 h_2}^{-1}(x, y) \, \overline{L}\left(f(x, y) \right) - \rho_{n,m} \frac{(1-x)^{h_1 - \alpha_{1i}}}{\Gamma\left(h_1 - \alpha_{1i} + 1\right)} \frac{(1-y)^{h_2 - \alpha_{2j}}}{\Gamma\left(h_2 - \alpha_{2j} + 1\right)} \tag{73}
$$

$$
+ \sum_{i=h_1}^{v_1} \sum_{j=h_2}^{v_2} \alpha_{h_1 h_2}^{-1}(x, y) \, \alpha_{ij}(x, y)
$$

$$
\cdot \left[D_{1-}^{(\alpha_{1i}, \alpha_{2j})} Q_{n,m}^{**}(x, y) - D_{1-}^{(\alpha_{1i}, \alpha_{2j})} f(x, y) + \rho_{n,m} D_{1-}^{(\alpha_{1i}, \alpha_{2j})} \left(\frac{x^{h_1}}{h_1!} \frac{y^{h_2}}{h_2!} \right) \right]
$$

$$
\overset{(71)}{\leq} -\rho_{n,m} \frac{(1-x)^{h_1 - \alpha_{1i}}}{\Gamma\left(h_1 - \alpha_{1i} + 1\right)} \frac{(1-y)^{h_2 - \alpha_{2j}}}{\Gamma\left(h_2 - \alpha_{2j} + 1\right)} \tag{74}
$$

$$
+ \left[\sum_{i=h_1}^{v_1} \sum_{j=h_2}^{v_2} l_{ij} \frac{2^{(i+j) - (\alpha_{1i} + \alpha_{2j})}}{\Gamma\left(i - \alpha_{1i} + 1\right) \Gamma\left(j - \alpha_{2j} + 1\right)} \frac{C}{n^{r-i} m^{p-j}} \omega_{s,q} \left(f^{(r,p)}; \frac{1}{n}, \frac{1}{m} \right) \right]
$$

$$
= \rho_{n,m} \left[1 - \frac{(1-x)^{h_1 - \alpha_{1i}}}{\Gamma\left(h_1 - \alpha_{1i} + 1\right)} \frac{(1-y)^{h_2 - \alpha_{2j}}}{\Gamma\left(h_2 - \alpha_{2j} + 1\right)} \right]
$$

$$
= \rho_{n,m} \left[\frac{\Gamma\left(h_1 - \alpha_{1i} + 1\right) \Gamma\left(h_2 - \alpha_{2j} + 1\right) - (1-x)^{h_1 - \alpha_{1i}} (1-y)^{h_2 - \alpha_{2j}}}{\Gamma\left(h_1 - \alpha_{1i} + 1\right) \Gamma\left(h_2 - \alpha_{2j} + 1\right)} \right]
$$

$$
\leq \rho_{n,m} \left[\frac{1 - (1-x)^{h_1 - \alpha_{1i}} (1-y)^{h_2 - \alpha_{2j}}}{\Gamma\left(h_1 - \alpha_{1i} + 1\right) \Gamma\left(h_2 - \alpha_{2j} + 1\right)} \right] \leq 0. \tag{75}
$$

Explanation: for $x, y \in [-1, 0]$ we get that $1 - x$, $1 - y \geq 1$, and $0 \leq h_1 - \alpha_{1 h_1}$, $h_2 - \alpha_{2 h_2} < 1$. Hence $(1-x)^{h_1 - \alpha_{1 h_1}}$, $(1-y)^{h_2 - \alpha_{2 h_2}} \geq 1$, and then

$$
(1-x)^{h_1 - \alpha_{1 h_1}} (1-y)^{h_2 - \alpha_{2 h_2}} \geq 1,
$$

so that

$$
1 - (1-x)^{h_1 - \alpha_{1 h_1}} (1-y)^{h_2 - \alpha_{2 h_2}} \leq 0. \tag{76}
$$

Hence again

$$
\overline{L}\left(Q_{n,m}^{**}(x, y) \right) \geq 0, \text{ for } (x, y) \in [-1, 0]^2. \tag{77}
$$

References

1. Anastassiou, G.A.: Monotone approximation by pseudopolynomials. In: Approximation Theory. Academic Press, New York (1991)
2. Anastassiou, G.A.: Bivariate monotone approximation. Proc. Amer. Math. Soc. **112**(4), 959–963 (1991)

3. Anastassiou, G.A., Shisha, O.: Monotone approximation with linear differential operators. J. Approx. Theory **44**, 391–393 (1985)
4. Gonska, H.H., Simultaneously approximation by algebraic blending functions. In: Alfred Haar Memorial Conference, Budapest, Coloquia Mathematica Soc. Janos Bolyai, vol. 49, pp. 363–382. North-Holand, Amsterdam (1985)
5. Gonska, H.H.: Personal communication with author, 2–24–2014
6. S. Iqbal, K. Krulic, J. Pecaric, On an inequality of H.G. Hardy, J. Inequal. Appl. Article ID 264347, 23pp. (2010)
7. A. Marchaud, Differences et deerivees d'une fonction de deux variables. C.R. Acad. Sci. **178**, 1467–1470 (1924)
8. Marchaud, A.: Sur les derivees et sur les differences des fonctions de variables reelles. J. Math. Pures Appl. **6**, 337–425 (1927)
9. Schumaker, L.L.: Spline Functions: Basic Theory. Wiley, New York (1981)
10. Shisha, O.: Monotone approximation. Pacific J. Math. **15**, 667–671 (1965)

Nonlinear Approximation: q-Bernstein Operators of Max-Product Kind

Oktay Duman

Abstract In this paper, we construct a certain family of nonlinear operators in order to approximate a function by these nonlinear operators. For construction, we use the linear q-Bernstein polynomials and also the max-product algebra.

1 Introduction

The main motivation for this study is the papers by Bede et al. [4, 5]. We construct a certain family of max-product type approximating operators based on q-integers having the property of pseudo-linearity which is a weaker notion than the classical linearity.

We first recall some basic concepts from approximating operators. The classical Bernstein polynomial and its q-generalization are given respectively by

$$B_n(f; x) = \sum_{k=0}^{n} \binom{n}{k} f\left(\frac{k}{n}\right) x^k (1-x)^{n-k} \tag{1}$$

and

$$B_n(f; x; q) = \sum_{k=0}^{n} \begin{bmatrix} n \\ k \end{bmatrix}_q f\left(\frac{[k]_q}{[n]_q}\right) x^k \prod_{s=0}^{n-k-1} (1 - q^s x), \tag{2}$$

where $n \in \mathbb{N}$, $f \in C[0, 1]$, $x \in [0, 1]$, $q \in (0, 1]$ (see [16, 17]). Recall that, a q-integer is given by

$$[n]_q := \frac{1 - q^n}{1 - q} \text{ if } q \neq 1, \text{ and } [n]_q = n \text{ if } q = 1;$$

O. Duman (✉)
Department of Mathematics, TOBB Economics and Technology University,
Ankara, Turkey
e-mail: oduman@etu.edu.tr; okitayduman@gmail.com

© Springer International Publishing Switzerland 2016   33
G.A. Anastassiou and O. Duman (eds.), *Intelligent Mathematics II:*
Applied Mathematics and Approximation Theory, Advances in Intelligent Systems
and Computing 441, DOI 10.1007/978-3-319-30322-2_3

and the q-factorial is defined by

$$[n]_q! := [n]_q \ldots [2]_q [1]_q \text{ with } [0]_q! := 1;$$

and also the q-binomial coefficient is defined by

$$\begin{bmatrix} n \\ k \end{bmatrix}_q := \frac{[n]_q!}{[k]_q! [n-k]_q!}.$$

It is well-known that the polynomials in (1) and (2) are positive and linear, and so their approximation properties can easily be obtain from the classical Korovkin theorem (see [15]). However, in recent years, a nonlinear modification of the classical Bernstein polynomial has been introduced by Bede and Gal [5] (see, also, [4]). Although, of course, the Korovkin theorem fails for this nonlinear operator, they showed in [5] that the new operator has a similar approximation behavior to the classical Bernstein polynomial. Later, they applied this idea for other well-known approximating operators, such as, Favard-Szá sz-Mirakjan operator, Meyer-König and Zeller operator, Baskakov operator, Bleimann-Butzer-Hahn operator, and etc. (see [1, 3, 7, 8]). These studies provide an important improvement on the approximation theory due to nonlinearity of operators. However, so far, there is no such an improvement on the q-Bernstein polynomial given by (2). The aim of the present paper is to fill in this gap in the literature.

This paper is organized as follows. In the second section, we construct a nonlinear q-Bernstein operator of max-product kind, and in the third section, we obtain an error estimation for these operators. In the last section, we give an statistical approximation theorem and discuss some concluding remarks.

2   Construction of the Operators

In this section, we construct a nonlinear approximation operator by modifying the q-Bernstein polynomial given by (2). For the construction, we mainly use the idea by Bede et al. in [4, 5].

We consider the operations "$\vee$" (maximum) and "$\cdot$" (product) over the interval $[0, +\infty)$. Then, $([0, +\infty), \vee, \cdot)$ has a semiring structure and is called "max-product algebra" (see, for instance, [5, 6]). Now let

$$C_+[0, 1] := \{f : [0, 1] \to [0, +\infty) : f \text{ is continuous on } [0, 1]\}.$$

Considering the identity

$$\sum_{k=0}^{n} \binom{n}{k} x^k (1-x)^{n-k} = 1,$$

the classical Bernstein polynomial can be written in the form that

$$B_n(f; x) = \frac{\sum_{k=0}^{n} \binom{n}{k} f\left(\frac{k}{n}\right) x^k (1-x)}{\sum_{k=0}^{n} \binom{n}{k} x^k (1-x)^{n-k}}.$$

In this case, replacing the sum operator $\sum$ by the maximum operator $\vee$, Bede and Gal [5] (see, also, [4]) introduce the following nonlinear approximation operators

$$B_n^{(M)}(f; x) = \frac{\bigvee_{k=0}^{n} \binom{n}{k} f\left(\frac{k}{n}\right) x^k (1-x)^{n-k}}{\bigvee_{k=0}^{n} \binom{n}{k} x^k (1-x)^{n-k}} \quad (f \in C_+[0, 1], \ n \in \mathbb{N}). \quad (3)$$

Now, by a similar idea, we define our operators as follows:

$$B_n^{(M)}(f; x; q) = \frac{\bigvee_{k=0}^{n} p_{n,k}(x; q) f\left(\frac{[k]_q}{[n]_q}\right)}{\bigvee_{k=0}^{n} p_{n,k}(x; q)}, \quad (4)$$

where $n \in \mathbb{N}$, $f \in C_+[0, 1]$, $x \in [0, 1]$, $q \in (0, 1)$, and $p_{n,k}(x; q)$ is given by

$$p_{n,k}(x; q) = \begin{bmatrix} n \\ k \end{bmatrix}_q x^k \prod_{s=1}^{n-k} (1 - q^s x). \quad (5)$$

Here, we assume that the empty product is one. Notice that, in (5), we prefer $\prod_{s=1}^{n-k}(1 - q^s x)$ rather than $\prod_{s=0}^{n-k-1}(1 - q^s x)$ due to some technical complexity which comes from the q-calculus. Actually, Lemma 5 and its proof explain why we need such a preference. Observe that when $q \to 1^-$, then our operators $B_n^{(M)}(f; x; q)$ reduce to the operators $B_n^{(M)}(f; x)$ given by (3).

It is easy to check that $\bigvee_{k=0}^{n} p_{n,k}(x; q) > 0$ for all $x \in [0, 1]$ and $q \in (0, 1)$; and hence, $B_n^{(M)}(f; x; q)$ is well-defined. Also, we see that, for any $f \in C_+[0, 1]$,

$$B_n^{(M)}(f; 0; q) = f(0).$$

Also, if $f \in C_+[0, 1]$ and nondecreasing on $[0, 1]$, then we have

$$B_n^{(M)}(f; 1; q) = f(1).$$

Indeed, we will see in Lemma 5 (the case of $j = n$) that

$$\bigvee_{k=0}^{n} p_{n,k}(1; q) = p_{n,n}(1; q) = 1,$$

which implies

$$B_n^{(M)}(f; 1; q) = \bigvee_{k=0}^{n} p_{n,k}(1; q) f \left(\frac{[k]_q}{[n]_q} \right).$$

Since f is nondecreasing on $[0, 1]$, we observe that

$$p_{n,k}(1; q) f \left(\frac{[k]_q}{[n]_q} \right) \leq p_{n,n}(1; q) f \left(\frac{[n]_q}{[n]_q} \right) = f(1),$$

which gives $B_n^{(M)}(f; 1; q) = f(1)$. But this property is not true for all $f \in C_+[0, 1]$. For example, consider the function $f(x) = 1 - x$ on $[0, 1]$. Then, for any $q \in (0, 1)$, we observe that

$$B_n^{(M)}(f; 1; q) > p_{n,0}(1; q) f(0) = \prod_{s=1}^{n}(1 - q^s) > 0 = f(1).$$

If $f \in C_+[0, 1]$, then it is not hard to see that $B_n^{(M)}(f; \cdot; q) \in C_+[0, 1]$ for any $n \in \mathbb{N}$ and $q \in (0, 1)$; and hence $B_n^{(M)}(f; x; q)$ is a positive operator. However, we will see that it is not linear over $C_+[0, 1]$.

Now let $f, g \in C_+[0, 1]$. Then, by the definition, we see that

$$f \leq g \Rightarrow B_n^{(M)}(f; x; q) \leq B_n^{(M)}(g; x; q). \tag{6}$$

So, $B_n^{(M)}(f; x; q)$ is increasing with respect to $f \in C_+[0, 1]$. Also, for any $f, g \in C_+[0, 1]$, we have

$$B_n^{(M)}(f + g; x; q) \leq B_n^{(M)}(f; x; q) + B_n^{(M)}(g; x; q). \tag{7}$$

We can find functions f and g such that the above inequality is strict. For example, define $f(t) = 1 - t$ and $g(t) = t$ for all $t \in [0, 1]$. Then, the above facts yield that

$$B_n^{(M)}(f + g; 1; q) = B_n^{(M)}(e_0; 1; q) = 1(e_0(t) := 1),$$
$$B_n^{(M)}(f; 1; q) > f(1) = 0,$$
$$B_n^{(M)}(g; 1; q) = g(1) = 1,$$

and hence

$$B_n^{(M)}(f + g; 1; q) < B_n^{(M)}(f; 1; q) + B_n^{(M)}(g; 1; q)$$

for such f and g. This clearly shows that our operators $B_n^{(M)}(f; x; q)$ is not linear over $C_+[0, 1]$.

As usual, let $\omega(f, \delta)$, $\delta > 0$, denote the modulus of continuity of $f \in C_+[0, 1]$, defined by

$$\omega(f, \delta) = \max_{|x-y| \leq \delta} |f(x) - f(y)|.$$

Now, using (6), (7), and the fact that $B_n^{(M)}(e_0; 1; q) = 1$, and also applying Corollary 3 in [5] (see also Corollary 2.3 in [4]), we immediately obtain the following estimation for the operators $B_n^{(M)}(f; x; q)$.

Corollary 1 *For all $f \in C_+[0, 1]$, $n \in \mathbb{N}$, $x \in [0, 1]$ and $q \in (0, 1)$, we have*

$$\left| B_n^{(M)}(f; x; q) - f(x) \right| \leq 2\omega(f, \delta_n(x; q)), \tag{8}$$

where

$$\delta_n(x; q) := B_n^{(M)}(\varphi_x; x; q) \text{ with } \varphi_x(t) = |t - x|. \tag{9}$$

In the next section, after the estimation of $B_n^{(M)}(\varphi_x; x; q)$ is computed, we will give the arranged version of Corollary 1.

3   An Error Estimation

Firstly, we estimate $B_n^{(M)}(\varphi_x; x; q)$ with $\varphi_x(t) = |t - x|$. To see this we need some notations and lemmas. In this section, we will mainly use the similar way to [4]; however, we have to adapt all items to the q-calculus. We should note that when computing all estimations, it is enough to consider $x \in (0, 1]$ since $B_n^{(M)}(f; 0; q) - f(0) = 0$ for any $f \in C_+[0, 1]$.

For $k, j \in \{0, 1, ..., n\}$ and $x \in \left[\dfrac{[j]_q}{[n+1]_q}, \dfrac{[j+1]_q}{[n+1]_q} \right]$, we define

$$M_{k,n,j}(x; q) = \frac{p_{n,k}(x; q) \left| \dfrac{[k]_q}{[n]_q} - x \right|}{p_{n,j}(x; q)} \tag{10}$$

and

$$m_{k,n,j}(x; q) = \frac{p_{n,k}(x; q)}{p_{n,j}(x; q)}. \tag{11}$$

In this case, (10) and (11) imply respectively that

$$k \geq j+1 \Rightarrow M_{k,n,j}(x; q) = \frac{p_{n,k}(x; q) \left(\dfrac{[k]_q}{[n]_q} - x \right)}{p_{n,j}(x; q)} \tag{12}$$

and

$$k \leq j - 1 \Rightarrow M_{k,n,j}(x;q) = \frac{p_{n,k}(x;q)\left(x - \dfrac{[k]_q}{[n]_q}\right)}{p_{n,j}(x;q)}, \tag{13}$$

where we use the following fact

$$\frac{[j-1]_q}{[n]_q} < \frac{[j]_q}{[n+1]_q}.$$

Now, for $k, j \in \{0, 1, ..., n\}$ with $k \geq j + 2$ and $x \in \left[\dfrac{[j]_q}{[n+1]_q}, \dfrac{[j+1]_q}{[n+1]_q}\right]$, we define

$$\overline{M}_{k,n,j}(x;q) = \frac{p_{n,k}(x;q)\left(\dfrac{[k]_q}{[n+1]_q} - x\right)}{p_{n,j}(x;q)} \tag{14}$$

and also; for $k, j \in \{0, 1, ..., n\}$ with $k \leq j - 2$ and $x \in \left[\dfrac{[j]_q}{[n+1]_q}, \dfrac{[j+1]_q}{[n+1]_q}\right]$, we consider

$$\underline{M}_{k,n,j}(x;q) = \frac{p_{n,k}(x;q)\left(x - \dfrac{[k]_q}{[n+1]_q}\right)}{p_{n,j}(x;q)}. \tag{15}$$

Then, we first get the next lemma.

Lemma 2 *Let* $q \in (0, 1)$, $j \in \{0, 1, ..., n\}$ *and* $x \in \left[\dfrac{[j]_q}{[n+1]_q}, \dfrac{[j+1]_q}{[n+1]_q}\right]$. *Then, we obtain the following inequalities:*
(i) *for all* $k \in \{0, 1, ..., n\}$ *with* $k \geq j + 2$,

$$\overline{M}_{k,n,j}(x;q) \leq M_{k,n,j}(x;q) \leq \left(1 + \frac{2}{q^{n+1}}\right)\overline{M}_{k,n,j}(x;q).$$

(ii) *for all* $k \in \{0, 1, ..., n\}$ *with* $k \leq j - 2$,

$$M_{k,n,j}(x;q) \leq \underline{M}_{k,n,j}(x;q) \leq \left(1 + \frac{2}{q^n}\right)M_{k,n,j}(x;q).$$

Proof (i) The inequality $\overline{M}_{k,n,j}(x;q) \leq M_{k,n,j}(x;q)$ follows from (12) and (14), immediately. Also, using the fact that $[n+1]_q = [n]_q + q^n$, we have

$$\frac{M_{k,n,j}(x;q)}{\overline{M}_{k,n,j}(x;q)} = \frac{\frac{[k]_q}{[n]_q} - x}{\frac{[k]_q}{[n+1]_q} - x} \leq \frac{\frac{[k]_q}{[n]_q} - \frac{[j]_q}{[n+1]_q}}{\frac{[k]_q}{[n+1]_q} - \frac{[j+1]_q}{[n+1]_q}}$$

$$= \frac{[k]_q[n+1]_q - [n]_q[j]_q}{[n]_q\left([k]_q - [j+1]_q\right)} = \frac{[k]_q[n]_q - [n]_q[j]_q + q^n[k]_q}{[n]_q\left([k]_q - [j+1]_q\right)}$$

$$\leq \frac{[k]_q[n]_q - [n]_q[j]_q + q^n[n]_q}{[n]_q\left([k]_q - [j+1]_q\right)} = \frac{[k]_q - [j]_q + q^n}{[k]_q - [j+1]_q}$$

$$= \frac{[k]_q - [j]_q + q^n}{[k]_q - [j]_q - q^j} = 1 + \frac{q^n + q^j}{[k]_q - [j]_q - q^j}$$

$$\leq 1 + \frac{2}{[k]_q - [j]_q - q^j}.$$

Now using the facts that $k \geq j + 2$ and $j \leq n$ we get

$$[k]_q - [j]_q - q^j \geq [j+2]_q - [j]_q - q^j$$
$$= q^{j+1}$$
$$\geq q^{n+1}.$$

Hence, we obtain that

$$\frac{M_{k,n,j}(x;q)}{\overline{M}_{k,n,j}(x;q)} \leq 1 + \frac{2}{q^{n+1}},$$

which proves (i).

(ii) From (13) and (15), we easily get $M_{k,n,j}(x;q) \leq \underline{M}_{k,n,j}(x;q)$. Also, using again the fact that $[n+1]_q = [n]_q + q^n$, we observe that

$$\frac{\underline{M}_{k,n,j}(x;q)}{M_{k,n,j}(x;q)} = \frac{x - \frac{[k]_q}{[n+1]_q}}{x - \frac{[k]_q}{[n]_q}} \leq \frac{\frac{[j+1]_q}{[n+1]_q} - \frac{[k]_q}{[n+1]_q}}{\frac{[j]_q}{[n+1]_q} - \frac{[k]_q}{[n]_q}}$$

$$= \frac{[n]_q\left([j+1]_q - [k]_q\right)}{[n]_q[j]_q - [k]_q[n+1]_q} = \frac{[n]_q\left([j+1]_q - [k]_q\right)}{[n]_q[j]_q - [k]_q[n]_q - q^n[k]_q}$$

$$\leq \frac{[n]_q\left([j+1]_q - [k]_q\right)}{[n]_q[j]_q - [k]_q[n]_q - q^n[n]_q} = \frac{[j+1]_q - [k]_q}{[j]_q - [k]_q - q^n}$$

$$= \frac{[j]_q - [k]_q + q^j}{[j]_q - [k]_q - q^n} = 1 + \frac{q^n + q^j}{[j]_q - [k]_q - q^n}$$

$$\leq 1 + \frac{2}{[j]_q - [k]_q - q^n}$$

Since $k \leq j - 2$ and $j \leq n$, we may write that

$$[j]_q - [k]_q - q^n \geq [j]_q - [j-2]_q - q^n$$
$$= q^{j-2} + q^{j-1} - q^n$$
$$\geq q^n$$

Therefore, we get

$$\frac{M_{k,n,j}(x;q)}{M_{k,n,j}(x;q)} \leq 1 + \frac{2}{q^n},$$

which completes the proof.

Lemma 3 *Let $q \in (0, 1)$. Then, for all $k, j \in \{0, 1, ..., n\}$ and*

$$x \in \left[\frac{[j]_q}{[n+1]_q}, \frac{[j+1]_q}{[n+1]_q} \right],$$

we get

$$m_{k,n,j}(x;q) \leq 1.$$

Proof We consider two possible cases: (*a*) $k \geq j$ and (*b*) $k < j$.

Case (*a*): Let $k \geq j$. Since $h(x) = \dfrac{1 - q^{n-k}x}{x}$ is nonincreasing on the interval $\left[\dfrac{[j]_q}{[n+1]_q}, \dfrac{[j+1]_q}{[n+1]_q} \right]$, it follows from (11) that

$$\frac{m_{k,n,j}(x;q)}{m_{k+1,n,j}(x;q)} = \frac{[k+1]_q}{[n-k]_q} \cdot \frac{1 - q^{n-k}x}{x}$$

$$\geq \frac{[k+1]_q}{[n-k]_q} \cdot \frac{1 - q^{n-k}\dfrac{[j+1]_q}{[n+1]_q}}{\dfrac{[j+1]_q}{[n+1]_q}}$$

$$= \frac{[k+1]_q}{[n-k]_q} \cdot \frac{[n+1]_q - q^{n-k}[j+1]_q}{[j+1]_q}$$

$$\geq \frac{[k+1]_q}{[j+1]_q} \cdot \frac{[n+1]_q - q^{n-k}[k+1]_q}{[n-k]_q}$$

Since $[k+1]_q \geq [j+1]_q$, we get

$$\frac{m_{k,n,j}(x;q)}{m_{k+1,n,j}(x;q)} \geq \frac{[n+1]_q - q^{n-k}[k+1]_q}{[n-k]_q}$$

$$= \frac{1 - q^{n+1} - q^{n-k}(1 - q^{k+1})}{1 - q^{n-k}}$$

$$= 1.$$

Then, we conclude that

$$1 = m_{j,n,j}(x;q) \geq m_{j+1,n,j}(x;q) \geq m_{j+2,n,j}(x;q) \geq \dots \geq m_{n,n,j}(x;q),$$

which completes the proof of case (a).

Case (b): Let $k < j$. Since $h(x) = \dfrac{x}{1 - q^{n-k+1}x}$ is nondecreasing on the interval

$$\left[\frac{[j]_q}{[n+1]_q}, \frac{[j+1]_q}{[n+1]_q} \right],$$

we have

$$\frac{m_{k,n,j}(x;q)}{m_{k-1,n,j}(x;q)} = \frac{[n-k+1]_q}{[k]_q} \cdot \frac{x}{1 - q^{n-k+1}x}$$

$$\geq \frac{[n-k+1]_q}{[k]_q} \cdot \frac{\dfrac{[j]_q}{[n+1]_q}}{1 - q^{n-k+1}\dfrac{[j]_q}{[n+1]_q}}$$

$$= \frac{[n-k+1]_q}{[k]_q} \cdot \frac{[j]_q}{[n+1]_q - q^{n-k+1}[j]_q}$$

$$\geq \frac{[j]_q}{[k]_q} \cdot \frac{[n-k+1]_q}{[n+1]_q - q^{n-k+1}[k]_q}$$

Since $[k]_q < [j]_q$, we may write that

$$\frac{m_{k,n,j}(x;q)}{m_{k-1,n,j}(x;q)} \geq \frac{[n-k+1]_q}{[n+1]_q - q^{n-k+1}[k]_q}$$

$$= \frac{1 - q^{n-k+1}}{1 - q^{n+1} - q^{n-k+1}(1 - q^k)}$$

$$= 1.$$

Then we easily get

$$1 = m_{j,n,j}(x;q) \geq m_{j-1,n,j}(x;q) \geq m_{j-2,n,j}(x;q) \geq \dots \geq m_{0,n,j}(x;q).$$

Therefore, the proof is completed.

Lemma 4 *Let $q \in (0, 1)$, $j \in \{0, 1, ..., n\}$ and $x \in \left[\dfrac{[j]_q}{[n+1]_q}, \dfrac{[j+1]_q}{[n+1]_q} \right]$. Then, we have:*

(i) If $k \in \{j+2, j+3, ..., n-1\}$ and $[k+1]_q - \sqrt{q^k [k+1]_q} \geq [j+1]_q$, then

$$\overline{M}_{k,n,j}(x; q) \geq \overline{M}_{k+1,n,j}(x; q).$$

(ii) If $k \in \{1, 2, ..., j-2\}$ and $[k]_q + \sqrt{q^{k-1} [k]_q} \leq [j]_q$, then

$$\underline{M}_{k,n,j}(x; q) \geq \underline{M}_{k-1,n,j}(x; q).$$

Proof (*i*) Let $k \in \{j+2, j+3, ..., n-1\}$ and $[k+1]_q - \sqrt{q^k [k+1]_q} \geq [j+1]_q$. First we may write that

$$\frac{\overline{M}_{k,n,j}(x; q)}{\overline{M}_{k+1,n,j}(x; q)} = \frac{[k+1]_q}{[n-k]_q} \cdot \frac{(1 - q^{n-k}x)}{x} \cdot \frac{\dfrac{[k]_q}{[n+1]_q} - x}{\dfrac{[k+1]_q}{[n+1]_q} - x}$$

$$= \frac{[k+1]_q}{[n-k]_q} \cdot \frac{(1 - q^{n-k}x)}{x} \cdot \frac{\left([k]_q - [n+1]_q x\right)}{\left([k+1]_q - [n+1]_q x\right)}.$$

Since $g_1(x) = \dfrac{(1 - q^{n-k}x)}{x} \cdot \dfrac{\left([k]_q - [n+1]_q x\right)}{\left([k+1]_q - [n+1]_q x\right)}$ is nonincreasing on the interval $\left[\dfrac{[j]_q}{[n+1]_q}, \dfrac{[j+1]_q}{[n+1]_q} \right]$, we get

$$\frac{\overline{M}_{k,n,j}(x; q)}{\overline{M}_{k+1,n,j}(x; q)} \geq \frac{[k+1]_q}{[n-k]_q} \cdot \frac{1 - q^{n-k}\dfrac{[j+1]_q}{[n+1]_q}}{\dfrac{[j+1]_q}{[n+1]_q}} \cdot \frac{[k]_q - [j+1]_q}{[k+1]_q - [j+1]_q}$$

$$= \frac{[n+1]_q - q^{n-k}[j+1]_q}{[n-k]_q} \cdot \frac{[k+1]_q}{[j+1]_q} \cdot \frac{[k]_q - [j+1]_q}{[k+1]_q - [j+1]_q}$$

$$\geq \frac{[n+1]_q - q^{n-k}[k+1]_q}{[n-k]_q} \cdot \frac{[k+1]_q}{[j+1]_q} \cdot \frac{[k]_q - [j+1]_q}{[k+1]_q - [j+1]_q}$$

$$= \frac{[k+1]_q}{[j+1]_q} \cdot \frac{[k]_q - [j+1]_q}{[k+1]_q - [j+1]_q}.$$

The hypothesis

$$[k+1]_q - \sqrt{q^k [k+1]_q} \geq [j+1]_q$$

is equivalent to

$$[k+1]_q - \sqrt{[k+1]_q^2 - [k]_q[k+1]_q} \geq [j+1]_q,$$

which implies that

$$[k+1]_q \left([k]_q - [j+1]_q\right) \geq [j+1]_q \left([k+1]_q - [j+1]_q\right).$$

Hence we obtain that

$$\frac{\overline{M}_{k,n,j}(x;q)}{\overline{M}_{k+1,n,j}(x;q)} \geq 1,$$

which proves (i).

(ii) Now let $k \in \{1, 2, ..., j-2\}$ and $[k]_q + \sqrt{q^{k-1}[k]_q} \leq [j]_q$. Then, we observe that

$$\frac{\underline{M}_{k,n,j}(x;q)}{\underline{M}_{k-1,n,j}(x;q)} = \frac{[n-k+1]_q}{[k]_q} \cdot \frac{x}{(1-q^{n-k+1}x)} \cdot \frac{x - \dfrac{[k]_q}{[n+1]_q}}{x - \dfrac{[k-1]_q}{[n+1]_q}}$$

$$= \frac{[n-k+1]_q}{[k]_q} \cdot \frac{x}{1-q^{n-k+1}x} \cdot \frac{[n+1]_q x - [k]_q}{[n+1]_q x - [k-1]_q}.$$

Since $g_2(x) = \dfrac{x}{1-q^{n-k+1}x} \dfrac{[n+1]_q x - [k]_q}{[n+1]_q x - [k-1]_q}$ is nondecreasing on the interval $\left[\dfrac{[j]_q}{[n+1]_q}, \dfrac{[j+1]_q}{[n+1]_q}\right]$, we get

$$\frac{\underline{M}_{k,n,j}(x;q)}{\underline{M}_{k-1,n,j}(x;q)} \geq \frac{[n-k+1]_q}{[k]_q} \cdot \frac{\dfrac{[j]_q}{[n+1]_q}}{1-q^{n-k+1}\dfrac{[j]_q}{[n+1]_q}} \cdot \frac{[j]_q - [k]_q}{[j]_q - [k-1]_q}$$

$$= \frac{[n-k+1]_q}{[n+1]_q - q^{n-k+1}[j]_q} \cdot \frac{[j]_q}{[k]_q} \cdot \frac{[j]_q - [k]_q}{[j]_q - [k-1]_q}$$

$$\geq \frac{[n-k+1]_q}{[n+1]_q - q^{n-k+1}[k]_q} \cdot \frac{[j]_q}{[k]_q} \cdot \frac{[j]_q - [k]_q}{[j] - [k-1]_q}$$

$$= \frac{[j]_q}{[k]_q} \cdot \frac{[j]_q - [k]_q}{[j]_q - [k-1]_q}.$$

The hypothesis $[k]_q + \sqrt{q^{k-1}[k]_q} \leq [j]_q$ is equivalent to

$$[k]_q + \sqrt{[k]_q^2 - [k]_q[k-1]_q} \le [j]_q,$$

which implies that

$$[j]_q \left([j]_q - [k]_q\right) \ge [k]_q \left([j]_q - [k-1]_q\right).$$

Then, we obtain that

$$\frac{M_{k,n,j}(x; q)}{M_{k-1,n,j}(x; q)} \ge 1,$$

which completes the proof of (ii).

Lemma 5 *Let $q \in (0, 1)$, $j \in \{0, 1, ..., n\}$ and $x \in \left[\dfrac{[j]_q}{[n+1]_q}, \dfrac{[j+1]_q}{[n+1]_q}\right]$. Then, we have*

$$\bigvee_{k=0}^{n} p_{n,k}(x; q) = p_{n,j}(x; q).$$

Proof Firstly, we claim that, for $0 \le k < k+1 \le n$,

$$0 \le p_{n,k+1}(x; q) \le p_{n,k}(x; q) \Leftrightarrow x \in \left[0, \frac{[k+1]_q}{[n+1]_q}\right]. \qquad (16)$$

Indeed, by definition, we have

$$0 \le p_{n,k+1}(x; q) \le p_{n,k}(x; q)$$

$$\Leftrightarrow 0 \le \begin{bmatrix} n \\ k+1 \end{bmatrix}_q x^{k+1} \prod_{s=1}^{n-k-1}(1 - q^s x) \le \begin{bmatrix} n \\ k \end{bmatrix}_q x^k \prod_{s=1}^{n-k}(1 - q^s x)$$

$$\Leftrightarrow 0 \le \left(\begin{bmatrix} n \\ k+1 \end{bmatrix}_q + q_q^{n-k}\begin{bmatrix} n \\ k \end{bmatrix}_q\right) x \le \begin{bmatrix} n \\ k \end{bmatrix}_q.$$

Since

$$\begin{bmatrix} n \\ k+1 \end{bmatrix}_q + q_q^{n-k}\begin{bmatrix} n \\ k \end{bmatrix}_q = \begin{bmatrix} n+1 \\ k+1 \end{bmatrix}_q, \qquad (17)$$

we get

$$0 \le p_{n,k+1}(x; q) \le p_{n,k}(x; q) \Leftrightarrow 0 \le x \le \frac{[k+1]_q}{[n+1]_q},$$

which corrects the claim (16). Hence, if we take $k = 0, 1, 2, ..., n - 1$, then we obtain that

$$p_{n,1}(x; q) \le p_{n,0}(x; q) \Leftrightarrow 0 \le x \le \frac{1}{[n + 1]_q},$$

$$p_{n,2}(x; q) \le p_{n,1}(x; q) \Leftrightarrow 0 \le x \le \frac{[2]_q}{[n + 1]_q},$$

...

$$p_{n,k+1}(x; q) \le p_{n,k}(x; q) \Leftrightarrow 0 \le x \le \frac{[k + 1]_q}{[n + 1]_q},$$

...

$$p_{n,n-1}(x; q) \le p_{n,n-2}(x; q) \Leftrightarrow 0 \le x \le \frac{[n - 1]_q}{[n + 1]_q},$$

$$p_{n,n}(x; q) \le p_{n,n-1}(x; q) \Leftrightarrow 0 \le x \le \frac{[n]_q}{[n + 1]_q}.$$

Using the above facts, we may write that

$$0 \le x \le \frac{1}{[n + 1]_q} \Rightarrow p_{n,k}(x; q) \le p_{n,0}(x; q)$$
(for all $k = 0, 1, ..., n$),

$$\frac{1}{[n + 1]_q} \le x \le \frac{[2]}{[n + 1]_q} \Rightarrow p_{n,k}(x; q) \le p_{n,1}(x; q)$$
(for all $k = 0, 1, ..., n$),

...

$$\frac{[n - 1]_q}{[n + 1]_q} \le x \le \frac{[n]_q}{[n + 1]_q} \Rightarrow p_{n,k}(x; q) \le p_{n,n-1}(x; q)$$
(for all $k = 0, 1, ..., n$),

$$\frac{[n]_q}{[n + 1]_q} \le x \le 1 \Rightarrow p_{n,k}(x; q) \le p_{n,n}(x; q)$$
(for all $k = 0, 1, ..., n$).

As a result, we get, for all $x \in \left[\frac{[j]_q}{[n + 1]_q}, \frac{[j + 1]_q}{[n + 1]_q} \right]$ ($j = 0, 1, 2, ..., n$), that

$$p_{n,k}(x; q) \le p_{n,j}(x; q) \text{ for all } k = 0, 1, ..., n,$$

which gives

$$\bigvee_{k=0}^{n} p_{n,k}(x;q) = p_{n,j}(x;q).$$

The proof is completed.

Note that to obtain the equality (17) we need the preference $\prod_{s=1}^{n-k}(1 - q^s x)$ rather than $\prod_{s=0}^{n-k-1}(1 - q^s x)$ in the definition of our operator.

Now, the next result gives an estimation for $B_n^{(M)}(\varphi_x; x; q)$ with $\varphi_x(t) = |t - x|$.

Corollary 6 *Let* $q \in (0, 1)$. *Then, for all* $x \in [0, 1]$ *and* $n \in \mathbb{N}$, *we have*

$$B_n^{(M)}(\varphi_x; x; q) \le \frac{1 + \frac{2}{q^{n+1}}}{\sqrt{[n]_q}}. \tag{18}$$

Proof By definition, we get

$$B_n^{(M)}(\varphi_x; x; q) = \frac{\bigvee_{k=0}^{n} p_{n,k}(x;q) \left| \frac{[k]_q}{[n]_q} - x \right|}{\bigvee_{k=0}^{n} p_{n,k}(x;q)}.$$

Firstly assume that $x \in \left[\dfrac{[j]_q}{[n+1]_q}, \dfrac{[j+1]_q}{[n+1]_q} \right]$ for a fixed $j \in \{0, 1, ..., n\}$. Then we have

$$B_n^{(M)}(\varphi_x; x; q) = \bigvee_{k=0}^{n} M_{k,n,j}(x;q), \; x \in \left[\dfrac{[j]_q}{[n+1]_q}, \dfrac{[j+1]_q}{[n+1]_q} \right], \tag{19}$$

where $M_{k,n,j}(x;q)$ is the same as in (10). We first claim that, for $k = 0, 1, 2, ..., n$ and $j = 0$,

$$M_{k,n,0}(x;q) \le \frac{1}{[n]_q} \text{for} x \in \left[0, \dfrac{1}{[n+1]_q} \right]. \tag{20}$$

Indeed, if $j = k = 0$, then $x \in \left[0, \dfrac{1}{[n+1]_q} \right]$, and so

$$M_{0,n,0}(x;q) = x \le \frac{1}{[n+1]_q} \le \frac{1}{[n]_q}.$$

Also, if $j = 0$ and $k \in \{1, 2, ..., n\}$, then we obtain, for $x \in \left[0, \dfrac{1}{[n+1]_q} \right]$, that

$$M_{k,n,0}(x;q) = \frac{p_{n,k}(x;q)\left(\frac{[k]_q}{[n]_q} - x\right)}{p_{n,0}(x;q)}$$

$$\leq \begin{bmatrix} n \\ k \end{bmatrix}_q \frac{[k]_q}{[n]_q} \cdot \frac{x^k}{(1 - q^{n-k+1}x)...(1 - q^n x)}$$

$$\leq \begin{bmatrix} n \\ k \end{bmatrix}_q \frac{[k]_q}{[n]_q} \cdot \frac{\frac{1}{[n+1]_q^k}}{\left(1 - \frac{q^{n-k+1}}{[n+1]_q}\right)...\left(1 - \frac{q^n}{[n+1]_q}\right)}$$

$$= \begin{bmatrix} n \\ k \end{bmatrix}_q \frac{[k]_q}{[n]_q} \cdot \frac{1}{\left([n+1]_q - q^{n-k+1}\right)...\left([n+1]_q - q^n\right)}$$

$$= \frac{[n]_q!}{[k]_q![n-k]_q!} \frac{[k]_q}{[n]_q} \cdot \frac{1}{\left([n+1]_q - q^{n-k+1}\right)...\left([n+1]_q - q^n\right)}$$

$$= \frac{[n-k+1]_q}{[n+1]_q - q^{n-k+1}}...\frac{[n]_q}{[n+1]_q - q^n} \cdot \frac{1}{[k-1]_q![n]_q}.$$

Here, for each $m = 1, 2, ..., k$, we observe that

$$\frac{[n-k+m]_q}{[n+1]_q - q^{n-k+m}} \leq 1.$$

Hence the above inequality gives that

$$M_{k,n,0}(x;q) \leq \frac{1}{[k-1]_q![n]_q} \leq \frac{1}{[n]_q},$$

which corrects (20). Now we claim that the inequality

$$M_{k,n,j}(x;q) \leq \frac{1 + \frac{2}{q^{n+1}}}{\sqrt{[n+1]_q}} \tag{21}$$

holds for $x \in \left[\frac{[j]_q}{[n+1]_q}, \frac{[j+1]_q}{[n+1]_q}\right]$ $(j = 1, ..., n; k = 0, 1, 2, ..., n)$. To see this we consider the following five possible cases:

(a) $k \in \{j - 1, j, j + 1\}$,
(b) $k \geq j + 2$ and $[k+1]_q - \sqrt{q^k[k+1]_q} < [j+1]_q$,
(c) $k \geq j + 2$ and $[k+1]_q - \sqrt{q^k[k+1]_q} \geq [j+1]$,
(d) $k \leq j - 2$ and $[k]_q + \sqrt{q^{k-1}[k]_q} \geq [j]_q$,
(e) $k \leq j - 2$ and $[k]_q + \sqrt{q^{k-1}[k]_q} < [j]_q$.

Case (a): If $k = j - 1$, then it follows from Lemma 3 that

$$M_{j-1,n,j}(x;q) = m_{j-1,n,j}(x;q)\left(x - \frac{[j-1]_q}{[n]_q}\right)$$

$$\leq \frac{[j+1]_q}{[n+1]_q} - \frac{[j-1]_q}{[n]_q} \leq \frac{[j+1]_q}{[n+1]_q} - \frac{[j-1]_q}{[n+1]_q}$$

$$= \frac{q^{j-1}(1+q)}{[n+1]_q} \leq \frac{2}{[n+1]_q}.$$

If $k = j$, then, by Lemma 3, we obtain that

$$M_{j,n,j}(x;q) = \left|\frac{[j]_q}{[n]_q} - x\right| \leq \frac{1}{[n+1]_q}$$

If $k = j+1$, then using again Lemma 3, we have

$$M_{j+1,n,j}(x;q) = m_{j+1,n,j}(x;q)\left(\frac{[j+1]_q}{[n]_q} - x\right)$$

$$\leq \frac{[j+1]_q}{[n]_q} - \frac{[j]_q}{[n+1]_q} \leq \frac{[n+1]_q[j+1]_q - [n]_q[j]_q}{[n]_q[n+1]_q}$$

$$= \frac{([n]_q + q^n)([j]_q + q^j) - [n]_q[j]_q}{[n]_q[n+1]_q}$$

$$= \frac{q^n[j]_q + q^j[n]_q + q^{n+j}}{[n]_q[n+1]_q}$$

$$\leq \frac{q^n[n]_q + q^j[n]_q + q^{n+j}[n]_q}{[n]_q[n+1]_q} = \frac{q^n + q^j + q^{n+j}}{[n+1]_q}$$

$$\leq \frac{3}{[n+1]_q}.$$

Case (b): Let $k \geq j+2$ and $[k+1]_q - \sqrt{q^k[k+1]_q} < [j+1]_q$. Then, we obtain from Lemma 3 that

$$\overline{M}_{k,n,j}(x;q) = m_{k,n,j}(x;q)\left(\frac{[k]_q}{[n+1]_q} - x\right) \leq \frac{[k]_q}{[n+1]_q} - \frac{[j]_q}{[n+1]_q}.$$

By hypothesis, since

$$q[j]_q > q[k]_q - \sqrt{q^k[k+1]_q},$$

we get

$$\overline{M}_{k,n,j}(x;q) \le \frac{[k]_q}{[n+1]_q} - \frac{[k]_q - \frac{1}{q}\sqrt{q^k[k+1]_q}}{[n+1]_q}$$

$$= \frac{\sqrt{q^{k-2}[k+1]_q}}{[n+1]_q} \le \frac{\sqrt{q^{k-2}[n+1]_q}}{[n+1]_q}.$$

Since $k - 2 \ge 0$, we conclude that

$$\overline{M}_{k,n,j}(x;q) \le \frac{1}{\sqrt{[n+1]_q}}$$

Also using Lemma 2 (i), we obtain that

$$M_{k,n,j}(x;q) \le \frac{1 + \frac{2}{q^{n+1}}}{\sqrt{[n+1]_q}}.$$

Case (c): Let $k \ge j + 2$ and $[k+1]_q - \sqrt{q^k[k+1]_q} \ge [j+1]_q$. In this case, we first show that the function $h(k) = [k+1]_q - \sqrt{q^k[k+1]_q}$ is increasing with respect to k. Indeed, we may write that

$$h(k+1) - h(k) = [k+2]_q - [k+1]_q + \sqrt{q^k[k+1]_q} - \sqrt{q^{k+1}[k+2]_q}$$

$$\ge [k+2]_q - [k+1]_q + \sqrt{q^k[k+1]_q} - \sqrt{q^k[k+2]_q}$$

$$= q^{k+1} - q^{k/2}\left(\sqrt{[k+2]_q} - \sqrt{[k+1]_q}\right)$$

$$= q^{k+1} - \frac{q^{k/2}q^{k+1}}{\sqrt{[k+2]_q} + \sqrt{[k+1]_q}}$$

$$= q^{k+1}\left(1 - \frac{q^{k/2}}{\sqrt{[k+2]_q} + \sqrt{[k+1]_q}}\right)$$

$$\ge q^{k+1}\left(1 - \frac{1}{\sqrt{[k+2]_q} + \sqrt{[k+1]_q}}\right)$$

$$> 0.$$

Hence, there exists $\bar{k} \in \{0, 1, 2, ..., n\}$, of maximum value, such that

$$[\bar{k}+1]_q - \sqrt{q^{\bar{k}}[\bar{k}+1]_q} < [j+1]_q.$$

Let $\tilde{k} = \bar{k} + 1$. Then, for all $k \geq \tilde{k}$, we get

$$[k+1]_q - \sqrt{q^k[k+1]_q} \geq [j+1]_q.$$

It is easy to check that $\tilde{k} \geq j + 2$. Indeed, observe that $h(j+1) < [j+1]_q$, but $h(\tilde{k}) \geq [j+1]_q$. Since h is increasing, $\tilde{k} > j+1$ and so $\tilde{k} \geq j+2$. Also, we get

$$\overline{M}_{\tilde{k},n,j}(x;q) = m_{\tilde{k},n,j}(x;q)\left(\frac{[\tilde{k}]_q}{[n+1]_q} - x\right) \leq \frac{[\tilde{k}+1]_q}{[n+1]_q} - x$$

$$\leq \frac{[\bar{k}+1]_q}{[n+1]_q} - \frac{[j]_q}{[n+1]_q}.$$

Since

$$[j]_q > [\bar{k}+1]_q - q^j - \sqrt{q^{\bar{k}}[\bar{k}+1]_q},$$

we see that

$$\overline{M}_{\tilde{k},n,j}(x;q) \leq \frac{[\bar{k}+1]_q}{[n+1]_q} - \frac{[\bar{k}+1]_q - q^j - \sqrt{q^{\bar{k}}[\bar{k}+1]_q}}{[n+1]_q}$$

$$= \frac{q^j + \sqrt{q^{\bar{k}}[\bar{k}+1]_q}}{[n+1]_q}$$

$$\leq \frac{1 + \sqrt{[n+1]_q}}{[n+1]_q}$$

$$\leq \frac{2}{\sqrt{[n+1]_q}}.$$

By Lemma 4 (i), we can write that

$$\overline{M}_{\tilde{k},n,j}(x;q) \geq \overline{M}_{\tilde{k}+1,n,j}(x;q) \geq \dots \geq \overline{M}_{n,n,j}(x;q).$$

Hence, we see that, for all $k \in \{\tilde{k}, \tilde{k}+1, \tilde{k}+2, \dots, n\}$

$$\overline{M}_{k,n,j}(x;q) \leq \frac{2}{\sqrt{[n+1]_q}}.$$

Thus, for the same k's, it follows from Lemma 2 (i) that

$$M_{k,n,j}(x;q) \le \frac{2\left(1 + \frac{2}{q^{n+1}}\right)}{\sqrt{[n+1]_q}}$$

Case (d): Let $k \le j - 2$ and $[k]_q + \sqrt{q^{k-1}[k]_q} \ge [j]_q$. Then, we get

$$\underline{M}_{k,n,j}(x;q) = m_{k,n,j}(x;q)\left(x - \frac{[k]_q}{[n+1]_q}\right) \le \frac{[j+1]_q}{[n+1]_q} - \frac{[k]_q}{[n+1]_q}$$

$$= \frac{[j] + q^j}{[n+1]_q} - \frac{[k]_q}{[n+1]_q}.$$

By hypothesis, we have

$$\underline{M}_{k,n,j}(x;q) \le \frac{[k]_q + q^j + \sqrt{q^{k-1}[k]_q}}{[n+1]_q} - \frac{[k]_q}{[n+1]_q}$$

$$= \frac{q^j + \sqrt{q^{k-1}[k]_q}}{[n+1]_q} \le \frac{1 + \sqrt{[n+1]_q}}{[n+1]_q}$$

$$\le \frac{2}{\sqrt{[n+1]_q}}.$$

Also using Lemma 2 (ii), we conclude that

$$M_{k,n,j}(x;q) \le \frac{2}{\sqrt{[n+1]_q}}.$$

Case (e): $k \le j - 2$ and $[k]_q + \sqrt{q^{k-1}[k]_q} < [j]_q$. Now define the function u as follows:

$$u(t) = \frac{1 - q^t}{1 - q} + \sqrt{q^{t-1}\left(\frac{1-q^t}{1-q}\right)}, \quad t \ge 0.$$

Let

$$A_q := \frac{\ln\left((1 - \sqrt{q})/2\right)}{\ln q}.$$

Then we observe the following facts:

(i) $u(k) = [k]_q + \sqrt{q^{k-1}[k]_q}$ for each $k \in \{0, 1, 2, ..., n\}$,
(ii) u is increasing on $[0, A_q)$,
(iii) u is decreasing on $(A_q, +\infty)$,
(iv) $u'(A_q) = 0$,
(v) $u = \frac{1}{1-q}$ is a horizontal asymptote of the graph of $u = u(t)$,
(vi) $u(A_q) > \frac{1}{1-q}$.

Let $\bar{k} \in \{0, 1, 2, ..., n\}$ be the minimum value such that $[\bar{k}]_q + \sqrt{q^{\bar{k}-1}[\bar{k}]_q} \geq [j]_q$. It follows from the above facts that $\bar{k} \in (0, A_q)$. Let $\tilde{k} = \bar{k} - 1$. Then, Since u is increasing on $[0, A_q)$, we may write that $[\bar{k} - 1]_q + \sqrt{q^{k-2}[k-1]_q} < [j]_q$. It is easy to check that $\tilde{k} \leq j - 2$. Indeed, observe that $u(j - 1) > [j]_q$, but $u\left(\tilde{k}\right) < [j]_q$. Since u is increasing, $\tilde{k} < j - 1$ and so $\tilde{k} \leq j - 2$. Also, we get

$$\underline{M}_{\tilde{k}-1,n,j}(x; q) = m_{\tilde{k}-1,n,j}(x; q)\left(x - \frac{[\bar{k} - 1]_q}{[n + 1]_q}\right) \leq x - \frac{[\bar{k} - 1]_q}{[n + 1]_q}$$

$$\leq \frac{[j + 1]_q}{[n + 1]_q} - \frac{[\bar{k} - 1]_q}{[n + 1]_q} = \frac{[j]_q + q^j}{[n + 1]_q} - \frac{[\bar{k} - 1]_q}{[n + 1]_q}.$$

Since

$$[j]_q \leq [\bar{k}]_q + \sqrt{q^{\bar{k}-1}[\bar{k}]_q},$$

we see that

$$\underline{M}_{\tilde{k}-1,n,j}(x; q) \leq \frac{[\bar{k}]_q + q^j + \sqrt{q^{\bar{k}-1}[\bar{k}]_q}}{[n + 1]} - \frac{[\bar{k} - 1]_q}{[n + 1]_q}$$

$$= \frac{q^j + q^{\bar{k}-1} + \sqrt{q^{\bar{k}-1}[\bar{k}]_q}}{[n + 1]_q} \leq \frac{2 + \sqrt{[n + 1]_q}}{[n + 1]_q}$$

$$\leq \frac{3}{\sqrt{[n + 1]_q}}.$$

By Lemma 4 (i), we can write that

$$\underline{M}_{\tilde{k}-1,n,j}(x; q) \geq \underline{M}_{\tilde{k}-2,n,j}(x; q) \geq ... \geq \underline{M}_{0,n,j}(x; q).$$

Hence, we see that, for all $k \in \{0, 1, ..., \tilde{k} - 1, \tilde{k}\}$

$$\underline{M}_{k,n,j}(x; q) \leq \frac{3}{\sqrt{[n + 1]_q}},$$

which implies, for the same k's, that

$$M_{k,n,j}(x; q) \leq \frac{3}{\sqrt{[n + 1]_q}}$$

due to Lemma 2 (ii). As a result, if we combine all results obtained above with (19) we conclude, for all $k, j \in \{0, 1, 2, ..., n\}$, $n \in \mathbb{N}$ and $x \in [0, 1]$, that

$$B_n^{(M)}(\varphi_x; x; q) \leq \frac{2\left(1 + \frac{2}{q^{n+1}}\right)}{\sqrt{[n+1]_q}}.$$

So the proof is completed.

Now we are ready to give an error estimation for the q-Bernstein operators of max-product kind.

Theorem 7 *If $f \in C_+[0, 1]$, $x \in [0, 1]$ and $n \in \mathbb{N}$, then we get*

$$\left|B_n^{(M)}(f; x; q) - f(x)\right| \leq 4\left(1 + \frac{2}{q^{n+1}}\right)\omega\left(f; \frac{1}{\sqrt{[n]_q}}\right).$$

Proof It is clear from Corollaries 1 and 6.

We should note that if we take $q \to 1^-$, then Theorem 7 reduces to Theorem 4.1 introduced by [4]. Furthermore, for the classical q-Bernstein polynomials given by (2), we know from [17] that

$$B_n(e_0; x; q) = 1, \quad B_n(e_1; x; q) = x, \quad B_n(e_2; x; q) = x^2 + \frac{x(1-x)}{[n]_q},$$

where $e_0(y) := 1$, $e_1(y) = y$ and $e_2(y) = y^2$ on $[0, 1]$. Hence, by positivity and linearity of $B_n(f; x; q)$ ($f \in C[0, 1]$), we may write that (see, for instance, Theorem 2.3 in [9])

$$|B_n(f; x; q) - f(x)| \leq 2\omega\left(f; \sqrt{\frac{x(1-x)}{[n]_q}}\right) \leq \omega\left(f; \frac{1}{\sqrt{[n]_q}}\right).$$

4 Approximation Results

In this section, we obtain an approximation theorem for the operators $B_n^{(M)}(f; x; q)$ defined by (4). But, to get such an approximation we have to replace a fixed $q \in (0, 1)$ considered in the previous sections with an appropriate sequence (q_n) whose terms are in the interval $(0, 1)$. Because, otherwise, $[n]_q$ goes to $\frac{1}{1-q}$ as $n \to \infty$ for a fixed q. Actually, this idea was first used by Phillips [17] for the q-Bernstein polynomials given by (2).

Assume now that (q_n) is a real sequence satisfying the following conditions:

$$0 < q_n < 1 \text{ for every } n \in \mathbb{N}, \tag{22}$$

$$st_A - \lim_n q_n = 1, \tag{23}$$

and

$$st_A - \lim q_n^n = 1. \tag{24}$$

Note that the notations in (23) and (24) denote the A-statistical limit of (q_n), where $A = [a_{jn}]$ $(j, n \in \mathbb{N})$ is an infinite nonnegative regular summability matrix, i.e., $a_{jn} \geq 0$ for every $j, n \in \mathbb{N}$ and $\lim_j \sum_{n=1}^{\infty} a_{jn} x_n = L$ whenever $\lim_n x_n = L$ provided that the series $\sum_{n=1}^{\infty} a_{jn} x_n$ is convergent for each $j \in \mathbb{N}$ (see, e.g. [14]). Recall that, for a given sequence (x_n), we say that (x_n) is A-statistically convergent to a number L if, for every $\varepsilon > 0$, $\lim_j \sum_{n:|x_n - L| \geq \varepsilon} a_{jn} = 0$ (see [12]). We should remark that this method of convergence generalizes both the classical convergence and the concept of statistical convergence which was first introduced by Fast [11]. Its usage in the classical approximation theory may be found in [2, 10, 13]. It is easy to find a sequence (q_n) satisfying (22)–(24); for example, define $q_n = \frac{1}{2}$ if $n = m^2$, $m \in \mathbb{N}$, and $q_n = e^{1/n}(1 - 1/n)$ otherwise; and also consider the Cesáro matrix $C_1 = [c_{jn}]$ defined to be $c_{jn} = 1/j$ if $1 \leq n \leq j$, and $c_{jn} = 0$ otherwise.

We get the following statistical approximation theorem.

Theorem 8 *Let $A = [a_{jn}]$ be a nonnegative regular summability matrix, and let (q_n) be a sequence satisfying (22)–(24). Then, for every $f \in C_+[0, 1]$, we have*

$$st_A - \lim_n \left\{ \sup_{x \in [0,1]} \left| B_n^M(f; x; q_n) - f(x) \right| \right\} = 0.$$

Proof Let $f \in C_+[0, 1]$. Replacing q with (q_n), and taking supremum over $x \in [0, 1]$, and also using the monotonicity of the modulus of continuity, we obtain from Theorem 7 that

$$E_n := \sup_{x \in [0,1]} |B_n(f; x; q_n) - f(x)| \leq 4 \left(1 + \frac{2}{q_n^{n+1}} \right) \omega \left(f; \frac{1}{\sqrt{[n]_{q_n}}} \right) \tag{25}$$

holds for every $n \in \mathbb{N}$. Then, it is enough to prove

$$st_A - \lim_n E_n = 0.$$

The hypotheses (22)–(24) imply that

$$st_A - \lim_n \frac{1}{\sqrt{[n]_{q_n}}} = 0 \quad \text{and} \quad st_A - \lim_n \frac{1}{q_n^{n+1}} = 1.$$

Also, by the right continuity of $\omega(f; \cdot)$ at zero, we may write that

$$st_A - \lim_n \omega \left(f; \frac{1}{\sqrt{[n]_{q_n}}} \right) = 0. \tag{26}$$

Hence, the proof follows from (22)–(26) immediately.

We should note that the A-statistical approximation result in Theorem 8 includes the classical approximation by choosing $A = I$, the identity matrix. Hence, the next result is an immediate consequence of Theorem 8.

Corollary 9 *Assume that (q_n) is a sequence for which (22), $\lim_n q_n = 1$ and $\lim_n q_n^n = 1$ hold. Then, for every $f \in C_+[0, 1]$, we have*

$$\lim_n \left\{ \sup_{x \in [0,1]} |B_n(f; x; q_n) - f(x)| \right\} = 0.$$

However, defining $q_n = \frac{1}{2}$ if $n = m^2$, $m \in \mathbb{N}$, and $q_n = e^{1/n}\left(1 - \dfrac{1}{n}\right)$ otherwise; and also considering the Cesáro matrix $C_1 = [c_{jn}]$ as stated before, we see that Theorem 8 works for the max-product operators constructed with this sequence (q_n) while Corollary 9 fails.

References

1. Anastassiou, G.A., Coroianu, L., Gal, S.G.: Approximation by a nonlinear Cardaliaguet-Euvrard neural network operator of max-product kind. J. Comput. Anal. Appl. **12**, 396–406 (2010)
2. Anastassiou, G.A., Duman, O.: Towards Intelligent Modeling: Statistical Approximation Theory, Intelligent Systems Reference Library, vol. 14. Springer, Berlin (2011)
3. Bede, B., Coroianu, L., Gal, S.G.: Approximation by truncated Favard-Szász-Mirakjan operator of max-product kind. Demonstratio Math. **44**, 105–122 (2011)
4. Bede, B., Coroianu, L., Gal, S.G.: Approximation and shape preserving properties of the Bernstein operator of max-product kind. Int. J. Math. Math. Sci. Art. ID 590589, 26pp. (2009)
5. Bede, B., Gal, S.G.: Approximation by nonlinear Bernstein and Favard-Szász-Mirakjan operators of max-product kind. J. Concr. Appl. Math. **8**, 193–207 (2010)
6. Bede, B., Nobuhara, H., Daňková, M., Di Nola, A.: Approximation by pseudo-linear operators. Fuzzy Sets and Systems **159**, 804–820 (2008)
7. Coroianu, L., Gal, S.G.: Approximation by nonlinear Lagrange interpolation operators of max-product kind on Chebyshev knots of second kind. J. Comput. Anal. Appl. **13**, 211–224 (2011)
8. Coroianu, L., Gal, S.G.: Approximation by nonlinear generalized sampling operators of max-product kind. Sampl. Theory Signal Image Process. **9**, 59–75 (2010)
9. DeVore, R.A.: The approximation of continuous functions by positive linear operators. In: Lecture Notes in Mathematics, vol. 293. Springer, Berlin (1972)
10. Duman, O.: Statistical convergence of max-product approximating operators. Turkish J. Math. **34**, 501–514 (2010)
11. Fast, H.: Sur la convergence statistique. Colloquium Math. **2**, 241–244 (1951)
12. Freedman, A.R., Sember, J.J.: Densities and summability. Pacific J. Math. **95**, 293–305 (1981)
13. Gadjiev, A.D., Orhan, C.: Some approximation theorems via statistical convergence. Rocky Mountain J. Math. **32**(1), 129–138 (2002)
14. Hardy, G.H.: Divergent Series. Clarendon Press, Oxford (1949)
15. Korovkin, P.P.: Linear Operators and Approximation Theory. Hindustan Publishing Corp. (India), Delhi (1960)

16. Lorentz, G.G.: Bernstein Polynomials, Mathematical Expositions, vol. 8. University of Toronto Press, Toronto (1953)
17. Phillips, G.M.: Bernstein polynomials based on the q-integers. Ann. Numer. Math. **4**, 511–518 (1997)

A Two Dimensional Inverse Scattering Problem for Shape and Conductive Function for a Dielectic Cylinder

Ahmet Altundag

Abstract The inverse problem under consideration is to simultaneously reconstruct the conductive function and shape of a coated homogeneous dielectric infinite cylinder from the far-field pattern for scattering of a time-harmonic E-polarized electromagnetic plane wave. We propose an inverse algorithm that combine the approaches suggested by Ivanyshyn et al. [1–3], and extend the approaches from the case of impenetrable scatterer to the case of penetrable scatterer. It is based on a system of non-linear boundary integral equation associated with a single-layer potential approach to solve the forward scattering problem. We present the mathematical foundations of the method and exhibit its feasibility by numerical examples.

1 Introduction

The problem is to determine simultaneously both the shape of the obstacle and the impedance function defined on the coated boundary from scattering of time-harmonic E-polarized electromagnetic plane waves. In the current paper we deal with dielectric scatterers covered by a thin boundary layer described by a impedance boundary condition and confine ourselves to the case of infinitely long coated homogeneous dielectric cylinders. This restriction provides us to reduce the problem into two dimension.

Let the simply connected bounded domain $D \subset \mathbb{R}^2$ with C^2 boundary Γ represent the cross section of an infinitely long coated homogeneous dielectric cylinder having constant wave number k_d with $Im\{k_d\}$, $\mathrm{Re}\{k_d\} \geqslant 0$ and denote the exterior wave number of background by $k_0 \in \mathbb{R}$. Denote by ν the outward unit normal vector to Γ. Then, given an incident plane wave $v^i = e^{ik_0 x.d}$ with incident direction given by the unit vector d, the direct scattering problem for E-polarized electromagnetic

A. Altundag (✉)
Istanbul Sabahattin Zaim University, Istanbul, Turkey
e-mail: ahmet.altundag@izu.edu.tr

© Springer International Publishing Switzerland 2016
G.A. Anastassiou and O. Duman (eds.), *Intelligent Mathematics II:*
Applied Mathematics and Approximation Theory, Advances in Intelligent Systems
and Computing 441, DOI 10.1007/978-3-319-30322-2_4

57

waves by a coated homogeneous dielectric is modelled by the following conductive-transmission boundary value problem for the Helmholtz equation: Find solutions $v \in H^1_{loc}(\mathbb{R}^2 \setminus \bar{D})$ and $v \in H^1(D)$ to the Helmholtz equations

$$\Delta v + k_0^2 v = 0 \quad \text{in } R^2 \setminus \bar{D}, \qquad \Delta w + k_d^2 w = 0 \quad \text{in } D \tag{1}$$

with the conductive-transmission boundary conditions

$$v = w, \qquad \frac{\partial v}{\partial \nu} = \frac{\partial w}{\partial \nu} + i \eta w \quad \text{on } \Gamma \tag{2}$$

for some continuous function defined in one continuously differentiable real valued function space $\eta \in C^1(\Gamma)$ with $\eta \le 0$ and where the total field is given by $v = v^i + v^s$ with the scattered wave v^s fulfilling the Sommerfeld radiation condition

$$\lim_{r \to \infty} r^{1/2} \left(\frac{\partial v^s}{\partial r} - i k_0 v^s \right) = 0, \quad r = |x|, \tag{3}$$

uniformly with respect to all directions. The latter is equivalent to an asymptotic behavior of the form

$$v^s(x) = \frac{e^{ik_0|x|}}{\sqrt{|x|}} \left\{ v_\infty \left(\frac{x}{|x|} \right) + O \left(\frac{1}{|x|} \right) \right\}, \quad |x| \to \infty, \tag{4}$$

uniformly in all directions, with the far field pattern v_∞ defined on the unit circle Ω in $\mathbb{R}^2$ (see [4]). In the above, v and w represent the electric field that is parallel to the cylinder axis, (1) corresponds to the time-harmonic Maxwell equations and the impedance-transmission conditions (2) model the continuity of the tangential components of the electric and magnetic field across the interface Γ.

The inverse obstacle problem we are interested in is, given the far field pattern v_∞ for one incident plane wave with incident direction $d \in \Omega$ to determine simultaneously both the boundary Γ of the scattering dielectric D and the impedance function η. More generally, we also consider the simultaneous reconstruction of Γ and η from the far field patterns for a small finite number of incident plane waves with different incident directions. This inverse problem is non-linear and ill-posed, since the solution of the scattering problem (1)–(3) is non-linear with respect to the boundary and impedance function, and since the mapping from the boundary and impedance function into the far field pattern is extremely smoothing.

For a stable solution of the inverse problem we propose an algorithm that combines the approaches suggested and investigated by Kress and Rundell [2, 3] and by Ivanyshyn and Kress [1], and extend the approaches to the case of an infinitely long coated homogeneous dielectric cylinder with arbitrarily shaped cross section embedded in a homogeneous background. Representing the solution w and v^s to the forward scattering problem in terms of single-layer potentials in D and in $\mathbb{R}^2 \setminus \bar{D}$ with densities ξ_d and ξ_0, respectively, the impedance-transmission boundary condition (2) provides a system of two boundary integral equations on Γ for the corresponding densities, that in the sequel we will denote as field equations. For the inverse problem, the required coincidence of the far field of the single-layer potential representing v^s and the given far field v_∞ provides a further equation that we denote as data equation. The system of the field and data equations can be viewed as three equations for four unknowns, i.e., the two densities, boundary of the scatterer Γ, and the conductive function η. They are linear with respect to the densities and non-linear with respect to the impedance function and the boundary.

In the spirit of [1–3], given approximations Γ_{approx}, η_{approx}, $\xi_{d_{approx}}$, and $\xi_{0_{approx}}$ for the boundary Γ, the impedance function η, the densities ξ_d and ξ_0 we linearise simultaneously both the field and the data equation with respect to all unknowns, i.e., the boundary curve, the impedance function, and the two densities. The linear equations are then solved to update the boundary curve, the impedance function, and the two densities. Because of the ill-posedness the solution of the update equations require stabilization, for example, by Tikhonov regularization. This procedure is then iterated until some suitable stopping criterion is satisfied.

At this point we note that uniqueness results for this inverse impedance—transmission problem are only available for the case of infinitely many incident waves (see [5]). A general uniqueness result based on the far field pattern for one or finitely many incident waves is still lacking.

To some extend, the inverse problem consists in solving a certain Cauchy problem, i.e., extending a solution to the Helmholtz equation from knowing their Cauchy data on some boundary curve. With this respect we also mention the related work of Ben Hassen et al. [6], Cakoni and Colton [7], Cakoni et al. [8], Eckel and Kress [9], Fang and Zeng [10], Ivanyshyn and Kress [11], Jakubik and Potthast [12]. For the simultaneous reconstruction of the shape and the impedance function for an impenetrable scatterers in a homogeneous background we refer to Kress and Rundell [2], Liu et al. [13, 14]. For the shape or the impedance reconstruction for penetrable scatterers, i.e., for dielectric obstacles we refer to Altundag and Kress [15, 16], Altundag [17–20], Akduman et al. [21], and Yaman [22].

The plan of the paper is as follows: In Sect. 2, as ingredient of our inverse algorithm we demonstrate the solution of the forward scattering problem via a single-layer approach followed by a corresponding numerical solution method in Sect. 3. In Sect. 4, we describe our inverse algorithm via simultaneous linearisation of the field and data equation in detail. In Sect. 5, we illustrate the feasibility of the method by some numerical examples.

2 The Direct Problem

The forward scattering problem (1)–(3) has at most one solution (see Gerlach and Kress [5]). Existence can be proven via boundary integral equations by a combined single- and double-layer approach (see Gerlach and Kress [5]).

Here, as one of the ingredients of our inverse algorithm, we follow [15] and suggest a single-layer approach. For this we denote by

$$\Phi_k(x, y) := \frac{i}{4} H_0^{(1)}(k|x - y|), \quad x \neq y,$$

the fundamental solution to the Helmholtz equation with wave number k in $\mathbb{R}^2$ in terms of the Hankel function $H_0^{(1)}$ of order zero and of the first kind. Adopting the notation of [4], in a Sobolev space setting, for $k = k_d$ and $k = k_0$, we introduce the single-layer potential operators

$$S_k : H^{-1/2}(\Gamma) \rightarrow H^{1/2}(\Gamma)$$

by

$$(S_k\xi)(x) := 2 \int_\Gamma \Phi_k(x, y)\xi(y) \, ds(y), \quad x \in \Gamma, \tag{5}$$

and the normal derivative operators

$$K_k' : H^{-1/2}(\Gamma) \rightarrow H^{-1/2}(\Gamma)$$

by

$$(K_k'\xi)(x) := 2 \int_\Gamma \frac{\partial \Phi_k(x, y)}{\partial \nu(x)} \xi(y) \, ds(y), \quad x \in \Gamma. \tag{6}$$

For the Sobolev spaces and the mapping properties of these operators we refer to [23, 24].

Then, from the jump relations it can be seen that the single-layer potentials

$$w(x) = \int_\Gamma \Phi_{k_d}(x, y)\xi_d(y) \, ds(y), \quad x \in D,$$

$$v^s(x) = \int_\Gamma \Phi_{k_0}(x, y)\xi_0(y) \, ds(y), \quad x \in \mathbb{R}^2 \backslash \bar{D}, \tag{7}$$

solve the scattering problem (1)–(3) provided the densities ξ_d and ξ_0 fulfil the system of integral equations

$$S_{k_d}\xi_d - S_{k_0}\xi_0 = 2v^i|_\Gamma,$$

$$\xi_d + \xi_0 + i\eta S_{k_d}\xi_d + K_{k_d}'\xi_d - K_{k_0}'\xi_0 = 2 \left.\frac{\partial v^i}{\partial \nu}\right|_\Gamma. \tag{8}$$

Provided k_0 is not a Dirichlet eigenvalue of the negative Laplacian for D, (7) has at most one solution. For the existence analysis and uniqueness of a solution, we refer to [19].

3 Numerical Solution

For the numerical solution of (8) and the presentation of our inverse algorithm we assume that the boundary curve Γ is given by a regular 2π-periodic parameterization

$$\Gamma = \{\zeta(s) : 0 \le s \le 2\pi\}. \tag{9}$$

Then, via $\chi = \xi \circ \zeta$. emphasizing the dependence of the operators on the boundary curve, we introduce the parameterized single-layer operator

$$\widetilde{S}_k : H^{-1/2}[0, 2\pi] \times C^2[0, 2\pi] \to H^{1/2}[0, 2\pi]$$

by

$$\widetilde{S}_k(\chi, \zeta)(s) := \frac{i}{2} \int_0^{2\pi} H_0^{(1)}(k|\zeta(s) - \zeta(\tau)|) \, |\zeta'(\tau)| \, \chi(\tau) \, d\tau$$

and the parameterized normal derivative operators

$$\widetilde{K}_k' : H^{-1/2}[0, 2\pi] \times C^2[0, 2\pi] \to H^{-1/2}[0, 2\pi]$$

by

$$\widetilde{K}_k'(\chi, \zeta)(s) := \frac{ik}{2} \int_0^{2\pi} \frac{[\zeta'(s)]^{\perp} \cdot [\zeta(\tau) - \zeta(s)]}{|\zeta'(s)| \, |\zeta(s) - \zeta(\tau)|}$$

$$\times H_1^{(1)}(k|\zeta(s) - \zeta(\tau)|) \, |\zeta'(\tau)| \, \chi(\tau) \, d\tau$$

for $s \in [0, 2\pi]$. Here we made use of $H_0^{(1)'} = -H_1^{(1)}$ with the Hankel function $H_1^{(1)}$ of order zero and of the first kind. Furthermore, we write $\zeta^{\perp} = (\zeta_2, -\zeta_1)$ for a vector $\zeta = (\zeta_1, \zeta_2)$, that is, $\zeta^{\perp}$ is obtained by rotating ζ clockwise by $90°$. Then the parameterized form of (8) is given by

$$\widetilde{S}_{k_d}(\chi_d, \zeta) - \widetilde{S}_{k_0}(\chi_0, \zeta) = 2\, v^i \circ \zeta,$$

$$\chi_d + \chi_0 + (\eta \cup \zeta)\widetilde{S}_{k_d}(\chi_d, \zeta) \tag{10}$$

$$+ \widetilde{K}_{k_d}'(\chi_d, \zeta) - \widetilde{K}_{k_0}'(\chi_0, \zeta) = \frac{2}{|\zeta'|} \, [\zeta']^{\perp} \cdot \operatorname{grad} v^i \circ \zeta.$$

The kernels

$$A(s, \tau) := \frac{i}{2} H_0^{(1)}(k|\zeta(t) - \zeta(\tau)|) |\zeta'(\tau)|$$

and

$$B(s, \tau) := \frac{ik}{2} \frac{[\zeta'(s)]^\perp \cdot [\zeta(\tau) - \zeta(s)]}{|\zeta'(s)| |\zeta(s) - \zeta(\tau)|} H_1^{(1)}(k|\zeta(s) - \zeta(\tau)|) |\zeta'(\tau)|$$

of the operators $\widetilde{S}_k$ and $\widetilde{K}'_k$ can be written in the form

$$A(s, \tau) = A_1(s, \tau) \ln \left(4 \sin^2 \frac{s-\tau}{2}\right) + A_2(s, \tau),$$

$$B(s, \tau) = B_1(s, \tau) \ln \left(4 \sin^2 \frac{s-\tau}{2}\right) + B_2(s, \tau),$$

(11)

where

$$A_1(s, \tau) := -\frac{1}{2\pi} J_0(k|\zeta(s) - \zeta(\tau)|)|\zeta'(\tau)|,$$

$$A_2(s, \tau) := A(s, \tau) - A_1(s, \tau) \ln \left(4 \sin^2 \frac{s-\tau}{2}\right),$$

$$B_1(s, \tau) := -\frac{k}{2\pi} \frac{[\zeta'(s)]^\perp \cdot [\zeta(\tau) - \zeta(s)]}{|\zeta'(s)| |\zeta(s) - \zeta(\tau)|} J_1(k|\zeta(s) - \zeta(\tau)|) |\zeta'(\tau)|,$$

$$B_2(s, \tau) := B(s, \tau) - B_1(s, \tau) \ln \left(4 \sin^2 \frac{s-\tau}{2}\right).$$

J_0 and J_1 denote the Bessel functions of order zero and one respectively. The functions A_1, A_2, B_1, and B_2 turn out to be analytic with diagonal terms

$$A_2(s, s) = \left[\frac{i}{2} - \frac{C}{\pi} - \frac{1}{\pi} \ln\left(\frac{k}{2}|\zeta'(s)|\right)\right] |\zeta'(s)|$$

in terms of Euler's constant C and

$$B_2(s, s) = -\frac{1}{2\pi} \frac{[\zeta'(s)]^\perp \cdot \zeta''(s)}{|\zeta'(s)|^2}.$$

For integral equations with kernels of the form (11) a combined collocation and quadrature methods based on trigonometric interpolation as described in Sect. 3.5 of [4] or in [25] is at our disposal. We refrain from repeating the details. For a related error analysis we refer to [23] and note that we have exponential convergence for smooth, i.e., analytic boundary curves Γ.

For a numerical example, we consider the scattering of a plane wave by a dielectric cylinder with a non-convex apple-shaped cross section with boundary Γ described by the parametric representation

$$\zeta(s) = \left\{ \frac{0.5 + 0.4 \cos s + 0.1 \sin 2s}{1 + 0.7 \cos s} (\cos s, \sin s) : s \in [0, 2\pi] \right\} \tag{12}$$

The following impedance functions are chosen in our experiments.

•

$$\eta_1 = -1 - 0.5 \sin(s) \tag{13}$$

•

$$\eta_2 = -2 - \cos(2s) - 0.5 \sin(s) \tag{14}$$

From the asymptotics for the Hankel functions, it can be deduced that the far field pattern of the single-layer potential v^s with density ξ_0 is given by

$$v_\infty(\hat{x}) = \gamma \int_\Gamma e^{-ik_0 \hat{x} \cdot y} \xi_0(y) \, ds(y), \quad \hat{x} \in \Omega, \tag{15}$$

where

$$\gamma = \frac{e^{i\frac{\pi}{4}}}{\sqrt{8\pi k_0}}.$$

The latter expression can be evaluated by the composite trapezoidal rule after solving the system of integral equations (8) for ξ_0, i.e., after solving (10) for χ_0. Table 1 gives some approximate values for the far field pattern $v_\infty(d)$ and $v_\infty(-d)$ in the forward direction d and the backward direction $-d$. The direction d of the incident wave is $d = (1, 0)$ and the wave numbers are $k_0 = 1$ and $k_d = 4 + 1i$, and the conductive function η_2 in (14) is chosen. Note that the exponential convergence is clearly exhibited.

Table 1 Numerical results for direct scattering problem

n	Re $u_\infty(d)$	Im $u_\infty(d)$	Re $u_\infty(-d)$	Im $u_\infty(-d)$
8	−0.9246701916	0.1927903437	−0.9330234793	0.2054859030
16	−0.9246830723	0.1927431421	−0.9330023078	0.2053750332
32	−0.9246871867	0.1927361525	−0.9330054172	0.2053781827
64	−0.9246871412	0.1927360822	−0.9330054392	0.2053782645

4 The Inverse Problem

We now proceed describing an iterative algorithm for approximately solving the inverse scattering problem by combining the method proposed by Ivanyshyn, Kress and Rundell [1–3] and by extending from the case of impenetrable obstacles to the case of penetrable scatterers. After introducing the far field operator $S_\infty : H^{-1/2}(\Gamma) \to L^2(\Omega)$ by

$$(S_\infty \varphi)(\hat{x}) := \gamma \int_\Gamma e^{-ik_0 \hat{x} \cdot y} \xi(y)\, ds(y), \quad \hat{x} \in \Omega, \tag{16}$$

from (7) and (15) we observe that the far field pattern for the solution to the scattering problem (1)–(3) is given by

$$v_\infty = S_\infty \xi_0 \tag{17}$$

in terms of the solution to (8). Thus as theoretical basis of our inverse algorithm we can state the following theorem.

Theorem 1 *For a given incident field v^i and a given far field pattern v_∞, assume that the boundary Γ, the impedance function η, and the densities ξ_d and ξ_0 satisfy the system of three integral equations*

$$S_{k_d} \xi_d - S_{k_0} \xi_0 = 2v^i,$$

$$\xi_d + \xi_0 + i\eta S_{k_d} \xi_d + K'_{k_d} \xi_d - K'_{k_0} \xi_0 = 2\frac{\partial v^i}{\partial \nu}, \tag{18}$$

$$S_\infty \xi_0 = v_\infty.$$

Then Γ and η solve the inverse scattering problem.

The ill-posedness of the inverse problem is reflected through the ill-posedness of the third integral equation, the far field equation that we denote as *data equation*. Note that (18) is linear with respect to the densities and nonlinear with respect to the boundary Γ and the impedance function η. This opens up a variety of approaches to solve (18) by linearization and iteration. In the current paper, we are going to proceed as follows: Given approximations Γ_{approx} and η_{approx} for the boundary Γ and the impedance function η, and approximations $\xi_{d_{approx}}$ and $\xi_{0_{approx}}$ for the densities ξ_d and ξ_0 we linearise simultaneously both the field and the data equations with respect to the boundary curve, the impedance function and the two densities. The linear equations are then solved to update the boundary curve, the conductive function and the two densities. Because of the ill-posedness the solution of the update equations require stabilization. For this, we use Tikhonov regularization. This procedure is then iterated until some suitable stopping criterion is achieved.

To describe this in more detail, we also require the parameterized version

$$\widetilde{S}_\infty : H^{-1/2}[0, 2\pi] \times C^2[0, 2\pi] \to L^2(\Omega)$$

of the far field operator as given by

$$\widetilde{S}_\infty(\chi, \zeta)(\hat{x}) := \gamma \int_0^{2\pi} e^{-ik_0 \hat{x} \cdot \zeta(\tau)} |\zeta'(\tau)| \chi(\tau) \, d\tau, \quad \hat{x} \in \Omega. \tag{19}$$

Then the parameterized form of (18) is given by

$$\widetilde{S}_{k_d}(\chi_d, \zeta) - \widetilde{S}_{k_0}(\chi_0, \zeta) = 2 \, v^i \circ \zeta,$$

$$\begin{aligned} \chi_d + \chi_0 + i(\eta \circ \zeta)\widetilde{S}_{k_d}(\chi_d, \zeta) \\ + \widetilde{K}'_{k_d}(\chi_d, \zeta) - \widetilde{K}'_{k_0}(\chi_0, \zeta) = \frac{2}{|\zeta'|} \, [\zeta']^\perp \cdot \operatorname{grad} v^i \circ \zeta, \end{aligned} \tag{20}$$

$$\widetilde{S}_\infty(\chi_0, z) = v_\infty.$$

For a fixed χ the Fréchet derivative of the operator $\widetilde{S}_k$ and $\widetilde{K}'_k$ with respect to the boundary ζ in the direction h are given by

$$\begin{aligned} \partial \widetilde{S}_k(\chi, \zeta; h)(s) = \frac{-ik}{2} \int_0^{2\pi} \frac{(\zeta(s) - \zeta(\tau)) \cdot (h(s) - h(\tau))}{|\zeta(s) - \zeta(\tau)|} \\ \times |\zeta'(\tau)| \, H_1^{(1)}(k|\zeta(s) - \zeta(\tau)|)\chi(\tau)d\tau \\ + \frac{i}{2} \int_0^{2\pi} \frac{\zeta'(\tau) \cdot h'(\tau)}{|\zeta'(\tau)|} H_0^{(1)}(k|\zeta(t) - \zeta(\tau)|)\chi(\tau)d\tau, \end{aligned}$$

and

$$\begin{aligned} & \partial \widetilde{K}'_k(\chi, \zeta; h)(s) \\ & = -\frac{ik}{2|\zeta'(s)|} \int_0^{2\pi} \frac{[\zeta'^\perp \cdot (h(s) - h(\tau)) + [h'^\perp \cdot (\zeta(s) - \zeta(\tau))}{|\zeta(s) - \zeta(\tau)|} \\ & \times |\zeta'(\tau)| H_1^{(1)}(k|\zeta(s) - \zeta(\tau)|)\chi(\tau)d\tau \\ & + \frac{ik}{|\zeta'(s)|} \int_0^{2\pi} \frac{[\zeta'^\perp \cdot (\zeta(s) - \zeta(\tau))(h(s) - h(\tau)) \cdot (\zeta(s) - \zeta(\tau))}{|\zeta(s) - \zeta(\tau)|^3} \\ & \times |\zeta'(\tau)| H_1^{(1)}(k|\zeta(s) - \zeta(\tau)|)\chi(\tau)d\tau \end{aligned} \tag{21}$$

$$-\frac{ik^2}{2|\zeta'(s)|}\int_0^{2\pi}\frac{[\zeta'^\perp\cdot(\zeta(s)-\zeta(\tau))(h(s)-h(\tau))\cdot(\zeta(s)-\zeta(\tau))}{|\zeta(s)-\zeta(\tau)|^2}$$

$$\times|\zeta'(\tau)|H_0^{(1)}(k|\zeta(s)-\zeta(\tau)|)\chi(\tau)d\tau$$

$$+\frac{ik}{2}\frac{h'(s)\cdot\zeta'(s)}{|\zeta'^2}\int_0^{2\pi}\frac{[\zeta'^\perp\cdot(\zeta(s)-\zeta(\tau))}{|\zeta(s)-\zeta(\tau)|}$$

$$\times|\zeta'(\tau)|H_1^{(1)}(k|\zeta(s)-\zeta(\tau)|)\chi(\tau)d\tau$$

$$-\frac{ik}{2|\zeta'(s)|}\int_0^{2\pi}\frac{[\zeta'^\perp\cdot(\zeta(s)-\zeta(\tau))h'(\tau)\cdot\zeta'(\tau)}{|\zeta(s)-\zeta(\tau)||\zeta'(\tau)|}$$

$$\times H_1^{(1)}(k|\zeta(s)-\zeta(\tau)|)\chi(\tau)d\tau.$$

Then the linearisation (20) with respect to all variables ζ, η, χ_d and χ_0 in the direction h, θ, μ_d and μ_0, respectively, reads

$$\widetilde{S}_{k_d}(\chi_d,\zeta)+\widetilde{S}_{k_d}(\mu_d,\zeta)+\partial\widetilde{S}_{k_d}(\chi_d,\zeta;h)-\widetilde{S}_{k_0}(\chi_0,\zeta)-\widetilde{S}_{k_0}(\mu_0,\zeta)$$

$$-\partial\widetilde{S}_{k_0}(\chi_0,\zeta;h)=2v^i\circ\zeta+2\mathrm{grad}v^i\circ\zeta\cdot h,$$

$$\chi_d+\mu_d+\chi_0+\mu_0+i(\theta\circ\zeta)\widetilde{S}_{k_d}(\chi_d,\zeta)+i(\eta\circ\zeta)\widetilde{S}_{k_d}(\chi_d,\zeta)$$
$$+\widetilde{K}'_{k_d}(\chi_d,\zeta)+\widetilde{K}'_{k_d}(\mu_d,\zeta)+\partial\widetilde{K}'_d(\chi_d,\zeta;h) \qquad (22)$$

$$-\widetilde{K}'_{k_0}(\chi_0,\zeta)-\widetilde{K}'_{k_0}(\mu_0,\zeta)-\partial\widetilde{K}'_{k_0}(\chi_0,\zeta;h)$$
$$=\frac{2}{|\zeta'|}[\zeta'^\perp\cdot\mathrm{grad}v^i\circ\zeta+2\sigma(\zeta)\cdot h,$$

$$\widetilde{S}_\infty(\chi_0,\zeta)+\widetilde{S}_\infty(\mu_0,\zeta)+\partial\widetilde{S}_\infty(\chi_0,\zeta;h)=v_\infty.$$

Here the term $\sigma(\zeta)\cdot h$ is the form (see [26])

$$\sigma(\zeta)\cdot h=-\frac{\partial v^i}{\partial\tau}\frac{v\cdot h'}{|\zeta'|}+(\frac{\partial^2 v^i}{\partial v\partial\tau}-H\frac{\partial v^i}{\partial\tau})\tau\cdot h+\frac{\partial^2 v^i}{\partial v^2}v\cdot h \qquad (23)$$

and τ and H stand for the tangential vector and the mean curvature respectively. They are given by

$$\tau=\frac{\zeta'}{|\zeta'|}, \quad\text{and } H=-\frac{\zeta''\cdot v}{|\zeta'^2}$$

and the matrix form of (22) can be written as

$$
\begin{bmatrix}
\left\{ \partial \widetilde{S}_{k_d}(\chi_d,\zeta;.) - \partial \widetilde{S}_{k_0}(\chi_0,\zeta;.) \atop -2\mathrm{grad}v^i \circ \zeta;\, Zeros \right\} & \widetilde{S}_{k_d}(.,\zeta) & -\widetilde{S}_{k_d}(.,\zeta) \\[2ex]
\left\{ \partial \widetilde{K}'_{k_d}(\chi_d,\zeta;.) - \partial \widetilde{K}'_{k_0}(\chi_0,\zeta;.) \atop -\xi(\zeta);\, i\widetilde{S}_{k_d}\chi_d \right\} & I + \widetilde{K}'_{k_d}(.,\zeta)\ I - \widetilde{K}'_{k_d}(.,\zeta) \\[2ex]
\partial \widetilde{S}_{\infty}(\chi_0,\zeta;.);\, Zeros & 0 & \widetilde{S}_{\infty}(.,\zeta)
\end{bmatrix}
\begin{bmatrix}
[h;\theta] \\[2ex]
\mu_d \\[2ex]
\mu_0
\end{bmatrix}
$$

$$
=
\begin{bmatrix}
2v^i \circ \zeta - \widetilde{S}_{k_d}(\chi_d,\zeta) + \widetilde{S}_{k_0}(\chi_0,\zeta) \\[2ex]
\left\{ \dfrac{2}{|\zeta'|}[\zeta'^{\perp} \cdot \mathrm{grad}v^i \circ \zeta - \chi_d - \widetilde{K}'_{k_d}(\chi_d,\zeta) - \chi_0 \atop + \widetilde{K}'_{k_0}(\chi_0,\zeta) - i(\eta \circ \zeta)\widetilde{S}_{k_d}(\chi_d,\zeta) \right\} \\[2ex]
v_{\infty} - \widetilde{S}_{\infty}(\chi_0,\zeta)
\end{bmatrix}
\tag{24}
$$

Now we can describe the method in a short form as follows:
Each iteration step of the proposed inverse algorithm consists of one part.

• Given an approximation ζ for the boundary, η for the conductive function, and χ_d, χ_0 for the densities, we solve the linearized sistem of integral equation (24), for h, θ, μ_d, and μ_0 to obtain updates $\zeta + h$, $\eta + \theta$, $\chi_d + \mu_d$, and $\chi_0 + \mu_0$. We continue this procedure until some stopping criteria is achieved. The stopping criterion for the iterative scheme is given by the relative error

$$
\frac{\| \, v_{\infty;N} - v_{\infty} \, \|}{\| \, v_{\infty} \, \|} \le \epsilon(\delta),
$$

where $v_{\infty;N}$ is the computed far field pattern after N iteration steps.

As a theoretical basis for the application of Tikhonov regularization from [27] we cite that, after the restriction to star-like boundaries, the operator $\partial \widetilde{S}_{\infty}$ is injective provided k_0^2 is not a Neumann eigenvalue for the negative Laplacian in D.

5 Numerical Examples

To avoid an inverse crime, in our numerical examples the synthetic far field data were obtained by a numerical solution of the boundary integral equations based on a combined single- and double-layer approach (see [28, 29]) using the numerical schemes as described in [4, 23, 30]. In each iteration step of the inverse algorithm for the solution of the matrix equation (24) we used the numerical method described in Sect. 3 using 64 quadrature points. The linearized matrix equation (24) were solved

Table 2 Boundary curves

Types	Representations
Apple-shaped	$\zeta(s) = \{\frac{0.5+0.4\cos s+0.1\sin 2s}{1+0.7\cos s}(\cos s, \sin s) : s \in [0, 2\pi]\}$
Kite-shaped	$\zeta(s) = \{(\cos s + 1.3\cos^2 s - 1.3, 1.5\sin s) : s \in [0, 2\pi]\}$
Peanut-shaped	$\zeta(s) = \{\sqrt{\cos^2 s + 0.25\sin s}(\cos s, \sin s) : s \in [0, 2\pi]\}$
Rounded triangle	$\zeta(s) = \{(2 + 0.3\cos 3s)(\cos s, \sin s) : s \in [0, 2\pi]\}$

by Tikhonov regularization with an H^2 penalty term, i.e., $p = 2$. The regularized equation is solved by Nyström's method with the composite trapezoidal rule again using 64 quadrature points.

In all our four examples we used an incident wave with the direction $d = (1, 0)$ and $J = 5$ as degree for the approximating trigonometric polynomials for the boundary curve, $P = 3$ as degree for the approximating trigonometric polynomials for the conductive function, $\alpha = 10^{-6}$ as regularization parameter for the data equation, and the wave numbers $k_0 = 1$ and $k_d = 5 + 1i$.

For simplicity, for the stopping rule we chose $\epsilon(\delta)$ the same for all noise levels since this already gave satisfactory reconstructions.

In according with the general convergence results on regularized Gauss–Newton method (see [31]) for the regularization parameters we used decreasing sequences

$$\lambda_{1,n} = \tau_1^{-n}\lambda_1$$
$$\lambda_{2,n} = \tau_2^{-n}\lambda_2$$

with λ_1, λ_2 positive and τ_1, $\tau_2 > 1$ chosen by trial and error. The iteration numbers and the regularization parameters λ_1 and λ_2 for the Tikhonov regularization of boundary ζ and conductive function η, respectively, were chosen by trial and error and their values are indicated in the following description of the individual examples.

In order to obtain noisy data, random errors are added point-wise to v_∞,

$$\tilde{v}_\infty = v_\infty + \delta\rho\frac{||v_\infty||}{|\rho|}$$

with the random variable $\rho \in \mathbb{C}$ and $\{\text{Re}\rho, \text{Im}\rho\} \in (0, 1)$. For all examples, 2 % noise level, i.e., $\delta = 0.02$ is added into the far-field pattern.

In the first example Fig. 1 illustrates reconstructions after 10 iterations with the regularization parameters $\lambda_1 = 0.7$, $\tau_1 = 1.1$ and $\lambda_2 = 1.2$, $\tau_2 = 1.15$.

In the second example Fig. 2 shows reconstructions after 15 iterations with the regularization parameter chosen as in the first example.

In the third example the reconstructions in Fig. 3 were obtained after 13 iterations with the regularization parameter chosen as in the first example.

In the fourth example the reconstructions in Fig. 4 were obtained after 15 iterations with the regularization parameters chosen as $\lambda_1 = 1.3$, $\tau_1 = 1.1$ and $\lambda_2 = 0.8$, $\tau_2 = 1.5$.

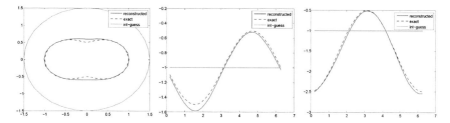

Fig. 1 Reconstruction of the peanut-shaped contour in Table 2 (*left*), conductive function η_1 in (13) (*middle*), and η_2 in (14) (*right*)

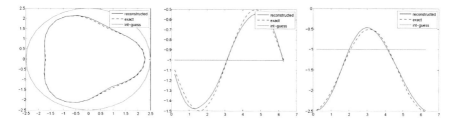

Fig. 2 Reconstruction of the rounded-shaped contour in Table 2 (*left*), conductive function η_1 in (13) (*middle*), and η_2 in (14) (*right*)

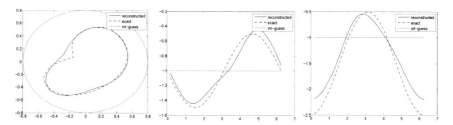

Fig. 3 Reconstruction of the apple-shaped contour in Table 2 (*left*), conductive function η_1 in (13) (*middle*), and η_2 in (14) (*right*)

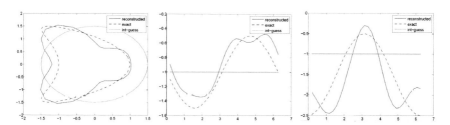

Fig. 4 Reconstruction of the kite-shaped contour in Table 2 (*left*), conductive function η_1 in (13) (*middle*), and η_2 in (14) (*right*)

Our examples clearly indicate the feasibility of the proposed algorithm with a reasonable stability against noise. An appropriate initial guess was important to ensure numerical convergence of the iterations. Further research will be directed towards applying the algorithm to real data, to extend the numerics to the three dimensional case.

Acknowledgments The author would like to thank Professor Rainer Kress for the helpful discussions and suggestions on the topic of this paper.

References

1. Ivanyshyn, O., Kress, R.: Nonlinear integral equations in inverse obstacle scattering. In: Fotiatis, D., Massalas, C. (eds.) Mathematical Methods in Scattering Theory and Biomedical Engineering. World Scientific, Singapore, pp. 39–50 (2006)
2. Kress, R., Rundell, W.: Inverse scattering problem for shape and impedance. Inverse Prob. **17**, 1075–1085 (2001)
3. Kress, R., Rundell, W.: Nonlinear integral equations and the iterative solution for an inverse boundary value problem. Inverse Prob. **21**, 1207–1223 (2005)
4. Colton, D., Kress, R.: Inverse Acoustic and Electromagnetic Scattering Theory, 2nd edn. Springer, Berlin (1998)
5. Gerlach, T., Kress, R.: Uniqueness in inverse obstacle scattering with conductive boundary condition. Inverse Prob. **12**, 619–625 (1996)
6. Hassen, M.F.B., Ivanyshyn, O., Sini, M.: The 3D acoustic scattering by complex obstacles. Inverse Prob. **26**, 105–118 (2010)
7. Cakoni, F., Colton, D.: The determination of the surface impedance of a partially coated obstacle from the far field data. SIAM J. Appl. Math. **64**, 709–723 (2004)
8. Cakoni, F., Colton, D., Monk, P.: The determination of boundary coefficients from far field measurements. J. Integr. Eqn. Appl. **22**, 167–191 (2010)
9. Eckel, H., Kress, R.: Nonlinear integral equations for the inverse electrical impedance problem. Inverse Prob. **23**, 475–491 (2007)
10. Fang, W., Zeng, S.: A direct solution of the Robin inverse problem. J. Integr. Eqn. Appl. **21**, 545–557 (2009)
11. Ivanyshyn, O., Kress, R.: Inverse scattering for surface impedance from phase-less far field data. J. Comput. Phys. **230**, 3443–3452 (2011)
12. Jakubik, P., Potthast, R.: Testing the integrity of some cavity-the Cauchy problem and the range test. Appl. Numer. Math. **58**, 899–914 (2008)
13. Liu, J., Nakamura, G., Sini, M.: Reconstruction of the shape and surface impedance from acoustic scattering data for arbitrary cylinder. SIAM J. Appl. Math. **67**, 1124–1146 (2007)
14. Nakamura, G., Sini, M.: Reconstruction of the shape and surface impedance from acoustic scattering data for arbitrary cylinder. SIAM J. Anal. **39**, 819–837 (2007)
15. Altundag, A., Kress, R.: On a two dimensional inverse scattering problem for a dielectric. Appl. Anal. **91**, 757–771 (2012)
16. Altundag, A., Kress, R.: An iterative method for a two-dimensional inverse scattering problem for a dielectric. J. Inverse Ill-Posed Prob. **20**, 575–590 (2012)
17. Altundag, A.: On a two-dimensional inverse scattering problem for a dielectric, Dissertation, Göttingen, Feb 2012
18. Altundag, A.: A hybrid method for inverse scattering problem for a dielectric. In: Advances in Applied Mathematics and Approximation Theory, vol. 41, pp. 185–203. Springer (2013)
19. Altundag, A.: Inverse obstacle scattering with conductive boundary condition for a coated dielectric cylinder. J. Concr. Appl. Math. **13**, 11 22 (2015)

20. Altundag, A.: A second degree Newton method for an inverse scattering problem for a dielectric. Hittite J. Sci. Eng. **2**, 115–125 (2015)
21. Akduman, I., Kress, R., Yaman, F., Yapar, A.: Inverse scattering for an impedance cylinder buried in a dielectric cylinder. Inverse Prob. Sci. Eng. **17**, 473–488 (2009)
22. Yaman, F.: Location and shape reconstructions of sound-soft obstacle buried in penetrable cylinders. Inverse Prob. **25**, 1–17 (2009)
23. Kress, R.: Integral Equation, 2nd edn. Springer, Berlin (1998)
24. McLean, W.: Strongly Elliptic Systems and Boundary Integral Equations. Cambridge University Press (2000)
25. Kress, R., Sloan, I.H.: On the numerical solution of a logarithmic integral equation of the first kind for the Helmholtz equation. Numerische Mathematik **66**, 199–214 (1993)
26. Serranho, P.: A hybrid method for inverse scattering for shape and impedance. Inverse Prob. **22**, 663–680 (2006)
27. Ivanyshyn, O., Johansson, T.: Boundary integral equations for acoustical inverse sound-soft scattering. J. Inverse Ill-Posed Prob. **15**, 1–14 (2007)
28. Colton, D., Kress, R.: Integral Equation Methods in Scattering Theory. Wiley-Interscience Publications, New York (1983)
29. Kress, R., Roach, G.F.: Transmission problems for the Helmholtz equation. J. Math. Phys. **19**, 1433–1437 (1978)
30. Kress, R.: On the numerical solution of a hypersingular integral equation in scattering theory. J. Comput. Appl. Math. **61**, 345–360 (1995)
31. Burger, M., Kaltenbacher, B., Neubauer, A.: Iterative solution methods. In: Scherzer (ed.) Handbook of Mathematical Methods in Imaging. Springer, Berlin, pp. 345–384 (2011)

Spinning Particle in Interaction with a Time Dependent Magnetic Field: A Path Integral Approach

Hilal Benkhelil and Mekki Aouachria

Abstract We consider a spin 1/2 particle interacting with a time dependent magnetic field using path integral formalism. The propagator is first of all written in the standard form by replacing the spin by two fermionic oscillators via the Schwinger's model; then it is determined exactly thanks to a simple transformations and the probability transition is then deduced.

1 Introduction

The applications of path-integral formalism have widely increased since a large class of potentials had been resolved [10]. However it is known that the most relativistic interactions are those where the spin, which is a very useful and very important notion in physics, is taken into account. In the framework of non-relativistic theory the phenomena of spin is automatically introduced by the Pauli equation which contains the Schrödinger Hamiltonian and a spin-field interaction. This motivates the research into the solvable Pauli equations which is inevitably useful in applied physics. For instance a well-known example of its direct application is the time-dependent field acting on an atom with two levels whose time-evolution is controlled by the Pauli-type equation. The solution for this equation has made clear the associated transition amplitudes [19]. This and similar [20, 21] types of interaction aside, there are little analytical and exact computations which treat the time-dependent spin-field interaction. Furthermore, if one replaces the time dependence of the exterior field by a space-time dependence or by only space dependence this becomes even more restrictive [5, 8, 9, 11, 15, 18].

Moreover, the problem becomes nearly unsolvable if we try to build these solutions by the path integral formalism because the spin is a discrete quantity. The difficulty

---

H. Benkhelil · M. Aouachria (✉)
University of Hadj lakhdar, Batna, Algeria
e-mail: mekkiaouachria@yahoo.fr

H. Benkhelil
e-mail: benkhelil_hilal@hotmail.fr

© Springer International Publishing Switzerland 2016
G.A. Anastassiou and O. Duman (eds.), *Intelligent Mathematics II:*
Applied Mathematics and Approximation Theory, Advances in Intelligent Systems
and Computing 441, DOI 10.1007/978-3-319-30322-2_5

73

here is associated to the fact that the path integral lacks some classical ideas such as trajectories and up to now one does not know how to deal with this kind of technique in this important case. Thus some effort has been made to find a partial solution using the Schwinger's model of spin and some explicit computations are then carried out [1–4, 6, 7, 12–14, 17].

In this paper we are devoted to this type of interaction; by considering a problem treats according to usual quantum mechanics [21]. It acts of a spin $1/2$ which interacts with a time dependent magnetic field.

$$\mathbf{B}(t) = \begin{pmatrix} \frac{B}{2} \sin\left(\frac{\omega t}{1+\omega^2 t^2}\right) \\ -\frac{B}{\cosh(\frac{\omega_0}{2})t} - B_0 \frac{1-\omega^2 t^2}{(1+\omega^2 t^2)^2} \\ \frac{B}{2} \cos\left(\frac{\omega t}{1+\omega^2 t^2}\right) \end{pmatrix}.$$

Its dynamics is described by the Hamiltonian

$$H = -\frac{g}{2}\sigma\mathbf{B},$$

where g is the gyromagnetic ratio. Then the Hamiltonian become

$$H = -\frac{B}{4} \sin\left(\frac{\omega t}{1+\omega^2 t^2}\right)\sigma_x \tag{1}$$
$$+ \left(\frac{\omega_0}{2\cosh(\frac{\omega_0}{2})t} + \frac{\omega}{2}\frac{1-\omega^2 t^2}{(1+\omega^2 t^2)^2}\right)\sigma_y$$
$$- \frac{\omega_0}{4} \cos\left(\frac{\omega t}{1+\omega^2 t^2}\right)\sigma_z,$$

where we have put $B = \frac{\omega_0}{g}$ and $B_0 = \frac{\omega}{g}$. The Pauli matrices are the following:

$$\sigma_z = \begin{pmatrix} 1 & 0 \\ 0 & -1 \end{pmatrix}, \sigma_x = \begin{pmatrix} 0 & 1 \\ 1 & 0 \end{pmatrix}, \sigma_y = \begin{pmatrix} 0 & -i \\ i & 0 \end{pmatrix}.$$

Considering this problem by the path integral approach, our motivation is the following. We show that for interaction with the coupling of spin-field type, the propagator is first, by construction, written in the standard form $\sum_{path} \exp(iS(path)/\hbar)$, where S is the action that describes the system, where the discrete variable relative to spin being inserted as the (continuous) path using fermionic coherent states. The knowledge of the propagator is essential to the determination of physical quantities such as the transition probability which is the aim of this paper.

The paper is organized as follows. In Sect. 2, we give some notation and the necessary spin coherent state path integral for spin $\frac{1}{2}$ system for our further computations. In Sect. 3, after setting up a path integral formalism for the propagator, we perform the direct calculations. The integration over the spin variables is easy to carry out thanks to simple transformations. The explicit result of the propagator is directly computed and the transition probability is then deduced. Finally, in Sect. 4, we present our conclusions.

2 Coherent States Formalism

Now, let us focus on some definitions, properties and notations needed for the further developments. As we are interested by the spin field interaction, we shall replace the Pauli matrices σ_i by a pair of fermionic operators (u, d) known as Schwinger fermionic model of spin following the recipe:

$$\sigma \longrightarrow \left(u^\dagger, d^\dagger\right) \sigma \begin{pmatrix} u \\ d \end{pmatrix},$$

where the pair (u, d) describes a two-dimensional fermionic oscillators.

Incidentally, the spin eigenstates $|\uparrow\rangle$ and $|\downarrow\rangle$ are generated from the fermionic vacuum state $|0, 0\rangle$ by the action of the fermionic oscillators u^+ and d^+ following the relations

$$u^+ |0, 0\rangle = |\uparrow\rangle \quad \text{and} \quad d^+ |0, 0\rangle = |\downarrow\rangle,$$

where the action of u and d on this vacuum state is given by the vanishing results

$$u |0, 0\rangle = 0 \quad \text{and} \quad d |0, 0\rangle = 0.$$

The pair of the fermionic oscillators (u, d) and its adjoint $\left(u^+, d^+\right)$ satisfy the usual fermionic algebra defined by the following anticommutator relations:

$$\left[u, u^+\right]_+ = 1, \quad \left[d, d^+\right]_+ = 1,$$

where all other anticommutators vanish.

The notation $[A, B]_+$ stands for

$$[A, B]_+ = AB + BA.$$

Let us now introduce coherent states relative to this fermionic oscillators algebra. These states are generally defined as eigenvectors of the fermionic oscillators u and d:

$$u |\alpha, \beta\rangle = \alpha |\alpha, \beta\rangle, \quad d |\alpha, \beta\rangle = \beta |\alpha, \beta\rangle,$$

where (α, β) is a pair of Grasmmann variables which are anticommuting with fermionic oscillators and with themselves, namely

$$\begin{cases} [\alpha, u]_+ = \left[\alpha, u^+\right]_+ = [\alpha, d]_+ = \left[\alpha, d^\dagger\right]_+ = 0 \\ [\beta, u]_+ = \left[\beta, u^+\right]_+ = [\beta, d]_+ = \left[\beta, d^+\right]_+ = 0 \end{cases}$$

and are commuting with vacuum states $|0, 0\rangle$, $\langle 0, 0|$:

$$\begin{cases} \alpha \, |0, 0\rangle = |0, 0\rangle \, \alpha, & \langle 0, 0| \, \alpha = \alpha \, \langle 0, 0| \\ \beta \, |0, 0\rangle = |0, 0\rangle \, \beta, & \langle 0, 0| \, \beta = \beta \, \langle 0, 0| \, . \end{cases}$$

The above definitions are equivalent to the fact that these states are generated from the vacuum state according to the following relation

$$|\alpha, \beta\rangle = \exp\left(-\alpha u^+ - \beta d^+\right) |0, 0\rangle$$

The main properties of these states are:

- the completeness relation

$$\int d\bar{\alpha} d\alpha d\bar{\beta} d\beta \, e^{-\bar{\alpha}\alpha - \bar{\beta}\beta} \, |\alpha, \beta\rangle \, \langle \alpha, \beta| = 1,$$

- non-orthogonality

$$\langle \alpha, \beta \mid \alpha', \beta'\rangle = e^{\bar{\alpha}\alpha' + \bar{\beta}\beta'}.$$

3   Path Integral Formulation

At this stage we shall provide a path integral expression for the propagator for the Hamiltonian given by the expression (1). This can be readily done by exploiting the above model of the spin by which this Hamiltonian converts to the following fermionic form

$$H = -\frac{\omega_0}{4} \sin\left(\frac{\omega t}{1 + \omega^2 t^2}\right) (u^\dagger d + d^\dagger u)$$

$$+ \left(\frac{\omega_0}{2 \cosh(\frac{\omega_0}{2}) t} + \frac{\omega}{2} \frac{1 - \omega^2 t_n^2}{(1 + \omega^2 t_n^2)^2}\right) (-i u^\dagger d + i d^\dagger u)$$

$$-\frac{\omega_0}{4} \cos\left(\frac{\omega t}{1 + \omega^2 t^2}\right) (u^\dagger u - d^\dagger d).$$

Moreover, it is convenient to choose the quantum state as $|\alpha, \beta\rangle$ where (α, β) describes the spin variables. According to the habitual construction procedure of the path integral, we define the propagator as the matrix element of the evolution operator between the initial state $|\alpha_i, \beta_i\rangle$ and final state $|\alpha_f, \beta_f\rangle$

$$\mathbf{K}(\alpha_f, \beta_f; \alpha_i, \beta_i; T) = \langle \alpha_f, \beta_f \mid U(T) \mid \alpha_i, \beta_i\rangle, \tag{2}$$

where

$$U(T) = \mathbf{T}_D \exp\left(-\frac{i}{\hbar}\int_0^T H(t)dt\right),$$

where $\mathbf{T}_D$ is the Dyson chronological operator.

To move to path integral representation, we first subdivide the time interval $[0, T]$ into $N + 1$ intervals of length ε, intermediate moments, by using the Trotter's formula and we then introduce the projectors according to these intermediate instants N, which are regularly distributes between 0 and T in (2), we obtain the discretized path integral form of the propagator

$$\mathbf{K}(\alpha_f, \beta_f, \alpha_i, \beta_i T) = \lim_{N\to\infty} \prod_{n=1}^{N} \int d\bar{\alpha}_n d\alpha_n d\bar{\beta}_n d\beta_n$$

$$\times e^{-\bar{\alpha}_n\alpha_n - \bar{\beta}_n\beta_n} \exp\sum_{n=1}^{N+1}\left[\bar{\alpha}_n\alpha_{n-1} + \bar{\beta}_n\beta_{n-1}\right.$$

$$i\varepsilon\frac{\omega_0}{4}\cos\left(\frac{\omega t_n}{1+\omega^2 t_n^2}\right)\left(\bar{\alpha}_n\alpha_{n-1} - \bar{\beta}_n\beta_{n-1}\right)$$

$$+i\varepsilon\frac{\omega_0}{4}\sin\left(\frac{\omega t_n}{1+\omega^2 t_n^2}\right)\left(\bar{\alpha}_n\beta_{n-1} + \bar{\beta}_n\alpha_{n-1}\right)$$

$$-i\varepsilon\left(\frac{\omega_0}{2\cosh(\frac{\omega_0}{2})t_n} + \frac{\omega}{2}\frac{1-\omega^2 t_n^2}{(1+\omega^2 t_n^2)^2}\right)$$

$$\left.\times\left(-i\bar{\alpha}_n\beta_{n-1} + i\bar{\beta}_n\alpha_{n-1}\right)\right]. \tag{3}$$

The formal continuous expression for the transition amplitude (3) is found by taking the limit $N \to \infty$.

$$\mathbf{K}(\alpha_f, \beta_f; \alpha_i, \beta_i; T) = \int \mathcal{D}\bar{\alpha}\mathcal{D}\alpha\mathcal{D}\bar{\beta}\mathcal{D}\beta e^{\frac{1}{2}\left(\bar{\alpha}_f\alpha_f + \bar{\beta}_f\beta_f + \bar{\alpha}_i\alpha_i + \bar{\beta}_i\beta_i\right)}$$

$$\times\exp\int_0^\infty dt\left[-\frac{1}{2}\left(\bar{\alpha}\dot{\alpha} + \bar{\beta}\dot{\beta} - \dot{\bar{\alpha}}\alpha - \dot{\bar{\beta}}\beta\right)\right.$$

$$+i\frac{\omega_0}{4}\left(\bar{\alpha}\beta + \bar{\beta}\alpha\right)\sin\left(\frac{\omega t}{1+\omega^2 t^2}\right)$$

$$+i\frac{\omega_0}{4}\left(\bar{\alpha}\alpha - \bar{\beta}\beta\right)\cos\left(\frac{\omega t}{1+\omega^2 t^2}\right)$$

$$i\left(\frac{\omega_0}{2\cosh(\frac{\omega_0}{2})t} + \frac{\omega}{2}\frac{1-\omega^2 t^2}{(1+\omega^2 t^2)^2}\right)$$

$$\left.\times\left(-i\bar{\alpha}\beta + i\bar{\beta}\alpha\right)\right],$$

with $(\alpha_0, \beta_0) = (\alpha_i, \beta_i)$ and $(\bar{\alpha}_{N+1}, \bar{\beta}_{N+1}) = (\bar{\alpha}_f, \bar{\beta}_f)$. This last expression represents the path integral of the propagator which has been the purpose subject of previous papers, and has the advantage that it permits us to perform explicitly some concrete calculations.

4 Calculation of the Propagator

To begin, we first introduce new Grassmann variables via an unitary transformation in spin coherent space which eliminates the angle $\dfrac{\omega t}{1 + \omega^2 t^2}$ present in the expression of the magnetic field:

$$
\begin{cases}
(\alpha_n, \beta_n) \mapsto (\eta_n, \xi_n) \\[2mm]
\begin{pmatrix} \alpha_n \\ \beta_n \end{pmatrix} = e^{-\frac{i}{2}\frac{\omega t_n}{1+\omega^2 t_n^2}\sigma_y} \begin{pmatrix} \eta_n \\ \xi_n \end{pmatrix}.
\end{cases}
$$

Then, it is easy to show that the measure and the infinitesimal action become respectively

$$
\prod_{n=1}^{N} \left(d\bar{\alpha}_n d\alpha_n d\bar{\beta}_n d\beta_n e^{-\bar{\alpha}_n \alpha_n - \bar{\beta}_n \beta_n} \right) = \prod_{n=1}^{N} \left(d\bar{\eta}_n d\eta_n d\bar{\xi}_n d\xi_n e^{-\bar{\eta}_n \eta_n d - \bar{\xi}_n \xi_n} \right)
$$

$$
\begin{aligned}
\bar{\alpha}_n \alpha_{n-1} + \bar{\beta}_n \beta_{n-1} &= \bar{\eta}_n \eta_{n-1} + \bar{\xi}_n \xi_{n-1} \\
&\quad + i \frac{\omega}{2} \varepsilon \frac{1 - \omega^2 t_n^2}{(1+\omega^2 t_n^2)^2} \left(-i\bar{\eta}_n \xi_{n-1} + i\bar{\xi}_n \eta_{n-1} \right) + O(\varepsilon^2) \\
&\quad i\varepsilon \left[\sin \tfrac{\omega t_n}{1+\omega^2 t_n^2} \left(\bar{\alpha}_n \beta_{n-1} + \bar{\beta}_n \alpha_{n-1} \right) \right. \\
&\quad \left. + \cos \tfrac{\omega t_n}{1+\omega^2 t_n^2} \left(\bar{\alpha}_n \alpha_{n-1} - \bar{\beta}_n \beta_{n-1} \right) \right] \\
&= i\varepsilon \left(\bar{\eta}_n \eta_{n-1} - \bar{\xi}_n \xi_{n-1} \right).
\end{aligned}
$$

The propagator in function of the new Grassmann variables η and ξ, becomes

$$
\mathbf{K}(f, i; T) = \lim_{N \to \infty} \prod_{n=1}^{N} d\bar{\eta}_n d\eta_n d\bar{\xi}_n d\xi_n e^{-\bar{\eta}_n \eta_n - \bar{\xi}_n \xi_n}
$$

$$
\prod_{n=1}^{N+1} \exp \left[\bar{\eta}_n \eta_{n-1} + \bar{\xi}_n \xi_{n-1} \right.
$$

$$
+ i\, \varepsilon \frac{\omega_0}{4} \left(\bar{\eta}_n \eta_{n-1} - \bar{\xi}_n \xi_{n-1} \right)
$$

$$
\left. - i\varepsilon \frac{\omega_0}{2 \cosh \frac{\omega_0 t_n}{2}} \left(-i\bar{\eta}_n \xi_{n-1} + i\bar{\xi}_n \eta_{n-1} \right) \right].
$$

Now using the following transformation:

$$\varepsilon = -\frac{\cosh \frac{\omega_0 t_n}{2}}{\omega_0} \tau \ \text{ with } \ \tau = s_n - s_{n-1},$$

where

$$s_n = 2 \arcsin \frac{1}{\cosh \frac{\omega_0 t_n}{2}},$$

$$Z_n = \begin{pmatrix} \eta_n \\ \xi_n \end{pmatrix} \ \text{ and } \ \bar{Z}_n = \left(\bar{\eta}_n, \bar{\xi}_n \right).$$

The propagator becomes

$$\mathbf{K}(f, i; T) = \lim_{N \to \infty} \prod_{n=1}^{N} \int d\bar{Z}_n d Z_n e^{-\bar{Z}_n Z_n}$$

$$\prod_{n=1}^{N+1} \exp \left[\bar{Z}_n Z_{n-1} + i\tau \bar{Z}_n Q(n) Z_{n-1} \right], \tag{4}$$

where

$$Q(n) = \begin{pmatrix} -\frac{1}{4 \sin \frac{s_n}{2}} & -\frac{i}{2} \\ \frac{i}{2} & \frac{1}{4 \sin \frac{s_n}{2}} \end{pmatrix}.$$

Then we introduce new Grassmann variables Ψ via an unitary transformation in spin coherent state space defined by

$$Z_n = U(n)\Psi_n \qquad \bar{Z}_n = \Psi_n U^{\dagger}(n)$$

with

$$U(n) = e^{-\frac{i}{2} \ln \tan \frac{s_n}{4} \sigma_z},$$

which modify the expression (4) to the following form

$$\mathbf{K}(\alpha_f, \beta_f; \alpha_i, \beta_i; T) = \lim_{N \to \infty} \prod_{n=1}^{N} \int d\Psi_n d\Psi_n e^{-\Psi_n \Psi_n}$$

$$\times \prod_{n=1}^{N+1} \exp \left[\overline{\Psi}_n \Psi_{n-1} + i\tau \overline{\Psi}_n Q_1(n) \Psi_{n-1} \right],$$

where

$$Q_1(n) = \begin{pmatrix} 0 & -\frac{i}{2}e^{+\frac{i}{2}\ln\tan\frac{s_n}{4}}\sigma_z \\ \frac{i}{2}e^{-\frac{i}{2}\ln\tan\frac{s_n}{4}}\sigma_z & 0 \end{pmatrix}.$$

The next step consists of taking the diagonal form for the action in order to be able to integrate. Thus, we set a unit transformation over the Grassmann variables

$$\begin{cases} \Psi \longrightarrow \Phi \\ \Psi = U_1(s)\Phi = \begin{pmatrix} A(s) & -B^*(s) \\ B(s) & A^*(s) \end{pmatrix}\Phi \end{cases} \tag{5}$$

$$\begin{cases} U_1(s)U_1^\dagger(s) = U_1^\dagger(s)U_1(s) = 1 \\ \det U_1(s) = 1 \end{cases}$$

and the initial conditions $A(t=0) = 1$, $B(t=0) = 0$. By means of a simple calculation including the following development

$$U_1(s_{n-1}) = U_1(s_n) - \tau\frac{dU_1(n)}{dt_n} + O(\tau^2)$$

$$U_1^\dagger(s_n)U_1(s_{n-1}) = \mathbf{I} - \tau U_1^\dagger(s_n)\frac{dU_1}{dt}(s_n),$$

we obtain

$$\mathbf{K}(\alpha_f, \beta_f; \alpha_i, \beta_i; T) = \lim_{N\to\infty}\prod_{n=1}^{N}\int d\Phi_n d\bar{\Phi}_n e^{-\bar{\Phi}_n\Phi_n}$$

$$\times \prod_{n=1}^{N+1}\exp\left[\bar{\Phi}_n\Phi_{n-1} + i\tau\bar{\Phi}_n Q_2(n)\Phi_{n-1}\right],$$

where

$$Q_2(n) = iU_1^\dagger(s_n)\frac{dU_1}{ds}(s_n) + U_1^\dagger(s_n)Q_1(n)U_1(s_n).$$

Now, we determine the unit transformation by fixing the diagonal form for the action, which leads us to the following condition

$$Q_2(n) = 0. \tag{6}$$

To be able to integrate, we have to write the expression in an appropriate form

$$\mathbf{K}(\alpha_f, \beta_f; \alpha_i, \beta_i; T) = \int d\xi^\dagger d\xi \exp\left[-\xi^\dagger\xi + \mathbf{V}^\dagger\xi + \xi^\dagger\mathbf{W}\right],$$

where

$$\mathbf{V}^{\dagger} = \left(0, ..., \overline{\Phi}_{N+1}\right), \quad \xi = \begin{pmatrix} \Phi_1 \\ \vdots \\ \Phi_1 \end{pmatrix}, \quad \mathbf{W} = \begin{pmatrix} \Phi_0 \\ \vdots \\ 0 \end{pmatrix}.$$

Now, we absorb the linear terms in ξ and $\xi^{\dagger}$ thanks to the shift

$$\xi \to \xi + \mathbf{W},$$
$$\xi^{\dagger} \to \xi^{\dagger} + \mathbf{V}^{\dagger},$$

and we integrate over the Grassmann variables. Our propagator relative to the spin 1/2 subject to time dependent magnetic field is finally written as follows:

$$\mathbf{K}(\alpha_f, \beta_f; \alpha_i, \beta_i; T) = e^{\overline{\Phi}_f \Phi_i}.$$

In terms of the old variables (α, β) it becomes

$$\mathbf{K}(\alpha_f, \beta_f, \alpha_i, \beta_i; T) = \exp\left(\bar{\alpha}_f, \bar{\beta}_f\right) \mathbf{R}(t) \begin{pmatrix} \alpha_i \\ \beta_i \end{pmatrix}, \tag{7}$$

where

$$\mathbf{R}(t) = e^{-\frac{i}{2}\frac{\omega t}{1+\omega^2 t^2}\sigma_y}$$

$$\times e^{-\frac{i}{2}\ln\tan\frac{s(t)}{4}\sigma_z} \begin{pmatrix} A(s(t)) & B(s(t)) \\ -B^*(s(t)) & A^*(s(t)) \end{pmatrix}.$$

5   The Transition Probability

Let us now turn to the calculation of this propagator (7), between the spin states. We just evaluate the matrix $K(\uparrow, \uparrow; T)$ only, and all the other matrices can be deduced following the same method. In fact, the propagator on the spin eigenstates is given by

$$K(\uparrow, \uparrow; T) = \langle \uparrow | U(T) | \uparrow \rangle$$

With the help of the completeness relations, this amplitude becomes

$$K(\uparrow, \uparrow; T) = \int d\bar{\alpha}_f d\alpha_f d\bar{\beta}_f d\beta_f d\bar{\alpha}_i d\alpha_i d\bar{\beta}_i d\beta_i$$
$$\times e^{-\bar{\alpha}_f \alpha_f - \bar{\beta}_f \beta_f} e^{-\bar{\alpha}_i \alpha_i - \bar{\beta}_i \beta_i}$$
$$\times \langle \uparrow | \alpha_f, \beta_f \rangle \langle \alpha_i, \beta_i | \uparrow \rangle \mathbf{K}(\alpha_f, \beta_f, \alpha_i, \beta_i; T).$$

Then

$$K(\uparrow,\uparrow;T) = \int d\bar{\alpha}_f d\alpha_f d\bar{\beta}_f d\beta_f d\bar{\alpha}_i d\alpha_i d\bar{\beta}_i d\beta_i$$
$$e^{-\bar{\alpha}_i\alpha_i-\bar{\beta}_i\beta_i}e^{-\bar{\alpha}_f\alpha_f-\bar{\beta}_f\beta_f}\langle\uparrow|\alpha_f,\beta_f\rangle\langle\alpha_i,\beta_i|\uparrow\rangle$$
$$\exp\left\{(\bar{\alpha}_f,\bar{\beta}_f)\begin{pmatrix} R_{11}(t) & R_{12}(t) \\ R_{21}(t) & R_{22}(t) \end{pmatrix}\begin{pmatrix} \alpha_i \\ \beta_i \end{pmatrix}\right\}. \tag{8}$$

Thanks to the features [16]

$$\langle\uparrow|\alpha_f,\beta_f\rangle = \alpha_f, \quad \langle\alpha_i,\beta_i|\uparrow\rangle = \bar{\alpha}_i \text{ and } \alpha_f\bar{\alpha}_i = e^{-\bar{\alpha}_i\alpha_f} - 1,$$

(8) takes the following form:

$$K(\uparrow,\uparrow;T) = \int dv^\dagger dv. \left[\exp v^\dagger M'v - \exp v^\dagger Mv\right],$$

where the matrices M and M' are, respectively

$$M = \begin{pmatrix} -1 & 0 & 0 & 0 \\ R_{11} & -1 & R_{11} & 0 \\ 0 & 0 & -1 & 0 \\ R_{21} & 0 & R_{22} & -1 \end{pmatrix}$$

and

$$M' = \begin{pmatrix} -1 & -1 & 0 & 0 \\ R_{11} & -1 & R_{11} & 0 \\ 0 & 0 & -1 & 0 \\ R_{21} & 0 & R_{22} & -1 \end{pmatrix}$$

and R_{nm} are the elements of the matrix R and

$$v = \begin{pmatrix} \alpha_i \\ \alpha_f \\ \beta_i \\ \beta_f \end{pmatrix}, \quad v^\dagger = (\bar{\alpha}_i\ \bar{\alpha}_f\ \bar{\beta}_i\ \bar{\beta}_f),$$

are the vectors gathering the old Grassmann variables.

The integration over the Grassmann variables is thus simple:

$$K(\downarrow,\uparrow;T) = \det M' - \det M.$$

As $\det M' = 1 + R_{11}$ and $\det M = 1$, the propagator following the states of the up–up spin is finally

$$K(\uparrow,\uparrow;T) = R_{11}(t).$$

Normally, if we repeat the calculations by considering all the initial and final states of the spin, the propagator will take the following matrix form:

$$K(m_f, m_i; T) = \begin{pmatrix} \mathbf{R}_{11}(t) & \mathbf{R}_{12}(t) \\ \mathbf{R}_{21}(t) & \mathbf{R}_{22}(t) \end{pmatrix}.$$

Hence the probability transition from down to up spin states is given by:

$$P_{\downarrow\uparrow} = |\mathbf{R}_{21}(t)|^2$$

$$P_{\downarrow\uparrow} = \left\| \left[A(t)e^{-\frac{i}{2} \ln \tan \frac{s(t)}{4}} \sin \frac{1}{2} \left(\frac{\omega t}{1+\omega^2 t^2} \right) \right. \right.$$

$$\left. \left. - B^*(t)e^{\frac{i}{2} \ln \tan \frac{s(t)}{4}} \cos \frac{1}{2} \left(\frac{\omega t}{1+\omega^2 t^2} \right) \right] \right\|^2.$$

Note that the matrix $U_1(s)$ introduced in (5) has been fixed by the condition (6), so it has to satisfy the following auxiliary equation

$$i \frac{dU_1}{ds} + Q_1(s)U_1(s) = 0,$$

i.e., a system of two coupled equations

$$\begin{cases} \frac{dA}{ds} = -\frac{1}{2} B^* e^{i \ln \tan \frac{s(t)}{4}} \\ \frac{dB^*}{ds} = \frac{1}{2} A e^{-i \ln \tan \frac{s(t)}{4}} \end{cases} \text{ with } \begin{cases} A(\pi) = 1 \\ B(\pi) = 0 \end{cases}$$

and whose solution determines the elements A and B of the matrix U_1. Let us uncouple this system

$$\frac{d^2 B^*}{ds^2} + \frac{i}{2 \sin \frac{s}{2}} \frac{dB^*}{ds} + \frac{1}{4} B^* = 0.$$

The solution of this equation is

$$B^*(s) = \frac{i}{2} \left(i + \cos \frac{s}{2} \right) h(s).$$

Then

$$A(s) = 2e^{i \ln \tan \frac{s(t)}{4}} \left[-\frac{i}{4} h(s) \sin \frac{s}{2} + \frac{i}{2} \left(i + \cos \frac{s}{2} \right) \frac{dh(s)}{ds} \right],$$

where

$$h(s) = \int_{\pi}^{s} \frac{e^{-i \ln \tan \frac{\tau}{4}}}{(i + \cos \frac{\tau}{2})^2} d\tau.$$

A straightforward calculation leads to the well known like Rabi formula.

$$P_{\downarrow\uparrow} = \frac{1 + \cos^2 \frac{s}{2}}{4} \left| \left[h(s) \sin \frac{1}{2} \left(\frac{\omega t}{1+\omega^2 t^2} \right) + \right. \right.$$
$$\left. \left. 2 \left[\frac{dh(s)}{ds} - \frac{\sin \frac{s}{2}}{2(i + \cos \frac{s}{2})} h(s) \right] \cos \frac{1}{2} \left(\frac{\omega t}{1+\omega^2 t^2} \right) \right] \right|^2 .$$

This result coincides with that in [7] and [21].

6 Conclusion

By using the formalism of the path integral and the fermionic coherent states approach, we have been able to calculate the explicit expression of the propagator relative to spin $1/2$ interacting with time dependent magnetic field. To treat the spin dynamics, we have used the Schwinger's recipe which replaces the Pauli matrices by a pair of fermionic oscillators. The introduction of a particular rotations in coherent state space has eliminated the rotation angle of the magnetic field and has then simplified somewhat the Hamiltonian of the considered system. As a consequence, we have been able to integrate over the spin variables described by fermionic oscillators. The exactness of the result is displayed in the evaluation of the transition probability formula.

References

1. Aouachria, M., Chetouani, L.: Rabi oscillations in gravitational fields: exact solution via path integral. Eur. Phys. J. C **25**, 333–338 (2002)
2. Aouachria, M., Chetouani, L.: Treatment of a damped two-level atom in an electromagnetic wave by the path integral formalism. Chin. J. Phys. **40**, 496–504 (2002)
3. Aouachria, M.: Spin coherent state path integral for a two-level atom in an electromagnetic wave of circular polarization. Chin. J. Phys. **49**, 689–698 (2011)
4. Aouachria, M.: Rabi oscillation in a damped rotating magnetic field: a path integral approach. J. Phys. Conf. Ser. **435**, 012021 (2013)
5. Barut, A.O., Beker, H.: Exact Solutions of spin (Rabi) precession, transmission and reflection in the Eckhart potential. Eur. Phys. Lett. **14**, 197–202 (1991)
6. Boudjedaa, T., Bounames, A., Nouicer, Kh, Chetouani, L., Hammann, T.F.: Path integral for the generalized Jaynes-Cummings model. Phys. Scr. **54**, 225–233 (1996)
7. Boudjedaa, T., Bounames, A., Nouicer, Kh, Chetouani, L., Hammann, T.F.: Coherent state path integral for the interaction of spin with external magnetic fields. Phys. Scr. **56**, 545–554 (1998)

8. Calvo, M., Codriansky, S.: A class of solvable Pauli-Schrödinger hamiltonians. J. Math. Phys. **24**, 553–559 (1983)
9. Codriansky, S., Cordero, P., Salamo, S.: On a class of solvable Pauli-Schrödinger hamiltonians. Z. Phys. A **353**, 341–343 (1995)
10. Grosche, C., Steiner, F.: Handbook of Feynman Path Integrals. Springer, Berlin (1998)
11. Lämmerzahl, C., Bordé, C.J.: Rabi oscillations in gravitational fields: exact solution. Phys. Lett. A **203**, 59–67 (1995)
12. Merdaci, A., Boudjedaa, T., Chetouani, L.: Exact path Integral for neutral spinning particle in interaction with helical magnetic field. Phys. Scr. **64**, 15–19 (2001)
13. Merdaci, A., Boudjedaa, T., Chetouani, L.: Path integral for a neutral spinning particle in interaction with a rotating magnetic field and a scalar potential. Czech. J. Phys. **51**, 865–881 (2001)
14. Merdaci, A., Boudjedaa, T., Chetouani, L.: A neutral spinning particle in interaction with a magnetic field and Poschl-Teller potential. Eur. Phys. J. C **22**, 585–592 (2001)
15. Mijatović, M., Ivanovski, C., Veljanoski, B., Trenčevski, K.: Scattering and bound states of a nonrelativistic neutral spin-1/2 particle in a magnetic field. Z. Phys. A **345**, 65–77 (1993)
16. Nakamura, M., Kazuo Kitahara, K.: Spin polarization of a quantum particle; fermionic path integral approach. J. Phys. Soc. Jpn. **60**, 1388–1397 (1991)
17. Nouicer, Kh, Chetouani, L.: Path integral approach to the supersymmetric generalized Jaynes-Cummings model. Phys. Lett. A **281**, 218–230 (2001)
18. Qiong-Gui, L.: Exact solutions for neutral particles in the field of a circularly polarized plane electromagnetic wave. Phys. Lett. A **342**, 67–76 (2005)
19. Rabi, I.I.: Space quantization in a gyrating magnetic field. Phys. Rev. **51**, 652–654 (1937)
20. Tahmasebi, M.J., Sobouti, Y.: Exact solutions of Schrödinger's equation for spin systems in a class of time-dependent magnetic fields. Mod. Phys. Lett. B **5**, 1919–1924 (1991)
21. Tahmasebi, M.J., Sobouti, Y.: Exact solutions of Schrödinger's equation for spin systems in a class of time-dependent magnetic fields II. Mod. Phys. Lett. B **6**, 1255–1261 (1992)

New Complexity Analysis of the Path Following Method for Linear Complementarity Problem

El Amir Djeffal, Lakhdar Djeffal and Farouk Benoumelaz

Abstract In this paper, we present an interior point algorithm for solving an optimization problem using the central path method. By an equivalent reformulation of the central path, we obtain a new search direction which targets at a small neighborhood of the central path. For a full-Newton step interior-point algorithm based on this search direction, the complexity bound of the algorithm is the best known for linear complementarity problem. For its numerical tests some strategies are used and indicate that the algorithm is efficient.

1 Introduction

Let us consider the linear complementarity problem (LCP): find vectors x and y in real space $\mathbb{R}^n$ that satisfy the following conditions

$$x \geq 0, \ y = Mx + q \geq 0 \text{ and } x^t y = 0, \tag{1}$$

where q is a given vector in $\mathbb{R}^n$ and M is a given $\mathbb{R}^{n \times n}$ real matrix. The linear complementarity problems have important applications in mathematical programming and various areas of engineering [1, 2]. Primal-dual path-following are the most attractive methods among interior point methods to solve a large wide of optimization problems because of their polynomial complexity and their simulation efficiency [3–8]. In this paper we deal with the complexity analysis and the numerical implementation of a primal-dual interior point algorithm based on new kernel function. These algorithms are based on the strategy of the central path and on a method for finding a new

E.A. Djeffal (✉) · L. Djeffal · F. Benoumelaz
Department of Mathematics, University of Hadj Lakhdar, Batna, Algeria
e-mail: djeffal_elamir@yahoo.fr

L. Djeffal
e-mail: lakdar_djeffal@yahoo.fr

F. Benoumelaz
e-mail: fbenoumelaz@yahoo.fr

© Springer International Publishing Switzerland 2016
G.A. Anastassiou and O. Duman (eds.), *Intelligent Mathematics II:*
Applied Mathematics and Approximation Theory, Advances in Intelligent Systems
and Computing 441, DOI 10.1007/978-3-319-30322-2_6

search directions. This technique was used first by Darvay for linear optimization, we reconsider this technique to the monotone linear complementarity problem case where we show also that this short-step algorithm deserves the best current polynomial complexity namely, which is analogous to linear optimization problem. Finally, the algorithm is applied on some monotone linear complementarity problems.

The paper is organized as follows. In the next section, the statement of the problem is presented. In Sect. 3, we deal with the new search directions and the description of the algorithm. In Sect. 4, we state its polynomial complexity. In Sect. 5, we present a numerical implementation. In Sect. 6, a conclusion and remarks are given.

The following notations are used throughout the paper. $\mathbb{R}^n$ denotes the space of real n-dimensional vectors and $\mathbb{R}^n_+$ the nonnegative orthant of $\mathbb{R}^n$. Let $u, v \in \mathbb{R}^n$, $u^t v = \sum_{i=1}^{n} u_i v_i$ is their inner product, $\|u\| = \sqrt{u^t u}$ is the Euclidean norm and $\|u\|_\infty = \max_{1 \le i \le n} |u_i|$ is the maximum norm, $e =' 1, 1, \ldots, 1)^t$ is the vector of ones in $\mathbb{R}^n$. Given vectors x and y in $\mathbb{R}^n$ $xy = (x_1 y_1, \ldots, x_n y_n)$ denotes the Hadamard coordinate-wise product of the two vectors x and y.

2 Presentation of the Problem

The feasible set and the strictly feasible set and the solution set of (1) are denoted, respectively by

$$S = \{(x, y) \in \mathbb{R}^{2n} : y = Mx + q, x \ge 0, y \ge 0\},$$
$$S_{str} = \{(x, y) \in S : x > 0, y > 0\},$$
$$\Omega = \{(x, y) \in S : x > 0, y > 0 \text{ and } x^t y = 0\}.$$

In this paper, we assume that the following assumptions hold

1. The strictly feasible set S_{str} is not empty
2. The matrix M is a positive semidefinite matrix

These assumptions imply that S_{str} is the relative interior of S and Ω is a non empty polyhedral convex and bounded set. In addition (1), is equivalent to the following convex quadratic problem, see, e.g., [6].

$$\min \{x^t y : y = Mx + q, x \ge 0, y \ge 0\}. \tag{2}$$

Hence, finding the solution of (1) is equivalent to find the minimizer of (2) with its objective value is zero.

In order to introduce an interior point method to solve (2), we associate with it the following barrier minimization problem

$$\min\left\{f_\mu(x, y) = x^t y - \mu \sum_{i=1}^{n} \ln(x_i y_i) : y = Mx + q, x > 0, y > 0\right\} \quad (3)$$

where $\mu > 0$ is a positive real number and it is called the barrier parameter.

The problem (2) is a convex optimization problem and then its first order optimality conditions are:

$$\begin{cases} Mx + q = y \\ xy = \mu e \\ x > 0, \ y > 0. \end{cases} \quad (4)$$

If the Assumptions (1) and (2) hold then for a fixed $\mu > 0$, the problem (3) and the system (4) have a unique solution [5, 6] denoted as $(x(\mu), y(\mu))$, with $x(\mu) > 0$ and $y(\mu) > 0$. We call $(x(\mu), y(\mu))$, with $\mu > 0$, the μ-centers of (3) or (4). The set of the μ-centers defines the so-called the central path of (1).

In the next section, we introduce a method for tracing the central path based a new class of search directions.

3 New Class of Search Directions

Now, following [9] the basic idea behind this approach is to replace the non linear equation:

$$\frac{xy}{\mu} = e$$

in (4) by an equivalent equation

$$\psi\left(\frac{xy}{\mu}\right) = \psi(e),$$

where ψ, is a real valued function on $[0, \infty)$ and differentiable on $[0, \infty)$ such that $\psi(t)$ and $\psi'(t) > 0$, for all $t > 0$. Then the system (4) can be written as the following equivalent form:

$$\begin{cases} Mx + q = y \\ \psi\left(\frac{xy}{\mu}\right) = \psi(e) \\ x > 0, \ y > 0. \end{cases} \quad (5)$$

Suppose that we have $(x, y) \in S_{str}$. Applying Newton's method for the system (5), we obtain a new class of search directions:

$$\begin{cases} M \triangle x = y \\ \frac{y}{\mu}\psi'\left(\frac{xy}{\mu}\right)\triangle x + \frac{x}{\mu}\psi'\left(\frac{xy}{\mu}\right)\triangle y = \psi(e) - \psi\left(\frac{xy}{\mu}\right). \end{cases} \quad (6)$$

Now, the following notations are useful for studying the complexity of the proposed algorithm.

The vectors

$$v = \sqrt{\frac{xy}{\mu}} \text{ and } d = \sqrt{\frac{xy^{-1}}{\mu}},$$

and observe that these notations lead to

$$d^{-1}x = dy = v. \tag{7}$$

Denote by

$$d_x = d^{-1}\Delta x, \ d_y = d\Delta y,$$

and hence, we have

$$\mu v(d_x + d_y) = y\Delta x + x\Delta y \tag{8}$$

and

$$d_x d_y = \frac{\Delta x \Delta y}{\mu}.$$

So using (7) and (8), the system (6) becomes

$$\begin{cases} \overline{M}d_x = d_y \\ d_x + d_y = p_v, \end{cases}$$

where $\overline{M} = DMD$ with $D = diag(d)$ and

$$p_v = \frac{\psi(e) - \psi(v^2)}{v\psi'(v^2)}.$$

As in [10], we shall consider the following function

$$\psi(t) = (m+1)t^2 - (m+2)t + \frac{1}{t^m} \text{ for all } t > 0, \text{ where } m > 4,$$

and

$$\psi'(t) = 2(m+1)t - (m+2) - m\frac{1}{t^{m+1}} \text{ for all } t > 0.$$

Hence, the Newton directions in (6) is

$$\begin{cases} M\Delta x = \Delta y \\ \frac{y}{\mu}\Delta x + \frac{x}{\mu}\Delta y = \dfrac{-(m+1)v^2 + (m+2)v - \frac{1}{v^m}}{2(m+1)v - (m+2) - m\frac{1}{v^{m+1}}}. \end{cases} \tag{9}$$

3.1   The Generic Interior-Point Algorithm for (LCP)

In this paper, we replace $\psi(t)$ with a new kernel function $\psi(t)$ which is defined in the previous section and assume that $\tau \geq 1$.

The new interior-point algorithm works as follows. Assume that we are given a strictly feasible point (x, y) which is in a τ-neighborhood of the given μ-center. Then we decrease μ to $\mu^{+} = (1 - \theta)\mu$, for some fixed $\theta \in (0, 1)$ and then we solve the Newton system (6) to obtain the unique search direction. The positivity condition of a new iterate is ensured with the right choice of the step size α which is defined by some line search rule. This procedure is repeated until we find a new iterate (x^{+}, y^{+}) that is in a τ-neighborhood of the μ^{+}-center. Then μ is again reduced by the factor $1 - \theta$ and we solve the Newton system targeting at the new μ^{+}-center, and so on. This process is repeated until μ is small enough, i.e., $n\mu \leq \varepsilon$.

The parameters τ, θ and the step size α should be chosen in such a way that the algorithm is optimized in the sense that the number of iterations required by the algorithm is as small as possible. The choice of the so-called barrier update parameter θ plays an important role both in theory and practice of $IPMs$.

The algorithm for our IPM for the (LCP) is given as follows:

IPM for the (LCP)

Begin algorithm
Input:
an accuracy parameter $\varepsilon > 0$,
an update parameter $\theta, 0 < \theta < 1$,
a threshold parameter $\tau, 0 < \tau < 1$,
a strictly feasible point (x^0, y^0) and $\mu^0 = \frac{(x^0)^t y^0}{n}$ such that $\delta(x^0 y^0, \mu^0) \leq \tau$.
begin
$x := x^0, y := y^0, \mu := \mu^0,$
  While $(n\mu) \geq \varepsilon$ do
   begin
  $\mu = (1 - \theta)\mu$
    While $(\delta(x^0 y^0, \mu^0) > \tau)$ do
    begin
      Solve system (3.2) to obtain $(\triangle x, \triangle y)$,
      Determine a step size α
      $x := x + \alpha \triangle x$
      $y := y + \alpha \triangle y$
    End While
  End While
End algorithm.

In the next section, we give the properties of the kernel function which are essential to our complexity analysis.

Lemma 1 *For $\psi(t)$, we have the following:*

(i) $\psi(t)$ *is exponentially convex for all $t > 0$,*
(ii) $\psi''(t)$ *is monotonically decreasing for all $t > 0$,*
(iii) $t\psi''(t) - \psi'(t) > 0$ *for all $t > 0$.*

Proof For (i), by Lemma 2.1.2 in [11], it suffices to show that the function $\psi(t)$ satisfies $t\psi''(t) + \psi'(t) \geq 0$, for all $t > 0$. We have

$$t\psi''(t) + \psi'(t) = t\left(2(m+1) - m(-m-1)t^{-m-2}\right)$$
$$+ \left(2(m+1)t - (m+2) - mt^{-m-1}\right)$$
$$= 4(m+1)t + m^2 t^{-m-1} - (m+2).$$

Let

$$g(t) = 4(m+1)t + m^2 t^{-m-1} - (m+2).$$

Then

$$g'(t) = 4(m+1) - m^2(m+1)t^{-m-2}$$

$$g''(t) = m^2(m+1)(m+2)t^{-m-3} > 0 \text{ for all } t > 0.$$

Let $g'(t) = 0$, we get $t = (\frac{m^2}{4})^{\frac{1}{m+2}}$. Since $g(t)$ is strictly convex and has a global minimum, $g((\frac{m^2}{4})^{\frac{1}{m+2}}) > 0$. And by Lemma 2.1.2 in [12], we have the result.

For (ii), we have $\psi'''(t) > 0$, so we have the result.

For (iii), we have

$$t\psi''(t) - \psi'(t) = m(m+2)t^{-m-1} + (m+2) > 0 \text{ for all } t > 0.$$

Lemma 2 *For $\psi(t)$, we have the following.*

$$(m+1)(t-1)^2 \leq \psi(t) \leq \frac{1}{4(m+1)}\psi'(t)^2, \text{ for all } t > 0, \tag{10}$$

$$\psi(t) \leq \frac{(m+1)(m+2)}{2}(t-1)^2, \text{ for all } t \geq 1. \tag{11}$$

Proof For (10), we have

$$\psi(t) = \int_1^t \int_1^\xi \psi''(\zeta)d\zeta\,d\xi \geq 2(m+1) \int_1^t \int_1^\xi d\zeta\,d\xi = (m+1)(t-1)^2,$$

also,

$$\psi(t) = \int_1^t \int_1^\xi \psi''(\zeta) d\zeta \, d\xi$$

$$\leq \frac{1}{2(m+1)} \int_1^t \int_1^\xi \psi''(\xi) \psi''(\zeta) d\zeta \, d\xi = \frac{1}{2(m+1)} \int_1^t \psi''(\xi) \psi'(\xi) \, d\xi$$

$$= \frac{1}{2(m+1)} \int_1^t \psi'(\xi) \, d(\psi'(\xi)) = \frac{1}{4(m+1)} \psi'(t)^2.$$

For (11), using Taylor's Theorem, we have

$$\psi(t) = \psi(1) + \psi'(1)(t-1) + \frac{1}{2}\psi''(1)(t-1)^2 + \frac{1}{6}\psi'''(\xi)(\xi-1)^3$$

$$= \frac{1}{2}\psi''(1)(t-1)^2 + \frac{1}{6}\psi'''(\xi)(\xi-1)^3$$

$$\leq \frac{1}{2}\psi''(1)(t-1)^2$$

$$= \frac{(m+1)(m+2)}{2}(t-1)^2.$$

This completes the proof.

Now, we define $\gamma : (0, \infty) \to (1, \infty)$ be the inverse function of $\psi(t)$ for all $t \geq 1$ and $\rho : (0, \infty) \to (0, 1)$ be the inverse function of $-\frac{1}{2}\psi'(t)$ for all $t \in (0, 1)$. then we have the following Lemma.

Lemma 3 *For $\psi(t)$, we have the following*

$$\sqrt{\frac{s}{m+1} + 1} \leq \gamma(s) \leq 1 + \sqrt{\frac{s}{m+1}}, \quad s \geq 0, \tag{12}$$

and

$$\rho(s) \geq \left(\frac{m}{2s+m}\right)^{\frac{1}{m+1}}, \quad s \geq 0. \tag{13}$$

Proof For (12), let $s = \psi(t)$, $t \geq 1$, i.e., $\gamma(s) = t, t \geq 1$, then we have

$$(m+1)t^2 = s + (m+2)t - t^{-m}.$$

Because $(m+2)t - t^{-m}$ is monotone increasing with respect to $t \geq 1$, we have

$$(m+1)t^2 \geq s + m + 1,$$

which implies that

$$t = \gamma(s) \geq \sqrt{\frac{s}{m+1}} + 1.$$

We have

$$s = \psi(t) \geq (m+1)(t-1)^2,$$

so

$$t = \gamma(s) \leq 1 + \sqrt{\frac{s}{m+1}}.$$

For (13), let $z = -\frac{1}{2}\psi'(t)$ for all $t \in (0,1)$. By the definition of ρ, we have $\rho(z) = t$ and $2z = -\psi'(t)$. Then

$$mt^{-m-1} = 2z + 2(m+1)t - (m+2).$$

Because $2(m+1)t - (m+2)$ is monotone increasing with respect to $t \in (0,1)$, we have

$$mt^{-m-1} \leq 2z + m,$$

which implies that

$$\rho(z) = t \geq \left(\frac{m}{2z+m}\right)^{\frac{1}{m+1}}.$$

This completes the proof.

Lemma 4 *Let $\gamma : (0,\infty) \to (1,\infty)$, be the inverse function of $\psi(t)$ for all $t \geq 1$. Then we have*

$$\delta(\beta v) \leq n\psi\left(\beta\gamma\left(\frac{\delta(v)}{n}\right)\right), \quad \beta \geq 1. \tag{14}$$

Proof Using Theorem 3.2 in [10], we get the result. This completes the proof.

Lemma 5 *Let $0 \leq \theta \leq 1$, $v^+ = \frac{1}{\sqrt{1-\theta}}v$. If $\delta(v) \leq \tau$, then we have*

$$\delta(v^+) \leq \frac{(m+1)(m+2)}{2(1-\theta)}\left(\sqrt{n\theta} + \sqrt{\frac{\tau}{m+1}}\right)^2. \tag{15}$$

Proof Since $\frac{1}{\sqrt{1-\theta}} \geq 1$ and $\gamma\left(\frac{\delta(v)}{n}\right) \geq 1$, we have $\frac{\gamma\left(\frac{\delta(v)}{n}\right)}{\sqrt{1-\theta}} \geq 1$. Using Lemma 4 with $\beta = \sqrt{1-\theta}$, (12), (13) and $\delta(v) \leq \tau$, we have

$$\delta(v^+) \le n\psi\left(\frac{1}{\sqrt{1-\theta}}\gamma\left(\frac{\delta(v)}{n}\right)\right)$$

$$\le n\frac{(m+1)(m+2)}{2}\left(\frac{1}{\sqrt{1-\theta}}\gamma\left(\frac{\delta(v)}{n}\right)-1\right)^2$$

$$= n\frac{(m+1)(m+2)}{2(1-\theta)}\left(\gamma\left(\frac{\delta(v)}{n}\right)-\sqrt{1-\theta}\right)^2$$

$$\le n\frac{(m+1)(m+2)}{2(1-\theta)}\left(1+\sqrt{\frac{\delta(v)}{(m+1)n}}-\sqrt{1-\theta}\right)^2$$

$$\le n\frac{(m+1)(m+2)}{2(1-\theta)}\left(\theta+\sqrt{\frac{\tau}{(m+1)n}}\right)^2$$

$$\le n\frac{(m+1)(m+2)}{2(1-\theta)}\left(\sqrt{n}\theta+\sqrt{\frac{\tau}{(m+1)n}}\right)^2.$$

This completes the proof.

Denote

$$\Psi_0 = L(n,\theta,\tau) = n\frac{(m+1)(m+2)}{2(1-\theta)}\left(\sqrt{n}\theta+\sqrt{\frac{\tau}{(m+1)n}}\right)^2, \qquad (16)$$

then Ψ_0 is an upper bound for $\Psi(V)$ during the process of the algorithm.

4 Analysis of Algorithm

The aim of this paper is to define a new kernel function and to obtain new complexity results for an (LCP) problem using the proximity function defined by the kernel function and following the approach of Bai et al. [10].

In the following, we compute a proper step size α and the decrease of the proximity function during an inner iteration and give the complexity results of the algorithm. For fixed $\mu > 0$.

4.1 Determining a Default Step Size

Taking a step size α, we have new iterates

$$x^+ = x + \alpha\triangle x, \ y^+ = y + \alpha\triangle y$$

Let

$$x^+ = x \left(e + \alpha \frac{\Delta x}{x} \right) = x \left(I + \alpha \frac{d_x}{v} \right) = \frac{x}{v} \left(v + \alpha d_x \right),$$

$$y^+ = y \left(e + \alpha \frac{\Delta y}{y} \right) = y \left(e + \alpha \frac{d_y}{v} \right) = \frac{y}{v} \left(v + \alpha d_y \right).$$

So, we have

$$v^+ = \left((v + \alpha d_x)^{\frac{1}{2}} \left(v + \alpha d_y \right) (v + \alpha d_x)^{\frac{1}{2}} \right)^{\frac{1}{2}}.$$

Since the proximity after one step is defined by

$$\delta(v^+) = \delta \left(\left((v + \alpha d_x)^{\frac{1}{2}} \left(v + \alpha d_y \right) (v + \alpha d_x)^{\frac{1}{2}} \right)^{\frac{1}{2}} \right).$$

By (i) in Lemma 2, we have

$$\delta(v^+) = \delta \left(\left((v + \alpha d_x) (v + \alpha d_y) \right)^{\frac{1}{2}} \right),$$

$$\leq \frac{1}{2} \left(\delta (v + \alpha d_x) + \delta(v + \alpha d_y) \right).$$

Define, for $\alpha > 0$,

$$f(\alpha) = \delta(V^+) - \delta(V).$$

Therefore, we have $f(\alpha) \leq f_1(\alpha)$, where

$$f_1(\alpha) = \frac{1}{2} \left(\delta (v + \alpha d_x) + \delta(v + \alpha d) \right) - \delta(v). \tag{17}$$

Obviously,

$$f(0) = f_1(0) = 0.$$

Throughout the paper, we assume that $\tau \geq 1$. Using Lemma 5 and the assumption that $\delta(v) \geq \tau$, we have $\delta(v) \geq \sqrt{(m + 1)}$.

By the definition of ρ, the largest step size of the worse case is given as follows:

$$\alpha^* = \frac{\rho(\delta) - \rho(2\delta)}{2\delta}. \tag{18}$$

Lemma 6 *Let the definition of ρ and α^* be as defined in (18), then we have*

$$\alpha^* \geq \frac{1}{(m + 1)(m + 2)^{\frac{m+2}{m+1}}}.$$

Proof Using Lemma 4.4 in [10], the definition of $\psi''(t)$, we have

$$\alpha^* \geq \frac{1}{\psi''(\rho(2\delta))} = \frac{1}{2(m+1) + \frac{m(m+1)}{\rho(2\delta)^{m+2}}}$$

$$\geq \frac{1}{2(m+1) + m(m+1)(\frac{4\delta+m}{m})^{\frac{m+2}{m+1}}}$$

$$\geq \frac{1}{2(m+1)\delta + 3m(m+2)\delta^{\frac{m+2}{m+1}}}$$

$$\geq \frac{1}{3(m+1)(m+2)\delta^{\frac{m+2}{m+1}}}.$$

This completes the proof.

For using $\bar{\alpha}$ as the default step size in the algorithm, define the $\bar{\alpha}$ as follows

$$\bar{\alpha} = \frac{1}{3(m+1)(m+2)\delta^{\frac{m+2}{m+1}}}. \tag{19}$$

4.2 Decrease of the Proximity Function During an Inner Iteration

Now, we show that our proximity function δ with our default step size $\bar{\alpha}$ is decreasing. It can be easily established by using the following result:

Lemma 7 (see [7]) *Let $h(t)$ be a twice differentiable convex function with $h(0) = 0$, $h'(0) < 0$ and let $h(t)$ attain its (global) minimum at $t > 0$. If $h''(t)$ is increasing for $t \in [0, t^*]$, then*

$$h(t) = \frac{th'(0)}{2}.$$

Let the univariate function h be such that

$$h(0) = f_1(0) = 0, \ h'(0) = f_1'(0) = -2\delta^2, \ h''(\alpha) = 2\delta^2\psi''(v - 2\alpha\delta).$$

Since $f_2(\alpha)$ holds the condition of the above lemma,

$$f(\alpha) \leq f_1(\alpha) \leq f_2(\alpha) \leq \frac{f_2'(0)}{2}\alpha, \ \text{for all } 0 \leq \alpha \leq \alpha^*.$$

We can obtain the upper bound for the decreasing value of the proximity in the inner iteration by the above lemma.

Theorem 8 *Let $\bar{\alpha}$ be a step size as defined in (19) and $\delta = \delta(v) \geq \tau = 1$. Then we* *have*

$$f(\bar{\alpha}) \leq -\frac{(m+1)^{\frac{-m-2}{2(m+1)}}}{3(m+2)}\Psi(V)^{\frac{m}{2(m+1)}}.$$

Proof For all $\bar{\alpha} \leq \alpha^*$, we have

$$f(\bar{\alpha}) \leq -\bar{\alpha}\delta^2 = -\frac{1}{3(m+1)(m+2)\delta^{\frac{m+2}{m+1}}}\delta^2$$

$$= -\frac{1}{3(m+1)(m+2)}\delta^{\frac{m}{m+1}}$$

$$\leq -\frac{1}{3(m+1)(m+2)}\left(\sqrt{(m+1)\delta(v)}\right)^{\frac{m}{m+1}}$$

$$\leq -\frac{(m+1)^{\frac{m}{2(m+1)}}}{3(m+1)(m+2)}\delta(v)^{\frac{m}{2(m+1)}}$$

$$\leq \frac{(m+1)^{\frac{-m-2}{2(m+1)}}}{3(m+2)}\delta(v)^{\frac{m}{2(m+1)}}.$$

This completes the proof.

4.3 Iteration Bound

We need to count how many inner iterations are required to return to the situation where $\delta(v) \leq \tau$ after a μ-update. We denote the value of $\delta(v)$ after μ-update as δ_0 the subsequent values in the same outer iteration are denoted as δ_k, $k = 1, \ldots$ If K denotes the total number of inner iterations in the outer iteration, then we have

$$\delta_0 \leq L = O(n, \theta, \tau), \ \delta_{K-1} > \tau, \ 0 \leq \delta_K \leq \tau.$$

and according to (14),

$$\delta_{k+1} \leq \delta_k - \frac{(m+1)^{\frac{-m-2}{2(m+1)}}}{3(m+2)}\delta_k^{\frac{m}{2(m+1)}}.$$

At this stage we invoke the following lemma from Lemma 14 in [7]

Lemma 9 (see [7]) *Let $t_0, t_1, \ldots, t_k$ be a sequence of positive numbers such that*

$$t_{k+1} \leq t_k - \beta t_k^{1-\nu}, \ k = 0, 1, \ldots, K-1,$$

where $\beta > 0$, $0 < v \leq 1$; then

$$K \leq \frac{t_0^v}{\beta v}.$$

Letting

$$t_k = \delta_k, \quad \beta = \frac{(m+1)^{\frac{-m-2}{2(m+1)}}}{3(m+2)} \quad \text{and} \quad v = \frac{m+2}{2(m+1)},$$

we can get the following lemma

Lemma 10 *Let K be the total number of inner iterations in the outer iteration. Then we have*

$$K \leq 6(m+1)^{\frac{3m+4}{2(m+1)}} \delta_0^{\frac{m+2}{2(m+1)}}.$$

Proof Using Lemma 9, we have

$$K \leq \frac{\delta_0^v}{\beta v} = 6(m+1)^{\frac{3m+4}{2(m+1)}} \delta_0^{\frac{m+2}{2(m+1)}}.$$

This completes the proof.

Now we estimate the total number of iterations of our algorithm.

Theorem 11 *If $\tau \geq 1$, the total number of iterations is not more than*

$$6(m+1)^{\frac{3m+4}{2(m+1)}} \delta_0^{\frac{m+2}{2(m+1)}} \frac{1}{\theta} \log \frac{n\mu^0}{\varepsilon}.$$

Proof In the algorithm, $n\mu \leq \varepsilon$, $\mu^k = (1-\theta)^k \mu^0$ and $\mu^0 = \frac{(x^0)^t y^0)}{n}$. By simple computation, we have

$$K \leq \frac{1}{\theta} \log \frac{n\mu^0}{\varepsilon}.$$

Therefore, the number of outer iterations is bounded above by $\frac{1}{\theta} \log \frac{n\mu^0}{\varepsilon}$. Multiplying the number of outer iterations by the number of inner iterations, we get an upper bound for the total number of iterations, namely,

$$6(m+1)^{\frac{3m+4}{2(m+1)}} \delta_0^{\frac{m+2}{2(m+1)}} \frac{1}{\theta} \log \frac{n\mu^0}{\varepsilon}.$$

This completes the proof.

5   Numerical Tests

In this section, we deal with the numerical implementation of this algorithm applied to some problems of monotone $LCPs$. Here we used **Iter** means the iterations number produced by the algorithm. The implementation is manipulated in DEV C++. Our tolerance is $\varepsilon = 10^{-6}$. For the update parameter we have vary $0 < \theta < 1$. Finally we note that the linear system of Newton in (6) is solved thanks to Gauss.

Example 12

$$M = \begin{pmatrix} 0 & 0 & 2 & 1 & 0 \\ 0 & 0 & 1 & 2 & 1 \\ -2 & -1 & 0 & 0 & 0 \\ -1 & -2 & 0 & 0 & 0 \\ 0 & -1 & 0 & 0 & 0 \end{pmatrix}, \ q = \begin{pmatrix} -4 & -5 & 8 & 7 & 3 \end{pmatrix},$$

The numerical results are presented in the following table.

Function	Large update	Short update	θ	Iter
$\frac{t^2-1}{2} - \log t$	$O(n \log \frac{n}{\varepsilon})$	$O(\sqrt{n} \log \frac{n}{\varepsilon})$	0.15	84
			0.30	75
			0.60	35
			0.95	24
$\frac{t^2-1}{2} + \frac{t^{1-q}}{q(q-1)} - \frac{q-1}{q}(t-1),$ $(q>1)$	$O\left(qn^{\frac{q+1}{2q}} \log \frac{n}{\varepsilon}\right)$	$O(q\sqrt{n} \log \frac{n}{\varepsilon})$	0.15	83
			0.30	77
			0.60	64
			0.95	28
$\frac{t^2-1}{2} + \frac{(e-1)^2}{e(e^t-1)} - \frac{e-1}{e}$	$O(n^{\frac{3}{4}} \log \frac{n}{\varepsilon})$	$O(\sqrt{n} \log \frac{n}{\varepsilon})$	0.15	79
			0.30	67
			0.60	56
			0.95	34
$\frac{1}{2}(t - \frac{1}{t})^2$	$O(n^{\frac{2}{3}} \log \frac{n}{\varepsilon})$	$O(\sqrt{n} \log \frac{n}{\varepsilon})$	0.15	82
			0.30	76
			0.60	45
			0.95	19
$\frac{t^{p+1}-1}{p+1} + \frac{t^{1-q}-1}{q-1}$ $p\in[0,1],q>1$	$O\left(qn^{\frac{p+q}{q(1+p)}} \log \frac{n}{\varepsilon}\right)$	$O(q^2 \sqrt{n} \log \frac{n}{\varepsilon})$	0.15	78
			0.30	75
			0.60	58
			0.95	27
$(m+1)t^2 - (m+2)t + \frac{1}{t^m},$ $t>0, m>4$	$O\left(m^{\frac{3m+1}{2m}} n^{\frac{m+1}{2m}} \log \frac{n}{\varepsilon}\right)$	$O\left(m^{\frac{3m+1}{2m}} \sqrt{n} \log \frac{n}{\varepsilon}\right)$	0.15	83
			0.30	63
			0.60	24
			0.95	12

Example 13

$$M = \begin{pmatrix} 0 & 0 & 0 & 0 & 0 & 3 & 0.8 & 0.32 & 1.128 & 0.0512 \\ 0 & 0 & 0 & 0 & 0 & 0 & 1 & 0.8 & 0.32 & 0.128 \\ 0 & 0 & 0 & 0 & 0 & 0 & 0 & 1 & 0.8 & 0.32 \\ 0 & 0 & 0 & 0 & 0 & 0 & 0 & 0 & 1 & 0.8 \\ 0 & 0 & 0 & 0 & 0 & 0 & 0 & 0 & 0 & 1 \\ -1 & 0 & 0 & 0 & 0 & 0 & 0 & 0 & 0 & 0 \\ -0.8 & -1 & 0 & 0 & 0 & 0 & 0 & 0 & 0 & 0 \\ -0.32 & -0.8 & -1 & 0 & 0 & 0 & 0 & 0 & 0 & 0 \\ -1.128 & -0.32 & -0.8 & -1 & 0 & 0 & 0 & 0 & 0 & 0 \\ -0.0512 & -1.128 & -0.32 & -0.8 & -1 & 0 & 0 & 0 & 0 & 0 \end{pmatrix}$$

and

$$q = \begin{pmatrix} -0.0256 & -0.064 & -0.16 & 5.59 & -1 & 1 & 1 & 1 & 1 & 1 \end{pmatrix}.$$

The numerical results are presented in the following table.

Function	Large update	Short update	θ Iter
$\frac{t^2-1}{2} - \log t$	$O(n \log \frac{n}{\varepsilon})$	$O(\sqrt{n} \log \frac{n}{\varepsilon})$	0.15 83
			0.30 77
			0.60 45
			0.95 14
$\frac{t^2-1}{2} + \frac{t^{1-q}}{q(q-1)} - \frac{q-1}{q}(t-1)$ $q>1$	$O\left(qn^{\frac{q+1}{2q}} \log \frac{n}{\varepsilon}\right)$	$O(q\sqrt{n} \log \frac{n}{\varepsilon})$	0.15 85
			0.30 78
			0.60 61
			0.95 23
$\frac{t^2-1}{2} + \frac{(e-1)^2}{e(e^t-1)} - \frac{e-1}{e}$	$O(n^{\frac{3}{4}} \log \frac{n}{\varepsilon})$	$O(\sqrt{n} \log \frac{n}{\varepsilon})$	0.15 79
			0.30 61
			0.60 33
			0.95 14
$\frac{1}{2}(t - \frac{1}{t})^2$	$O(n^{\frac{2}{3}} \log \frac{n}{\varepsilon})$	$O(\sqrt{n} \log \frac{n}{\varepsilon})$	0.15 80
			0.30 78
			0.60 41
			0.95 17
$\frac{t^{p+1}-1}{p+1} + \frac{t^{1-q}-1}{q-1}$ $p\in[0,1],q>1$	$O\left(qn^{\frac{p+q}{q(1+p)}} \log \frac{n}{\varepsilon}\right)$	$O(q^2\sqrt{n} \log \frac{n}{\varepsilon})$	0.15 76
			0.30 75
			0.60 57
			0.95 17
$(m+1)t^2 - (m+2)t + \frac{1}{t^m}$ $t>0, m>4$	$O\left(m^{\frac{3m+1}{2m}} n^{\frac{m+1}{2m}} \log \frac{n}{\varepsilon}\right)$	$O\left(m^{\frac{3m+1}{2m}} \sqrt{n} \log \frac{n}{\varepsilon}\right)$	0.15 81
			0.30 43
			0.60 14
			0.95 5

Example 14 Let $M \in \mathfrak{R}^{n \times n}$ and $q \in \mathfrak{R}^n$ defined by:

$$M = \begin{pmatrix} 1 & 2 & 2 & \ldots & 2 \\ 0 & 1 & 2 & \ldots & 2 \\ 0 & 0 & . & . & . & . \\ . & . & . & . & . & . & . \\ . & . & . & . & . & . \\ . & . & . & . & . & 2 \\ 0 & 0 & 0 & . & . & 0 & 1 \end{pmatrix}, \quad q = (-1 \ldots -1).$$

Case 1: $n = 10$. The numerical results are presented in the following table

Function	Large update	Short update	θ Iter
$\frac{t^2-1}{2} - \log t$	$O(n \log \frac{n}{\varepsilon})$	$O(\sqrt{n} \log \frac{n}{\varepsilon})$	0.15 81
			0.30 72
			0.60 44
			0.95 11
$\frac{t^2-1}{2} + \frac{t^{1-q}}{q(q-1)} - \frac{q-1}{q}(t-1)$ $q>1$	$O\left(qn^{\frac{q+1}{2q}} \log \frac{n}{\varepsilon}\right)$	$O(q\sqrt{n} \log \frac{n}{\varepsilon})$	0.15 80
			0.30 71
			0.60 61
			0.95 21
$\frac{t^2-1}{2} + \frac{(e-1)^2}{e(e^t-1)} - \frac{e-1}{e}$	$O(n^{\frac{3}{4}} \log \frac{n}{\varepsilon})$	$O(\sqrt{n} \log \frac{n}{\varepsilon})$	0.15 70
			0.30 51
			0.60 32
			0.95 12
$\frac{1}{2}(t - \frac{1}{t})^2$	$O(n^{\frac{2}{3}} \log \frac{n}{\varepsilon})$	$O(\sqrt{n} \log \frac{n}{\varepsilon})$	0.15 81
			0.30 79
			0.60 40
			0.95 13
$\frac{t^{p+1}-1}{p+1} + \frac{t^{1-q}-1}{q-1}$ $p \in [0,1], q>1$	$O\left(qn^{\frac{p+q}{q(1+p)}} \log \frac{n}{\varepsilon}\right)$	$O(q^2\sqrt{n} \log \frac{n}{\varepsilon})$	0.15 73
			0.30 74
			0.60 45
			0.95 27
$(m+1)t^2 - (m+2)t + \frac{1}{t^m}$ $t>0, m>4$	$O\left(m^{\frac{3m+1}{2m}} n^{\frac{m+1}{2m}} \log \frac{n}{\varepsilon}\right)$	$O\left(m^{\frac{3m+1}{2m}} \sqrt{n} \log \frac{n}{\varepsilon}\right)$	0.15 63
			0.30 23
			0.60 11
			0.95 4

Case 2: $n = 15$.

The numerical results are presented in the following table.

Function	Large update	Short update	θ Iter
$\frac{t^2-1}{2} - \log t$	$O(n \log \frac{n}{\varepsilon})$	$O(\sqrt{n} \log \frac{n}{\varepsilon})$	0.15 60
			0.30 52
			0.60 24
			0.95 13
$\frac{t^2-1}{2} + \frac{t^{1-q}}{q(q-1)} - \frac{q-1}{q}(t-1)$, $q>1$	$O\left(qn^{\frac{q+1}{2q}} \log \frac{n}{\varepsilon}\right)$	$O(q\sqrt{n} \log \frac{n}{\varepsilon})$	0.15 74
			0.30 51
			0.60 44
			0.95 16
$\frac{t^2-1}{2} + \frac{(e-1)^2}{e(e^t-1)} - \frac{e-1}{e}$	$O(n^{\frac{3}{4}} \log \frac{n}{\varepsilon})$	$O(\sqrt{n} \log \frac{n}{\varepsilon})$	0.15 63
			0.30 51
			0.60 32
			0.95 13
$\frac{1}{2}(t - \frac{1}{t})^2$	$O(n^{\frac{2}{3}} \log \frac{n}{\varepsilon})$	$O(\sqrt{n} \log \frac{n}{\varepsilon})$	0.15 81
			0.30 79
			0.60 40
			0.95 13
$\frac{t^{p+1}-1}{p+1} + \frac{t^{1-q}-1}{q-1}$, $p\in[0,1], q>1$	$O\left(qn^{\frac{p+q}{q(1+p)}} \log \frac{n}{\varepsilon}\right)$	$O(q^2\sqrt{n} \log \frac{n}{\varepsilon})$	0.15 73
			0.30 74
			0.60 45
			0.95 27
$(m+1)t^2 - (m+2)t + \frac{1}{t^m}$, $t>0, m>4$	$O\left(m^{\frac{3m+1}{2m}} n^{\frac{m+1}{2m}} \log \frac{n}{\varepsilon}\right)$	$O\left(m^{\frac{3m+1}{2m}} \sqrt{n} \log \frac{n}{\varepsilon}\right)$	0.15 45
			0.30 34
			0.60 9
			0.95 6

6 Concluding Remarks

We propose a new barrier function and primal–dual interior point algorithms for (LCP) problems and analyze the iteration complexity of the algorithm based on the kernel function. We have

$$O\left(m^{\frac{3m+1}{2m}} n^{\frac{m+1}{2m}} \log \frac{(x^0)^t y^0)}{\varepsilon}\right)$$

for large-update methods and

$$O\left(m^{\frac{3m+1}{2m}}\sqrt{n}\log\frac{(x^0)^t y^0}{\varepsilon}\right)$$

for small-update methods which are the best known iteration bounds for such methods. Future research might focus on the extension to symmetric cone optimization. Finally, the numerical tests some strategies are used and indicate that our kernel function used in the algorithm is efficient.

References

1. de Klerk, E.: Aspects of Semidefinite Programming: Interior Point Methods and Selected Applications. Kluwer Academic Publishers, Dordrecht (2002)
2. Mansouri, H., Roos, C.: A new full-Newton step $O(n)$ infeasible interior-point algorithm for semidefinite optimization. Numer. Algorithms **52**(2), 225–255 (2009)
3. Lustig, I.J.: Feasible issues in a primal-dual interior point method for linear programming. Math. Program. **49**, 145–162 (1990–1991)
4. Tanabe, K.: Centered Newton method for linear programming: interior and 'exterior' point method. In: Tone, K. (ed.) New Methods for Linear Programming, vol. 3, pp. 98–100. The Institute of Statistical Mathematics, Tokyo (1990), in Japanese
5. Kojima, M., Megiddo, N., Mizuno, S.: A primal-dual infeasible-interior-point algorithm for linear programming. Math. Program. **61**, 263–280 (1993)
6. Zhang, Y.: On the convergence of a class of infeasible-interior-point methods for the horizontal linear complementarity problem. SIAM J. Optim. **4**, 208–227 (1994)
7. Mansouri, H., Roos, C.: Simplified $O(nL)$ infeasible interior-point algorithm for linear optimization using full Newton steps. Optim. Methods Softw. **22**(3), 519–530 (2007)
8. Bai, Y.Q., El Ghami, M., Roos, C.: A new efficient large-update primal-dual interior-point method based a finite barrier. SIAM J. Optim. **13**(3), 766–782 (2002)
9. Bellman, R.: Introduction to matrix analysis. In: Classics in Applied Mathematics, vol. 12. SIAM, Philadelphia (1995)
10. Bai, Y.Q., El Ghami, M., Roos, C.: A comparative study of kernel functions for primal-dual interior-point algorithms in linear optimization. SIAM J. Optim. **15**(1), 101–128 (2014)
11. Nesterov, Y.E., Nemirovskii, A.S.: Interior point polynomial algorithms in convex programming. In: SIAM Studies in Applied Mathematics, vol. 13. SIAM, Philadelphia (1994)
12. Peng, J., Roos, C., Terlaky, T.: Self-Regularity: A New Paradigm for Primal-Dual Interior-Point Algorithms. Princeton University Press, Princeton (2002)

Branch and Bound Method to Resolve the Non-convex Quadratic Problems

R. Benacer and Boutheina Gasmi

Abstract In this paper, we present a new rectangle Branch and Bound approach for solving non convex quadratic programming problems in which we construct a new lower approximate convex quadratic function of the objective quadratic function over an n-rectangle $S^k = \left[a^k, b^k \right]$ or $S^k = \left[L^k, U^k \right]$. This quadratic function (the approximate one) is given to determine a lower bound of the global optimal value of the original problem (NQP) over each rectangle. In the other side, we apply a simple two-partition technique on rectangle, as well as, the tactics on reducing and deleting subrectangles are used to accelerate the convergence of the proposed algorithm. This proposed algorithm is proved to be convergent and shown to be effective with some examples.

1 Introduction

We consider the non-convex quadratic programming problems below:

$$\begin{cases} \min f(x) = \frac{1}{2} x^T Q x + d^T x \\ x \in S \cap (X_f) \end{cases} \qquad (1)$$

where

$$S^0 = \left\{ x \in \mathbb{R}^n : L_i^0 \le x_i \le U_i^0 : i = \overline{1, n} \right\},$$
$$(D_f) = \left\{ x \in \mathbb{R}^n : Ax \le b; x \ge 0 \right\},$$
$$Q : \text{ is a real } (n \times n) \text{ non-positive symmetric matrix,}$$
$$A : \text{ is a real } (n \times n) \text{ symmetric matrix,}$$

R. Benacer · B. Gasmi (✉)
University of Hadj Lakhder Batna, Batna, Algeria
e-mail: gasmi.boutheina@gmail.com

R. Benacer
e-mail: r.benacer@hotmail.fr

© Springer International Publishing Switzerland 2016
G.A. Anastassiou and O. Duman (eds.), *Intelligent Mathematics II:*
Applied Mathematics and Approximation Theory, Advances in Intelligent Systems
and Computing 441, DOI 10.1007/978-3-319-30322-2_7

$$d^T = (d_1, d_2, \ldots, d_n) \in \mathbb{R}^n,$$
$$b^T = (b_1, b_2, \ldots, b_m) \in \mathbb{R}^m.$$

In our life, every things, every problems are created as a mathematical problems [1]. We can also take the quotes of Galilee: "The word is created at mathematical problems"; especially "quadratic problems" that have worthy of study. Because they frequently appear in many applied field of science and technology as well as the convergent of many other nonlinear problems into this form of problems; for example: bilinear programming, quadratic 0–1 programming, modularization of product, and etc., which can be interpreted as quadratic problems.

In this paper, we present a new rectangle Branch and Bound approach for solving non-convex quadratic programming problems, where we proposed a new lower approximate convex quadratic functions of the objective quadratic function f over an n-rectangle. This lower approximate is given to determine a lower bound of the global optimal value of the original problem (1) over each rectangle.

To accelerate the convergence of the proposed algorithm we used a simple two-partition technique on rectangle and the tactics on reducing and deleting subrectangles from [2].

The paper is organized as follows:

In Sect. 1, we give a simple introduction of our studies in which we give and define the standard form of our problem. In Sect. 2, a new equivalent form of the objective function is proposed as an lower approximate linear functions of the quadratic form over an n-rectangle [3]. We can also propose as an upper approximate linear functions, but we must respect the procedure of calculate the lower and the upper bound of the original principal rectangle S^0 which is noted by $S^k = \left[L^k, U^k\right] \subseteq \mathbb{R}^n$ in the k-step [4]. In Sect. 3, we define a new lower approximate quadratic functions of the quadratic no convex function over an n-rectangle with respect to a rectangle to calculate a lower bound on the global optimal value of the original no convex problem (1). We also give a new simple rectangle partitioning method and describe rectangle reducing tactics from [2]. In Sect. 4, we give a new Branch and Reduce Algorithm in order to solving the original non-convex problem (1) [5]. In Sect. 5, we study the convergence of the proposed algorithm and we give a simple comparison between the linear approximate and the quadratic one. Finally, a conclusion of the paper is given to show and explain the effective of the proposed algorithm.

2 Equivalent Forms of f

In this section, we construct and define the equivalent form of the non-convex quadratic function which proposed as a lower approximate linear functions over an n-rectangle $S^k = \left[L^k, U^k\right]$. This work is proposed to determine the lower bound of the global optimal value of the original problem (1).

Let $\lambda_{\min}$ and $\lambda_{\max}$ be the minimum eigenvalue and the maximum eigenvalue of the matrix Q respectively, and we show the number θ that $\theta \geq |\lambda_{\min}|$.

The equivalent linear form of the objective function f is given by:

$$f(x) = \left(x - L^K\right)^T (Q + \theta I) \left(x - L^K\right) + d^T x - \theta \sum_{i=1}^{n} x_i^2$$
$$+ 2\left(L^K\right)^T (Q + \theta I) x - \left(L^K\right)^T (Q + \theta I) L^K$$

by the use of the lower bound L^k, and is given by:

$$f(x) = \left(x - U^K\right)^T (Q + \theta I) \left(x - U^K\right) + d^T x - \theta \sum_{i=1}^{n} x_i^2$$
$$+ 2\left(U^K\right)^T (Q + \theta I) x - \left(U^K\right)^T (Q + \theta I) U^K$$

by the use of the upper bound U^k of the rectangle S^k.

On the other hand, the convex envelope of the function $h(x) = (-x_j^2)$ over the interval $S_j^k = \left[L_j^k, U_j^k\right]$ is given by the function

$$\overline{h}(x) = -\left(U_j^k + L_j^k\right) x_i + L_j^k U_j^k,$$

which is a linear function, then we get the best linear lower bound of $h(x) = \sum_{j=1}^{n}(-x_j^2)$ given by:

$$\varphi_{S^k}(x) = -\left(U^k + L^k\right)^T x + \left(L^k\right)^T U^k$$

by definition, the initial rectangle S^0 is given by:

$$S^0 = \left\{x \in \mathbb{R}^n : L_i^0 \leq x_i \leq U_i^0 : i = \overline{1, n}\right\}.$$

We subdivide this rectangle into two sub-rectangles defined by:

$$S_{+1} = \left\{x \in \mathbb{R}^n : L_s^0 \leq x_s \leq h_s^0 : L_j^0 \leq x_j \leq U_j^0 : j = \overline{1, n} : j \neq s\right\}$$
$$S_{+2} = \left\{x \in \mathbb{R}^n : h_s^0 \leq x_s \leq U_s^0 : L_j^0 \leq x_j \leq U_j^0 : j = \overline{1, n} : j \neq s\right\},$$

where we calculate the point h_s by a normal rectangular subdivision (ω-subdivision).

The lower approximate linear function of f over the rectangle S^K:

Gao et al. [2] determine the best lower approximate linear function of the objective non-convex function f over the rectangle S^K, which is given by the following theorem:

Theorem 1 *Consider the function $f : C \subseteq \mathbb{R}^n \longrightarrow \mathbb{R}$ and the rectangle $S^0 \subseteq \mathbb{R}^n$, where $C \subseteq S^0 \subseteq \mathbb{R}^n$. Then, the lower approximate linear function of f is given by:*

$$L_{S^K}(x) = (a_{S^K})^T x + b_{S^K}$$
$$U_{S^K}(x) = (\overline{a_{S^K}})^T x + \overline{b_{S^K}},$$

where

$$a_{S^K} = d + 2(Q + \theta I)L^K - \theta \left(L^K + U^K\right),$$
$$b_{S^K} = -\left(L^K\right)^T (Q + \theta I)L^K + \theta \left(L^K\right)^T \left(U^K\right),$$
$$\overline{a_{S^K}} = d + 2(Q + \theta I)U^K - \theta \left(L^K + U^K\right),$$
$$\overline{b_{S^K}} = -\left(U^K\right)^T (Q + \theta I)U^K + \theta \left(L^K\right)^T \left(U^K\right).$$

The new lower approximate quadratic convex function of f over the rectangle S^K:

By using the president lower approximate linear function of f over the rectangle S^K, we can define the new lower approximate quadratic convex function of f over the same rectangle by:

$$L_{quad}(x) = L_{S^K}(x) - \frac{1}{2}K\left(U^K - x\right)\left(x - L^K\right)$$

and

$$U_{quad}(x) = U_{S^K}(x) - \frac{1}{2}K\left(U^K - x\right)\left(x - L^K\right),$$

where

- K is a positive real number given by the spectral radius of the matrix $(Q + \theta I)$,
- $\theta \geq |\lambda_{\min}|$,
- $L_{S^K}(x)$ the best lower approximate linear function of f over the rectangle S^K.

The new lower approximate linear function of f over the rectangle S^K:

By using the president new lower approximate quadratic function of f over the rectangle S^K, we can define the new lower approximate linear function of f over the same rectangle by:

$$\widetilde{L}_{quad}(x) = L_{S^K}(x) - \frac{1}{2}Kh^2$$

and

$$\widetilde{U}_{quad}(x) = U_{S^K}(x) - \frac{1}{2}Kh^2$$

with $h = \left\| U^K - L^K \right\|$.

The relation between the convex quadratic approximation and the linear one
We have the following theorem.

Theorem 2 *The following inequalities are satisfied:*

$$\widetilde{L}_{quad}(x) = L_{S^K}(x) - \frac{1}{8}Kh^2 \le L_{quad}(x) \le f(x)$$

$$\widetilde{U}_{quad}(x) = U_{S^K}(x) - \frac{1}{8}Kh^2 \le U_{quad}(x) \le f(x)$$

for all $x \in (D_f) \cap S^K$ and $h = \|U^K - L^K\|$.

Approximation error

We can estimate the approximation error by the distance between the non-convex objective function f and here lower approximation functions.

(1) *The linear approximation error* is presented by the distance between the function f and here new lower approximate linear function $\widetilde{L}_{quad}$ over the rectangle S^K. Then we have the following proposition.

Proposition 3 *Let the function $f : C \subseteq \mathbb{R}^n \longrightarrow \mathbb{R}$, where $C \subseteq S^0 \subseteq \mathbb{R}^n$ and $\theta \ge |\lambda_{\min}|$ for this the matrix $(Q + \theta I)$, be semi-positive. Then we have*

$$\max_{x \in S^K \cap (D_f)} \left\{ \left| f(x) - \widetilde{L}_{quad}(x) \right| \right\} \le \left(\rho(Q + \theta I) + \theta + \frac{1}{8}Kh^2 \right) \|U^K - L^K\|^2$$

$$\max_{x \in S^K \cap (D_f)} \left\{ \left| f(x) - \widetilde{U}_{quad}(x) \right| \right\} \le \left(\rho(Q + \theta I) + \theta + \frac{1}{8}Kh^2 \right) \|U^K - L^K\|^2.$$

(2) *The quadratic approximation error* is presented by the distance between the function f and here lower approximate quadratic function $\widetilde{L}_{quad}$ over the rectangle S^K. Then we have the following proposition.

Proposition 4 *Let the function $f : C \subseteq \mathbb{R}^n \longrightarrow \mathbb{R}$, where $C \subseteq S^0 \subseteq \mathbb{R}^n$ and $\theta \ge |\lambda_{\min}|$ for this the matrix $(Q + \theta I)$, be semi-positive. Then we have*

$$\max_{x \in S^K \cap (D_f)} \left\{ \left| f(x) - L_{quad}(x) \right| \right\} \le \left(\rho(Q + \theta I) + \theta + \frac{1}{2}K \right) \|U^K - L^K\|^2$$

$$\max_{x \in S^K \cap (D_f)} \left\{ \left| f(x) - U_{quad}(x) \right| \right\} \le \left(\rho(Q + \theta I) + \theta + \frac{1}{2}K \right) \|U^K - L^K\|^2.$$

Proof By the definition of the function $L_{quad}(x)$ as well as the meaning of $\varphi_{S^k}(x)$, we have

$$f(x) - L_{quad}(x) = f(x) - L_{S^k}(x) + \frac{1}{2}K\left(U^K - x\right)\left(x - L^K\right)$$

$$- \left(x - L^K\right)^T (Q + \theta I)\left(x - L^K\right)$$

$$+ \left(\frac{1}{2}K + \theta\right)\left(U^K - x\right)\left(x - L^K\right).$$

Then, we get

$$
\begin{aligned}
\left\| f(x) - L_{quad}(x) \right\|_\infty &= \max \left\{ f(x) - L_{quad}(x) : x \in S^K \cap (D_f) \right\} \\
&\leq \left\| (x - L^K)^T (Q + \theta I)(x - L^K) \right\|_\infty \\
&\quad + \left\| \left(\frac{1}{2} K + \theta \right)(U^K - x)(x - L^K) \right\|_\infty \\
&\leq \left(\rho(Q + \theta I) + \theta + \frac{1}{2} K \right) \left\| U^K - L^K \right\|^2 .
\end{aligned}
$$

Using the same thing for the lower bound $U_{quad}(x)$ with the equivalent linear form of the objective function f, we obtain that

$$
\left\| f(x) - U_{quad}(x) \right\|_\infty \leq \left(\rho(Q + \theta I) + \theta + \frac{1}{2} K \right) \left\| U^K - L^K \right\|^2 .
$$

The quadratic approximate problem (QAP)
Construction of the interpolate problem (IP):
It is clear that

$$
f(x) \geq \max \left\{ L_{quad}(x), U_{quad}(x) : \forall x \in (X_f) \cap S^K \right\} = \gamma(x).
$$

This function presents the best quadratic lower bound of f. Similarly, we construct the interpolation problem as follows:

$$
\begin{cases}
\alpha_h = \max \widehat{x} \\
\widehat{x} \in \left\{ L_{quad}(x), U_{quad}(x) \right\} : \forall x \in (X_f) \cap S^K
\end{cases}
\tag{2}
$$

and the convex quadratic problem is defined by

$$
\begin{cases}
\min \alpha_h \\
\forall x \in (X_f) \cap S^K .
\end{cases}
\tag{3}
$$

Suppose that $f^* = f(x^*)$ is the global optimal value of the original problem (1) and $\widetilde{x}$ be the optimal solution of (3).

Question 1 *What is the relation between $f(\widetilde{x})$, $f(x^*)$ and $L_{quad}(\widetilde{x})$ the optimal value of (3)?*

Proposition 5 *Let $f : C \subseteq \mathbb{R}^n \longrightarrow \mathbb{R}$ and $S^0 \subseteq \mathbb{R}^n$, where $C \subseteq S^0 \subseteq \mathbb{R}^n$. Then we have*

$$
0 \leq f(\widetilde{x}) - f(x^*) \leq \left(\rho(Q + \theta I) + \theta + \frac{1}{2} K \right) \left\| U^K - L^K \right\|^2
$$

and

$$
L_{quad}(\widetilde{x}) \leq f^* \leq f(\widetilde{x}).
$$

Proof From Proposition 4, we have:

$$f(x) - L_{quad}(x) \leq \left(\rho \left(Q + \theta I \right) + \theta + \frac{1}{2} K \right) \left\| U^K - L^K \right\|^2 : x \in S^K \cap (D_f)$$

and, for $x = \tilde{x}$,

$$f(\tilde{x}) - L_{quad}(\tilde{x}) \leq \left(\rho \left(Q + \theta I \right) + \theta + \frac{1}{2} K \right) \left\| U^K - L^K \right\|^2.$$

Thus

$$f(\tilde{x}) - f^* + f^* - L_{quad}(\tilde{x}) \leq \left(\rho \left(Q + \theta I \right) + \theta + \frac{1}{2} K \right) \left\| U^K - L^K \right\|^2$$

and

$$f(\tilde{x}) - f^* \leq \left(\rho \left(Q + \theta I \right) + \theta + \frac{1}{2} K \right) \left\| U^K - L^K \right\|^2 + \left(L_{quad}(\tilde{x}) - f^* \right),$$

as well as $L_{quad}(\tilde{x}) - f^* \leq 0$. We obtain that:

$$0 \leq f(\tilde{x}) - f^* \leq \left(\rho \left(Q + \theta I \right) + \theta + \frac{1}{2} K \right) \left\| U^K - L^K \right\|^2.$$

On the other hand, we have

$$\begin{cases} L_{quad}(\tilde{x}) - f^* \leq 0 \\ f(\tilde{x}) - f^* \geq 0 \end{cases} \implies \left(L_{quad}(\tilde{x}) \leq f^* \leq f(\tilde{x}) \right).$$

Then, the proof is completed.

Question 2 *Is the solution $\tilde{x}$ presented the best lower bound of the global optimal solution of* (1)?

Let the estimate function be noted by

$$E(x) = f(x) - L_{quad}(x)$$

for all $x \in S^K \cap (D_f)$, and verified the inequality:

$$E(\tilde{x}) \geq f(\tilde{x}) - f^*.$$

Lemma 6 *If $E(\tilde{x})$ is a small value, then $f(\tilde{x})$ is an acceptable approximative value for the global optimum $f^* = f(x^*)$ over S^K. Similarly, we find that the point $\tilde{x}$ is the global approximative solution of the global solution x^* over the rectangle S^K.*

Remarks

1. In this method, the range of the non-convex function f over the new rectangle (sub-rectangle) S^K is small than the range over the initial rectangle S°. Immediately, we obtain that the value $E(\widetilde{x})$ will be very small.
2. From Remark (1), we find that the point $\widetilde{x}$ is a global approximative solution for the global optimal solution x^* over the rectangle S^K.

3   The Reduction Technique (Eliminate Technique)

We get to describe the rectangle partition as follows:

step (0) Let $S^K = \{x^k \in \mathbb{R}^n : L_i^K \le x_i^k \le U_i^K : i = \overline{1, n}\}$ with $x^k \in S^K$;
step (1) we find a partition information point:

$$h_s = \max \left\{ (x_i - L_i^K)(U_i^K - x_i) : i = \overline{1, n} \right\};$$

step (2) if $h_s \ne 0$, then we divide the rectangle S^K into two subrectangle on edge $\left[L_s^K, U_s^K \right]$ by the point h_s, **else**, we divide the rectangle S^K into two subrectangle on the longest edge $\left[L_s^K, U_s^K \right]$ by the midpoint $\frac{L^K + U^K}{2}$ which is yet noted as h_s;
step (3) the rest rectangle is yet noted as S^K.
Now, we describe the rectangle reducing tactics from [2] to accelerate the convergence of the proposed global optimization Algorithm (ARSR).

Remarks

1. All linear constraints of the problem (1) are expressed by $\sum_{j=1}^{n} a_{ij} x_j \le b_i : i = 1, n$.
2. The rectangle S^K be also recorded as constrain to be added to the problem (1).
3. The minimum and the maximum of each function $\psi(x_i) = a_{is} x_s$ for $i = \overline{1, n}$ over the interval $\left[L_s^K, U_s^K \right]$ are obtained at the extreme points of the same interval.

Linearity Based Range Reduction Algorithm
We describe this algorithm to reduce and delete the rectangle S^K.
program (LBRRA)
Let $I_k' := \{1, 2, 3, \ldots, n\}$ the set of the index, $P_k := P$,
for $1 \le i \le n$ **do**
compute $rU_i := \sum_{j=1}^{n} \max \left\{ a_{ij} L_j^k, a_{ij} U_j^k \right\}$

compute $rL_i := \sum_{j=1}^{n} \min \left\{ a_{ij} L_j^k, a_{ij} U_j^k \right\}$

if $rL_i > b_i$ **then**
stop. The problem (1) is infeasible over S^K (there are no solution of (1) over S^K, because S^K is deleted from the subrectangle set produced through partitioning of the rectangle S°)

else
  if $rU_i < b_i$ **then**
   the constraint is redundant.
   $I'_k := I'_k - \{i\}$
   $P_k := P_k - \{x \in \mathbb{R}^n : (a_i)^T x \le b_i\}$
  else
   for $1 \le j \le n$ **do**
   if $a_{ij} > 0$ **then**

$$U_j^k := \min \left\{ U_j^k, \frac{b_i - rL_i + \min\left\{a_{ij}L_j^k, a_{ij}U_j^k\right\}}{a_{ij}} \right\}$$

   else

$$L_j^k := \max \left\{ L_j^k, \frac{b_i - rU_i + \max\left\{a_{ij}L_j^k, a_{ij}U_j^k\right\}}{a_{ij}} \right\}$$

   end if
   enddo
  end if
  end if
enddo
end program

4 Algorithm (ARSR): Branch and Bound

Algorithm (ARSR): Branch and Bound
  program (ARSR) initialization: determine the initial rectangle S^0 where (χ_f)

$\subset S^0$ and suppose that

$$QLBP_{S^0} := S^0 \cap (\chi_f)$$

  iteration k:
  if $QLBP_{S^0} \ne \phi$ **then**
  solve the quadratic problem (2) when $k = 0$
  Let x^0 be an optimal solution of (2) and $\alpha(S^0)$ be the optimal value accompanied
to x^0
  $H := \{S^0\}$ (the set of the subrectangle of the initial rectangle S^0)
  $\alpha_0 := \min\{\alpha(S^0)\}$, $\beta_0 := f(x^0)$ (the upper bound of $f(x^*)$)
  $k := 0$
  while Stop=false do
   if $\alpha_k = \beta_k$ **then**
   Stop=true (x^k is a global optimal solution of the problem (1))
   else

we subdivide the rectangle S^k into two sub-rectangle $\{S_j^k : j = 1, 2\}$
by the proposed algorithm.

for $j = 1, 2$ **do**

applied the Linearity Based Range Reduction Algorithm over the two sub-rectangle $\{S_j^k\}$

the obtained set is yet noted as le rectangle S_j^k

if $S_j^k \neq \phi$ **then**

$(QLBP)_{S_j^k} := \{x \in \mathbb{R}^n : x \in S_j^k \cap (\chi_f)\}$,

solve the quadratic problem $(QLBP)$ when $S^k := S_j^k$

let x^{k_j} be the optimal solution and $\alpha(S_j^k)$ be the optimal value

$H := H \cup \{S_j^k\}$

$\beta_{k+1} := \min\{f(x^k), f(x^{k_j})\}$

$x^k := arg \min \beta_{k+1}$

end if

end for

$H := H - \{S^k\}$

$\alpha_{k+1} := \min_{S \in H}\{\alpha(S)\}$;choose an rectangle $S^{k+1} \in H$

such that $\alpha_{k+1} = \alpha(S^{k+1})$

$k \leftarrow k + 1$;

end if

end do

end if

end program

5   The Convergence of the Algorithm (ARSR)

In this section, we study the convergence of the proposed Algorithm (ARSR) and we give a simple comparison between the linear approximation and the quadratic one. On the other hand, we give some example to explain the proposed Algorithm.

The convergence of the proposed algorithm:

The proposed algorithm in Sect. 4 is different from the one in [2] in lower-bounding (quadratic approximation), and added to the rectangle-reducing strategy. We will prove that the proposed algorithm is convergent.

Theorem 7 *If the proposed algorithm terminates in finite steps, then a global optimal solution of the problem (1) is obtained when the algorithm terminates.*

Proof Assume that when the algorithm terminates, the result is coming to x^k. Then, immediately, we have $a_x = B_k$ when terminating at the k-step, and so x^k is a global optimal solution of the problem (1).

Theorem 8 *If the algorithm generates an infinite sequence* $\{x^k\}_{k \in \mathbb{N}^*}$*, then every accumulation point* x^* *of this sequence is a global optimal solution of the problem (1) (i.e., the global optimal solution is not unique).*

Proof Let x^* be an accumulation point of the sequence $\{x^k\}_{k \in \mathbb{N}^*}$ and $\{x_p^k\}_{k \in \mathbb{N}^*, p \in \mathbb{N}^*}$ be a subsequence of the sequence $\{x^k\}_{k \in \mathbb{N}^*}$ converging to x^*. Obviously, in the proposed algorithm, the lower sequence $\{a_k\}_{k \in \mathbb{N}^*}$ is monotone increasing and the upper sequence $\{B_k\}_{k \in \mathbb{N}^*}$ is monotone decreasing, and we have

$$\alpha_k = l_{quad}\left(x^k\right), \ B_k = f\left(x^k\right).$$

Then, we can write that

$$\alpha_k = l_{quad}\left(x^k\right) \leq \min_{x \in S_k} f(x) \leq B_k = f\left(x^k\right).$$

So, both $\{x_k\}_{k \in \mathbb{N}^*}$ and $\{B_k\}_{k \in \mathbb{N}^*}$ are convergent, and

$$\lim_{k \to \infty} B_k = \lim_{q \to \infty} B_{k_q} = \lim_{k \to \infty} f\left(x^k\right) = \lim_{q \to \infty} f\left(x^{k_q}\right) = f\left(x^*\right).$$

Without loss of generality, we assume that x^{k_q} is the solution of the problem $(QLBP)$ on S_{k_q}, which satisfies $S_{k_{q+1}} \subset S_{k_q}, q \geq 1$. By the properties of the proposed rectangle partition, which is exhaust, i.e. $\lim_{q \to \infty} S_{k_q} = x^*$, and from Theorem 7 again, we know that

$$0 \leq B_{k_q} - \alpha_{k_q} = f\left(x_q^k\right) - l_{quad}\left(x_q^k\right) \leq \left(\rho\left(Q + \theta I\right) + \theta + \frac{1}{2}K\right) \|U_q^K - L_q^K\|^2.$$

Then

$$\lim_{q \to 0}\left(f\left(x_q^k\right) - l_{quad}\left(x_q^k\right)\right) = \lim_{q \to 0}\left(B_{k_q} - \alpha_{k_q}\right) = 0.$$

Thus, we have

$$\lim_{q \to 0}\left(B_{k_q} - \alpha_{k_q}\right) = \lim_{q \to 0}\left(\alpha_{k_q} - B_{k_q} - \left(B_{k_q} - \alpha_{k_q}\right)\right) = 0,$$

and so

$$\lim_{k \to 0} \alpha_k = \lim_{q \to 0} \alpha_{k_q} = \lim_{q \to 0}\left(B_{k_q} - \left(B_{k_q} - \alpha_{k_q}\right)\right) = \lim_{q \to 0} B_{k_q}.$$

Then we have:

$$\lim_{k \to 0} \alpha_k = \lim_{q \to 0} B_{k_q} = \lim_{q \to \infty} f\left(x^{k_q}\right) = f\left(x^*\right).$$

Therefore, the point x^* is an global optimal solution of the problem (1).

Fig. 1 Graphs

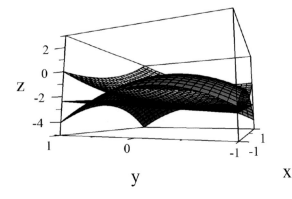

Example 9 Let the non-convex quadratic function be defined by

$$f(x) = (x_1 + 1)^2 + (x_2 + 1)^2 - \frac{5}{2}(x_1 + x_2) - 3\left(x_1^2 + x_2^2\right) - 2.$$

In Fig. 1, we see the graphs of the non-convex quadratic function f, the linear approximate function and the convex quadratic lower bound function over the rectangle $[-1, 0] \subseteq \mathbb{R}^n$. It is clear that the convex quadratic approximate function is between the objective function and the linear approximate one of the same function over he rectangle $S^0 = [-1, 0] \subseteq \mathbb{R}^n$.

6   Conclusion

In this paper, we present a new rectangle Branch and Bound approach for solving non-convex quadratic programming problems, where we proposed a new lower approximate convex quadratic functions of the objective quadratic function f over an n-rectangle. This lower approximate is given to determine a lower bound of the global optimal value of the original problem (1) over each rectangle. To accelerate the convergence of the proposed algorithm we used a simple two-partition technique on rectangle and the tactics on reducing and deleting subrectangles from [2].

References

1. Horst, R., Pardalos, P.M., Thoai, N.V.: Introduction to Global Optimization. Kluwer Academic Publishers, Dordrecht (1995)
2. Gao, Y., Xue, H., Shen, P.: A new rectangle branch and bound reduce approach for solving non convex quadratic programming problems. Appl. Math. Comput. **168**, 1409–1418 (2005)

3. Honggang, X., Chengxian, X.: A branch and bound algorithm for solving a class of DC-programming. Appl. Math. Comput. **165**, 291–302 (2005)
4. Pardalos, P.M.: Global optimization algorithms for linearly constrained indefinite quadratic problems. Comput. Math. Appl. Lic. **21**(6–7), 87–97 (1991)
5. Jiao, H.: A branch and bound algorithm for globally solving a class of non convex programming problems. Nonlinear Anal. **70**, 1113–1123 (2009)

Rogue Wave Solutions for the Myrzakulov-I Equation

Gulgassyl Nugmanova

Abstract In this paper, we consider the (2+1)-dimensional generalization of the Landau–Lifshitz equation, so-called the Myrzakulov-I (M-I) equation, which describes a two-dimensional dynamics of magnetization in ferromagnetics. The Darboux transformation (DT) for the M-I equation is constructed. Using the DT the solution of the type of destructive waves for the M-I equation is found.

1 Introduction

Among the nonlinear evolution equations, integrable ones are of special interest, since only in this case it is possible to carry out detailed and in-depth theoretical study, using the methods of the theory of solitons. Spin systems are convenient model systems for the study of nonlinear phenomena in magnetics. The (2+1)-dimensional Myrzakulov-I (M-I) equation, which is considered in this paper, has a rich internal structure. Various algebraic and geometric properties of the M-I equation with the scalar potential, and its integrable reductions were studied based on the theory of solitons and differential geometry in the papers [1–5]. In the recent literature the problem of describing the phenomenon of destructive waves is widely discussed. For example, in [6] the solution for the type of destructive waves with dimensional Heisenberg ferromagnet was found. These waves are not only applied, but also of theoretical interest. The problem of finding solutions of the type destructive waves of the M-I equation is the main objective of this work. To construct the solution of this type, we will focus on the method of Darboux transformation (DT).

The paper is organized as follows. In Sect. 2, the M-I equation, and its Lax representation are introduced. In Sect. 3, we derived the DT of the M-I equation. Using the DT, some exact soliton solutions are derived. Section 4 is devoted to conclusion.

G. Nugmanova (✉)
L.N. Gumilyov Eurasian National University, Astana, Kazakhstan
e-mail: nugmanovagn@gmail.com

© Springer International Publishing Switzerland 2016   119
G.A. Anastassiou and O. Duman (eds.), *Intelligent Mathematics II:*
Applied Mathematics and Approximation Theory, Advances in Intelligent Systems
and Computing 441, DOI 10.1007/978-3-319-30322-2_8

2 The M-I Equation

Let us consider the M-I equation. It looks like [1]

$$i S_t + \frac{1}{2}[S, S_{xy}] + iu S_x = 0, \tag{1}$$

$$u_x - \frac{i}{4}tr(S[S_x, S_y]) = 0, \tag{2}$$

where

$$S = \begin{pmatrix} S_3 & S^- \\ S^+ & -S_3 \end{pmatrix}.$$

The M-I equation is integrable by the IST. Its Lax representation reads as [5]

$$\Phi_x = U\Phi, \tag{3}$$
$$\Phi_t = 2\lambda\Phi_y + V\Phi, \tag{4}$$

where

$$\Phi = \begin{pmatrix} \psi_1 \\ \psi_2 \end{pmatrix},$$
$$U = -i\lambda S,$$
$$V = \frac{\lambda}{2}([S, S_y] + 2iuS).$$

3 Darboux Transformation for the M-I Equation

To construct the DT for the M-I equation (1)–(2), we associate desired solution of its corresponding linear system (3)–(4) Φ' with the known solution Φ at the form

$$\Phi' = L\Phi,$$

where we assume that
$$L = \lambda N - K.$$

Here
$$N = \begin{pmatrix} n_{11} & n_{12} \\ n_{21} & n_{22} \end{pmatrix}.$$

The matrix function Φ' satisfies the same Lax representation as (3)–(4), so that

$$\Phi'_x = U'\Phi',$$
$$\Phi'_t = 2\lambda\Phi'_y + V'\Phi'.$$

The matrix function L obeys the following equations

$$L_x + LU = U'L, \tag{5}$$
$$L_t + LV = 2\lambda L_y + V'L. \tag{6}$$

From the first equation of this system we obtain

$$N : \lambda_x = 0$$
$$\lambda^0 : K_x = 0 \tag{7}$$
$$\lambda^1 : N_x = iS'K - iKS,$$
$$\lambda^2 : 0 = -iS'N + iNS. \tag{8}$$

Hence we have the following DT for the matrix S:

$$S' = NSN^{-1}.$$

The second equation of the system (5)–(6) gives us

$$N : \lambda_t = 2\lambda\lambda_y,$$
$$\lambda^0 : K_t = 0, \tag{9}$$
$$\lambda^1 : N_t = -2K_y - \frac{1}{2}([S', S'_y] + 2iU'S') + \frac{1}{2}([S, S_y] + 2iUS),$$
$$\lambda^2 : N_y = -\frac{1}{4}([S', S'_y] + 2iU'S')N + \frac{1}{4}N([S, S_y] + 2iUS).$$

From (7) and (9) we can see that K is constant. So that we can put

$$K = \begin{pmatrix} 1 & 0 \\ 0 & 1 \end{pmatrix}.$$

Then from (8) we get the form of the DT for S as

$$S' = S - iN_x,$$

After some cumbersome calculations we get that

$$N_y N^{-1} = iuNSN^{-1} - NSN^{-1}N_y SN^{-1} - iu'NSN^{-1}.$$

This equation can be simplified as

$$i(u' - u)NSN^{-1} = -(NSN^{-1}N_ySN^{-1} + N_yN^{-1}).$$

Hence we get the DT for the potential u:

$$u' = u + itr(SN^{-1}N_y).$$

Thus, in general we have constructed the DT for the M-I with the following form:

$$S' = NSN^{-1},$$
$$u' = u + itr(SN^{-1}N_y).$$

It is not difficult to verify that the matrix N has the form

$$N = \begin{pmatrix} n_{11} & n_{12} \\ -n_{12}^* & n_{11}^* \end{pmatrix},$$

so that we have

$$N^{-1} = \frac{1}{n} \begin{pmatrix} n_{11}^* & -n_{12} \\ n_{12}^* & n_{11} \end{pmatrix},$$

Here

$$n = \det N = |n_{11}|^2 + |n_{12}|^2.$$

Finally we have the DT in terms of the elements of N as:

$$S' = \frac{1}{n} \begin{pmatrix} S_3(|n_{11}|^2 - |n_{12}|^2) & S^-n_{11}^2 - S^+n_{12}^2 \\ +S^-n_{11}n_{12}^* + S^+n_{11}^*n_{12} & -2S_3n_{11}n_{12} \\ S^+n_{11}^{*2} - S^-n_{12}^{*2} & S_3(|n_{12}|^2 - |n_{11}|^2) \\ -2S_3n_{11}^*n_{12}^* & -S^-n_{11}n_{12}^* - S^+n_{11}^*n_{12} \end{pmatrix}, \tag{10}$$

$$u' = u + \frac{i}{n}[(n_{11y}n_{11}^* + n_{12y}^*n_{12} - n_{12y}n_{12}^* - n_{11y}^*n_{11})S_3$$
$$+ (n_{11y}n_{12}^* - n_{12y}^*n_{11})S^- + (n_{12y}n_{11}^* - n_{11y}^*n_{12})S^+].$$

At last, we give the other form of the DT of S as:

$$S' = S - iN_x = S - i \begin{pmatrix} n_{11x} & n_{12x} \\ -n_{12x}^* & n_{11x}^* \end{pmatrix}.$$

To construct exact solutions of the M-I equation, we must find the explicit expressions of n_{ij}. To do that, we assume that [5]

$$N = H\Lambda^{-1}H^{-1},$$

where

$$H = \begin{pmatrix} \psi_1(\lambda_1; t, x, y) & \psi_1(\lambda_2; t, x, y) \\ \psi_2(\lambda_1; t, x, y) & \psi_2(\lambda_2; t, x, y) \end{pmatrix}.$$

Here

$$\Lambda = \begin{pmatrix} \lambda_1 & 0 \\ 0 & \lambda_2 \end{pmatrix}$$

and $det\ H \neq 0$, where λ_1 and λ_2 are complex constants. It is easy to show that H satisfies the following equations

$$H_x = -iSH\Lambda,$$
$$H_t = 2H_y\Lambda + \frac{1}{2}([S, S_y] + 2iuS)H\Lambda.$$

From these equations it follows that N obeys the equations

$$N_x = iNSN^{-1} - iS,$$
$$N_y = [H_yH^{-1}, N],$$
$$N_t = \frac{1}{2}(([S, S_y] + 2iuS) - 2Z' - ([S, S_y] + 2iuS)).$$

In order to satisfy the constraint of S, the S and the matrix solution of the Lax equations obey the condition

$$\Phi^\dagger = \Phi^{-1}, \quad S^\dagger = S,$$

which follows from the equations

$$\Phi_x^\dagger = i\lambda\Phi^\dagger S^\dagger, \quad (\Phi^{-1})_x = i\lambda\Phi^{-1}S^{-1},$$

where $\dagger$ denote the Hermitian conjugate. After some calculations we come to the formulas

$$\lambda_2 = \lambda_1^*, \quad H = \begin{pmatrix} \psi_1(\lambda_1; t, x, y) & -\psi_2^*(\lambda_1; t, x, y) \\ \psi_2(\lambda_1; t, x, y) & \psi_1^*(\lambda_1; t, x, y) \end{pmatrix},$$

$$H^{-1} = \frac{1}{\Delta}\begin{pmatrix} \psi_1^*(\lambda_1; t, x, y) & \psi_2^*(\lambda_1; t, x, y) \\ -\psi_2(\lambda_1; t, x, y) & \psi_1(\lambda_1; t, x, y) \end{pmatrix},$$

where

$$\Delta = |\psi_1|^2 + |\psi_2|^2.$$

So, for the matrix N we have

$$N = \frac{1}{\Delta} \begin{pmatrix} \lambda_1^{-1}|\psi_1|^2 + \lambda_2^{-1}|\psi_2|^2 & (\lambda_1^{-1} - \lambda_2^{-1})\psi_1\psi_2^* \\ (\lambda_1^{-1} - \lambda_2^{-1})\psi_1^*\psi_2 & \lambda_1^{-1}|\psi_2|^2 + \lambda_2^{-1}|\psi_1|^2) \end{pmatrix}.$$

4 Rogue Wave Solution

The solution of the linear system corresponding to the M-I equation will be sought in the form

$$\Psi_1 = [i(2x - y - 1) - 2t]e^{-0,5it},$$
$$\Psi_2 = [2x - y + 1 + 2it]e^{0,5it}$$

Then, from the Eq. (10), we obtain the solution of the M-I in the following form

$$S_3 = -\frac{2(2x - y)}{(2x - y)^2 + 4t^2 + 1},$$
$$S^+ = \frac{[(2x - y)^2 i - 4t^2 - 1]i - 4t(2x - y)}{2[(2x - y)^2 + 4t^2 + 1]}.$$

These solutions are behave as destructive waves within a certain range (see Figs. 1 and 2); accordingly, they are called rogue waves.

Fig. 1 Rogue wave solution when $t = 0.5$

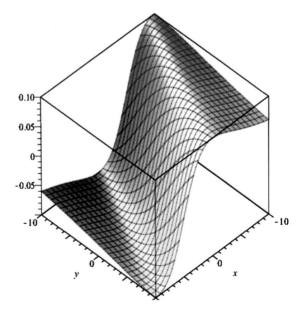

Fig. 2 Rogue wave solution when $t = 5$

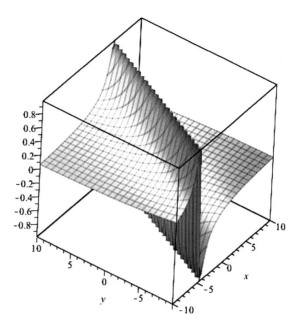

5 Conclusion

One of the interesting areas of research in the theory of solitons is the development of methods for constructing exact solutions of nonlinear differential equations in partial derivatives. In this paper, the method of DT is extended to the (2+1)-dimensional spin systems for the M-I equation. By building the DT for the M-I equation, we found its rogue waves type solutions.

References

1. Myrzakulov, R., Vijayalakshmi, S., Nugmanova, G., Lakshmanan, M.: A (2+1)-dimensional integrable spin model: geometrical and gauge equivalent counterpart, solitons and localized coherent structures. Phys. Lett. A **233**, 391–396 (1997)
2. Myrzakulov, R., Nugmanova, G.N., Danlybaeva, A.K.: Geometry and multidimensional spliton equations. Theor. Math. Phys. **118**(13), 441–451 (1999)
3. Myrzakulov, R., Nugmanova, G., Syzdykova, R.: Gauge equivalence between (2+1)-dimensional continuous Heisenberg ferromagnetic models and nonlinear Schrödinger-type equations. J. Phys. A: Math. Theor. **31**(147), 9535–9545 (1998)
4. Zhang, Zh.-H., Deng, M., Zhao, W.-Zh., Ke, W.: On the integrable inhomogeneous Myrzakulov I equation. arXiv:nlin/0603069v1 [nlin.SI] 30 March 2006
5. Chen, Ch., Zhou, Zi.-X.: Darboux transformation and exact solutions of the Myrzakulov I equation. Chin. Phys. Lett. **26**(8), 080504 (2009)
6. Zhang, Y., Nie, X.-J., Zha, Q.-L.: Rogue wave solutions for the Heisenberg ferromagnet equations. Chin. Phys. Lett. **31**(6), 060201 (2014)

Fuzzy Bilevel Programming with Credibility Measure

Hande Günay Akdemir

Abstract This paper considers a hierarchical decision making problem in which the wholesaler is considered as the leader and decides the price, while the retailer is the decision maker of the second level and decides the ordering quantity from the wholesaler. Both decision makers try to maximize their profit in case of the fuzzy demand of the product depends on the price. To deal with uncertainty, credibility measure theory is applied. Examples are given to illustrate the proposed model.

1 Introduction

Supply chain is a system consists of suppliers, manufacturers, distributors, retailers, transporters who act in a coordinated manner to accomplish product development, marketing, distribution and warehousing tasks in order to provide a competitive advantage. In most real world situations, the possible values of model parameters are only determined ambiguously by using linguistic terms, i.e. cannot be described precisely by the experts. Besides, probability distributions of model parameters cannot be known due to lack of historical data. Supply chain management related problems can also include imprecise parameters due to possible failures in production line, fluctuations in unit costs, prices and demands. The impact of vagueness can be captured by using fuzzy set theory. In fuzzy context, the degree of membership expresses how much a particular element belongs to a fuzzy set.

Bilevel programming (BP) is a nested hierarchical system where two decision makers act in a cooperative/noncooperative and sequential manner to optimize their possibly conflicting individual objective functions. It is closely related to game theory and economic equilibrium problems. Each unit optimizes its own objective function, but is affected by the action or reaction of the other unit. BP has been applied

H. Günay Akdemir (✉)
Department of Mathematics, Giresun University, Giresun, Turkey
e-mail: hande.akdemir@giresun.edu.tr

© Springer International Publishing Switzerland 2016   127
G.A. Anastassiou and O. Duman (eds.), *Intelligent Mathematics II:*
Applied Mathematics and Approximation Theory, Advances in Intelligent Systems
and Computing 441, DOI 10.1007/978-3-319-30322-2_9

in fields which involve hierarchical relationship between two classes of decision makers, for example in pricing and fare optimization problems in airline industry or in other transportation networks, management of multi-divisional firms, economic planning, optimal design, engineering, chemistry, environmental sciences, etc. For a detailed literature review on BP and its generalization: mathematical programs with equilibrium constraints, reader may refer to [4]. For example, a government in a distribution network, or a service provider in a communication market or a manufacturer in a supply chain acts as a dominant player and makes his decision first. As followers, users of those networks, or competitors or retailers use that decision as an input to form their strategy.

Cao and Chen [2] discussed a capacitated plant selection problem in decentralized manufacturing environment. Ji and Shao [7] formulated a BP model with more than one retailer for the newsboy problem with fuzzy demands and quantity discounts. Zhu et al. [17] considered fuzzy bilevel expected value model for a supply chain composed of a supplier producing a single period product in the upper level and multiple retailers in the lower level. In [9], a bilevel multi-objective supply chain model, where market demand, production capacity, and supplies are fuzzy numbers, is considered. Yao et al. [16] developed a possibilistic bilevel multi-objective stone resource assignment model between a government and a plant. Hashim et al. [6] presented a fuzzy bilevel decision making problem with the lower level corresponding to a plant planning problem, while the upper level to a distribution network problem. In [14], the authors investigated a supply chain optimization problem in a bi-level structure in which customer demand and supply chain costs were considered to be fuzzy.

In this paper, a hierarchical supply chain consisting of one wholesaler and one retailer is studied and fuzzy customer demand is supposed to be price sensitive. Three types of membership functions are considered for fuzzy demand, namely triangular, trapezoidal, and "close to" type. This paper is organized as follows. Section 2 gives brief information on credibility theory. In Sect. 3, BP models and a solution technique, namely branch and bound algorithm are given. Illustrative examples for three types of fuzzy numbers are considered in Sect. 4. Finally, it is concluded with some remarks in the last section.

2 Fuzzy Set Theory and Credibility Measures

In this section, some knowledge of credibility theory is introduced. A fuzzy event must hold if its credibility is 1, and fail if its credibility is 0 [7]. In literature, possibility measures are often used for optimistic decisions, necessity measures are for pessimistic decisions [8]. So, credibility measures are suitable for risk-neutral fuzzy decision problems.

Definition 1 Let Θ be a nonempty set, and 2^{Θ} be the power set of Θ. Each element of 2^{Θ} is called an event. Let ξ be a fuzzy variable with the membership function μ, and t be a real number. The possibility, necessity and credibility of the fuzzy event, characterized by $\{\xi \geq t\}$, can be given respectively as

$$Pos\{\xi \geq t\} = \sup_{u \geq t}\mu(u),$$

$$Nec\{\xi \geq t\} = 1 - \sup_{u < t}\mu(u),$$

$$Cr\{\xi \geq t\} = \frac{1}{2}[Pos\{\xi \geq t\} + Nec\{\xi \geq t\}].$$

The set function Cr is normalized, monotonic, and self-dual fuzzy measure [12]. For axiomatic definition of credibility and further details, reader can refer to [13].

Example 2 A triangular fuzzy variable ξ can be determined by a triplet (a, b, c) with $a < b < c$, whose membership function is

$$\mu(t) = Pos\{\xi = t\} = \begin{cases} \frac{t-a}{b-a}, & \text{if } a < t \leq b \\ \frac{c-t}{c-b}, & \text{if } b < t \leq c \\ 0, & \text{otherwise.} \end{cases}$$

It is easy to obtain that

$$Cr\{\xi \geq t\} = \begin{cases} 1, & \text{if } t \leq a \\ \frac{2b-a-t}{2(b-a)}, & \text{if } a < t \leq b \\ \frac{c-t}{2(c-b)}, & \text{if } b < t \leq c \\ 0, & \text{if } t > c \end{cases}.$$

Example 3 Credibility measure of the trapezoidal fuzzy variable ξ which is determined by a quadruplet (a, b, c, d) with $a < b < c < d$, is:

$$Cr\{\xi \geq t\} = \begin{cases} 1, & \text{if } t \leq a \\ \frac{2b-a-t}{2(b-a)}, & \text{if } a < t \leq b \\ \frac{1}{2}, & \text{if } b < t \leq c \\ \frac{d-t}{2(d-c)}, & \text{if } c < t \leq d \\ 0, & \text{if } t > d, \end{cases}.$$

where membership function is

$$\mu(t) = Pos\{\xi = t\} = \begin{cases} \frac{t-a}{b-a}, & \text{if } a < t \leq b \\ 1, & \text{if } b < t \leq c \\ \frac{d-t}{d-c}, & \text{if } c < t \leq d \\ 0, & \text{otherwise.} \end{cases}$$

Example 4 For a fixed real number m, the membership function of the linguistic fuzzy set of real numbers "close to m" or "about m" can be defined as $\mu(t) = \frac{1}{1+(t-m)^2}$.

So,

$$Cr\{\xi \geq t\} = \begin{cases} 1 - \frac{1}{2(1+(t-m)^2)}, & \text{if } t \leq m \\ \frac{1}{2(1+(t-m)^2)}, & \text{if } t > m \end{cases} \tag{1}$$

3   Branch and Bound Algorithm for Bilevel Programming

The first level decision maker (the leader) controls over the vector x and the second level decision maker (the follower) controls over the vector y. The follower's problem (the lower level/inner problem) is a constraint of the leader's problem (the upper level/outer problem).

The general formulation of a BP problem is:

$$\begin{aligned} \min_{x} \ & F(x, y) \\ & G(x, y) \leq 0 \\ & H(x, y) = 0 \\ & \text{where } y \text{ solves} \\ & \begin{cases} \min_{y} \ f(x, y) \\ \quad g(x, y) \leq 0 \\ \quad h(x, y) = 0 \end{cases} \end{aligned} \tag{2}$$

where $x \in \mathbb{R}^m$ is a vector for upper-level variables, $y \in \mathbb{R}^n$ is a vector for lower-level variables, $F : \mathbb{R}^m \times \mathbb{R}^n \to \mathbb{R}$ is upper-level objective function, $f : \mathbb{R}^m \times \mathbb{R}^n \to \mathbb{R}$ is lower-level objective function, the vector-valued functions $G : \mathbb{R}^m \times \mathbb{R}^n \to \mathbb{R}^r$ and $H : \mathbb{R}^m \times \mathbb{R}^n \to \mathbb{R}^{r'}$ give upper-level constraints, and the vector-valued functions $g : \mathbb{R}^m \times \mathbb{R}^n \to \mathbb{R}^s$ and $h : \mathbb{R}^m \times \mathbb{R}^n \to \mathbb{R}^{s'}$ give lower-level constraints.

Rather than working with hierarchical form (2), we convert the model into a standard mathematical program by replacing the second level problem with its Karush-Kuhn-Tucker (KKT) optimality conditions. This operation reduces the original problem to a single level program involving nonlinear complementary constraints [3]. Equivalent problem is a nonlinear programming problem whose constraints include the first order necessary optimality conditions of the second level problem. The equivalent single level program of the BP problem follows as:

$$\min_{x,y,u,v} \ F(x, y)$$
$$G(x, y) \leq 0$$
$$H(x, y) = 0$$
$$g(x, y) \leq 0 \tag{3}$$
$$h(x, y) = 0$$
$$\nabla_y f(x, y) + u^T \nabla_y g(x, y) + v^T \nabla_y h(x, y) = 0$$
$$u_i g_i(x, y) = 0, \quad i = 1, \ldots, s$$
$$u_i \geq 0, \qquad i = 1, \ldots, s$$

where u and v are the vectors of KKT multipliers and Lagrange multipliers, respectively.

In branch and bound algorithm [1], the complementary constraints $u_i g_i(x, y) = 0$ are suppressed and removed to construct the relaxed program. Supposing that the solution of the relaxed program does not satisfy some complementary constraints, branching is performed by separating two subproblems one with $u_i = 0$ as an additional constraint, and the other with the constraint $g_i(x, y) = 0$, selecting i for which $|u_i g_i(x, y)|$ is the largest. Branching is repeated until an infeasible solution is obtained or all complementary constraints are satisfied. In doing so, we examine all of the complementary slackness conditions one by one within a tolerance value. ($|u_i g_i(x, y)| < 10^{-6}$) Resulting feasible solutions are labeled as candidate solutions.

4   Illustrative Examples

The goal of this study is to apply fuzzy BP notion to the two-echelon supply chain, consisting of a manufacturer as the leader and a retailer as the follower. It is assumed that the decision makers know the distribution of the uncertain and price-sensitive customer demand. This problem is adapted from [10]. The authors considered stochastic versions of this lunch vendor problem. In this paper, we consider fuzzy bilevel chance constrained programming model for supply chain management problem.

- The retailer buy the product from the company at the unit wholesale price $x \in [a, b]$ which is determined by the company, where a and b are positive constants.
- Then, for a given wholesale price x of the leader, the retailer responses to the decision of the wholesaler company by the amount y of product that he purchases from the company and it is no less than c.
- The retailer sells the product at the unit retail price $2x$ to the market.
- The decisions are taken before demand realization. The demand ξ is supposed to be a fuzzy variable, and varies according to the price.
- The retailer can sell the amount $t = \min\{y, \xi\}$ of product, so $t \leq y$ and $t \leq \xi$. The objectives of the company and the retailer are to maximize profits xy and $(2xt - xy)$, respectively.

- If there are any unsold product, the retailer cannot return them to the company, he/she will dispose them with no cost.

We can formulate the problem as the following fuzzy BP problem with fuzzy chance constraint [5, 15]:

$$
\begin{aligned}
&\max_{x} \ (xy)\\
&\quad a \leq x \leq b\\
&\quad \text{where } y \text{ solves}\\
&\left\{
\begin{array}{l}
\max_{y} \ (2xt - xy) \qquad\qquad\qquad (4)\\
\quad y \geq c\\
\quad t \leq y\\
\quad Cr\{\xi \geq t\} \geq \alpha
\end{array}
\right.
\end{aligned}
$$

where $\alpha \in (0, 1)$ is pre-determined satisfactory level for credibility to ensure that the constraint should hold at some confidence level [11].

Example 5 Parameters are taken as $a = 10, b = 20, c = 100$, and demand is a price dependent triangular fuzzy number $\xi = (300 - 12x, 350 - 10x, 400 - 4x)$, $\alpha = 0.70$. Then, the credibility measure of demand can be calculated as:

$$
Cr\{\xi \geq t\} =
\begin{cases}
1, & \text{if } t \leq 300 - 12x\\
\frac{400-8x-t}{100+4x}, & \text{if } 300 - 12x < t \leq 350 - 10x\\
\frac{400-4x-t}{100+12x}, & \text{if } 350 - 10x < t \leq 400 - 4x\\
0, & \text{if } t > 400 - 4x
\end{cases}
\qquad (5)
$$

One way to represent the piecewise credibility function is to see it as a separable function. So,

$$
Cr\{\xi \geq t\} = \lambda_0 + \lambda_1 + 0.5\lambda_2 + 0\lambda_3 + 0\lambda_4
$$

is the linear approximation denoting 0, $(300 - 12x)$, $(350 - 10x)$, $(400 - 4x)$, and 400 are grid points, where

$$
\lambda_0 + \lambda_1 + \lambda_2 + \lambda_3 + \lambda_4 = 1, \ \lambda_0, \lambda_1, \lambda_2, \lambda_3, \lambda_4 \geq 0,
$$

and

$$
t = 0\lambda_0 + (300 - 12x)\,\lambda_1 + (350 - 10x)\,\lambda_2 + (400 - 4x)\,\lambda_3 + 400\lambda_4.
$$

The initial two-level programming problem (4) is transformed into an equivalent crisp two-level programming problem:

$$\max_{x} (xy)$$
$$10 \le x \le 20$$
where $y, t, \lambda_0, \lambda_1, \lambda_2, \lambda_3, \lambda_4$ solves

$$\begin{cases} \min_{y,t,\lambda_0,\dots,\lambda_4} (xy - 2xt) \\ 100 - y \le 0 \\ t - y \le 0 \\ -\lambda_0, -\lambda_1, -\lambda_2, -\lambda_3, -\lambda_4, -t \le 0 \\ 0.70 - \lambda_0 - \lambda_1 - 0.5\lambda_2 \le 0 \\ t - (300 - 12x)\lambda_1 - (350 - 10x)\lambda_2 \\ \quad - (400 - 4x)\lambda_3 - 400\lambda_4 = 0 \\ \lambda_0 + \lambda_1 + \lambda_2 + \lambda_3 + \lambda_4 - 1 = 0 \end{cases} \quad (6)$$

Then, by using (3), the equivalent single level programming model of (6) can be given as:

$$\max_{x,y,t,\lambda,u,v} (xy)$$
$$x \ge 10, x \le 20,$$
$$100 - y \le 0, t - y \le 0,$$
$$-\lambda_0, -\lambda_1, -\lambda_2, -\lambda_3, -\lambda_4, -t \le 0,$$
$$0.70 - \lambda_0 - \lambda_1 - 0.5\lambda_2 \le 0,$$
$$t - (300 - 12x)\lambda_1 - (350 - 10x)\lambda_2$$
$$\quad - (400 - 4x)\lambda_3 - 400\lambda_4 = 0,$$
$$\lambda_0 + \lambda_1 + \lambda_2 + \lambda_3 + \lambda_4 - 1 = 0,$$
$$x - u_1 - u_2 = 0, -2x + u_2 - u_8 + v_1 = 0,$$
$$-u_3 - u_9 + v_2 = 0, -u_4 - u_9 - v_1(300 - 12x) + v_2 = 0,$$
$$-u_5 - 0.5u_9 - v_1(350 - 10x) + v_2 = 0,$$
$$-u_6 - v_1(400 - 4x) + v_2 = 0, -u_7 - 400v_1 + v_2 = 0,$$
$$u_1(100 - y) = u_2(t - y) = 0,$$
$$u_3\lambda_0 = u_4\lambda_1 = u_5\lambda_2 = u_6\lambda_3 = u_7\lambda_4 = 0,$$
$$u_8 t = u_9(0.70 - \lambda_0 - \lambda_1 - 0.5\lambda_2) = 0,$$
$$u_1, \dots, u_9 \ge 0, v_1, v_2 \text{ urs.}$$

(7)

Based on the branch and bound algorithm, Fig. 1 demonstrates the solution process for the model (7).

Example 6 Parameters in Example 6 are the same as those in Example 5 except the credibility function of the demand. $a = 10$, $b = 20$, $c = 100$, demand is a price dependent trapezoidal fuzzy number

$$\xi = (300 - 12x, 300, 400 - 4x, 400), \quad \alpha = 0.70.$$

So,

$$Cr\{\xi \ge t\} = \lambda_0 + \lambda_1 + 0.5\lambda_2 + 0.5\lambda_3 + 0\lambda_4 + 0\lambda_5$$

Fig. 1 Branching and
backtracking for Example 5

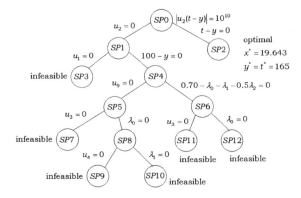

where

$$\lambda_0 + \lambda_1 + \lambda_2 + \lambda_3 + \lambda_4 + \lambda_5 = 1, \lambda_0, \lambda_1, \lambda_2, \lambda_3, \lambda_4, \lambda_5 \geq 0,$$

$$t = 0\lambda_0 + (300 - 12x)\,\lambda_1 + 300\lambda_2 + (400 - 4x)\,\lambda_3 + 400\lambda_4 + 1000\lambda_5.$$

The equivalent single level program for Example 6 can be given as follows:

$$
\begin{aligned}
\max_{x,y,t,\lambda,u,v} \quad & (xy) \\
& x \geq 10, x \leq 20, \\
& 100 - y \leq 0, t - y \leq 0, \\
& -\lambda_0, -\lambda_1, -\lambda_2, -\lambda_3, -\lambda_4, -\lambda_5, -t \leq 0, \\
& 0.70 - \lambda_0 - \lambda_1 - 0.5\lambda_2 - 0.5\lambda_3 \leq 0, \\
& t - (300 - 12x)\,\lambda_1 - 300\lambda_2 - (400 - 4x)\,\lambda_3 \\
& \quad -400\lambda_4 - 1000\lambda_5 = 0, \\
& \lambda_0 + \lambda_1 + \lambda_2 + \lambda_3 + \lambda_4 + \lambda_5 - 1 = 0, \\
& x - u_1 - u_2 = 0, -2x + u_2 - u_9 + v_1 = 0, \\
& -u_3 - u_{10} + v_2 = 0, -u_4 - v_1(300 - 12x) + v_2 = 0, \\
& -u_5 - 300v_1 + v_2 = 0, -u_6 - v_1(400 - 4x) + v_2 = 0, \\
& -u_7 - 400v_1 + v_2 = 0, -u_8 - 1000v_1 + v_2 = 0, \\
& u_1(100 - y) = u_2(t - y) = 0, \\
& u_3\lambda_0 = u_4\lambda_1 = u_5\lambda_2 = u_6\lambda_3 = u_7\lambda_4 = u_8\lambda_5 = 0, \\
& u_9 t = u_{10}(0.70 - \lambda_0 - \lambda_1 - 0.5\lambda_2) = 0, \\
& u_1, \ldots, u_{10} \geq 0, v_1, v_2 \text{ urs.}
\end{aligned}
\qquad (8)
$$

Branching and backtracking procedure for the model (8) is shown in Fig. 2.

Example 7 Parameters in Example 7 are the same as those in Examples 5 and 6 except
the credibility function of the demand. For Example 7, $a = 10$, $b = 20$, $c = 100$,

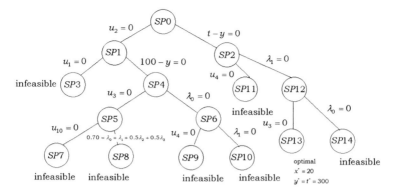

Fig. 2 Branching and backtracking for Example 6

demand is a price dependent fuzzy number "*close to* $(350 - 8x)$", $\alpha = 0.70$. Then, the credibility measure of demand can be calculated as:

$$Cr\{\xi \geq t\} = \begin{cases} 1 - \frac{1}{2\left(1+(t-350+8x)^2\right)}, & \text{if } t \leq 350 - 8x \\ \frac{1}{2\left(1+(t-350+8x)^2\right)}, & \text{otherwise.} \end{cases}$$

For Example 7, the equivalent single level program is given as:

$$\max_{x,y,t,\lambda,u,v} \ (xy)$$

$$x \geq 10, \ x \leq 20, \ 100 - y \leq 0, \ t - y \leq 0, \ -\lambda_1, -\lambda_2, -\lambda_3, -t \leq 0,$$

$$0.70 - \lambda_1 \left(1 - \frac{1}{2\left(1+(8x-160)^2\right)}\right) - 0.5\lambda_2$$

$$-\lambda_3 \left(\frac{1}{2\left(1+(8x-80)^2\right)}\right) \leq 0,$$

$$t - 190\lambda_1 - (350 - 8x)\lambda_2 - 270\lambda_3 = 0,$$

$$\lambda_1 + \lambda_2 + \lambda_3 - 1 = 0,$$

$$x - u_1 - u_2 = 0, \ -2x + u_2 - u_6 + v_1 = 0,$$

$$-u_3 - u_7 \left(1 - \frac{1}{2\left(1+(8x-160)^2\right)}\right) - 190v_1 + v_2 = 0,$$

$$-u_4 - 0.5u_7 - v_1(350 - 8x) + v_2 = 0,$$

$$-u_5 - u_7 \left(\frac{1}{2\left(1+(8x-80)^2\right)}\right) - 270v_1 + v_2 = 0,$$

$$u_1(100 - y) = u_2(t - y) = 0,$$

$$u_3\lambda_1 = u_4\lambda_2 = u_5\lambda_3 = u_6 t = 0,$$

$$u_7(0.70 - \lambda_1 \left(1 - \frac{1}{2\left(1+(8x-160)^2\right)}\right) - 0.5\lambda_2$$

$$-\lambda_3 \left(\frac{1}{2\left(1+(8x-80)^2\right)}\right) = 0,$$

$$u_1, \ldots, u_7 \geq 0, \ v_1, v_2 \text{ urs.}$$

The resulting optimal values of decision variables are summarized in Table 1.

Table 1 Optimal Solutions

	x	y	t			
Example 5	19.643	165.000	165.000			
Example 6	20.000	300.000	300.000			
Example 7	19.580	211.630	211.630			
	λ_0	λ_1	λ_2	λ_3	λ_4	λ_5
Example 5	0.00	0.70	0.00	0.00	0.30	–
Example 6	0.70	0.00	0.00	0.00	0.00	0.30
Example 7	–	0.73	0.00	0.27	–	–
	u_1	u_2	u_3	u_4	u_5	
Example 5	0.000	19.643	1262.755	0.000	1543.367	
Example 6	0.000	20.000	0.000	18800.000	14000.000	
Example 7	0.000	19.580	0.000	684.250	0.000	
	u_6	u_7	u_8	u_9	u_{10}	
Example 5	1543.367	0.000	0.000	6594.388	–	
Example 6	13600.000	12000.000	0.000	0.000	20000.000	
Example 7	0.000	1632.859	–	–	–	
	v_1	v_2	Objective			
Example 5	19.643	7857.143	3241.071			
Example 6	20.000	20000.000	6000.000			
Example 7	19.580	5285.657	4143.656			

5 Concluding Remarks

In this study, a two-echelon hierarchical supply chain is considered that consists of a manufacturer and a retailer when the retailer faces fuzzy and price-dependent customer demand. As the price of product influence customers to some extend, so it is vital factor for both decision makers. The models for different membership functions of demand such as triangular, trapezoidal, and "close to" types are analyzed separately. Different fuzzification techniques lead to different optimal solutions. For fuzzification process, it is advisable for the user to perform experimental and sensitivity analysis on selected fuzzy sets to determine the impact of varying membership values. Only demand uncertainty is considered by using credibility theory. The main advantage of using credibility measures is that they can easily be represented as separable functions in the models.

References

1. Bard, J.F., Moore, J.T.: A branch and bound algorithm for the bilevel programming problem. SIAM J. Sci. Stat. Comput. **11**(2), 281–292 (1990)
2. Cao, D., Chen, M.: Capacitated plant selection in a decentralized manufacturing environment: a bilevel optimization approach. Eur. J. Oper. Res. **169**(1), 97–110 (2006)
3. Colson, B., Marcotte, P., Savard, G.: A trust-region method for nonlinear bilevel programming: algorithm and computational experience. Comput. Optim. Appl. **30**, 211–227 (2007)
4. Dempe, S.: Annotated bibliography on bilevel programming and mathematical programs with equilibrium constraints. Optim.: J. Math. Program. Oper. Res. **52**(3), 333–359 (2003)
5. Gao, J., Liu, B.: Fuzzy multilevel programming with a hybrid intelligent algorithm. Comput. Math. Appl. **49**(9–10), 1539–1548 (2005)
6. Hashim, M., Nazim, M., Nadeem, A.H.: Production-distribution planning in supply chain management under fuzzy environment for large-scale hydropower construction projects. In: Proceedings of the Sixth International Conference on Management Science and Engineering Management, Lecture Notes in Electrical Engineering, vol. 185, pp. 559–576 (2013)
7. Ji, X., Shao, Z.: Model and algorithm for bilevel newsboy problem with fuzzy demands and discounts. Appl. Math. Comput. **172**(1), 163–174 (2006)
8. Katagiri, H., Kato, K., Uno, T.: Possibilistic Stackelberg solutions to bilevel linear programming problems with fuzzy parameters. In: IFSA World Congress and NAFIPS Annual Meeting (IFSA/NAFIPS), Joint, pp. 134–139. IEEE (2013)
9. Li, Y., Yang, S.: Fuzzy Bi-level multi-objective programming for supply chain. In: IEEE International Conference on Automation and Logistics 2007, pp. 2203–2207, 18–21 Aug 2007
10. Lin, G.H., Chen, X., Fukushima, M.: Smoothing implicit programming approaches for stochastic mathematical programs with linear complementarity constraints (revised version, March 2004), Technical Report 2003–2006. Department of Applied Mathematics and Physics , Kyoto University (May, 2003)
11. Liu, B., Iwamura, K.: Chance constrained programming with fuzzy parameters. Fuzzy Sets Syst. **94**, 227–237 (1998)
12. Liu, B., Liu, Y.K.: Expected value of fuzzy variable and fuzzy expected value model. IEEE Trans. Fuzzy Syst. **10**, 445–450 (2002)
13. Liu, B., Liu, B.: Theory and Practice of Uncertain Programming. Physica, Heidelberg (2002)
14. Taran, M., Roghanian, E.: A fuzzy multi-objective multi-follower linear bi-level programming problem to supply chain optimization. Uncertain Supply Chain Manag. **1**(4), 193–206 (2013)
15. Yang, L., Liu, L.: Fuzzy fixed charge solid transportation problem and algorithm. Appl. Soft Comput. **7**(3), 879–889 (2007)
16. Yao, L., Xu, J., Guo, F.: A stone resource assignment model under the fuzzy environment. Math. Probl. Eng., vol. 2012, Article ID 265837, 26 p. (2012)
17. Zhu, L., Zhao, R., Tang, W.: Fuzzy single-period product problem with return policy. Int. J. Manag. Sci. Eng. Manag. **2**(2), 126–137 (2007)

A New Approach of a Possibility Function Based Neural Network

George A. Anastassiou and Iuliana F. Iatan

Abstract The paper presents a new type of fuzzy neural network, entitled Possibility Function based Neural Network (PFBNN). Its advantages consist in that it not only can perform as a standard neural network, but can also accept a group of possibility functions as input. The PFBNN discussed in this paper has novel structures, consisting in two stages: the first stage of the network is a fuzzy based and it has two parts: a Parameter Computing Network (PCN), followed by a Converting Layer (CL); the second stage of the network is a standard backpropagation based neural network (BPNN). The PCN in a possibility function based network can also be used to predict functions. The CL is used to convert the possibility function to a value. This layer is necessary for data classification. The network can still function as a classifier using only the PCN and the CL or only the CL. Using only the PCN one can perform a transformation from one group of possibility functions to another.

1 Introduction

Scientists have proved a considerable interest in the study of the Artificial Neural Networks (ANNs) [2] and especially the Fuzzy Neural Networks (FNNs) [13] during the last decade. Their interest in the FNN applications was generated [3] by the following two events:

1. first, the success of Japanese fuzzy logic technology applications in the consumer products;
2. second, in some ANN applications sufficient data for training are not available.

G.A. Anastassiou
Department of Mathematical Sciences, University of Memphis,
Memphis, TN 38152, USA
e-mail: ganastss@memphis.edu

I.F. Iatan (✉)
Department of Mathematics and Computer Science, Technical University
of Civil Engineering, Bucharest, Romania
e-mail: iuliafi@yahoo.com

© Springer International Publishing Switzerland 2016   139
G.A. Anastassiou and O. Duman (eds.), *Intelligent Mathematics II:*
Applied Mathematics and Approximation Theory, Advances in Intelligent Systems
and Computing 441, DOI 10.1007/978-3-319-30322-2_10

In such situations, the fuzzy logic systems [5, 9, 13] are often workable.

Today, a lot of neural approaches [4, 6] are established [8] as fixed parts of machine learning, and a rigorous theoretical investigation for these approaches is available.

Many researchers of this academic world agree on the fact that the statistical notion is often the right language to formalize the learning algorithms and to investigate their mathematical properties. Nevertheless, according to the widespread models, tasks, and application areas of the neural approaches [10, 12, 14, 15] , the mathematical tools ranging from approximation theory [16], complexity theory, geometry, statistical physics, statistics, linear and nonlinear optimization, control theory and many more fields can be found in the literature dedicated of the ANNs.

Correspondingly, the role of Mathematics in the ANN literature is [7, 8] diverse:

(A) **development and presentation of algorithms**: Most neural algorithms are designed in mathematical terms and some learning schemes are even mainly motivated by abstract mathematical considerations such as support vector machines;

(B) **foundation of tools**: A fixed canon of mathematical questions has been identified for most network models and application areas which is to be answered in order to establish the models as well founded and reliable tools in the literature. Interestingly, many mathematical questions are thereby not yet solved satisfactorily also for old network models and constitute still open topics of ongoing research such as the loading problem of feed-forward networks or the convergence problem of the self-organizing map in its original formulation;

(C) **application of tools**: Mathematical formalization establishes standards for the assessment of the performance of methods and application in real-life scenarios (although these standards are not always followed and real life would sometimes be better characterized by slightly different descriptions than the mathematics).

We will in the following consider mathematical questions which are to be answered to justify standard models [8] as reliable machine learning tools. Thereby, we will focus on classical models used for machine learning: feed-forward networks. Depending on the output function of such type networks, feed-forward networks can be used for classification of patterns if the output set is finite and discrete, or approximation of functions if the output set is contained in a real-vector space.

Probabilistic neural network [11] is a kind of feed-forward neural network involved from the radial basis function networks. Its theoretical basis is the Bayesian minimum risk criteria. In pattern classification, its advantage is to substitute nonlinear learning algorithm with linear learning algorithm.

In recent years, the probabilistic neural networks are used [1, 11] in the field of face recognition for its structure, good approximation, fast training speed and good real-time performance.

The remainder of the paper is organized as follows. In Sect. 2 we analyze the architecture of the PFBNN. We follow with the training algorithm of the PBFNN in Sect. 3. We conclude in Sect. 4.

2   Architecture of the PFBNN

From the Fig. 1 which shows the block diagram of the proposed fuzzy neural network we can notice that there are two stages:

(1) the first stage of the network is a fuzzy based and it has two parts: a Parameter Computing Network (PCN), followed by a Converting Layer (CL);
(2) the second stage of the network is a standard back-propagation based neural network (BPNN).

This PBFNN can be segmented and still perform useful functions. The network can still function as a classifier using only the PCN and the CL or only the CL. Using only the PCN one can perform a transformation from one group of possibility functions to another.

There are three types of weight variables used between connecting nodes of the three networks in Fig. 1:

a. the first type called a λ-weight is used for connection weights between nodes in the PCN;
b. the second type called a r-weight is used for connection weights between the output of the PCN and the CL;
c. the third type called a w-weight is used for connection weights between the neurons in a standard BPNN.

As the w-weights are adjusted according to standard back-propagation algorithms, we shall discuss only the setting and adjustment of the λ- and r-weights.

The PCN accepts as input, a vector representing a group of possibility functions and generates a group of possibility functions as output.

In the PCNN, the weights associated of the neurons corresponding to the λ layers are

$$\{\lambda_{ij}^{(k)}\}_{i=\overline{1,L_{k-1}},\ j=\overline{1,L_k}}$$

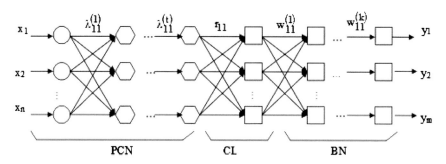

Fig. 1 The framework of a possibility based fuzzy neural network

where:

- k ($k = \overline{1, t}$) represents the order of the PCN layer,
- i is the index of the neuron from the $(k - 1)$ layer,
- j means the index of the neuron from the k layer,
- L_k is the number of the neurons of the kth layer, $L_0 = n$,

with

$$\lambda_{ij}^{(k)} : [0, 1] \to [-1, 1]$$

and they are always positive or always negative.

Each $\lambda_{ij}^{(k)}$ is represented as a binary tuple $(\rho_{ij}^{(k)}, \omega_{ij}^{(k)})$, where:

- $\rho_{ij}^{(k)}$ is a transformation function from $[0, 1]$ to $[0, 1]$;
- $\omega_{ij}^{(k)}$ is a constant real number in $[-1, 1]$.

One can use a fuzzy normal distribution function

$$f(x) = e^{-\frac{(x-\mu)^2}{2\sigma^2}} \tag{1}$$

for each element x of the crisp input vector X to obtain the fuzzified input data of the PCN.

We shall compute the outputs of the neurons from the kth layer of PCN using the relation:

$$y_j^{(k)}(u) = \sum_{i=1}^{L_{k-1}} \omega_{ij}^{(k)} y_i^{(k-1)} \left(\rho_{ij}^{(k)}(u)^{-1} \right), \ u \in [0, 1] \tag{2}$$

or shortly

$$y_j^{(k)} = \sum_{i=1}^{L_{k-1}} \omega_{ij}^{(k)} \left(y_i^{(k-1)} \circ \rho_{ij}^{(k)^{-1}} \right), \ (\forall) \ j = \overline{1, L_k}, \tag{3}$$

where:

- $Y^{(k)} = (y_1^{(k)}, \ldots, y_{L_k}^{(k)})$ is the output vector of the kth layer of PCN,
- $Y^{(k-1)} = (y_1^{(k-1)}, \ldots, y_{L_{k-1}}^{(k)})$ constitutes the input vector of the kth layer (namely the output vector of the $(k - 1)$th layer),
- "$\circ$" means the composite of two functions,
- $\rho_{ij}^{(k)^{-1}}$ is the inverse function of $\rho_{ij}^{(k)}$.

The CL accepts as input a possibility function (that represents a possibility vector) generated by the PCN and transforming it into a real number vector.

Each weight of this layer is a function

$$r_{ij} : [0, 1] \to [-1, 1], \ i = \overline{1, L_t}, \ j = \overline{1, M},$$

where L_t is the number of the neurons from the layer t of PCN and M is the number of the output neurons of CL.

Similar to λ in the PCN, r_{ij} is always positive or always negative. The r-weights of the CL can be also represented as a binary tuple $r_{ij} = (\gamma_{ij}, \tau_{ij}), i = \overline{1, L_t}, j = \overline{1, M}$ where

$$\gamma_{ij} : [0, 1] \to [0, 1]$$

is a possibility function, which is different from ρ (the transformation function in the PCN) and τ_{ij} is a constant real number in $[-1, 1]$.

The output $Z = (z_1, \ldots, z_M)$ of the CL is a vector having real numbers as components, see the following formula:

$$z_j = \sum_{i=1}^{L_t} \tau_{ij} \left(\max_{u \in [0,1]} \left(\min(y_i^{(t)}(u), \tau_{ij}(u)) \right) \right), \ j = \overline{1, M} \tag{4}$$

$Y^{(t)} = (y_1^{(t)}, \ldots, y_{L_t}^{(t)})$ being the fuzzy input vector of CL, which constitutes the output vector of PCN.

We shall use $y_i^{(t)}(u) \cdot \tau_{ij}(u)$ instead of $\min(y_i^{(t)}(u), \tau_{ij}(u))$ in order to compute easier the outputs of the CL using (4).

3 Training Algorithm of the PBFNN

We shall build the training algorithm of the PBFNN in the hypothesis that the PCN has three layers (namely $t = 3$):

1. the input layer which contains a number of $L_0 = n$ neurons
2. a hidden layer having L_1 neurons
3. an output layer with L_2 neurons.

Step 1. Initialize the weights of the PCN and CL in the following way:

(a) choose a linear function as the initial weight function for each ρ:

$$\lambda_{ij}^{(k)}(u_i) = v_j, \ i = \overline{1, L_t}, \ j = \overline{1, M} \tag{5}$$

and design a genetic algorithm to search for optimal ω's;

(b) let each weight function γ as a possibility function:

$$\gamma_{ij}(u) = e^{-\frac{(u-u_0)^2}{2\sigma^2}}, \ u_0 \in [0, 1], \ i = \overline{1, L_t}, \ j = \overline{1, M} \tag{6}$$

assigning usually $\sigma = 1$ and design a genetic algorithm to search for optimal τ's.

Let $Y^{(0)} = (y_1^{(0)}, \ldots, y_{L_0}^{(0)})$ be the input fuzzy vector of the PCN corresponding to the training vector by the index p.

Step 2. Compute the fuzzy output vector $Y^{(1)} = (y_1^{(1)}, \ldots, y_{L_1}^{(1)})$ of the hidden layer of PCN using the relation:

$$y_j^{(1)} = \sum_{i=1}^{L_0} \omega_{ij}^{(1)} \left(y_i^{(0)} \circ \rho_{ij}^{(1)-1} \right), \ j = \overline{1, L_1} \tag{7}$$

Step 3. Compute the fuzzy output vector $Y^{(2)} = (y_1^{(2)}, \ldots, y_{L_2}^{(1)})$ of the output layer of PCN using the relation:

$$y_k^{(2)} = \sum_{j=1}^{L_1} \omega_{jk}^{(2)} \left(y_j^{(1)} \circ \rho_{jk}^{(2)-1} \right), \ k = \overline{1, L_2}. \tag{8}$$

Step 4. Apply to the input of the CL the fuzzy vector $Y^{(2)} = (y_1^{(2)}, \ldots, y_{L_2}^{(1)})$, which one obtains at the output of the PCN.

Step 5. Determine the output vector $Z = (z_1, \ldots, z_M)$ of the CL, having each component a real number:

$$z_j = \sum_{i=1}^{L_2} \tau_{ij} \left(\max_{u \in [0,1]} \left(y_i^{(2)}(u), \tau_{ij}(u) \right) \right), \ j = \overline{1, M}, \tag{9}$$

where M is the number of the output neurons of the CL.

Step 6. Adjust the weights of the output layer of the PCN:

$$\begin{cases} \rho_{jk}^{(2)}(u_j) \leftarrow \rho_{jk}^{(2)}(u_j) + \mu_\rho \cdot \frac{\partial E}{\partial \rho_{jk}^{(2)}(u_j)}, \\ \omega_{jk}^{(2)}(u_j) \leftarrow \omega_{jk}^{(2)}(u_j) + \mu_\omega \cdot \frac{\partial E}{\partial \omega_{jk}^{(2)}(u_j)}, \end{cases} \tag{10}$$

$j = \overline{1, L_1}$, $k = \overline{1, L_2}$, μ_ρ, μ_ω being two constants with the meaning of learning rates, and

$$E = \frac{1}{|S_T|} \sum_{p=1}^{|S_T|} E_p \tag{11}$$

defines the performance of the system, where:

- $|S_T|$ represents the number of the vectors from the training lot,
- E_p is the output error of the PCN for the pth training sample, defined by:

$$E_p = \sum_{k=1}^{L_2} \left(\int_0^1 E_p(v_k) dv_k \right) \tag{12}$$

and

$$E_p(v_k) = \frac{1}{2} \left(T_k(v_k) - y_k^{(2)}(v_k) \right)^2, \tag{13}$$

$T_k = (T_1, \ldots, T_{L_2})$ being the ideal output vector (the target vector) of the input vector by the index p applied to the PCN.

We shall have

$$
\begin{aligned}
\frac{\partial E}{\partial \rho_{jk}^{(2)}(u_j)} &= \sum_{k=1}^{L_2} \left(\int_0^1 \frac{\partial E_p(v_k)}{\partial \rho_{jk}^{(2)}(u_j)} dv_k \right) \\
&= \int_0^1 \frac{\partial E_p(v_k)}{\partial \rho_{jk}^{(2)}(u_j)} dv_k \\
&= \int_0^1 \left(T_k(v_k) - y_k^{(2)}(v_k) \right) \cdot \frac{\partial y_k^{(2)}(v_k)}{\partial \rho_{jk}^{(2)}(u_j)} dv_k
\end{aligned} \tag{14}
$$

and

$$\frac{\partial y_k^{(2)}(v_k)}{\partial \rho_{jk}^{(2)}(u_j)} = \omega_{jk}^{(2)} \cdot \frac{\partial y_j^{(1)}}{\partial \rho_{jk}^{(2)-1}(v_k)} \cdot \frac{\partial \rho_{jk}^{(2)-1}(v_k)}{\partial \rho_{jk}^{(2)}(u_j)}. \tag{15}$$

Substituting (14) and (15) into (10) it will results:

$$\rho_{jk}^{(2)}(u_j) \leftarrow \rho_{jk}^{(2)}(u_j) - \mu_\rho \omega_{jk}^{(2)} \int_0^1 \left(T_k(v_k) - y_k^{(2)}(v_k) \right) \cdot \frac{\partial y_j^{(1)}}{\partial \rho_{jk}^{(2)-1}(v_k)} \cdot \frac{\partial \rho_{jk}^{(2)-1}(v_k)}{\partial \rho_{jk}^{(2)}(u_j)} dv_k, \tag{16}$$

where $i = \overline{1, L_1}, k = \overline{1, L_2}$.

Similarly,

$$\frac{\partial E}{\partial \omega_{jk}^{(2)}} = \sum_{k=1}^{L_2} \left(\int_0^1 \frac{\partial E_p(v_k)}{\partial \omega_{jk}^{(2)}(u_j)}) dv_k \right) = \int_0^1 \frac{\partial E_p(v_k)}{\partial \omega_{jk}^{(2)}} dv_k \tag{17}$$

namely

$$\frac{\partial E}{\partial \omega_{jk}^{(2)}} = -\int_0^1 \left(T_k(v_k) - y_k^{(2)}(v_k) \right) \cdot \frac{\partial y_k^{(2)}(v_k)}{\partial \omega_{jk}^{(2)}} dv_k, \tag{18}$$

where

$$\frac{\partial y_k^{(2)}(v_k)}{\partial \omega_{jk}^{(2)}} = y_j^{(1)} \left(\rho_{jk}^{(2)-1}(v_k) \right). \tag{19}$$

Substituting (18) and (19) into (10) we obtain:

$$\omega_{jk}^{(2)}(u_j) \leftarrow \omega_{jk}^{(2)}(u_j) - \mu_\omega \int_0^1 \left(T_k(v_k) - y_k^{(2)}(v_k) \right) \cdot y_j^{(1)} \left(\rho_{jk}^{(2)-1}(v_k) \right) dv_k \tag{20}$$

where $i = \overline{1, L_1}, k = \overline{1, L_2}$.

Step 7. Adjust the weights of the hidden layer of the PCN:

$$\begin{cases} \rho_{ij}^{(1)}(u_i) \leftarrow \rho_{ij}^{(1)}(u_i) + \mu_\rho \cdot \frac{\partial E}{\partial \rho_{ij}^{(1)}(u_i)}, \\ \omega_{ij}^{(1)}(u_i) \leftarrow \omega_{ij}^{(1)}(u_i) + \mu_\omega \cdot \frac{\partial E}{\partial \omega_{ij}^{(1)}(u_i)}, \end{cases} \tag{21}$$

$i = \overline{1, L_0}, j = \overline{1, L_1}$, where:

$$\frac{\partial E}{\partial \rho_{ij}^{(1)}(u_i)} = \sum_{k=1}^{L_2} \left(\int_0^1 \frac{\partial E_p(v_k)}{\partial \rho_{ij}^{(1)}(u_i)} \right) dv_k \tag{22}$$

namely

$$\frac{\partial E}{\partial \rho_{ij}^{(1)}(u_i)} = -\sum_{k=1}^{L_2} \left(\int_0^1 (T_k(v_k) - y_k^{(2)}(v_k)) \cdot \frac{\partial y_k^{(2)}(v_k)}{\partial \rho_{ij}^{(1)}(u_i)} dv_k \right) \tag{23}$$

where

$$y_k^{(2)}(v_k) = \sum_{j=1}^{L_1} \omega_{jk}^{(2)} \cdot y_j^{(1)} \left(\rho_{jk}^{(2)-1}(v_k) \right), \tag{24}$$

namely

$$y_k^{(2)}(v_k) = \sum_{j=1}^{L_1} \omega_{jk}^{(2)} \cdot \sum_{i=1}^{L_0} \omega_{ij}^{(1)} \cdot y_i^{(0)} \left(\rho_{ij}^{(1)-1}(\rho_{jk}^{(2)-1}(v_k)) \right) \qquad (25)$$

and

$$\frac{\partial y_k^{(2)}(v_k)}{\partial \rho_{ij}^{(1)}(u_i)} = \omega_{jk}^{(2)} \cdot \omega_{ij}^{(1)} \cdot \frac{\partial y_i^{(0)}}{\partial \rho_{ij}^{(1)-1}(\rho_{jk}^{(2)-1}(v_k))} \cdot \frac{\partial \rho_{ij}^{(1)-1}(\rho_{jk}^{(2)-1}(v_k))}{\rho_{ij}^{(1)}(u_i)}, \qquad (26)$$

where

$$\frac{\partial \rho_{ij}^{(1)-1}(\rho_{jk}^{(2)-1}(v_k))}{\rho_{ij}^{(1)}(u_i)} = \frac{\partial \rho_{ij}^{(1)-1}}{\partial \rho_{jk}^{(2)-1}(v_k)} \cdot \frac{\partial \rho_{jk}^{(2)-1}(v_k)}{\partial \rho_{ij}^{(1)}(u_i)} \qquad (27)$$

Substituting (23), (26), (27) into the first formula from the relation (21) we achieve:

$$\rho_{ij}^{(1)}(u_i) \leftarrow \rho_{ij}^{(1)}(u_i) - \mu_\rho \cdot \omega_{ij}^{(1)}$$
$$\cdot \sum_{k=1}^{L_2} \omega_{jk}^{(2)} \left(T_k(v_k) - y_k^{(2)}(v_k) \right) \cdot \frac{\partial y_i^{(0)}}{\partial \rho_{ij}^{(1)-1}(\rho_{jk}^{(2)-1}(v_k))} \qquad (28)$$
$$\cdot \frac{\partial \rho_{ij}^{(1)-1}}{\partial \rho_{jk}^{(2)-1}(v_k)} \cdot \frac{\partial \rho_{jk}^{(2)-1}(v_k)}{\partial \rho_{ij}^{(1)}(u_i)} dv_k,$$

$(\forall)\ i = \overline{1, L_0},\ j = \overline{1, L_1}.$
  Analogically,

$$\frac{\partial E}{\partial \omega_{ij}^{(1)}(u_i)} = \sum_{k=1}^{L_2} \left(\int_0^1 \frac{\partial E_p(v_k)}{\partial \omega_{ij}^{(1)}(u_i)} dv_k \right) \qquad (29)$$

namely

$$\frac{\partial E}{\partial \omega_{ij}^{(1)}(u_i)} = -\sum_{k=1}^{L_2} \left(\int_0^1 (T_k(v_k) - y_k^{(2)}(v_k)) \cdot \frac{\partial y_k^{(2)}(v_k)}{\partial \omega_{ij}^{(1)}(u_i)} dv_k \right), \qquad (30)$$

where

$$\frac{\partial y_k^{(2)}(v_k)}{\partial \omega_{ij}^{(1)}(u_i)} = \omega_{jk}^{(2)} \cdot y_i^{(0)} \left(\rho_{ij}^{(1)-1}(\rho_{jk}^{(2)-1}(v_k)) \right). \qquad (31)$$

Substituting (30), (31) into the second formula from the relation (21) we achieve:

$$\omega_{ij}^{(1)}(u_i) \leftarrow \omega_{ij}^{(1)}(u_i) - \mu_\omega \cdot$$

$$\sum_{k=1}^{L_2} \omega_{jk}^{(2)} \int_0^1 \left(T_k(v_k) - y_k^{(2)}(v_k) \right) \cdot y_i^{(0)} \left(\rho_{ij}^{(1)-1}(\rho_{jk}^{(2)-1}(v_k)) \right) dv_k, \qquad (32)$$

$(\forall)\, i = \overline{1, L_2},\, j = \overline{1, M}.$

   Step 8. Adjust the weights of the CL:

$$\begin{cases} \gamma_{ij}(u) \leftarrow \gamma_{ij}(u) + \mu_\gamma \cdot \frac{\partial E}{\partial \gamma_{ij}(u)}, \\ \tau_{ij} \leftarrow \tau_{ij} + \mu_\tau \cdot \frac{\partial E}{\partial \tau_{ij}}, \end{cases} \qquad (33)$$

$u \in [0, 1], (\forall)\, i = \overline{1, L_2},\, j = \overline{1, M},\, \mu_\gamma$ and μ_τ being two constants with the meaning of learning rates and E is the performance of the system (defined as in (11)), E_p being in this case the output error of the CL for the pth training sample and it is defined by:

$$E_p = \frac{1}{2} \sum_{j=1}^{M} (U_j - z_j)^2, \qquad (34)$$

$U = (U_1, \ldots, U_M)$ being the ideal output vector (the target vector) of the input vector by the index p applied to the CL.

   We shall have:

$$\frac{\partial E}{\partial \gamma_{ij}(u)} = \frac{\partial E_p}{\partial \gamma_{ij}(u)} = -(U_j - z_j)\frac{\partial z_j}{\partial \gamma_{ij}(u)}. \qquad (35)$$

Let $u_{\max}$ the point for which $y_i^{(2)}(u)\gamma_{ij}(u)$ has maximum value. Hence:

$$\frac{\partial z_j}{\partial \gamma_{ij}(u)} = \sum_{k=1}^{M} \tau_{kj} \cdot \frac{\partial \left(y_k^{(2)}(u) \cdot \gamma_{kj}(u) \right)}{\partial \gamma_{ij}(u_{\max})}, \qquad (36)$$

namely

$$\frac{\partial z_j}{\partial \gamma_{ij}(u)} = \sum_{k=1}^{M} \tau_{kj} \left(\gamma_{kj}(u_{\max}) \cdot \frac{\partial y_k^{(2)}}{\partial \gamma_{ij}(u_{\max})} + y_k^{(2)}(u_{\max}) \cdot \frac{\partial \gamma_{kj}}{\partial \gamma_{ij}(u_{\max})} \right), \qquad (37)$$

where

$$\frac{\partial y_k^{(2)}}{\partial \gamma_{ij}(u_{\max})} = 0 \qquad (38)$$

and

$$\frac{\partial \gamma_{kj}}{\partial \gamma_{ij}(u_{\max})} = \begin{cases} 1, & \text{if } k = i \\ 0, & \text{otherwise.} \end{cases} \tag{39}$$

Introducing the relations (36), (37)–(39) into the first formula from (33) one obtains:

$$\gamma_{ij}(u_{\max}) \leftarrow \gamma_{ij}(u_{\max}) - \mu_\gamma \cdot \tau_{ij} \cdot (U_j - z_j) \cdot y_i^{(2)}(u_{\max}), \tag{40}$$

$(\forall)\ i = \overline{1, L_2},\ j = \overline{1, M}.$

Similarly, we shall have:

$$\frac{\partial E}{\partial \tau_{ij}} = \frac{\partial E_p}{\partial \tau_{ij}} = -(U_j - z_j)\frac{\partial z_j}{\partial \tau_{ij}}, \tag{41}$$

where

$$\frac{\partial z_j}{\partial \tau_{ij}} = \max_{u \in [0,1]} \left(y_i^{(2)}(u) \cdot \gamma_{ij}(u) \right) = y_i^{(2)}(u_{\max}) \cdot \gamma_{ij}(u_{\max}). \tag{42}$$

Substituting (41) and (42) into the second formula from (33) we shall achieve:

$$\tau_{ij} \leftarrow \tau_{ij} - \mu_\tau \cdot \tau_{ij} \cdot (U_j - z_j) \cdot y_i^{(2)}(u_{\max}) \cdot \gamma_{ij}(u_{\max}), \tag{43}$$

$(\forall)\ i = \overline{1, L_2},\ j = \overline{1, M}.$

Step 9. Compute the PCN error because of the pth training vector with (12).

Step 10. Compute the CL error because of the pth training vector, using (34).

Step 11. If the training algorithm has not applied for all the training vectors, then go to the next vector. Otherwise, test the stop condition. For example, we can stop the algorithm after a fixed training epoch numbers.

4   Conclusion

The paper describes a new type of fuzzy neural network, entitled possibility function based neural network. Its advantages consist in that it not only can perform as a standard neural network, but can also accept a group of possibility functions as input.

The possibility function based network discussed in this paper has novel structures. The parameter computing network in a possibility function based network can also be used to predict functions. The converting layer is used to convert the possibility function to a value. This layer is necessary for data classification.

References

1. Araghi, L.F., Khaloozade, H., Arvan, M.R.: Ship identification using probabilistic neural networks (PNN). In: Proceedings of the International Multi Conference of Engineers and Computer Scientists (IMECS 2009), vol II, 18–20, Hong Kong, 2009
2. Basu, J.K., Bhattacharyya, D., Kim, T.H.: Use of artificial neural network in pattern recognition. Int. J. Softw. Eng. Appl. **4**(2), 22–34 (2010)
3. Chen, L., Cooley, D.H., Zhang, J.: Possibility-based fuzzy neural networks and their application to image processing. IEEE Trans. Syst. Man Cybernetics **29**(1), 119–126 (1999)
4. Cherkassky, V., Mulier, F.: Learning from Data: Concepts, Theory, and Methods. Wiley-IEEE Press (2007)
5. Cooley, D.H., Zhang, J., Chen, L.: Possibility function based fuzzy neural networks: case study. IEEE Trans. Syst. Man Cybernetics **1**, 73–78 (1994)
6. Dutt, V., Chadhury, V., Khan, I.: Different approaches in pattern recognition. Comput. Sci. Eng. **1**(2), 32–35 (2011)
7. Hammer, B., Villmann, T.: Mathematical aspects of neural networks. In: 11th European Symposium on Artificial Neural Networks (ESANN' 2003), 59–72. Brussels, Belgium (2003)
8. Hastie, T., Tibshirani, R., Friedman, J.: The Elements of Statistical Learning: Data Mining, Inference and Prediction. Springer-Verlag, Berlin Heidelberg (2009)
9. Lin, C.J., Wang, M., Lee, C.Y.: Pattern recognition using neural-fuzzy networks based on improved particle swam optimization. Expert Syst. Appl. **36**, 5402–5410 (2009)
10. Maraqua, M., Al-Zboun, F., Dhyabat, M., Zitar, R.A.: Recognition of Arabic sign language (ArSL) using recurrent neural networks. J. Intell. Learn. Syst. Appl. **4**, 41–52 (2012)
11. Ni, Q., Guo, C., Yang, J.: Research of face image recognition based on probabilistic neural networks. In: IEEE Control and Decision Conference, pp. 3885–3888 (2012)
12. Pajares, G., Guijarro, M., Ribeiro, A.: A hopfield neural network for combining classifiers applied to textured images. Neural Networks **23**, 144–153 (2010)
13. Rubio, J.J.: Stability analysis for an online evolving neuro-fuzzy recurrent network in evolving intelligent systems: methodology and applications. In: Angelov, P., Filev, D.P., Kasabov, N. (eds.), pp. 173–199. Wiley-IEEE Press (2010)
14. Tsang, E.C.C., Qiu, S.S., Yeung, D.S.: Convergence analysis of a discrete hopfield neural network with delay and its application to knowledge refinement. Int. J. Pattern Recogn. Artif. Intell. **21**(3), 515–541 (2007)
15. Xhemali, D., Hinde, C.J., Stone, R.G.: Naïve Bayes vs. decision trees vs. neural networks in the classification of training web pages. Int. J. Comput. Sci. Issues **4**(1), 16–23 (2009)
16. Yin, H., Allinson, N.M.: Self-organizing mixture networks for probability density estimation. IEEE Trans. Neural Networks **12**(2), 405–411 (2001)

Elementary Matrix Decomposition Algorithm for Symmetric Extension of Laurent Polynomial Matrices and Its Application in Construction of Symmetric M-Band Filter Banks

Jianzhong Wang

Abstract In this paper, we develop a novel and effective algorithm for the construction of perfect reconstruction filter banks (PRFBs) with linear phase. In the algorithm, the key step is the symmetric Laurent polynomial matrix extension (SLPME). There are two typical problems in the construction: (1) For a given symmetric finite low-pass filter **a** with the polyphase, to construct a PRFBs with linear phase such that its low-pass band of the analysis filter bank is **a**. (2) For a given dual pair of symmetric finite low-pass filters, to construct a PRFBs with linear phase such that its low-pass band of the analysis filter bank is **a**, while its low-pass band of the synthesis filter bank is **b**. In the paper, we first formulate the problems by the SLPME of the Laurent polynomial vector(s) associated to the given filter(s). Then we develop a symmetric elementary matrix decomposition algorithm based on Euclidean division in the ring of Laurent polynomials, which finally induces our SLPME algorithm.

1 Introduction

The main purpose of this paper is to develop a novel and effective algorithm for the construction of perfect reconstruction filter banks (PRFBs) with linear phase. In the algorithm, the key step is the symmetric Laurent polynomial matrix extension (SLPME).

PRFBs have been widely used in many areas such as signal and image processing, data mining, feature extraction, and compressive sensing [3, 8, 11–14]. A PRFB consists of two sub-filter banks: an analysis filter bank, which decomposes a signal into different bands, and a synthesis filter bank, which composes a signal from its different band components. Either an analysis filter bank or a synthesis one consists of several band-pass filters. Assume that an analysis filter bank consists of the filter set $\{H_0, H_1, \ldots, H_{M-1}\}$ and a synthesis filter bank consists of the set

J. Wang (✉)
Department of Mathematics and Statistics, Sam Houston State University,
Huntsville, TX, USA
e-mail: jzwang@shsu.edu

© Springer International Publishing Switzerland 2016   151
G.A. Anastassiou and O. Duman (eds.), *Intelligent Mathematics II:*
Applied Mathematics and Approximation Theory, Advances in Intelligent Systems
and Computing 441, DOI 10.1007/978-3-319-30322-2_11

$\{B_0, B_1, \ldots, B_{M-1}\}$, where H_0 and B_0 are low-pass filters. Then they form an M-band PRFB if and only if the following condition holds:

$$\sum_{j=0}^{M-1} B_j(\uparrow M)(\downarrow M)\overline{H}_j = I, \tag{1}$$

where $\downarrow M$ is the M-downsampling operator, $\uparrow M$ is the M-upsampling operator, I is the identity operator, and $\overline{H}_j$ denotes the conjugate filter of H_j. Note that the conjugate of a real filter $\mathbf{a} = (\ldots, a_{-1}, a_0, a_1, \ldots)$ is defined as $\bar{\mathbf{a}} = (\ldots, a_1, a_0, a_{-1}, \ldots)$. In signal processing, a filter H having only finite non-zero entries is called a finite impulse response (FIR). Otherwise it is called an infinite impulse response (IIR). Since FIR is much more often used than IIR, in this paper we only study FIR with real entries. Recall that the z-transform of a FIR H is a Laurent polynomial (LP) and the z-transform of the conjugate filter of H is $\bar{H}(z) = H(1/z)$. We define the M-polyphase form of a signal (or a filter) $\mathbf{x}$ by the LP vector $[a^{[M,0]}(z), \ldots, a^{[M,M-1]}(z)\}]$, where

$$a^{[M,k]}(z) = \sum_j a(Mj+k)z^j, \quad 0 \le k \le M - 1.$$

For convenience, we will simplify $a^{[M,k]}$ to $a^{[k]}$ if it does not cause confusion. The polyphase form of a M-band filter bank $\{H_0, \ldots, H_{M-1}\}$ is the following LP matrix.

$$\mathbf{H}(z) = \begin{bmatrix} H_0^{[0]}(z) & H_0^{[1]}(z) & \cdots & H_0^{[M-1]}(z) \\ H_1^{[0]}(z) & H_1^{[1]}(z) & \cdots & H_1^{[M-1]}(z) \\ \vdots & \vdots & \cdots & \vdots \\ H_{M-1}^{[0]}(z) & H_{M-1}^{[1]}(z) & \cdots & H_{M-1}^{[M-1]}(z) \end{bmatrix}$$

Using polyphase form, we represent (1) as a LP matrix identity in the following theorem.

Theorem 1 *The filter bank pair of $\{H_0, \ldots, H_{M-1}\}$ and $\{B_0, \ldots, B_{M-1}\}$ realizes a PRFB if and only if the following identity holds:*

$$\mathbf{H}(z)\mathbf{B}^*(z) = \frac{1}{M}I, \tag{2}$$

where both $\mathbf{H}(z)$ and $\mathbf{B}(z)$ are LP matrices, and $\mathbf{B}^*(z)$ denotes the conjugate transpose matrix of $\mathbf{B}(z)$.

We denote by $\mathscr{L}$ the ring of all Laurent polynomials, and call a LP matrix is $\mathscr{L}$-invertible, if its inverse is a LP matrix too. Since $M\mathbf{B}^*(z) = \mathbf{H}^{-1}(z)$ in a PRFB, the polyphases of its analysis filter bank and its synthesis one are $\mathscr{L}$-invertible. By (2), we also have

$$\sum_{j=0}^{M-1} M H_0^{[j]}(z) \bar{B}_0^{[j]}(z) = 1.$$

In general, we will call a LP vector $\mathbf{a}(z) = [a_1(z), \ldots, a_M(z)]$ a *prime* one if there is a LP vector $\mathbf{b}(z) = [b_1(z), \ldots, b_M(z)]$ such that $\mathbf{a}(z)\mathbf{b}^T(z) = 1$. More details of the theory of PRFBs are referred to [10, 15].

The filters with symmetry (also called with linear phases) are more desirable in application [10]. They are formally defined as follows:

Definition 2 Let c be an integer. A filter (or signal) $\mathbf{x}$ is called symmetric or anti-symmetric about $c/2$ if $x(k) = x(c - k)$ or $x(k) = -x(c - k)$, $k \in \mathbb{Z}$, respectively.

Later, for simplification, we will use the term *symmetric* to mention both symmetric and antisymmetric. Thus, $\mathbf{x}$ is symmetric if and only if $x(k) = \varepsilon x(c - k)$, where ε ($= 1$ or -1) is the symbol of the symmetry-type. Note that we can always shift a signal/filter $\mathbf{x}$ such that the shifted one has the symmetric center at $c = 0$ or $c = \pm 1$. Hence, without loss of generality, in this paper we always assume that a symmetric filter has the center at $c = 0$, $c = 1$, or $c = -1$, and simply call it 0-symmetric, 1-symmetric, or (-1)-symmetric, respectively. Correspondingly, the set of all 0-symmetric filters (1-symmetric, or (-1)-symmetric ones) is denoted by $\mathcal{V}_0$ ($\mathcal{V}_1$ or $\mathcal{V}_{-1}$). Besides, when we need to stress on the symmetry-type, we denote by $\mathcal{V}_1^+, \mathcal{V}_0^+, \mathcal{V}_{-1}^+$ for $\varepsilon = +1$ and $\mathcal{V}_1^-, \mathcal{V}_0^-, \mathcal{V}_{-1}^-$ for $\varepsilon = -1$. It is clear that if $\mathbf{x} \in \mathcal{V}_0$, then so is $\bar{\mathbf{x}}$, and if $\mathbf{x} \in \mathcal{V}_1$, then $\bar{\mathbf{x}} \in \mathcal{V}_{-1}$. We also have the following: $\mathbf{x} \in \mathcal{V}_0$ if and only if $x(z) = \varepsilon x(1/z)$, $\mathbf{x} \in \mathcal{V}_1$ if and only if $x(z) = \varepsilon z x(1/z)$, and $\mathbf{x} \in \mathcal{V}_{-1}$ if and only if $x(z) = \varepsilon / z x(1/z)$. In addition, if $a(z) = \varepsilon b(1/z)$ ($a(z) = \varepsilon z b(1/z)$, $a(z) = \varepsilon / z b(1/z)$), we call $[a(z), b(z)]$ a $\mathcal{V}_0$ ($\mathcal{V}_1$, $\mathcal{V}_{-1}$) pair. For a symmetric filter H, we modify its M-polyphase to the following:

$$H^{[k]}(z) = \sum_j H(Mj + k)z^j, \quad -m \le k \le M - m - 1, \quad m = \left[\frac{M-1}{2}\right].$$

Later, a LP vector is called a *S-LP one* if it is a polyphase form of a symmetric filter. Similarly, we will call a LP matrix *Sr-LP matrix* (*Sc-LP matrix*) if its rows (columns) are S-LP vectors. They will be simply called S-LP matrices if row and column are not stressed. Similarly, a PRFB is called *symmetric* if all of its band-filters are symmetric. Two fundamental problems in the construction of symmetric PRFBs are the following:

Problem 1 Assume that a given symmetric low-band filter H_0 has a prime polyphase. How to construct a symmetric PRFB, in which the first band of its analysis filter bank is H_0?

Problem 2 Assume that a dual pair of symmetric low-band filters H_0 and B_0 are given. How to find other symmetric components $H_1, \ldots, H_{M-1}$ and $B_1, \ldots, B_{M-1}$ so that they form a symmetric PRFB?

By Theorem 1, we have the following:

Corollary 3 *The symmetric filter banks* $\{H_0, \ldots, H_{M-1}\}$ *and* $\{B_0, \ldots, B_{M-1}\}$ *form a symmetric PRFB if and only if the following identity holds:*

$$\mathbf{H}(z)\mathbf{B}^*(z) = \frac{1}{M}I, \tag{3}$$

where both $\mathbf{H}(z)$ *and* $\mathbf{B}(z)$ *are Sr-LP matrices.*

Ignoring the factor $\frac{1}{M}$ on the right-hand side of (3) in Corollary 3, we can see that the two fundamental problems are equivalent the following symmetric Laurent polynomial Matrix extension (SLPME) problems:

SLPME Problem 1 Assume that a given S-LP row vector $\mathbf{a}(z) \in \mathscr{L}^M$ is prime. To find an $\mathscr{L}$-invertible Sr-LP matrix $A(z)$ such that $A(1, :) = \mathbf{a}$.

SLPME Problem 2 Assume that a given pair of S-LP row vectors $[\mathbf{a}(z), \mathbf{b}(z)]$ satisfies $\mathbf{a}(z)\mathbf{b}^T(z) = 1$. To find an $\mathscr{L}$-invertible Sr-LP matrix $A(z)$ such that $A(1, :) = \mathbf{a}$ and $A^{-1}(:, 1) = \mathbf{b}^T$.

Laurent polynomial matrix extension (LPME) has been discussed in [1, 4, 9]. Having the aid of LPME technique, several algorithms have been developed for the construction of PRFBs [2, 5–7, 15, 16]. Unfortunately, the methods for constructing LPME usually do not produce SLPME. The main difficulty in SLPME is how to preserve the symmetry. Recently, Chui, Han, and Zhuang in [2] proposed a bottom-up algorithm for solving **SPLME Problem** 2 based on the properties of dual filters.

In this paper, we solve the problem in the framework of the algebra of Laurent polynomials. Our approach to SLPME is based on the decomposition of $\mathscr{L}$-invertible S-LP matrix in the LP ring [15]. To make the paper more readable, we restrict our discussion for $M = 2, 3, 4$. The readers can find that our algorithms can be extended for any integer M without essential difficulty.

The paper is organized as follows. In Sect. 2, we discuss the properties of S-LP vectors and the symmetric Euclidean division in the LP ring. In Sect. 3, we introduce the elementary S-LP matrix decomposition technique and apply it in the development of the SLPME algorithms. Finally, two illustrative examples are presented in Sect. 4.

2   S-LP Vectors and Symmetric Euclidean Division

For simplification, in the paper, we only discuss LP with real coefficients. Readers will find that our results can be trivially generalized to LP with coefficients in the complex field or other number fields. Let the ring of all polynomials be denoted by $\mathscr{P}$ and write $\mathscr{P}_h = \mathscr{P} \setminus \{0\}$. Similarly, let the ring of all Laurent polynomials be denoted by $\mathscr{L}$ and write $\mathscr{L}_h = \mathscr{L} \setminus \{0\}$. If $a \in \mathscr{L}_h$, we can write $a(z) = \sum_{k=m}^n a_k z^k$, where $n \geq m$ and $a_m a_n \neq 0$. We define the highest degree

Table 1 The symmetry of the components in a S-LP vector

$M = 2$	$c = 0$	$x^{[0]}(z) = \varepsilon x^{[0]}(1/z), x^{[1]}(z) = \varepsilon/z x^{[1]}(1/z)$
	$c = 1$	$x^{[0]}(z) = \varepsilon x^{[1]}(1/z)$
$M = 3$	$c = 0$	$x^{[0]}(z) = \varepsilon x^{[0]}(1/z), x^{[1]}(z) = \varepsilon x^{[-1]}(1/z)$
	$c = 1$	$x^{[0]}(z) = \varepsilon x^{[1]}(1/z), x^{[-1]}(z) = \varepsilon z x^{[-1]}(1/z)$
$M = 4$	$c = 0$	$x^{[0]}(z) = \varepsilon x^{[0]}(1/z), x^{[1]}(z) = \varepsilon x^{[-1]}(1/z), x^{[2]}(z) = \varepsilon/z x^{[2]}(1/z)$
	$c = 1$	$x^{[0]}(z) = \varepsilon x^{[1]}(1/z), x^{[-1]}(z) = \varepsilon x^{[2]}(1/z)$

and the lowest degree of $a \in \mathscr{L}_h$ by $\deg^+(a) = n$ and $\deg^-(a) = m$ respectively. When $a = 0$, we agree that $\deg^+(0) = -\infty$ and $\deg^-(0) = \infty$. We define the support length of a by $\mathrm{supp}(a) = \deg^+(a) - \deg^-(a)$. Particularly, when $a(z) \in \mathscr{L}_h$ is 0-symmetric, 1-symmetric, or (-1)-symmetric, we have $\deg^-(a) = -\deg^+(a)$, $\deg^-(a) = -\deg^+(a) + 1$, or $\deg^-(a) = -\deg^+(a) - 1$, respectively.

Let the semi-group $\mathscr{G} \subset \mathscr{P}_h$ be defined by $\mathscr{G} = \{p \in \mathscr{P}_h : p(0) \neq 0\}$. Then, the power mapping $\pi : \mathscr{L}_h \to \mathscr{G}, \pi(a(z)) = z^{-\deg^-(a)} a(z)$, defines an equivalent relation "$\frown$" in $\mathscr{L}_h$, i.e., $a \frown b$ if and only if $\pi(a) = \pi(b)$. For convenience, we agree that $\pi(0) = 0$. Let $\mathscr{L}_m$ denote the group of all non-vanished Laurent monomials: $\mathscr{L}_m = \{m \in \mathscr{L}_h; m = cz^\ell, c \neq 0, \ell \in \mathbb{Z}\}$. Then, we have $\pi(m) = c$. For a LP vector $\mathbf{a} = [a_1, \ldots, a_s]$, we define $\pi(\mathbf{a}) = [\pi(a_1), \ldots, \pi(a_s)]$. Then the greatest common divisor (gcd) of a nonzero row (or column) LP vector $\mathbf{a} \in \mathscr{L}^s$ is defined by $\gcd_{\mathscr{L}}(\mathbf{a}) = \gcd(\pi(\mathbf{a})) \in \mathscr{G}$. A LP $a(z) \in \mathscr{L}_h$ is said to be in the subset $\mathscr{L}_d$ if $a(z) = \varepsilon a(1/z)$ and $\gcd_{\mathscr{L}}(a(z), a(1/z)) = 1$. A LP matrix $A(z) \in \mathscr{L}^{s \times s}$ is said to be $\mathscr{L}$-invertible if $A(z)$ is invertible and $A^{-1}(z) \in \mathscr{L}^{s \times s}$ too. It is obvious that $A(z)$ is $\mathscr{L}$-invertible if and only if $\det(A(z)) \in \mathscr{L}_m$.

We now discuss the properties of S-LP vectors. Recall that an M dimensional S-LP vector is defined as the M-polyphase form of a symmetric filter. Let $\mathbf{x}(z)$ be an M-dimensional S-LP vector. We list its symmetric properties for $M = 2, 3, 4$, in Table 1.

Let $m = \left[\frac{M-1}{2}\right]$. We can verify that, when M is even and $c = 0$, $x^{[0]}(z) \in \mathscr{V}_0, x^{[m]}(z) \in \mathscr{V}_{-1}$, and $[x^{[i]}(z), x^{[-i]}(z)], i = 1, \ldots, M - 1$, are $\mathscr{V}_0$ pairs; when M is even and $c = 1$, $(x^{[i]}(z), x^{[-i+1]}(z)), i = 1, \ldots, M$, are $\mathscr{V}_0$ pairs; when M is odd and $c = 0$, $x^{[0]}(z) \in \mathscr{V}_0$, and $[x^{[i]}(z), x^{[-i]}(z)], i = 1, \ldots, M$, are $\mathscr{V}_0$ pairs; when M is odd and $c = 1$, $[x^{[i]}(z), x^{[-i+1]}(z)], i = 1, \ldots, M$, are $\mathscr{V}_0$ pair and $x^{[-M]}(z) \in \mathscr{V}_1$.

We need the following $\mathscr{L}$-Euclid's division theorem [15] in our discussion.

Theorem 4 *Let $(a, b) \in \mathscr{L}_h \times \mathscr{L}_h$ and $\mathrm{supp}(a) \geq \mathrm{supp}(b)$. Then there exists a unique pair $(q, r) \in \mathscr{L} \times \mathscr{L}$ such that $a(z) = q(z)b(z) + r(z)$ with*

$$\mathrm{supp}(r) + \deg^-(a) \leq \deg^+(r) < \mathrm{supp}(b) + \deg^-(a), \tag{4}$$

which implies that $\mathrm{supp}(q) \leq \mathrm{supp}(a) - \mathrm{supp}(b)$ and $\mathrm{supp}(r) < \mathrm{supp}(b)$.

By Theorem 4, it is also clear that if $\deg^-(a) = \deg^-(b)$, then $q \in \mathscr{P}_h$ and $\deg(q) \leq \deg^+(a) - \deg^+(b)$. From Theorem 4, we derive the symmetric $\mathscr{L}$-Euclid's division theorem to deal with S-LP vectors.

Theorem 5 *Let $a(z) \in \mathscr{V}_0$ with $supp(a) = 2m$, $b(z) \in \mathscr{V}_{-1}$ with $supp(b) = 2k - 1$, $c(z) \in \mathscr{V}_1$ with $supp(c) = 2s - 1$, and $d(z) \in \mathscr{L}_h$ with $supp(d) = \ell$ be given. Then we have the following:*

1. *If $m \geq k$, then there is $p(z) \in \mathscr{V}_1^+$ with $supp(p) \leq 2(m - k)$ and $a_1(z) \in \mathscr{V}_0$ with $supp(a_1) < supp(b)$ such that $a(z) = b(z)p(z) + a_1(z)$. If $m < k$, then there is $q(z) \in \mathscr{V}_{-1}^+$ with $supp(q) \leq 2(k - m) - 1$ and $b_1(z) \in \mathscr{V}_{-1}$ with $supp(b_1) < supp(a)$ such that $b(z) = q(z)a(z) + b_1(z)$.*
2. *If $m \geq s$, then there is $q(z) \in \mathscr{V}_{-1}^+$ with $supp(q) \leq 2(m - k)$ and $a_1(z) \in \mathscr{V}_0$ with $supp(a_1) < supp(c)$ such that $a(z) = c(z)q(z) + a_1(z)$. If $m < s$, then there is $p(z) \in \mathscr{V}_1^+$ with $supp(p) \leq 2(k - m) - 1$ and $c_1(z) \in \mathscr{V}_1$ with $supp(c_1) < supp(a)$ such that $c(z) = p(z)a(z) + c_1(z)$.*
3. *If $supp(a) > supp(d)$, there is a $p(z) \in \mathscr{P}_h$ with $\deg(p) \leq m - \left[\frac{\ell+1}{2}\right]$ and $a_1(z) \in \mathscr{V}_0$ with $supp(a_1) \leq \ell$ such that $a(z) = p(z)d(z) + \varepsilon p(1/z)d(1/z) + a_1(z)$.*
4. *If $supp(b) > supp(d)$, there is a $q(z) \in \mathscr{P}_h$ with $\deg(q) \leq k - 1 - \left[\frac{\ell+1}{2}\right]$ and $b_1(z) \in \mathscr{V}_{-1}$ with $supp(b_1) \leq \ell$ such that $b(z) = q(z)d(z) + \varepsilon/zq(1/z)d(1/z) + b_1(z)$. Similarly, if $supp(c) > supp(d)$, there is a $p(z) \in \mathscr{P}_h$ with $\deg(p) \leq c - 1 - \left[\frac{\ell+1}{2}\right]$ and $c_1(z) \in \mathscr{V}_1$ with $supp(c_1) \leq \ell$ such that $c(z) = p(z)d(z) + \varepsilon z p(1/z)d(1/z) + c_1(z)$.*

Proof To prove (1), we write $a(z) = \sum_{j=-m}^{m} a_j z^j$ and set $a^t(z) = \sum_{j=k}^{m} a_j z^j + \frac{1}{2}\sum_{j=-k+1}^{k-1} a_j z^j$ so that $a^t(z) + \varepsilon a^t(1/z) = a(z)$. By Theorem 4, we can find a $\hat{p}(z) \in \mathscr{P}_h$ with $\deg(\hat{p}) \leq m - k$ such that $a^t(z) = \hat{p}(z)b(z) + r(z)$, where $r \in \mathscr{L}$ with $\deg^+(r) < k$, $\deg^-(r) > -k$. It leads to

$$a(t) = \hat{p}(z)b(z) + \varepsilon \hat{p}(1/z)b(1/z) + r(z) + \varepsilon r(1/z).$$

Since $b(z) \in \mathscr{V}_{-1}$, we have $b(z) = \varepsilon/zb(1/z)$, which yields

$$a(z) = \big(\hat{p}(z) + z\hat{p}(1/z)\big) b(z) + (r(z) + \varepsilon r(1/z)).$$

Write $p(z) = \hat{p}(z) + z\hat{p}(1/z)$, $a_1(z) = r(z) + \varepsilon r(1/z)$. It is obvious that $p(z) \in \mathscr{V}_1^+$ with $supp(p) \leq 2(m - k) - 1$ and $a_1(z) \in \mathscr{V}_0$ with $supp(r) < supp(b)$. The proof of the first statement of (1) is completed. The proofs of the remains are similar.

3 SLPME Algorithms Based on Elementary S-LP Matrix Decomposition

We now discuss SLPME algorithms for $M = 2, 3, 4$, respectively.

3.1 The Case of $M = 2$

We say $\mathbf{a}(z) = [a_1(z), a_2(z)] \in \mathcal{V}^{0,2}$ if $a_1(z) \in \mathcal{V}_0, a_2(z) \in \mathcal{V}_{-1}$; and say $\mathbf{a}(z) \in \mathcal{V}^{1,2}$ if it is a $\mathcal{V}_0$ pair. We also say $\mathbf{b}(z) \in \mathcal{V}*^{0,2}$ if $b_1(z) \in \mathcal{V}_0, b_2(z) \in \mathcal{V}_1$. Define

$$\mathcal{S}^{0,2} = \left\{ S(z) = [s_{ij}(z)]_{i,j=1}^2; \quad s_{ii}(z) \in \mathcal{V}_0^+, i = 1, 2, s_{21}(z) \in \mathcal{V}_1^+, s_{12}(z) \in \mathcal{V}_{-1}^+ \right\}.$$

Thus, if $S(z) \in \mathcal{S}^{0,2}$, then $S(1, :)(z) \in \mathcal{V}^{0,2}$ and $S(:, 1)(z) \in \mathcal{V}*^{0,2}$.

3.1.1 The Case of $\mathbf{a} \in \mathcal{V}^{0,2}$

To develop our SLPME algorithm, we give the following:

Definition 6 Let $s(z) \in \mathcal{V}_{-1}^+, t(z) \in \mathcal{V}_1^+, k \in \mathbb{Z}$, and $r \in \mathbb{R} \setminus \{0\}$. Then the following matrices

$$E_u(s) = \begin{bmatrix} 1 & s(z) \\ 0 & 1 \end{bmatrix} \quad E_l(t) = \begin{bmatrix} 1 & 0 \\ t(z) & 1 \end{bmatrix}, \quad D(r, k) = \begin{bmatrix} r z^k & 0 \\ 0 & 1 \end{bmatrix} \tag{5}$$

are called the elementary $\mathcal{S}^{0,2}$ matrices, and their product is called a $\mathcal{S}^{0,2}$-fundamental matrix.

It can verify that all of the matrices in (5) are $\mathcal{L}$-invertible and their inverses are also in $\mathcal{S}^{0,2}$. Indeed, we have

$$(E_u(s))^{-1} = E_u(-s) \quad (E_l(t))^{-1} = E_l(-t) \quad (D(r, k))^{-1} = D(1/r, -k). \tag{6}$$

Later, we simply denote by E_u, E_l, D for the matrices in (6).

We now return the SLPME for $\mathbf{a} \in \mathcal{V}^{0,2}$. WLOG, we assume supp $(a_1) > \text{supp}(a_2) \geq 1$. Since $\gcd_{\mathcal{L}}(\mathbf{a}) = 1$, By Theorem 5, we can use elementary $\mathcal{S}^{0,2}$ matrices to make the following:

$$\mathbf{a}^0 \xrightarrow{E_l(-p_1)} \mathbf{a}^1 \xrightarrow{E_u(-q_1)} \mathbf{a}^2 \cdots \xrightarrow{E_l(-p_n)} \mathbf{a}^{2n-1} \xrightarrow{E_u(-q_n)} \mathbf{a}^{2n} \xrightarrow{D(r,k)} [1, 0],$$

where $\mathbf{a}^0 - \mathbf{a}$, $\mathbf{a}^{2i} E_l(-p_{i+1}) = \mathbf{a}^{2i+1}$, $\mathbf{a}^{2i+1} E_u(-q_{i+1}) = \mathbf{a}^{2i+2}$, $i = 1, \ldots, n - 1$. Let

$$E_a = E_l(-p_1)E_u(-q_1) \cdots E_l(-p_n)E_u(-q_n)D(r, k). \tag{7}$$

Then $E_a \in \mathscr{S}^{0,2}$, $\mathbf{a}E_a = [1,0]$, and its inverse

$$A_a(z) = E_a^{-1}(z) \equiv D(1/r, -k)E_u(q_n)E_l(p_n)\cdots E_u(q_1)E_l(p_1) \tag{8}$$

provides a solution for SLPME Problem 1.

We now consider the SPLME Problem 2. WLOG, assuming that the symmetric dual pair $[\mathbf{a}(z), \mathbf{b}(z)] \in \mathscr{V}^{0,2} \times \mathscr{V}*^{0,2}$ is given. Let $E_a(z)$, $A_a(z)$ be the matrices given in (7) and (8). By $\mathbf{a}(z)\mathbf{b}^T(z) = 1$ and $A_a(1,:) = \mathbf{a}$, we have $A_a(z)\mathbf{b}^T(z) = [1, w(z)]^T$ with $w(z) \in \mathscr{V}_1$, which yields $E_a(z)[1, w(z)]^T = \mathbf{b}^T(z)$. Then the matrices $\tilde{A}(z) = E_l(-w)(z)A_a(z)$ and $\tilde{B}(z) = E_a(z)E_l(w)(z)$ give the solution.

3.1.2 The Case of $\mathbf{a}(z) \in \mathscr{V}^{1,2}$

If its dual $\mathbf{b}(z)$ is not given, then by the extended Euclidean algorithm in [15], we can find LP vector $\mathbf{s}(z) = [s_1(z), s_2(z)]$, such that $\mathbf{as}^T = 1$. We define $b_1(z) = \frac{1}{2}(s_1(z) + \varepsilon s_2(1/z))$, $b_2(z) = \varepsilon b_1(1/z)$. The vector $\mathbf{b}(z)$ is a $\mathscr{V}_0$ pair and $\mathbf{ab}^T = 1$. We now define

$$A(z) = \begin{bmatrix} a_1(z) & a_2(z) \\ -b_2(z) & b_1(z) \end{bmatrix},$$

which is $\mathscr{L}$-invertible and its inverse is

$$B(z) = A^{-1}(z) = \begin{bmatrix} b_1(z) & -a_2(z) \\ b_2(z) & a_1(z) \end{bmatrix}.$$

Then $A(z)$ provides the solution of SPLME Problem 1, and the pair $[A(z), B(z)]$ gives the solution of SPLME Problem 2.

3.2 The Case of $M = 3$

We say $\mathbf{a}(z) = [a_1(z), a_2(z), a_3(z)] \in \mathscr{V}^{0,3}$ if $a_2(z) \in \mathscr{V}_0$ and $[a_1(z), a_3(z)]$ is a $\mathscr{V}_0$ pair, and say $\mathbf{a}(z) \in \mathscr{V}^{1,3}$ if $a_2(z) \in \mathscr{V}_1$ and $[a_1(z), a_3(z)]$ is a $\mathscr{V}_0$ pair. We also say $\mathbf{b}(z) \in \mathscr{V}*^{1,3}$ if $b_2(z) \in \mathscr{V}_{-1}$ and $[b_1(z), b_3(z)]$ is a $\mathscr{V}_0$ pair. Note that, in the case of $M = 3$, $c = 1$, the polyphase form $\mathbf{x}(z)$ is not in $\mathscr{V}^{1,3}$, but $[x^{[0]}(z), x^{[-1]}(z), x^{[1]}(z)] \in \mathscr{V}^{1,3}$.

3.2.1 The Case of $\mathbf{a} \in \mathscr{V}^{0,3}$

Definition 7 Let $q(z) \in \mathscr{L}$. The matrices of the following two types are called elementary $\mathscr{S}^{0,3}$ matrices:

$$E_v(q) = \begin{bmatrix} 1 & 0 & 0 \\ q(z) & 1 & q(1/z) \\ 0 & 0 & 1 \end{bmatrix}, \quad E_h(q) = \begin{bmatrix} 1 & q(z) & 0 \\ 0 & 1 & 0 \\ 0 & q(1/z) & 1 \end{bmatrix}.$$

In general, we simply denote by E an elementary $\mathscr{S}^{0,3}$ matrix, and call their product a Fundamental $\mathscr{S}^{0,3}$ one.

It is clear that $E_v^{-1}(q) = E_v(-q)$ and $E_h^{-1}(q) = E_h(-q)$. By the same argument for $M = 2$, using the elementary $\mathscr{S}^{0,3}$ matrices, we can obtain the following chain:

$$\mathbf{a}^0 \xrightarrow{E_1} \mathbf{a}^1 \xrightarrow{E_2} \mathbf{a}^2 \cdots \xrightarrow{E_n} \mathbf{a}^n$$

where $\mathbf{a}^n$ has the same symmetry as $\mathbf{a}$ and $\gcd_{\mathscr{L}}(\mathbf{a}^n) = 1$. Therefore,

$$\mathbf{a}^n = [p(z), 0, \varepsilon p(1/z)], \ p(z) \in \mathscr{L}_d.$$

Besides, when $\varepsilon = 1$, it may have another form $\mathbf{a}^n = [0, r, 0]$. Writing $E = E_1 E_2 \cdots E_n$, we have $\mathbf{a}(z) = \mathbf{a}^n(z) E^{-1}(z)$. Let $q(z) \in \mathscr{L}_d$ satisfy $p(z)q(z) + p(1/z)q(1/z) = 1$ and set $\mathbf{q}(z) = [q(z), 0, \varepsilon q(1/z)]$. We define

$$Q_1(z) = \begin{bmatrix} 0 & 1/2 & -1/2 \\ r & 0 & 0 \\ 0 & 1/2 & 1/2 \end{bmatrix} \quad Q_2(z) = \begin{bmatrix} q(z) & 0 & -\varepsilon p(1/z) \\ 0 & 1 & 0 \\ \varepsilon q(1/z) & 0 & p(z) \end{bmatrix}, \quad (9)$$

whose inverses are

$$Q_1^{-1}(z) = \begin{bmatrix} 0 & \frac{1}{r} & 0 \\ 1 & 0 & 1 \\ -1 & 0 & 1 \end{bmatrix} \quad Q_2^{-1}(z) = \begin{bmatrix} p(z) & 0 & \varepsilon p(1/z) \\ 0 & 1 & 0 \\ -\varepsilon q(1/z) & 0 & q(z) \end{bmatrix}.$$

Finally, we define

$$A(z) = \begin{cases} Q_1^{-1}(z) E^{-1}(z), & \text{if } \mathbf{a}^n = [0, r, 0], \\ Q_2^{-1}(z) E^{-1}(z), & \text{if } \mathbf{a}^n = [p(z), 0, \varepsilon p(1/z)]. \end{cases}$$

It is clear that $A(z)$ is a SLPME of $\mathbf{a}(z)$.

We now return to SPLME Problem 2. Assume a symmetric dual pair $[\mathbf{a}(z), \mathbf{b}(z)] \in \mathscr{V}^{0,3} \times \mathscr{V}^{0,3}$ is given so that $\mathbf{a}(z)\mathbf{b}^T(z) = 1$. Let $E(z)$ be the LP matrix above. Define $\mathbf{w}(z) = \mathbf{b}(z)(E^{-1})^T(z) \in \mathscr{V}^{0,3}$. Then $\mathbf{a}^n \mathbf{w}^T = \mathbf{a}^n E^{-1} \mathbf{b}^T = \mathbf{a}\mathbf{b}^T = 1$. Hence,

$$\mathbf{w}(z) = \begin{cases} [u(z), 1/r, \varepsilon u(1/z)], & \text{if } \mathbf{a}^n = [0, r, 0], \\ [v(z), v_c(z), \varepsilon v(1/z)], & \text{if } \mathbf{a}^n = [p(z), 0, \varepsilon p(1/z)], \end{cases}$$

where $p(z)v(z) + p(1/z)v(1/z) = 1$. Write $q_+(z) = u(z) + \varepsilon u(1/z)$, $q_-(z) = u(z) - \varepsilon u(1/z)$. Define

$$Q_1(z) = \begin{bmatrix} u(z) & 1 & 1 \\ 1/r & 0 & 0 \\ \varepsilon u(1/z) & 1 & -1 \end{bmatrix}, \quad Q_2(z) = \begin{bmatrix} v(z) & 0 & -\varepsilon p(1/z) \\ v_c(z) & 1 & 0 \\ \varepsilon v(1/z) & 0 & p(z) \end{bmatrix}, \quad (10)$$

whose inverses are

$$Q_1^{-1}(z) = \frac{1}{2} \begin{bmatrix} 0 & 2r & 0 \\ 1 & -rq_+(z) & 1 \\ 1 & -rq_-(z) & -1 \end{bmatrix}, \quad Q_2^{-1}(z) = \begin{bmatrix} p(z) & 0 & \varepsilon p(1/z) \\ -v_c(z)p(z) & 1 & -\varepsilon v_c(z)p(1/z) \\ -\varepsilon v(1/z) & 0 & v(z) \end{bmatrix}.$$

We now define

$$A(z) = \begin{cases} Q_1^{-1}(z)E^{-1}(z), & \text{if } \mathbf{a}^n = [0, r, 0], \\ Q_2^{-1}(z)E^{-1}(z), & \text{if } \mathbf{a}^n = [p(z), 0, \varepsilon p(1/z)]. \end{cases}$$

Then, the pair $[A(z), A^{-1}(z)]$ is a SLPME of the pair $[\mathbf{a}(z), \mathbf{b}(z)]$.

3.2.2   The Case of $\mathbf{a} \in \mathcal{V}^{1,3}$

The discussion is very similar to the case of $\mathbf{a} \in \mathcal{V}^{0,3}$.

Definition 8 Let $q(z) \in \mathcal{L}$. The matrices of the following two types are called elementary $\mathcal{S}^{1,3}$ matrices:

$$E_v(q) = \begin{bmatrix} 1 & 0 & 0 \\ q(z) & 1 & zq(1/z) \\ 0 & 0 & 1 \end{bmatrix}, \quad E_h(q) = \begin{bmatrix} 1 & q(z) & 0 \\ 0 & 1 & 0 \\ 0 & zq(1/z) & 1 \end{bmatrix}.$$

The product of elementary $\mathcal{S}^{1,3}$ matrices is called a Fundamental $\mathcal{S}^{1,3}$ one.

Using the elementary $\mathcal{S}^{1,3}$ matrices, we can obtain the following chain:

$$\mathbf{a}^0 \xrightarrow{E_1} \mathbf{a}^1 \xrightarrow{E_2} \mathbf{a}^2 \cdots \xrightarrow{E_n} \mathbf{a}^n$$

where $\mathbf{a}^n$ has the same symmetry as $\mathbf{a}$ and $\gcd_{\mathcal{L}}(\mathbf{a}^n) = 1$. Therefore,

$$\mathbf{a}^n = [p(z), 0, \varepsilon p(1/z)], \ p(z) \in \mathcal{L}_d.$$

Note that, because $a_2(z) \in \mathcal{S}^{1,3}$, $\mathbf{a}^n$ does not have other forms. Writing $E = E_1 E_2 \cdots E_n$, we have $\mathbf{a}(z) = \mathbf{a}^n(z)E^{-1}(z)$. Let $Q_2(z)$ be the LP matrix in (9). Then $A(z)$ is a SLPME of $\mathbf{a}(z)$.

We now consider SPLME Problem 2. In the given dual pair, $\mathbf{b}(z) \in \mathcal{V}*^{1,3}$. Let $E(z)$ be the LP matrix above. Then the LP vector $\mathbf{w}(z) = \mathbf{b}(z)(E^{-1})^T(z) \in \mathcal{V}*^{1,3}$ too. Hence, it has the only form of

$$\mathbf{w}(z) = [v(z), v_c(z), \varepsilon v(1/z)], \quad v_c(z) \in \mathscr{V}_{-1},$$

where $p(z)v(z) + p(1/z)v(1/z) = 1$. Let $Q_2(z)$ be the LP matrix in (10), and $A(z) = Q_2^{-1}(z)E^{-1}(z)$. Then $[(E(z)Q_2(z))^{-1}, E(z)Q_2(z)]$ is a SLPME of the dual pair $[\mathbf{a}(z), \mathbf{b}(z)]$.

3.3 The Case of $M = 4$

In this case, we say $\mathbf{a}(z) \in \mathscr{V}^{0,4}$ if $a_1(z) \in \mathscr{V}_0$, $[a_2(z), a_3(z)]$ is a $\mathscr{V}_0$ pair, and $a_4(z) \in \mathscr{V}_1$; say $\mathbf{a}(z) \in \mathscr{V}^{1,4}$ if both $[a_1(z), a_4(z)]$ and $[a_2(z), a_3(z)]$ are $\mathscr{V}_0$ pairs. We also say $\mathbf{b}(z) \in \mathscr{V}*^{0,4}$ if $\mathbf{b}(1/z) \in \mathscr{V}_{0,4}$. Note that, if $\mathbf{x}(z)$ is the polyphase form of an asymmetry filter in the case of $M = 4, c = 0$, then $[x^{[0]}(z), x^{[-1]}(z), x^{[1]}(z), x^{[2]}(z)] \in \mathscr{V}^{0,4}$.

3.3.1 The Case of a $\in \mathscr{V}^{0,4}$

Definition 9 Let $q(z) \in \mathscr{L}, s(z) \in \mathscr{V}_1, t(z) \in \mathscr{V}_{-1}$. The followings are called elementary $\mathscr{S}^{0,4}$ matrices:

$$E_h^0(q) = \begin{bmatrix} 1 & q(z) & q(1/z) & 0 \\ 0 & 1 & 0 & 0 \\ 0 & 0 & 1 & 0 \\ 0 & 0 & 0 & 1 \end{bmatrix}, \quad E_v^0(q) = \left(E_h^0(q)\right)^T, \quad E_t(s) = \begin{bmatrix} 1 & 0 & 0 & s(z) \\ 0 & 1 & 0 & 0 \\ 0 & 0 & 1 & 0 \\ 0 & 0 & 0 & 1 \end{bmatrix},$$

$$E_h^1(q) = \begin{bmatrix} 1 & 0 & 0 & 0 \\ 0 & 1 & 0 & 0 \\ 0 & 0 & 1 & 0 \\ 0 & q(z) & zq(1/z) & 1 \end{bmatrix}, \quad E_v^1(q) = \left(E_h^1(q)\right)^T, \quad E_b(t) = \begin{bmatrix} 1 & 0 & 0 & 0 \\ 0 & 1 & 0 & 0 \\ 0 & 0 & 1 & 0 \\ t(z) & 0 & 0 & 1 \end{bmatrix}.$$

Using the elementary $\mathscr{S}^{0,4}$ matrices, we can obtain the following chain:

$$\mathbf{a}^0 \xrightarrow{E_1} \mathbf{a}^1 \xrightarrow{E_2} \mathbf{a}^2 \cdots \xrightarrow{E_n} \mathbf{a}^n,$$

where $\mathbf{a}^n$ has the same symmetry as $\mathbf{a}$ and $\gcd_{\mathscr{L}}(\mathbf{a}^n) = 1$. Therefore,

$$\mathbf{a}^n = [0, p(z), \varepsilon p(1/z), 0], p(z) \in \mathscr{L}_d.$$

If $\varepsilon = 1$, it possibly can also have the form $(\mathbf{a})^n(z) = [r, 0, 0, 0], r \neq 0$. Let $q(z) \in \mathscr{L}_d$ satisfy $p(z)q(z) + p(1/z)q(1/z) = 1$. Write $E = E_1 E_2 \cdots E_n$ and define

$$Q_1(z) = \begin{bmatrix} 1/r & 0 & 0 & 0 \\ 0 & 1/2 & 1/2 & 0 \\ 0 & -1/2 & 1/2 & 0 \\ 0 & 0 & 0 & 1 \end{bmatrix} \quad Q_2(z) = \begin{bmatrix} 0 & 1 & 0 & 0 \\ q(z) & 0 & 0 & -\varepsilon p(1/z) \\ \varepsilon q(1/z) & 0 & 0 & p(z) \\ 0 & 0 & 1 & 0 \end{bmatrix},$$

whose inverses are

$$Q_1^{-1}(z) = \begin{bmatrix} r & 0 & 0 & 0 \\ 0 & 1 & -1 & 0 \\ 0 & 1 & 1 & 0 \\ 0 & 0 & 0 & 1 \end{bmatrix} \quad Q_2^{-1}(z) = \begin{bmatrix} 0 & p(z) & \varepsilon p(z) & 0 \\ 1 & 0 & 0 & 0 \\ 0 & 0 & 0 & 1 \\ 0 & -\varepsilon q(1/z) & q(z) & 0 \end{bmatrix}.$$

Set13sps3,

$$A(z) = \begin{cases} Q_1^{-1}(z)E^{-1}(z), & \text{if } \mathbf{a}^n = [r, 0, 0, 0], \\ Q_2^{-1}(z)E^{-1}(z), & \text{if } \mathbf{a}^n = [0, p(z), \varepsilon p(1/z), 0]. \end{cases} \tag{11}$$

Then $A(z)$ is a SLPME of $\mathbf{a}$.

We now consider SPLME Problem 2. In the given dual pair, $\mathbf{b}(z)$ is now in $\mathscr{V}*^{0,4}$. Let $E(z)$ be the LP matrix above. Then the LP vector $\mathbf{w} = \mathbf{b}(z)(E^{-1})^T(z)$ is in $\mathscr{V}*^{0,4}$ too. Hence, if $\mathbf{a}^n = [r, 0, 0, 0]$,

$$\mathbf{w}(z) = [1/r, v(z), \varepsilon v(1/z), v_{-1}(z)], \quad v_{-1}(z) \in \mathscr{V}_{-1},$$

else if $\mathbf{a}^n = [0, p(z), \varepsilon p(1/z), 0]$,

$$\mathbf{w}(z) = [v_0(z), v(z), \varepsilon v(1/z), v_{-1}(z)], \quad v_0(z) \in \mathscr{V}_0, v_{-1}(z) \in \mathscr{V}_{-1},$$

where $p(z)v(z) + p(1/z)v(1/z) = 1$. Let

$$Q_1(z) = \begin{bmatrix} 1/r & 0 & 0 & 0 \\ v(z) & 1/2 & 1/2 & 0 \\ \varepsilon v(1/z) & -1/2 & 1/2 & 0 \\ v_{-1}(z) & 0 & 0 & 1 \end{bmatrix} \quad Q_2(z) = \begin{bmatrix} v_0(z) & 1 & 0 & 0 \\ v(z) & 0 & 0 & -\varepsilon p(1/z) \\ \varepsilon v(1/z) & 0 & 0 & p(z) \\ v_{-1}(z) & 0 & 1 & 0 \end{bmatrix},$$

whose inverses are

$$Q_1^{-1}(z) = \begin{bmatrix} r & 0 & 0 & 0 \\ -w^+(z) & 1 & -1 & 0 \\ -w^-(z) & 1 & 1 & 0 \\ -v_{-1}(z) & 0 & 0 & 1 \end{bmatrix}, \quad Q_2^{-1}(z) = \begin{bmatrix} 0 & p(z) & \varepsilon p(z) & 0 \\ 1 & -p(z)v_0(z) & -\varepsilon p(1/z)v_0(z) & 0 \\ 0 & -p(z)v_{-1}(z) & -\varepsilon p(1/z)v_{-1}(z) & 1 \\ 0 & -\varepsilon v(1/z) & v(z) & 0 \end{bmatrix},$$

where $w^+(z) = s(z) + \varepsilon s(1/z)$ and $w^-(z) = s(z) - \varepsilon s(1/z)$. Let $A(z)$ be given by (11). Then $[A(z), A^{-1}(z)]$ is a SLPME of the dual pair $[\mathbf{a}(z), \mathbf{b}(z)]$.

3.3.2 The Case of a $\in \mathscr{V}^{1,4}$

Let $P = \begin{bmatrix} 0 & 1 & 0 & 0 \\ 1 & 0 & 0 & 0 \\ 0 & 0 & 0 & 1 \\ 0 & 0 & 1 & 0 \end{bmatrix}$ be the permutation matrix.

Definition 10 Let $q(z) \in \mathscr{L}$. The matrix with the form of

$$E_1(q) = \begin{bmatrix} 1 & q(z) & q(1/z) & 0 \\ 0 & 1 & 0 & 0 \\ 0 & 0 & 1 & 0 \\ 0 & 0 & 0 & 1 \end{bmatrix}$$

and $E_2(q) = P E_1(q) P$ are called elementary $\mathscr{S}^{1,4}$ matrices.

Using the elementary $\mathscr{S}^{1,4}$ matrices, we can obtain the following chain:

$$\mathbf{a}^0 \xrightarrow{E_1} \mathbf{a}^1 \xrightarrow{E_2} \mathbf{a}^2 \cdots \xrightarrow{E_n} \mathbf{a}^n,$$

where $\mathbf{a}_n(z) = [p(z), 0, 0, \varepsilon p(1/z)]$ with $p(z) \in \mathscr{L}_d$ or

$$\mathbf{a}^n = [0, p(z), \varepsilon p(1/z), 0], \ p(z) \in \mathscr{L}_d.$$

Since $[p(z), 0, 0, \varepsilon p(1/z)] = [0, p(z), \varepsilon p(1/z), 0]P$, we only discuss the first case. Let $q(z) \in \mathscr{L}_d$ satisfy $p(z)q(z) + p(1/z)q(1/z) = 1$. Write $E = E_1 E_2 \cdots E_n$ and define

$$Q(z) = \begin{bmatrix} q(z) & 0 & 0 & -\varepsilon p(z) \\ 0 & 1/2 & 1/2 & 0 \\ 0 & -1/2 & 1/2 & 0 \\ \varepsilon q(z) & 0 & 0 & p(z) \end{bmatrix},$$

whose inverse is

$$Q^{-1}(z) = \begin{bmatrix} p(z) & 0 & 0 & \varepsilon p(1/z) \\ 0 & 1 & -1 & 0 \\ 0 & 1 & 1 & 0 \\ -\varepsilon q(1/z) & 0 & 0 & q(z) \end{bmatrix}.$$

The matrix $A(z) = Q^{-1}(z)E^{-1}(z)$ is a SLPME of $\mathbf{a}(z)$.

We now consider SPLME Problem 2. In the given dual pair, $\mathbf{b}(z) \in \mathscr{V}^{1,4}$. Let $E(z)$ be the LP matrix above. We still assume that $\mathbf{a}_n = [p(z), 0, 0, \varepsilon p(1/z)], p(z) \in \mathscr{L}_d$. Then the LP vector $\mathbf{w}(z) = \mathbf{b}(z)(E^{-1})^T(z)$ has the form of

$$\mathbf{w}(z) = [v(z), s(z), \varepsilon s(1/z), \varepsilon v(1/z)], \quad v(z) \subset \mathscr{L}_d,$$

where $p(z)v(z) + p(1/z)v(1/z) = 1$.

Let $w^+(z) = s(z) + \varepsilon s(1/z)$, $w^-(z) = s(z) - \varepsilon s(1/z)$, and

$$Q(z) = \begin{bmatrix} v(z) & 0 & 0 & -\varepsilon p(1/z) \\ -v(z)w^+(z) & 1/2 & -1/2 & \varepsilon p(1/z)w^+(z) \\ v(z)w^-(z) & 1/2 & 1/2 & -\varepsilon p(1/z)w^-(z) \\ \varepsilon v(1/z) & 0 & 0 & p(z) \end{bmatrix},$$

whose inverse is

$$Q^{-1}(z) = \begin{bmatrix} p(z) & 0 & 0 & \varepsilon p(1/z) \\ s(z) & 1 & 1 & 0 \\ \varepsilon s(1/z) & -1 & 1 & 0 \\ -\varepsilon v(1/z) & 0 & 0 & p(z) \end{bmatrix}.$$

Define $A(z) = Q^{-1}(z)E^{-1}(z)$. Then $[A(z), A^{-1}(z)]$ is a SLPME of the dual pair $[\mathbf{a}(z), \mathbf{b}(z)]$.

4   Illustrative Examples

In this section, we present two examples to the readers for demonstrating the SLPME algorithm we developed in the previous section.

Example 11 (Construction of 3-band symmetric PRFB)

Let H_0 and B_0 be two given low-pass symmetric filters with the z-transforms

$$H_0(z) = \left(\frac{z^{-1} + 1 + z}{3} \right)^2$$

and

$$B_0(z) = \frac{1}{27}(z^{-1} + 1 + z)^2(-4z + 11 - 4z^{-1})$$

We want to construct the 3-band symmetric PRFB $\{H_0, H_1, H_2\}$, $\{B_0, B_1, B_2\}$, which satisfies

$$\sum_{j=0}^{2} B_j(\uparrow 3)(\downarrow 3)\overline{H}_j = frac13I,$$

Their polyphase forms are the following:

$$[\bar{H}_0^{[0]}(z), \bar{H}_0^{[1]}(z), \bar{H}_0^{[2]}(z)] = \frac{1}{9}[2 + z, 3, 2 + 1/z]$$

$$[B_0^{[0]}(z), B_0^{[1]}(z), B_0^{[2]}(z)] = \left[\frac{2}{9} + \frac{1}{9z}, \frac{-4z + 17 - 4z^{-1}}{27}, \frac{2 + z}{9} \right]$$

To normalize them, we set

$$\mathbf{a} = [\overline{H}_0^{[0]}, \overline{H}_0^{[1]}, \overline{H}_0^{[2]}] = \frac{1}{3}[2 + z, 3, 2 + 1/z]$$

and

$$\mathbf{b} = 3[B_0^{[0]}, B_0^{[1]}, B_0^{[2]}] = \left[\frac{2}{3} + \frac{1}{3z}; \frac{-4z + 17 - 4z^{-1}}{9}; \frac{2 + z}{3}\right]$$

so that $\mathbf{a}\mathbf{b}^T = 1$. We now use elementary $\mathscr{S}^{0,3}$ matrix decomposition technique. Let

$$E(z) = \begin{bmatrix} 1 & 0 & 0 \\ -\frac{2+z}{3} & 1 & -\frac{2+1/z}{3} \\ 0 & 0 & 1 \end{bmatrix}.$$

We have $\mathbf{a}^1(z) = \mathbf{a}(z)E(z) = [0, 1/3, 0]$. To make the SLPME for $\mathbf{a}(z)$, we set

$$Q(z) = \begin{bmatrix} 0 & 1/2 & -1/2 \\ 3 & 0 & 0 \\ 0 & 1/2 & 1/2 \end{bmatrix}.$$

Then the LP matrix

$$A(z) = Q^{-1}(z)E^{-1}(z) = \begin{bmatrix} \frac{2+z}{9} & 1/3 & \frac{2+1/z}{9} \\ 1 & 0 & 1 \\ -1 & 0 & 1 \end{bmatrix}$$

is a SLPME for $\mathbf{a}$. To obtain the SLPME for the dual pair $[\mathbf{a}, \mathbf{b}]$, we compute $\mathbf{w}(z) = \mathbf{b}(z)(E^{-1}(z))^T = \left[\frac{2+1/z}{3}, 3, \frac{2+z}{3}\right]$, which yields

$$Q(z) = \begin{bmatrix} \frac{2+1/z}{3} & 1 & 1 \\ 3 & 0 & 0 \\ \frac{2+z}{3} & -1 & 1 \end{bmatrix},$$

where $Q(:, 1) = \mathbf{w}^T$. Finally, we have

$$A(z) = (E(z)Q(z))^{-1} = \begin{bmatrix} \frac{2+z}{9} & \frac{1}{3} & \frac{2+1/z}{9} \\ -\frac{-2+26z+2z^2+z^3}{54z} & -\frac{-1+z^2}{18z} & -\frac{1+2z+26z^2-2z^3}{54z^2} \\ -\frac{2-18z+6z^2+z^3}{54z} & -\frac{1+4z+z^2}{18z} & -\frac{1+6z-18z^2+2z^3)}{54z^2} \end{bmatrix}$$

and

$$B(z) = E(z)Q(z) = \begin{bmatrix} \frac{2+1/z}{3} & 1 & 1 \\ \frac{4/z+17-4z}{9} & \frac{-1/z+z}{3} & -\frac{1/z+4+z}{3} \\ \frac{2+z}{3} & -1 & 1 \end{bmatrix},$$

which are SLPME of $(\mathbf{a}(z), \mathbf{b}(z))$.

Recovering the filters from their polyphases and applying the normalization factor $\frac{1}{3}$ to $B(z)$, we have the following z-transforms for the 3-band PRFB:

$$H_0(z) = \left(\frac{z^{-1} + 1 + z}{3}\right)^2,$$

$$H_1(z) = \frac{(1 - z)^3(1 + 5z + 15z^2 + 33z^3 + 33z^4 + 15z^5 + 5z^6 + z^7)}{54z^5},$$

$$H_2(z) = -\frac{(z - 1)^2(1 + 4z + 10z^2 + 22z^3 + 16z^4 + 22z^5 + 10z^6 + 4z^7 + z^8)}{54z^5},$$

and

$$B_0(z) = \frac{1}{27}(z^{-1} + 1 + z)^2(-4z + 11 - 4z^{-1}),$$

$$B_1(z) = -\frac{1}{9z^3} + \frac{1}{3z} - \frac{z}{3} - \frac{z^3}{9},$$

$$B_2(z) = -\frac{1}{9z^3} + \frac{1}{3z} - \frac{4}{9} + \frac{z}{3} - \frac{z^3}{9}.$$

Example 12 (Symmetric LP matrix extension of 4 × 4 matrix)

Let

$$\mathbf{a} = \frac{1}{16}\left[-\frac{2}{z} + 12 - 2z, -\frac{1}{z} + 8 + z, \frac{1}{z} + 8 - z, 4 + \frac{4}{z}\right] \in \mathcal{V}^{0,4}$$

be a given LP independent vector. We first consider the **SLPME Problem** 1. The symmetric Euclidean divisions yields the matrices

$$S_1 = \begin{bmatrix} 1 & 0 & 0 & 0 \\ 0 & 1 & 0 & 0 \\ 0 & 0 & 1 & 0 \\ \frac{1}{2}(1+z) & 0 & 0 & 1 \end{bmatrix} \quad S_2 = \begin{bmatrix} 1 & 0 & 0 & 0 \\ 0 & 1 & 0 & 0 \\ 0 & 0 & 1 & 0 \\ 0 & -\frac{1}{4}(7+z) & -\frac{1}{4}(1+7z) & 1 \end{bmatrix}$$

$$S_3 = \begin{bmatrix} 1 & \frac{1}{2z} & \frac{2}{z} & 0 \\ 0 & 1 & 0 & 0 \\ 0 & 0 & 1 & 0 \\ 0 & 0 & 0 & 1 \end{bmatrix} \quad S_4 = \begin{bmatrix} 1 & 0 & 0 & -\frac{1+z}{4z} \\ 0 & 1 & 0 & 0 \\ 0 & 0 & 1 & 0 \\ 0 & 0 & 0 & 1 \end{bmatrix}.$$

and $\mathbf{a} = [1, 0, 0, 0]$. Let $E = S_1 S_2 S_3 S_4 =$

$$\begin{bmatrix} 1 & \frac{1}{2z} & \frac{z}{2} & -\frac{1+z}{4z} \\ 0 & 1 & 0 & 0 \\ 0 & 0 & 1 & 0 \\ \frac{1+z}{2} & \frac{1-6z-z^2}{4z} & \frac{-1-6z+z^2}{4} & \frac{-1+6z-z^2}{8z} \end{bmatrix}$$

and

$$Q = \begin{bmatrix} 1 & 0 & 0 & 0 \\ 0 & 1/2 & 1/2 & 0 \\ 0 & -1/2 & 1/2 & 0 \\ 0 & 0 & 0 & 1 \end{bmatrix}.$$

Then the SLPME of **a** is

$$A(z) = \begin{bmatrix} \frac{-1+6z-z^2}{8z} & \frac{-1+8z+z^2}{16z} & -\frac{1+8z-z^2}{16z} & \frac{1+z}{4z} \\ 0 & 1 & -1 & 0 \\ 0 & 1 & 1 & 0 \\ -\frac{1+z}{2} & \frac{7+z}{4z} & \frac{1+7z}{4} & 1 \end{bmatrix}.$$

We now solve the **SLPME Problem** 2. Let

$$\mathbf{b} = \frac{1}{16}\left[-\frac{1}{z} + 10 - z, -\frac{2}{z} + 6, 2z + 6, 4 + 4z\right].$$

Then $(\mathbf{a}, \mathbf{b})$ is a dual pair. We have $\mathbf{w} = \mathbf{b}(E^{-1})^T = \left[1, \frac{3z+1}{8z}, \frac{3+z}{8}, \frac{(1+z)^3}{4z}\right]$. Set

$$Q(z) = \begin{bmatrix} 1 & 0 & 0 & 0 \\ \frac{1+3z}{8} & 1/2 & 1/2 & 0 \\ \frac{3+z}{8} & -1/2 & 1/2 & 0 \\ \frac{(1+z)^3}{4z} & 0 & 0 & 1 \end{bmatrix}.$$

Then, the SLPME for the dual pair is
$A(z) = (E(z)Q(z))^{-1} =$

$$\begin{bmatrix} \frac{-1+6z-z^2}{8z} & \frac{-1+8z+z^2}{16z} & \frac{1+8z-z^2}{16z} & \frac{1+z}{4z} \\ \frac{1-6z+6z^3-z^4}{64z^2} & \frac{1-8z+126z^2+8z^3+z^4}{128z^2} & -\frac{1+8z+126z^2-8z^3+z^4}{128z^2} & \frac{(z-1)(z+1)^2}{32z^2} \\ \frac{1-32z^2+z^4}{64z^2} & -\frac{1-2z+80z^2-14z^3+z^4}{128z^2} & \frac{-1-14z+80z^2-2z^3+z^4}{128z^2} & -\frac{1+7z+7z^2+z^3}{32z^2} \\ \frac{1-3z-30z^2-30z^3-3z^4+z^5}{32z^2} & \frac{1-5z+90z^2-10z^3-11z^4-z^5}{64z^2} & \frac{-1-11z-10z^2+90z^3-5z^4+z^5}{64z^2} & -\frac{(z-1)^2(1+6z+z^2)}{16z^2} \end{bmatrix}$$

and

$$B(z) = E(z)Q(z) = \begin{bmatrix} \frac{-1+10z-z^2}{16z} & \frac{1-z^2}{4z} & \frac{1+z^2}{4z} & -\frac{1+z}{4z} \\ \frac{3z+1}{8z} & \frac{1}{2} & \frac{1}{2} & 0 \\ \frac{3+z}{8} & -1/2 & 1/2 & 0 \\ \frac{1+z}{4} & \frac{1-5z+5z^2-z^3}{8z} & \frac{1-7z-7z^2+z^3}{8z} & \frac{-1+6z-z^2}{8z} \end{bmatrix}.$$

References

1. Chui, C.K., Lian, J.-A.: Construction of compactly supported symmetric and antisymmetric orthonormal wavelets with scale = 3. Appl. Comput. Harmon. Anal. **2**, 21–51 (1995)
2. Chui, C., Han, B., Zhuang, X.: A dual-chain approach for bottom-up construction of wavelet filters with any integer dilation. Appl. Comput. Harmon. Anal. **33**(2), 204–225 (2012)
3. Crochiere, R., Rabiner, L.R.: Multirate Digital Signal Processing. Prentice-Hall, Englewood Cliffs (1983)
4. Goh, S., Yap, V.: Matrix extension and biorthogonal multiwavelet construction. Linear Algebra Appl. **269**, 139–157 (1998)
5. Han, B.: Matrix extension with symmetry and applications to symmetric orthonormal complex m-wavelets. J. Fourier Anal. Appl. **15**, 684–705 (2009)
6. Han, B., Zhuang, X.: Matrix extension with symmetry and its applications to symmetric orthonormal multiwavelets. SIAM J. Math. Anal. **42**, 2297–2317 (2010)
7. Han, B., Zhuang, Z.: Algorithms for matrix extension and orthogonal wavelet filter banks over algebraic number fields. Math. Comput. **12**, 459–490 (2013)
8. Mallat, S.: A Wavelet Tour of Signal Processing. Academic Press, San Diego (1998)
9. Shi, X., Sun, Q.: A class of m-dilation scaling function with regularity growing proportionally to filter support width. Proc. Amer. Math. Soc. **126**, 3501–3506 (1998)
10. Strang, G., Nguyen, T.: Wavelets and Filter Banks. Wellesley, Cambrige (1996)
11. Vaidyanathan, P.: Theory and design of M-channel maximally decimated quadrature mirror filters with arbitrary M, having the perfect-reconstruction property, IEEE Trans. Acoust. Speech. Signal Process. **35**(4), 476–492 (1987)
12. Vaidyanathan, P.: How to capture all FIR perfect reconstruction QMF banks with unimodular matrices. Proceedings of IEEE International Symposium on Circuits Systems, vol. 3, pp. 2030–2033 (1990)
13. Vaidyanathan, P.: Multirate Systems and Filter Banks. Prentice-Hall, Englewood Cliffs (1993)
14. Vetterli, M.: A theory of multirate filter banks. IEEE Trans. Acoust. Speech Signal Process. **35**(3), 356–372 (1987)
15. Wang, J.Z.: Euclidean algorithm for Laurent polynomial matrix extension. Appl. Comput. Harmon. Anal. (2004)
16. Zhuang, X.: Matrix extension with symmetry and construction of biorthogonal multiwavelets with any integer dilation. Appl. Comput. Harmon. Anal. **33**, 159–181 (2012)

Solution of Equation for Ruin Probability of Company for Some Risk Model by Monte Carlo Methods

Kanat Shakenov

Abstract The classical process of risk and the equation for ruin probability of some model (model S. Anderson) is considered. This model is solved by classical numerical methods and Monte Carlo methods.

1 Introduction

Risk. We are interested not so much in the outcome of a process as the associated quantitative characteristics. The risk can be described by a random variable. The meaning of the word "risk" is probabilistic in nature, therefore, we shall call the risk as arbitrary random variable. The set of all risks is denoted X (see [6–10]).

The risk portfolio. The risk portfolio P is said to be an arbitrary subset of X.
Insurance. Insurance is a transfer of risk from one carrier (the insured) to another (the insurance company, the insurer) for a fee, called the cost of insurance, tariff rates or premiums. The essence of insurance is to redistribute risk among multiple carriers; relatively homogeneous set of risks will be called *insurance portfolio*.

Insurance portfolio. The simplest insurance portfolio. The simplest insurance portfolio

$$P = \{X_1, \ldots, X_N\}$$

consists of N risks (random variables)

$$X_1, \ldots, X_N,$$

K. Shakenov (✉)
Al Farabi Kazakh National University, Almaty, Kazakhstan
e-mail: shakenov2000@mail.ru

© Springer International Publishing Switzerland 2016
G.A. Anastassiou and O. Duman (eds.), *Intelligent Mathematics II:*
Applied Mathematics and Approximation Theory, Advances in Intelligent Systems
and Computing 441, DOI 10.1007/978-3-319-30322-2_12

169

which are independent and identically distributed; X has a Bernoulli distribution

$$X = \begin{cases} 1, & \text{with probability } p, \\ 0, & \text{with probability } 1 - p. \end{cases}$$

For each risk insurance event may occur with probability of p, and loss due to the occurrence of the insured event is 1, and the same for all risks. The risk of portfolio

$$X = \sum_{i=1}^{N} X_i$$

has a binomial distribution with parameters

$$N, p : \mathbf{P}\{X = k\} = \binom{N}{k} p^k (1 - p)^{N-k}, k = 0, 1, \ldots, N.$$

The main numerical characteristics (expectation and variance) of this distribution is equal

$$\mathbf{E}X = Np \quad \text{and} \quad \mathbf{Var}X = Np(1 - p).$$

Simple insurance portfolio. Simple insurance portfolio $P = \{X_1, \ldots, X_N\}$ also includes N independent risks $X_1, \ldots, X_N$ and their distributions has form

$$X = \begin{cases} S_i, & \text{with probability } p, \\ 0, & \text{with probability } 1 - p. \end{cases}$$

For the i-th risk the insured event occurs with probability p, and the amount of loss due to the occurrence of this event is S_i, and not the same for different risks. An example is whole life insurance to the value of S_i, determined by the insurance sum of i-th contract in portfolio. The main numerical characteristics (expectation and variance) of this distribution is equal

$$\mathbf{E}X = Np\overline{S}_N \quad \text{and} \quad \mathbf{Var}X = Np(1 - p)\widehat{S}_N^2,$$

where

$$\overline{S}_N = N^{-1} \sum_{i=1}^{N} S_i, \; \widehat{S}_N^2 = N^{-1} \sum_{i=1}^{N} S_i^2.$$

Real insurance portfolio. Real insurance portfolio is a complication of a simple portfolio; arbitrary losses from the range $[0, S_i]$ are allowed. It consists of N independent risks $P = \{X_1, \ldots, X_N$, the probability of occurrence of the insured event of i-th risk is p, and the amount of loss caused by the insured event is described by a random variable

$$X_i = \xi_i r_i S_i,$$

where

$$\xi_i = \begin{cases} 1, & \text{with probability } p, \\ 0, & \text{with probability } 1 - p. \end{cases}$$

is an indicator of the insured event for i-th risk, S_i is the insured amount (liability) by i-th risk, and $r_1, \ldots, r_N$ is a collection of independent and identically distributed random variables with distribution function $F_r(u) = \mathbf{P}\{r_1 \le u\}$. Here the distribution of portfolio risk has no simple explicit expression, but with known parameters of distribution r_1, $m = \mathbf{E}r_1$, $\tau^2 = \mathbf{Var}r_1$ you can calculate basic numerical characteristics of portfolio risk

$$\mathbf{E}X = Npm\overline{S}_N \quad \text{and} \quad \mathbf{Var}X = Npm^2\widehat{S}_N(1 - p + \tau^2/m^2).$$

The price of insurance. One of the main objectives of risk theory is to determine the price that must be paid when the transfer of risk from one carrier to another is executed. The insurance premium is usually expressed as a fraction of the sum insured (liability) S_i of the corresponding risk. Thus, when the amount of the premium T (same for all portfolio risks) the absolute amount of the premium of i-th risk is equal to TS_i, and the total premium of portfolio is $Q = TNS_N$.

The principles for determining the price. The principle of non-risk. Let's attempt to put a price such that the insurance company is non-risk, that is to collect premiums Q which is enough to cover all insurance claims in portfolio with probability 1. In the simplest case the maximum size of the portfolio loss is equal to N, and the probability of its occurrence: $p^N > 0$, so the following equality $Q = TN = N$ is required, hence $T = 1$, that is, the absolute size of the award coincides with the responsibility for the risk. It is clear that such insurance is quite unattractive to insurers, and its consideration is meaningless. It is easy to see that this conclusion is valid for more complex risk portfolios. The conclusion: non-risk management of insurance business is impossible. In particular, the premium must satisfy inequality $T < 1$.

The principle of justice. Now lets provide "justice" of the transfer of risks, that is the equivalence of financial liabilities of partners. Due that fact that the size of the insurance premium (financial liability of insurer) determined, and the amount of liabilities of the insurer (reimbursable insurance loss) is random, we will understand the equality of these liabilities on average within the portfolio: $TN = \mathbf{E}X$, hence, taking into account $\mathbf{E}X = Np$, obtain $T = p$. Such amount of the premium is too low, because the multiple reproduction of the insurance portfolio with a probability of 1 causes ruin the insurance company. Let's get an illustration of this fact. The question arises: what will be the profit of the insurer after m reproductions of portfolio formed by justice principles. Profit of j-th portfolio is a random variable

$$Z^{(j)} = Q - X^{(j)}$$

with

$$\mathbf{E}Z^{(j)} = 0 \quad \text{and} \quad \mathbf{Var}\,Z^{(j)} = \sigma^2 > 0.$$

Therefore, needed profit is $Z_m = \sum_{j=1}^{m} Z^{(j)}$, and $\mathbf{E}Z^{(m)} = 0$ and (in the case of independent portfolios) $\mathbf{Var}\,Z_m = m\sigma^2$, i.e. profit of m portfolios on the average is 0, but the uncertainty in its value increases with m, in particular, it can achieve an arbitrarily small value, leading to ruin of the company. Thus, the premium must satisfy inequality $T > p$. For more complex portfolios premium should exceed the size of average relative loss of portfolio $\mathbf{E}(X/R)$, where $R = \sum Z_i$ is overall liability for the portfolio.

The principle of sufficient coverage. We have proved that the last two of the principles of calculation of the premium are inefficient. We should look for other principles that lead to values $T \in (0, 1)$ (for the simplest portfolio). Now consider the principle of sufficient coverage, the essence of which is as follows: as a unit probability of covering of future losses in portfolio X with premiums Q fails, try to provide the given value of this probability: fix number $\alpha \in (0, 1)$ and will determine the premium T from equation $\mathbf{P}\{X \le Q\} = \alpha$. Let F is the distribution function of portfolio risk: $F(x) = \mathbf{P}\{X \le x\}$, F_0 is distribution function corresponding to the centered and normalized random variable $(X - \mathbf{E}X)/\sqrt{\mathbf{Var}\,X}$. Then the preceding equation reduces to

$$F_0\left((Q - \mathbf{E}X)/\sqrt{\mathbf{Var}\,X}\right) = \alpha,$$

where, taking into account $Q = TN\overline{S}_N$ and $\mathbf{E}X = Np$, $\mathbf{Var}\,X = Np(1-p)$. We get

$$T = p + \sqrt{\frac{p(1-p)}{N}}\,F_0^{-1}(\alpha) = p\left(1 + \sqrt{\frac{1-p}{pN}}\,F_0^{-1}(\alpha)\right).$$

In the case of a large volume of portfolio N the reference to the central limit theorem allows us to rewrite the last expression in the form

$$T = p + \sqrt{\frac{p(1-p)}{N}}\,\varPhi^{-1}(\alpha) = p\left(1 + \sqrt{\frac{1-p}{pN}}\,\varPhi^{-1}(\alpha)\right),$$

where $\varPhi$ is standard normal distribution function. From

$$F_0\left((Q - \mathbf{E}X)/\sqrt{\mathbf{Var}\,X}\right) = \alpha$$

using

$$\mathbf{E}X = Np\overline{S}_N, \ \mathbf{Var}X = Np(1-p)\widehat{S}_N^2,$$

and

$$\mathbf{E}X = Npm\overline{S}_N, \ \mathbf{Var}X = Npm^2\widehat{S}_N(1-p+\tau^2/m^2).$$

Similarly, we obtain the expression of the insurance premium for a simple portfolio

$$T = p + \frac{\widehat{S}_N}{\overline{S}_N}\sqrt{\frac{p(1-p)}{N}}F_0^{-1}(\alpha) = p\left(1 + \frac{\widehat{S}_N}{\overline{S}_N}\sqrt{\frac{1-p}{pN}}F_0^{-1}(\alpha)\right)$$

and real portfolio

$$T = pm\left(1 + \frac{\widehat{S}_N}{\overline{S}_N}\sqrt{\frac{1-p+\tau^2/m^2}{pN}}F_0^{-1}(\alpha)\right).$$

Here, the distribution function F_0 may also be replaced by a function of the standard normal distribution in the case of large volume portfolio. It should be noted that this last formula is recommended to the Russian and Kazakhstan insurers by corresponding standards for the calculation of insurance premiums for all types of insurance other than life insurance.

2 Risk Process

Classic risk process. Classic risk process has been studied for over a century [5]. The equation of this process is described by a dynamic portfolio of insurance companies, banks, other financial institutions, that are redistributors of financial flows in a risky environment. Consider the definition of risk on the example of an insurance company. Let premiums received forms steady stream with the intensity $\widetilde{c}$, and at random times $0 < T_1 < T_2 < \ldots$ insurance events occur, causing damage to a random size $Z_1, Z_2, \ldots$, respectively. Then capital size in time t, provided that the initial capital (at time $T_0 = 0$) is equal to x, described by the expression

$$\widetilde{X}(t) = x + \widetilde{c}t - \sum_{i=1}^{N(t)} \widetilde{Z}_i, \tag{1}$$

where $N(t) = \max\{k : T_k \leq t\}$ is the number of insured events occurring during the time interval $[0, t]$. Due that the moments of time $T_i, \ i = 1, 2, \ldots$ are random,

we can consider time intervals between consecutive insurance events also random
$\theta_i = T_{i+1} - T_i \geq 0$.

Stochastic process $\widetilde{X}(t)$ it called **classical risk process**, if the random variables
$\theta_i, i = 1, 2, \ldots$ are independent, identically distributed and have exponential distribution with parameter $\lambda > 0$:

$$F_\theta(u) = \mathbf{P}\{\theta_1 \leq u\} = 1 - \exp(-\lambda u), \ u \geq 0,$$

random variables $\widetilde{Z}_i, i = 1, 2, \ldots$ are also independent, identically distributed and
have a distribution function

$$\widetilde{F}_{\widetilde{Z}}(u) = \mathbf{P}\{\widetilde{Z}_1 \leq u\}, \ u \geq 0; \ \widetilde{F}_{\widetilde{Z}}(0) = 0.$$

The number of insurance events $N(t)$ has a Poisson distribution with a parameter λt:

$$\mathbf{P}\{N(t) = k\} = \frac{(\lambda t)^k}{k!} \exp(-\lambda t), k = 0, 1, \ldots,$$

and accumulated amount of insurance losses $\widetilde{Z}_{[0,t]} = \sum_{i=1}^{N(t)} \widetilde{Z}_i$ on time interval $[0, t]$
is a random variable with so-called compound Poisson distribution, the distribution
function has the form

$$\mathbf{P}\{\widetilde{Z}_{[0,t]} \leq u\} = \sum_{k=0}^{\infty} \frac{(\lambda t)^k}{k!} \exp(-\lambda t) \widetilde{F}_{\widetilde{Z}}^{*k}(u), \ u \geq 0,$$

where $\widetilde{F}_{\widetilde{Z}}^{*k}$ means k-times convolution of the distribution function $\widetilde{F}_{\widetilde{Z}}$ with itself, that
is distribution function of k independent and identically distributed random variables
with distribution function $\widetilde{F}_{\widetilde{Z}}$. When it is necessary to emphasize the dependence of
the process value $\widetilde{X}(t)$ from random argument $\omega \in \Omega$, use designation $\widetilde{X}(\omega, t)$, in
particular, separate trajectory of the process with fixed ω designated as

$$\widetilde{X}_\omega = \{\widetilde{X}(\omega, t), \ t \geq 0\}.$$

Classical risk process is completely determined by the values of the four parameters
$(x, \widetilde{c}, \lambda, \widetilde{F}_{\widetilde{Z}})$, satisfying $x \geq 0, \widetilde{c} > 0, \lambda > 0, \widetilde{F}_{\widetilde{Z}}(0) = 0$. Arbitrary classical risk
process with fixed parameter values that satisfy the above conditions is denoted by

$$\widetilde{X} = \widetilde{X}(x, \widetilde{c}, \lambda, \widetilde{F}_{\widetilde{Z}}),$$

and the set of all classical processes with such parameters is

$$\widetilde{\aleph} = \{\widetilde{X}(x, \widetilde{c}, \lambda, \widetilde{F}_{\widetilde{Z}}) : x \geq 0, \widetilde{c} > 0, \lambda > 0, \widetilde{F}_{\widetilde{Z}}(0) = 0\}.$$

(see [6–10]).

Ruin process. The **ruin** of process $\widetilde{X}(t)$ is understood as achieving to level 0, that is an event

$$\widetilde{\Re}(x) = \Re\left(x, \widetilde{c}, \lambda, \widetilde{F_{\widetilde{Z}}}\right) = \{\omega \in \Omega : \exists t \geq 0, \ \widetilde{X}(x,t) \leq 0\}.$$

The moment of ruin is random variable

$$\widetilde{\tau}(x) = \widetilde{\tau}\left(x, \widetilde{c}, \lambda, \widetilde{F_{\widetilde{Z}}}\right) = \min\{t : \widetilde{X}(t) \leq 0\}.$$

This random value depends on the process parameters $\widetilde{X}(t)$ and it may be improper, with positive probability of taking the value ∞; this situation corresponds to trajectories that do not bankrupt on the entire time $[0, \infty)$ [5].
The probability of ruin $\widetilde{X}(t)$ it is the value

$$\mathbf{P}\{\widetilde{\tau}(x) < \infty\} = \mathbf{P}\left(\widetilde{\Re}(x)\right),$$

that is a probability measure of the set of trajectories that go bankrupt in a finite time. This value is also a function of process parameters that emphasized by designation

$$\widetilde{R}(x) = \widetilde{R}\left(x, \widetilde{c}, \lambda, \widetilde{F_Z}\right) = \mathbf{P}\{\widetilde{\tau}(x) < \infty\}.$$

In some cases, more convenient characteristic of the process risk is the probability of **survival** of the process

$$\widetilde{S}(x) = \widetilde{S}\left(x, \widetilde{c}, \lambda, \widetilde{F_Z}\right) = 1 - \widetilde{R}(x).$$

The dependence of the probability of ruin on the parameters of the process. Note the properties of monotony of ruin probability as a function of process parameters. 1. $\widetilde{R}$ is a non-increasing function of x ; 2. $\widetilde{R}$ it is a non-increasing function of $\widetilde{c}$; 3. It is a decreasing function of λ; 4. $\widetilde{R}$ it is a non-increasing function of $\widetilde{F_{\widetilde{Z}}}$, if the order on the set of distribution functions specified as

$$F_1 \preceq F_2 \Longleftrightarrow F_1(x) \leq F_2(x), \ \forall x.$$

Aggregate risk process. The operation of aggregation. Let's consider the classical risk process $\widetilde{X}(t)$, fix a number $\delta > 0$ and divide the positive ray $\mathbb{R}^+$ on intervals

$$\Delta_i = [(i-1)\delta, i\delta)$$

length δ. Further, we group all premium incomes and insurance losses that have occurred in these time intervals. Then the size of the premium income for any period Δ_i is equal to $c = \widetilde{c}\delta$. Size Z_i of insurance losses accumulated on the interval Δ_i is calculated as follows. For each i size Z_i is a random variable, and for $i = 1$ its distribution has already been calculated and is equal to

$$\mathbf{P}\{\widetilde{Z}_{[0,t]} \le u\} = \sum_{k=0}^{\infty} \frac{(\lambda t)^k}{k!} \exp(-\lambda t)\widetilde{F}_{\widetilde{Z}}^{*k}(u), \ u \ge 0,$$

we need only to substitute the value of the length of the time interval $t = \delta$:

$$F_Z(u) = \mathbf{P}\{Z_1 \le u\} = \mathbf{P}\{\widetilde{Z}_{[0,\delta)} \le u\} = \sum_{k=0}^{\infty} \frac{(\lambda\delta)^k}{k!} \exp(-\lambda\delta)\widetilde{F}_{\widetilde{Z}}^{*k}(u),$$

$u \ge 0$. Further, due to the steady-state flow of insurance events and independence and identically distributed loss of classical risk process

$$\{\widetilde{Z}_k, \ k = 1, 2, \ldots\}$$

amount of damages $\{Z_i, \ i = 1, 2, \ldots\}$ at intervals $\varDelta_i$ are also independent and identically distributed random variables with distribution function

$$F_Z(u) = \mathbf{P}\{Z_1 \le u\} = \mathbf{P}\{\widetilde{Z}_{[0,\delta)} \le u\} = \sum_{k=0}^{\infty} \frac{(\lambda\delta)^k}{k!} \exp(-\lambda\delta)\widetilde{F}_{\widetilde{Z}}^{*k}(u),$$

$u \ge 0$. Thus, the importance of classical risk process $\widetilde{X}(t)$ at time $n\delta$ for integer n is

$$X(n) = x + cn - \sum_{i=1}^{n} Z_i, \ n = 0, 1, \ldots. \tag{2}$$

This process is called **aggregate risk process** and is approximating classic model of risk in $\delta \to 0$. Here, this process depends on three parameters (x, c, F_Z) calculated on the parameters of the original classical process $(x, \widetilde{c}, \lambda, \widetilde{F}_{\widetilde{Z}})$. We denote $X = X(x, c, F_Z)$ aggregate risk process defined by parameters x, c, F_Z, and $\aleph$ is the set of all aggregated risk processes at all admissible values of the parameters:

$$\aleph = \{X(x, c, F_Z) \, ; x \ge 0, c > 0, F_Z(0) = 0\}.$$

(see [5, 8, 10]).

   Ruin. Let note

$$\Re(x, c, F_Z) = \Re(x) = \{\omega : \exists n > 0, X_\omega(n) \le 0\}$$

is a ruin event of aggregated process,

$$\tau(x) = \tau(x, c, F_Z) = \min\{n : X(n) \le 0\}$$

is moment of ruin

$$R(x) = R(x, c, F_Z) = \mathbf{P}\{\tau < \infty\},$$
$$S(x) = S(x, c, F_Z) = \mathbf{P}\{\tau = \infty\} = 1 - R(x)$$

is the probability of ruin (bankruptcy) and survival of the process, respectively. Considering the relation of inclusion of the ruin (survival) events for various values of the parameters of risk, we conclude that the probability of survival is a decreasing function of the initial capital x, the flow rate of the premium c and distribution function loss F_Z, if on the set of distribution functions we consider a natural partial order

$$F_1 \preceq F_2 \iff F_1(u) \le F_2(u), \ \forall u.$$

Random walk. The aggregate risk process by replacement

$$Y_i = c - Z_i, \ i = 1, 2, \ldots,$$

can be represented as a random walk, [5], $X(n) = x + \sum_{i=1}^{n} Y_i$ with independent and identically distributed "steps" Y_i, $i = 1, 2, \ldots$. Hence, using the theorem of [5], we deduce

Theorem 1 *The aggregate risk process*

$$X(n) = x + cn - \sum_{i=1}^{n} Z_i, \ n = 0, 1, \ldots$$

may belong to one and only one from three types depending on the relation between its parameters:

1. $c = \mathbf{E}Z_1$: *oscillatory type; processes of this type with probability one reach any given level; more precisely:*

$$\mathbf{P}\left\{\inf_{n \in \mathbb{N}} X(n) = -\infty\right\} = 1 \quad and \quad \mathbf{P}\left\{\inf_{n \in \mathbb{N}} X(n) = \infty\right\} = 1.$$

2. $c < \mathbf{E}Z_1$: *ruining type; for this type of process*

$$\mathbf{P}\left\{\inf_{n \in \mathbb{N}} X(n) = -\infty\right\} = 1$$

and with probability 1 *there is a finite maximum* $\mathbf{M}(x) = \max_{n \in \mathbb{N}} X(n)$.

3. $c > \mathbf{E}Z_1$: *surviving type; for this type of process*

$$\mathbf{P}\left\{\inf_{n \in \mathbb{N}} X(n) = \infty\right\} = 1$$

and with probability 1 *there is a finite minimum* $\mathbf{m}(x) = \min_{n \in \mathbb{N}} X(n)$.

Proof Similar theorem is proved in [5].

According to the theorem it is clear that the processes of the first type

$$\mathbf{P}\left\{\inf_{n\in\mathbb{N}} X(n) = -\infty\right\} = 1$$

destructed with probability 1. Because the processes of the second type

$$\mathbf{P}\left\{\inf_{n\in\mathbb{N}} X(n) = -\infty\right\} = 1$$

is also true, they are busting with probability 1. For the third type of process the probability of survival

$$S(x) = \mathbf{P}\{\mathbf{m}(x) > 0\}$$

and can be, in general, positive. Moreover, since for arbitrary $x, y \geq 0$ holds

$$\mathbf{m}(x) = \mathbf{m}(y) + (x - y),$$

for G_x – distribution function of $\mathbf{m}(x)$ – we get

$$G_x(u) = \mathbf{P}\{\mathbf{m}(x) \leq u\} = \mathbf{P}\{\mathbf{m}(y) - y \leq u - x\}$$
$$= G_y(u + y - x) \to 0$$

at $x \to \infty$ and fixed u, y so the following is true.

Theorem 2 *Suppose that $c > \mathbf{E}Z_1$. Then* $\lim_{x\to\infty} S(x) = 1$.

Proof Similar theorem is proved in [5].

The condition $c > \mathbf{E}Z_1$ is called the condition of positive risk premium.

3 The Equation for the Probability of Ruin

Using the formula of total probability, it is easy to deduce the integral equation for the probability of survival

$$X(n) = x + cn - \sum_{i=1}^{n} Z_i, \ n = 0, 1, \ldots,$$

as a function of the initial capital. Let's fix parameters c, F_Z, and define the event "survival, with condition that initial capital is x":

$$S(x) = \{\omega \in \Omega : X(n) > 0, \ n = 0, 1, \ldots; \ X(0) = x\},$$

and for the interval I similar event of "survival, provided that the initial capital belong to the interval I ":

$$S(I) = \{\omega \in \Omega : X(n) > 0, \ n = 0, 1, \ldots; \ X(0) = I\}.$$

It is clear that $S(x) \subseteq S_1 = \{Z_1 < x + c\}$. For an arbitrary integer $m > 0$ we separate the interval $[0, x + c)$ on the m equal parts I_k of length

$$\gamma = (x + c)/m : I_k = \big[(k - 1)\gamma, \ k\gamma\big], \ k = 1, \ldots, m.$$

Then by the formula of total probability we have

$$\mathbf{P}\{S(x)\} = \sum_{k=1}^{m} S(x + c - I_k)\mathbf{P}\{Z_1 \in I_k\}$$

$$= \sum_{k=1}^{m} S(x + c - I_k) \left(F_Z(k\gamma) - F_Z((k - 1)\gamma)\right).$$

The right side of the last expression is an integral sum for the integral

$$\int_{0}^{x+c} S(x + c - u)d F_Z(u),$$

and converges to it when $m \to \infty$, and the left side is independent from m and is $S(x)$, so

$$S(x) = \int_{0}^{x+c} S(x + c - u)d F_Z(u). \tag{3}$$

This integral equation allows us to study many of the properties of the aggregate risk, expressed in terms of probability of survival or ruin (see also [8, 10]).

Example 3 The integral equation of survival for the simplest risk process. Let $c = 1$, $p \in (0, 1)$ and

$$F_Z(u) = \begin{cases} 0, \ u < 0, \\ p, \ 0 \le u < 2, \\ 1, \ u \ge 2. \end{cases}$$

Note that aggregate risk process

$$X(n) = x + cn - \sum_{i=1}^{n} Z_i, \ n = 0, 1, \ldots$$

turns into a simple process of risk $X_n = z + \sum_{i=1}^{n} Z_i$, where Z_i are independent and identically distributed random variables and the equation for the probability of survival takes the form

$$S(x) = S(x + 1) + (1 - p)S(x - 1).$$

The last equation is a difference equation of the second order and can be easily solved.

Remark 4 Type of the equation for ruin probability be up to risk process

$$X(n) = x + cn - \sum_{i=1}^{n} Z_i, \ n = 0, 1, \ldots,$$

i.e. from distribution function $F_Z(u)$ of random variables Z (from damages distribution).

4 Numerical Solutions of the Equation for Ruin Probability of Some Risk Model

In classical risk model (Model Anderson S.) with Poisson damages stream that enter an insurance company with intensity's λ and accumulation speed of payments c and with damages distribution $D(u)$ under condition $\dfrac{\lambda b}{c} < 1$, the equation is given as

$$R(x) = \frac{\lambda}{c} \int_0^x D(x - u) R(u) du + \frac{\lambda}{c} F(x), \tag{4}$$

where

$$F(x) = \int_x^\infty D(t) dt, \ b = \int_0^\infty t\, dD(t),$$

$R(x)$ stands for probability of ruin as a function of the initial capital $x \geq 0$, [1]. It is known that the function $R(x)$ is monotone decreasing to 0 as $x \to \infty$. If $F(x) x \geq 0$ is continuous then due to condition $\lambda b/c < 1$ and Contraction – Mapping Principle the integral equation for probability of ruin has continuous unique solution in $\mathbb{C}$ function class on interval $[0, \infty)$ (in space $\mathbb{C} [0, \infty)$). This is the case. The integral operator K is defined in form

$$(KR)(x) = \frac{\lambda}{c} \int_0^\infty R(u)D(x-u)du.$$

Then

$$\|KR\|_{\mathbb{C}} \le \|R\|_{\mathbb{C}} \sup_{0 \le x < \infty} \frac{\lambda}{c} \int_0^x D(x-u)du = \|R\|_{\mathbb{C}} \frac{\lambda b}{c}$$

from this it follows that $\|K\|_{\mathbb{C}} \le \dfrac{\lambda b}{c} < 1$.

This integral equation is not easy to be solved by "classical" methods when $D(x)$ is Pareto distribution,

$$D(x) \equiv P(x) = 0,\ 0 \le u \le \frac{\alpha - 1}{\alpha};$$

$$P(u) = 1 - \left(\frac{\alpha - 1}{\alpha u}\right)^\alpha,\ u > \frac{\alpha - 1}{\alpha}$$

with parameter $\alpha = 3$, with accuracy at most 0.03, $\lambda = 0.7$, $c = 1$, for $x > 500$.

It is known that equation (4) can be converted to Cauchy problem for second-order ODE. It can be solved by different classical numerical methods, for example, Euler's, Runge–Kutta and so on. But in this case, for large value of x ($x \gg 500$), error would grow. That is why we solve this integral equation for large x ($x \gg 500$) in one point by Monte Carlo methods (see [2–4, 11]. The calculating experiment was realized for parameters $\lambda = 0.8$, $c = 1$,

$$F(x) = \begin{cases} 1, & \text{for } x \le k, \\ \left(\dfrac{k}{x}\right)^\alpha, & \text{for } x > k \end{cases}$$

and also for Pareto distribution with parameters $\alpha = 2, 3, 5, 7$ and $k > 0$, $k = \dfrac{\alpha - 1}{\alpha}$. Results of numerical experiments are completely satisfactory.

References

1. Cincinashvili, G.S., Skvarnik, E.S.: Numerical solution of the mean-value problems in classical model of risk. In: Proceedings FAM 2002, Krasnoyarsk, ICM SB RAS, pp. 295–301 (2002) (In Russian)
2. Ermakov, S.M.: Method Monte Carlo and Adjacent Questions. Nauka, Moscow (1975). (In Russian)
3. Ermakov, S.M., Shakenov, K.K.: On the applications of the Monte Carlo method to Navier-Stokes equations. Bull. Leningrad State University, Series Mathematics, Mechanics, Astronomy, Deposited 6267–B86, pp. 1–14. (1986) (In Russian)

4. Ermakov, S.M., Shakenov, K.K.: On the applications of the Monte Carlo method to difference analogue Navier-Stokes equations. Bull. Leningrad State University, Series Mathematics, Mechanics, Astronomy, Deposited 8576–B86, pp. 1–9. (1986) (In Russian)
5. Feller, W.: An introduction to probability theory and its applications, vol. 1, John Wiley, New York (1970), vol. 2, John Wiley, New York (1971)
6. Lundberg, F.I.: Approximerad Framställning av Sannolikhetsfunktionen. II. Aterförsäkring av Kollektivrisker, Almqvist and Wiksell, Uppsala (1903)
7. MacMinn, R.D.: Notes on Pratt's risk aversion in the small and in the large, Lecture note, p. 6 http://kiwiclub.bus.utexas.edu/uncertainty/expected-utility/pratt.pdf
8. Novoselov, A.A.: Modeling of financial risks, the series of lectures for students of institute of mathematics Siberian federative university, Archives, Krasnoyarsk (1998) (In Russian)
9. Pratt, J.W.: Risk aversion in the small and in the large. J. Econometrica **32**, 122–136 (1964)
10. Shakenov, K.K.: Modeling of Financial Risks. The series of lectures. Al-Farabi Kazakh National University, Almaty (2015)
11. Sobol', I.M.: Monte Carlo Numerical Methods. Nauka, Moscow (1973). (In Russian)

Determinant Reprentation of Dardoux Transformation for the (2+1)-Dimensional Schrödinger-Maxwell-Bloch Equation

K.R. Yesmahanova, G.N. Shaikhova, G.T. Bekova
and Zh.R. Myrzakulova

Abstract In this article, we consider a (2+1)-dimensional Schrödinger-Maxwell-Bloch equation (SMBE). The (2+1)-dimensional SMBE is integrable by the inverse scattering method. We constructed Darboux transformation (DT) of this equation. Also, we derive determinant representation of one-fold, two-fold and n-fold DT of (2+1)-dimensional SMBE. As an application of these conversion of the (2+1)-dimensional SMBE, soliton solutions will get from trivial "seed" solutions.

1 Introduction

It is well known that the nonlinear nature of the real system is considered to be fundamental in modern science. Nonlinearity is the fascinating subject which has many applications in almost all areas of science. Usually nonlinear phenomena are modeled by nonlinear ordinary and/or partial differential equations. Many of these nonlinear differential equations (NDE) are completely integrable. This means that these integrable NDE have some classes of interesting exact solutions such as solitons, dromions, rogue waves, similaritons and so on. They are of great mathematical as well as physical interest and the investigation of the solitons and other its sisters have become one of the most exciting and extremely active areas of research in modern science and technology in the past several decades. In particular, many of the completely integrable NDE have been found and studied [1–15].

K.R. Yesmahanova (✉) · G.N. Shaikhova · G.T. Bekova · Zh. Myrzakulova
L. N. Gumilyov Eurasian National University, Astana, Kazakhstan
e-mail: kryesmakhanova@gmail.com

G.N. Shaikhova
e-mail: g.shaikhova@gmail.com

G.T. Bekova
e-mail: bekovaguldana@yahoo.com

Zh. Myrzakulova
e-mail: jaydary@mail.ru

© Springer International Publishing Switzerland 2016
G.A. Anastassiou and O. Duman (eds.), *Intelligent Mathematics II:
Applied Mathematics and Approximation Theory*, Advances in Intelligent Systems
and Computing 441, DOI 10.1007/978-3-319-30322-2_13

183

Among of such integrable nonlinear systems, the Schrödinger-Maxwell-Bloch equation (SMBE) plays an important role. The SMBE describes a soliton propagation in fibres with resonant and erbium-doped systems [16] and has the (1+1)-dimensions. The (1+1)-dimensional SMBE has been studied by Darboux transformation (DT) in [17], where the authors obtained the soliton and periodic solutions from the different "seeds".

Recently, the authors in [18, 19] presented the (2+1)-dimensional SMBE. In this paper, our aim is to construct the DT for the (2+1)-dimensional SMBE and to find its soliton solutions. It is well known that the DT is an efficient way to find different solutions for integrable equations [20]. In [21, 22], rogue waves, position solutions were obtained by the DT for one and two component Hirota-Maxwell-Bloch equation (HMBE). The determinant representation of the DT for the inhomogeneous HMBE was given in [23, 24], where the authors found the soliton, position solutions.

The paper is organized as follows. In Sect. 2, we present Lax representation of the (2+1)-dimensional SMBE. One-fold DT of the (2+1)-dimensional SMBE is constructed in Sect. 3. In Sect. 4, we give the determinant representation of the (2+1)-dimensional SMBE. In Sect. 5 the soliton solutions are obtained from the "seed" solutions. Here, in particular, we present one-soliton solutions. Finally, in Sect. 6 we present our conclusion.

2 Lax Representation of (2+1)-Dimensional Schrödinger-Maxwell-Bloch Equation

Here we consider the (2+1)-dimensional SMBE, which reads as

$$iq_t + q_{xy} - vq - 2ip = 0, \tag{1}$$

$$ir_t - r_{xy} + vr - 2ik = 0, \tag{2}$$

$$v_x + 2(rq)_y = 0, \tag{3}$$

$$p_x - 2i\omega p - 2\eta q = 0 \tag{4}$$

$$k_x + 2i\omega k - 2\eta r = 0, \tag{5}$$

$$\eta_x + rp + kq = 0, \tag{6}$$

where q, k, r, p are complex functions, η, v is a real function, ω is a real constant and subscripts x, y, t denote partial derivatives with respect to the variables. This system (1)–(6) is integrable by the Inverse Scattering Method [18].

Corresponding Lax representation for the (2+1)-dimensional SMBE (1)–(6) is given by

$$\Psi_x = A\Psi, \tag{7}$$

$$\Psi_t = 2\lambda\Psi_y + B\Psi, \tag{8}$$

where A and B matrices have the form

$$A = -i\lambda\sigma_3 + A_0,$$

$$B = B_0 + \frac{i}{\lambda + \omega}B_{-1}.$$

Here σ_3, A_0, B_0, B_1 are 2×2 matrices:

$$A_0 = \begin{pmatrix} 0 & q \\ -r & 0 \end{pmatrix}, \quad \sigma_3 = \begin{pmatrix} 1 & 0 \\ 0 & -1 \end{pmatrix},$$

$$B_0 = \begin{pmatrix} -0.5iv & iq_y \\ ir_y & 0.5iv \end{pmatrix}, \quad B_{-1} = \begin{pmatrix} \eta & -p \\ -k & -\eta \end{pmatrix}.$$

Let us consider the reductions $r = \delta q^*$, $k = \delta p^*$, where the asterisk symbol "$*$" means a complex conjugate and δ is the real constant. Then the system (1)–(6) takes the form

$$iq_t + q_{xy} - vq - 2ip = 0,$$

$$v_x + 2\delta(|q|^2)_y = 0,$$

$$p_x - 2i\omega p - 2\eta q = 0,$$

$$\eta_x + \delta(q^*p + p^*q) = 0,$$

where we can assume that the $\delta = \pm 1$. So that $\delta = +1$ corresponds to a attractive interaction and $\delta = -1$ corresponds to a repulsive interaction.

In the next section, we construct one-fold DT of the (2+1)-dimensional SMBE.

3 One-Fold Darboux Transformation for the (2+1)-Dimensional Schrödinger-Maxwell-Bloch Equation

We consider the following transformation of the system of Eqs. (7)–(8)

$$\Psi^{[1]} = T\Psi = (\lambda I - M)\Psi \tag{9}$$

such that

$$\Psi_x^{[1]} = A^{[1]}\Psi^{[1]}, \tag{10}$$

$$\Psi_t^{[1]} = 2\lambda \Psi_y^{[1]} + B^{[1]}\Psi^{[1]}, \tag{11}$$

where $A^{[1]}$ and $B^{[1]}$ depend on $q^{[1]}$, $v^{[1]}$, $p^{[1]}$, $\eta^{[1]}$ and λ. Here M and I are matrices have the form

$$M = \begin{pmatrix} m_{11} & m_{12} \\ m_{21} & m_{22} \end{pmatrix}, \quad I = \begin{pmatrix} 1 & 0 \\ 0 & 1 \end{pmatrix}. \tag{12}$$

The relation between $q^{[1]}$, $v^{[1]}$, $p^{[1]}$, $\eta^{[1]}$, λ and $A^{[1]} - B^{[1]}$ is the same as the relation between q, v, p, η, λ and $A - B$. In order the Eqs. (10)–(11) to hold, T must satisfy the following equations

$$T_x + TA = A^{[1]}T, \tag{13}$$

$$T_t + TB = 2\lambda T_y + B^{[1]}T. \tag{14}$$

Then the relation between q, v, p, η and $q^{[1]}$, $v^{[1]}$, $p^{[1]}$, $\eta^{[1]}$ can be reduced from these equations, which is in fact the DT of the (2+1)-dimensional SMBE. Comparing the coefficients of λ^i of the two sides of the Eq. (13), we get

$$\lambda^0 : \quad M_x = A_0^{[1]}M - MA_0, \tag{15}$$

$$\lambda^1 : \quad A_0^{[1]} = A_0 + i[M, \sigma_3], \tag{16}$$

$$\lambda^2 : \quad iI\sigma_3 = i\sigma_3 I. \tag{17}$$

Finally, from (15)–(17) we obtain

$$q^{[1]} = q - 2im_{12}, \quad r^{[1]} = r - 2im_{21}. \tag{18}$$

Hence we get $m_{21} = -m_{12}^*$ in our attractive interaction case, that is when $\delta = +1$. Then, comparing the coefficients of λ^i of the two sides of the Eq. (14) gives us

$$\lambda^0 : \quad -M_t = iB_{-1}^{[1]} - B_0^{[1]}M - iB_{-1} + MB_0, \tag{19}$$

$$\lambda^1 : \quad 2M_y = B_0^{[1]} - B_0, \tag{20}$$

$$(\lambda + \omega)^{-1} : \quad 0 = -i\omega B_{-1}^{[1]} - iB_{-1}^{[1]}M + i\omega B_{-1} + iMB_{-1}. \tag{21}$$

Then the system of Eqs. (19)–(21) gives

$$B_{-1}^{[1]} = (M + \omega I)B_{-1}(M + \omega I)^{-1}. \tag{22}$$

These Eqs. (18), (22) give one-fold transformation of the (2+1)-dimensional SMBE. We now assume that

$$M = H\Lambda H^{-1},$$

where

$$H = \begin{pmatrix} \psi_1(\lambda_1; t, x, y) & \psi_1(\lambda_2; t, x, y) \\ \psi_2(\lambda_1; t, x, y) & \psi_2(\lambda_2; t, x, y) \end{pmatrix} := \begin{pmatrix} \psi_{1,1} & \psi_{1,2} \\ \psi_{2,1} & \psi_{2,2} \end{pmatrix},$$

$$\Lambda = \begin{pmatrix} \lambda_1 & 0 \\ 0 & \lambda_2 \end{pmatrix} \tag{23}$$

and $\det H \neq 0$, where λ_1 and λ_2 are complex constants. In order to satisfy the constraints of A_0 and $B_{-1}^{[1]}$ as mentioned above, we first note that if $\delta = +1$, then

$$\Psi^+ = \Psi^{-1}, \quad A_0^+ = -A_0,$$

$$\lambda_2 = \lambda_1^*, \quad H = \begin{pmatrix} \psi_1(\lambda_1; t, x, y) & -\psi_2^*(\lambda_1; t, x, y) \\ \psi_2(\lambda_1; t, x, y) & \psi_1^*(\lambda_1; t, x, y) \end{pmatrix},$$

$$H^{-1} = \frac{1}{\Delta} \begin{pmatrix} \psi_1^*(\lambda_1; t, x, y) & \psi_2^*(\lambda_1; t, x, y) \\ -\psi_2(\lambda_1; t, x, y) & \psi_1(\lambda_1; t, x, y) \end{pmatrix}.$$

In the following section we give the determinant representation of the DT for the (2+1)-dimensional SMBE.

4 Determinant Representation of Darboux Transformation for the (2+1)-Dimensional Schrödinger-Maxwell-Bloch Equation

Here the determinant representation is constructed for the one-fold, two-fold and n-fold DT of the (2+1)-dimensional SMBE. The reduction condition on the eigenfunctions are $\Psi_{2,2i} = \Psi_{1,2i-1}^*$, $\Psi_{2,2i-1} = -\Psi_{1,2i}^*$ and for the eigenvalues are $\lambda_{2i} = -\lambda_{2i-1}^*$.

The determinant representation of the one-fold DT of the (2+1)-dimensional SMBE formulate the following theorem (as [21–24]).

Theorem 1 *The one-fold DT of the (2+1)-dimensional SMBE is*

$$T_1(\lambda, \lambda_1, \lambda_2) = \lambda I - M = \lambda I + t_0^{[1]} = \frac{1}{\Delta_1} \begin{pmatrix} (T_1)_{11} & (T_1)_{12} \\ (T_1)_{21} & (T_1)_{22} \end{pmatrix}, \tag{24}$$

where

$$t_0^{[1]} = \frac{1}{\Delta_1} \begin{pmatrix} \begin{vmatrix} \Psi_{2,1} & \lambda_1\Psi_{1,1} \\ \Psi_{2,2} & \lambda_2\Psi_{1,2} \end{vmatrix} - \begin{vmatrix} \Psi_{1,1} & \lambda_1\Psi_{1,1} \\ \Psi_{1,2} & \lambda_2\Psi_{1,2} \end{vmatrix} \\ \\ \begin{vmatrix} \Psi_{2,1} & \lambda_1\Psi_{2,1} \\ \Psi_{2,2} & \lambda_2\Psi_{2,2} \end{vmatrix} - \begin{vmatrix} \Psi_{1,1} & \lambda_1\Psi_{2,1} \\ \Psi_{1,2} & \lambda_2\Psi_{2,2} \end{vmatrix} \end{pmatrix},$$

$$\Delta_1 = \begin{vmatrix} \Psi_{1,1} & \Psi_{1,2} \\ \Psi_{2,1} & \Psi_{2,2} \end{vmatrix},$$

$$(T_1)_{11} = \begin{vmatrix} 1 & 0 & \lambda \\ \Psi_{1,1} & \Psi_{2,1} & \lambda_1\Psi_{1,1} \\ \Psi_{1,2} & \Psi_{2,2} & \lambda_2\Psi_{1,2} \end{vmatrix},$$

$$(T_1)_{12} = \begin{vmatrix} 0 & 1 & 0 \\ \Psi_{1,1} & \Psi_{2,1} & \lambda_1\Psi_{1,1} \\ \Psi_{1,2} & \Psi_{2,2} & \lambda_2\Psi_{1,2} \end{vmatrix},$$

$$(T_1)_{21} = \begin{vmatrix} 1 & 0 & 0 \\ \Psi_{1,1} & \Psi_{2,1} & \lambda_1\Psi_{2,1} \\ \Psi_{1,2} & \Psi_{2,2} & \lambda_2\Psi_{2,2} \end{vmatrix},$$

$$(T_1)_{22} = \begin{vmatrix} 0 & 1 & \lambda \\ \Psi_{1,1} & \Psi_{2,1} & \lambda_1\Psi_{2,1} \\ \Psi_{1,2} & \Psi_{2,2} & \lambda_2\Psi_{2,2} \end{vmatrix}.$$

T_1 satisfies the following equations

$$T_{1x} + T_1 A = A^{[1]}T_1,$$

$$T_{1t} + T_1 B = 2\lambda T_{1y} + B^{[1]}T_1.$$

$$A_0^{[1]} = A_0 + \left[\sigma_3, t_0^{[1]}\right], \quad B_{-1}^{[1]} = T_1\big|_{\lambda=-\mu} B_{-1} T_1^{-1}\big|_{\lambda=-\mu}.$$

Then the solutions of the system (1)–(6) have the form

$$q^{[1]} = q - 2i\frac{(T_1)_{12}}{\Delta_1},$$

$$v^{[1]} = v + 4i\left(\frac{(T_1)_{11}}{\Delta_1}\right)_y,$$

$$\eta^{[1]} = \frac{\left(|\omega+(T_1)_{11}|^2 - |(T_1)_{12}|^2\right)\eta + p(T_1)_{21}(\omega+(T_1)_{11}) - p^*(T_1)_{12}(\omega+(T_1)_{22})}{W},$$

$$p^{[1]} = \frac{p\left[(\omega+(T_1)_{11})^2 - p^*(T_1)_{12}^2\right] + 2\eta(T_1)_{12}(\omega+(T_1)_{11})}{W},$$

$$p^{*[1]} = \frac{p^*\left((\omega+(T_1)_{22})^2 + p(T_1)_{21}^2\right) - 2\eta(T_1)_{21}(\omega+(T_1)_{22})}{W},$$

where

$$W = (T_1)_{11}(T_1)_{22} - (T_1)_{12}(T_1)_{21}.$$

We can find the transformation T_1 which has the following property

$$T_1(\lambda, \lambda_1, \lambda_2)\big|_{\lambda=\lambda_i} \begin{pmatrix} \Psi_{1,i} \\ \Psi_{2,i} \end{pmatrix} = 0, \quad i = 1, 2.$$

Now we prove the theorem.

Proof of the Main Theorem From the formulae (13), (23) it follows that

$$M = \frac{1}{\Delta_1} \begin{pmatrix} \lambda_1 \Psi_{11} \Psi_{22} - \lambda_2 \Psi_{12} \Psi_{21} & (\lambda_2 - \lambda_1) \Psi_{11} \Psi_{12} \\ (\lambda_1 - \lambda_2) \Psi_{21} \Psi_{22} & \lambda_1 \Psi_{12} \Psi_{21} + \lambda_2 \Psi_{11} \Psi_{22} \end{pmatrix}.$$

From Eq. (24), we have

$$T_1(\lambda, \lambda_1, \lambda_2) = \lambda I - M$$

$$= \frac{1}{\Delta_1} \begin{pmatrix} \lambda \cdot \Delta_1 - \begin{vmatrix} \Psi_{2,1} & \lambda_1 \Psi_{1,1} \\ \Psi_{2,2} & \lambda_2 \Psi_{1,2} \end{vmatrix} & - \begin{vmatrix} \Psi_{1,1} & \lambda_1 \Psi_{1,1} \\ \Psi_{1,2} & \lambda_2 \Psi_{1,2} \end{vmatrix} \\[4mm] \begin{vmatrix} \Psi_{2,1} & \lambda_1 \Psi_{2,1} \\ \Psi_{2,2} & \lambda_2 \Psi_{2,2} \end{vmatrix} & \lambda \cdot \Delta_1 + \begin{vmatrix} \Psi_{1,1} & \lambda_1 \Psi_{2,1} \\ \Psi_{1,2} & \lambda_2 \Psi_{2,2} \end{vmatrix} \end{pmatrix},$$

$$\lambda I + t_0^{[1]} = \lambda I - M$$

$$= \frac{1}{\Delta_1} \begin{pmatrix} \lambda \cdot \Delta_1 - \begin{vmatrix} \Psi_{2,1} & \lambda_1 \Psi_{1,1} \\ \Psi_{2,2} & \lambda_2 \Psi_{1,2} \end{vmatrix} & - \begin{vmatrix} \Psi_{1,1} & \lambda_1 \Psi_{1,1} \\ \Psi_{1,2} & \lambda_2 \Psi_{1,2} \end{vmatrix} \\[4mm] \begin{vmatrix} \Psi_{2,1} & \lambda_1 \Psi_{2,1} \\ \Psi_{2,2} & \lambda_2 \Psi_{2,2} \end{vmatrix} & \lambda \cdot \Delta_1 + \begin{vmatrix} \Psi_{1,1} & \lambda_1 \Psi_{2,1} \\ \Psi_{1,2} & \lambda_2 \Psi_{2,2} \end{vmatrix} \end{pmatrix},$$

and the elements of the matrix are as follows:

$$(T_1)_{11} = \begin{vmatrix} 1 & 0 & \lambda \\ \Psi_{1,1} & \Psi_{2,1} & \lambda_1 \Psi_{1,1} \\ \Psi_{1,2} & \Psi_{2,2} & \lambda_2 \Psi_{1,2} \end{vmatrix} = \begin{vmatrix} \Psi_{2,1} & \lambda_1 \Psi_{1,1} \\ \Psi_{2,2} & \lambda_2 \Psi_{1,2} \end{vmatrix} - \lambda \cdot \Delta_1,$$

$$(T_1)_{12} = \begin{vmatrix} 0 & 1 & 0 \\ \Psi_{1,1} & \Psi_{2,1} & \lambda_1 \Psi_{1,1} \\ \Psi_{1,2} & \Psi_{2,2} & \lambda_2 \Psi_{1,2} \end{vmatrix} = - \begin{vmatrix} \Psi_{1,1} & \lambda_1 \Psi_{1,1} \\ \Psi_{1,2} & \lambda_2 \Psi_{1,2} \end{vmatrix},$$

$$(T_1)_{21} = \begin{vmatrix} 1 & 0 & 0 \\ \Psi_{1,1} & \Psi_{2,1} & \lambda_1 \Psi_{2,1} \\ \Psi_{1,2} & \Psi_{2,2} & \lambda_2 \Psi_{2,2} \end{vmatrix} = \begin{vmatrix} \Psi_{2,1} & \lambda_1 \Psi_{2,1} \\ \Psi_{2,2} & \lambda_2 \Psi_{2,2} \end{vmatrix},$$

$$(T_1)_{22} = \begin{vmatrix} 0 & 1 & \lambda \\ \Psi_{1,1} & \Psi_{2,1} & \lambda_1\Psi_{2,1} \\ \Psi_{1,2} & \Psi_{2,2} & \lambda_2\Psi_{2,2} \end{vmatrix} = - \begin{vmatrix} \Psi_{1,1} & \lambda_1\Psi_{2,1} \\ \Psi_{1,2} & \lambda_2\Psi_{2,2} \end{vmatrix} - \lambda \cdot \Delta_1.$$

$$
\begin{aligned}
& t_{0x}^{[1]} + \lambda A_0 - i\lambda^2\sigma_3 - \lambda t_0^{[1]}\sigma_3 + t_0^{[1]}A_0 \\
& = -i\lambda^2\sigma_3 + \lambda A_0^{[1]} - i\lambda\sigma_3 t_0^{[1]} + A_0^{[1]}t_0^{[1]}.
\end{aligned}
\tag{25}
$$

Comparing the coefficients of λ^i of the two sides of the Eq. (25), we get

$$\lambda^0: \quad t_{0x}^{[1]} + t_0^{[1]}A_0 = A_0^{[1]}t_0^{[1]},$$

$$\lambda^1: \quad A_0 - it_0^{[1]}\sigma_3 = A_0^{[1]} - i\sigma_3 t_0^{[1]},$$

$$\lambda^2: \quad i\sigma_3 = i\sigma_3.$$

In same way Theorem 1 we can formulate the next theorem.

Theorem 2 *The two-fold DT of the* (2+1)-*dimensional SMBE is*

$$T_2(\lambda, \lambda_1, \lambda_2, \lambda_3, \lambda_4) = \lambda^2 I + \lambda t_1^{[2]} + t_0^{[2]} = \frac{1}{\Delta_2}\begin{pmatrix} (T_2)_{11} & (T_2)_{12} \\ (T_2)_{21} & (T_2)_{22} \end{pmatrix},$$

where

$$\Delta_2 = \begin{vmatrix} \Psi_{1,1} & \Psi_{2,1} & \lambda_1\Psi_{1,1} & \lambda_1\Psi_{2,1} \\ \Psi_{1,2} & \Psi_{2,2} & \lambda_2\Psi_{1,2} & \lambda_2\Psi_{2,2} \\ \Psi_{1,3} & \Psi_{2,3} & \lambda_3\Psi_{1,3} & \lambda_3\Psi_{2,3} \\ \Psi_{1,4} & \Psi_{2,4} & \lambda_4\Psi_{1,4} & \lambda_4\Psi_{2,4} \end{vmatrix},$$

$$(T_2)_{11} = \begin{vmatrix} 1 & 0 & \lambda & 0 & \lambda^2 \\ \Psi_{1,1} & \Psi_{2,1} & \lambda_1\Psi_{1,1} & \lambda_1\Psi_{2,1} & \lambda_1^2\Psi_{1,1} \\ \Psi_{1,2} & \Psi_{2,2} & \lambda_2\Psi_{1,2} & \lambda_2\Psi_{2,2} & \lambda_2^2\Psi_{1,2} \\ \Psi_{1,3} & \Psi_{2,3} & \lambda_3\Psi_{1,3} & \lambda_3\Psi_{2,3} & \lambda_3^2\Psi_{1,3} \\ \Psi_{1,4} & \Psi_{2,4} & \lambda_4\Psi_{1,4} & \lambda_4\Psi_{2,4} & \lambda_4^2\Psi_{1,4} \end{vmatrix},$$

$$(T_2)_{12} = \begin{vmatrix} 0 & 1 & 0 & \lambda & 0 \\ \Psi_{1,1} & \Psi_{2,1} & \lambda_1\Psi_{1,1} & \lambda_1\Psi_{2,1} & \lambda_1^2\Psi_{1,1} \\ \Psi_{1,2} & \Psi_{2,2} & \lambda_2\Psi_{1,2} & \lambda_2\Psi_{2,2} & \lambda_2^2\Psi_{1,2} \\ \Psi_{1,3} & \Psi_{2,3} & \lambda_3\Psi_{1,3} & \lambda_3\Psi_{2,3} & \lambda_3^2\Psi_{1,3} \\ \Psi_{1,4} & \Psi_{2,4} & \lambda_4\Psi_{1,4} & \lambda_4\Psi_{2,4} & \lambda_4^2\Psi_{1,4} \end{vmatrix},$$

$$(T_2)_{21} = \begin{vmatrix} 1 & 0 & \lambda & 0 & 0 \\ \Psi_{1,1} & \Psi_{2,1} & \lambda_1\Psi_{1,1} & \lambda_1\Psi_{2,1} & \lambda_1^2\Psi_{2,1} \\ \Psi_{1,2} & \Psi_{2,2} & \lambda_2\Psi_{1,2} & \lambda_2\Psi_{2,2} & \lambda_2^2\Psi_{2,2} \\ \Psi_{1,3} & \Psi_{2,3} & \lambda_3\Psi_{1,3} & \lambda_3\Psi_{2,3} & \lambda_3^2\Psi_{2,3} \\ \Psi_{1,4} & \Psi_{2,4} & \lambda_4\Psi_{1,4} & \lambda_4\Psi_{2,4} & \lambda_4^2\Psi_{2,4} \end{vmatrix},$$

$$(T_2)_{22} = \begin{vmatrix} 0 & 1 & 0 & \lambda & \lambda^2 \\ \Psi_{1,1} & \Psi_{2,1} & \lambda_1\Psi_{1,1} & \lambda_1\Psi_{2,1} & \lambda_1^2\Psi_{2,1} \\ \Psi_{1,2} & \Psi_{2,2} & \lambda_2\Psi_{1,2} & \lambda_2\Psi_{2,2} & \lambda_2^2\Psi_{2,2} \\ \Psi_{1,3} & \Psi_{2,3} & \lambda_3\Psi_{1,3} & \lambda_3\Psi_{2,3} & \lambda_3^2\Psi_{2,3} \\ \Psi_{1,4} & \Psi_{2,4} & \lambda_4\Psi_{1,4} & \lambda_4\Psi_{2,4} & \lambda_4^2\Psi_{2,4} \end{vmatrix},$$

T_2 satisfies the following equations

$$T_{2x} + T_2 A = A^{[2]}T_2,$$

$$T_{2t} + T_2 B = 2\lambda T_{2y} + B^{[2]}T_1.$$

$$A_0^{[2]} = A_0 + \left[\sigma_3, t_0^{[2]}\right], \quad B_{-1}^{[2]} = T_2\big|_{\lambda=-\mu} B_{-1} T_2^{-1}\big|_{\lambda=-\mu}.$$

Then the solutions of the system (1)–(6) have the form

$$q^{[2]} = q - 2i\frac{(T_2)_{12}}{\Delta_2},$$

$$v^{[2]} = v + 4i\left(\frac{(T_2)_{11}}{\Delta_2}\right)_y,$$

$$\eta^{[2]} = \frac{\left(|\omega+(T_2)_{11}|^2 - |(T_2)_{12}|^2\right)\eta + p(T_2)_{21}(\omega+(T_2)_{11}) - p^*(T_2)_{12}(\omega+(T_2)_{22})}{W},$$

$$p^{[2]} = \frac{p\left[(\omega+(T_2)_{11})^2 - p^*(T_2)_{12}^2\right] + 2\eta(T_2)_{12}(\omega+(T_2)_{11})}{W},$$

$$p^{*[2]} = \frac{p^*\left((\omega+(T_2)_{22})^2 + p(T_2)_{21}^2\right) - 2\eta(T_2)_{21}(\omega+(T_2)_{22})}{W},$$

where

$$W = (T_2)_{11}(T_2)_{22} - (T_2)_{12}(T_2)_{21}.$$

We can find the transformation T_1 which has the following property

$$T_2(\lambda, \lambda_1, \lambda_2, \lambda_3, \lambda_4)\big|_{\lambda=\lambda_i} \begin{pmatrix} \Psi_{1,i} \\ \Psi_{2,i} \end{pmatrix} = 0, \quad i = 1, 2, 3, 4.$$

Theorem 3 *The n-fold DT of the (2+1)-dimensional SMBE is*

$$T_n(\lambda, \lambda_1, \lambda_2, \lambda_3, ..., \lambda_{2n}) = \lambda^n I + \lambda^{n-1} t_{n-1}^{[n]} + \cdots + \lambda t_1^{[n]} + t_0^{[n]}$$

$$= \frac{1}{\Delta_n} \begin{pmatrix} (T_n)_{11} & (T_n)_{12} \\ (T_n)_{21} & (T_n)_{22} \end{pmatrix},$$

where

Δ_n

$$= \begin{vmatrix} \Psi_{1,1} & \Psi_{2,1} & \lambda_1 \Psi_{1,1} & \lambda_1 \Psi_{2,1} & \cdots & \lambda_1^{n-1} \Psi_{1,1} & \lambda_1^{n-1} \Psi_{2,1} \\ \Psi_{1,2} & \Psi_{2,2} & \lambda_2 \Psi_{1,2} & \lambda_2 \Psi_{2,2} & \cdots & \lambda_2^{n-1} \Psi_{1,2} & \lambda_2^{n-1} \Psi_{2,2} \\ \Psi_{1,3} & \Psi_{2,3} & \lambda_3 \Psi_{1,3} & \lambda_3 \Psi_{2,3} & \cdots & \lambda_3^{n-1} \Psi_{1,3} & \lambda_3^{n-1} \Psi_{2,3} \\ \Psi_{1,4} & \Psi_{2,4} & \lambda_4 \Psi_{1,4} & \lambda_4 \Psi_{2,4} & \cdots & \lambda_4^{n-1} \Psi_{1,4} & \lambda_1^{n-1} \Psi_{2,4} \\ \vdots & \vdots & \vdots & \vdots & & \vdots & \vdots \\ \Psi_{1,2n-1} & \Psi_{2,2n-1} & \lambda_{2n-1} \Psi_{1,2n-1} & \lambda_{2n-1} \Psi_{2,2n-1} & \cdots & \lambda_{2n-1}^{n-1} \Psi_{1,2n-1} & \lambda_{2n-1}^{n-1} \Psi_{2,2n-1} \\ \Psi_{1,2n} & \Psi_{2,2n} & \lambda_{2n} \Psi_{1,2n} & \lambda_{2n} \Psi_{2,2n} & \cdots & \lambda_{2n}^{n-1} \Psi_{1,2n} & \lambda_{2n}^{n-1} \Psi_{2,2n} \end{vmatrix},$$

$(T_n)_{11}$

$$= \begin{vmatrix} 1 & 0 & \lambda & \cdots & \lambda^{n-1} & 0 & \lambda^n \\ \Psi_{1,1} & \Psi_{2,1} & \lambda_1 \Psi_{1,1} & \cdots & \lambda_1^{n-1} \Psi_{1,1} & \lambda_1^{n-1} \Psi_{2,1} & \lambda_1^n \Psi_{1,1} \\ \Psi_{1,2} & \Psi_{2,2} & \lambda_2 \Psi_{1,2} & \cdots & \lambda_2^{n-1} \Psi_{1,2} & \lambda_2^{n-1} \Psi_{2,2} & \lambda_2^n \Psi_{1,2} \\ \Psi_{1,3} & \Psi_{2,3} & \lambda_3 \Psi_{1,3} & \cdots & \lambda_3^{n-1} \Psi_{1,3} & \lambda_3^{n-1} \Psi_{2,3} & \lambda_3^n \Psi_{1,3} \\ \vdots & \vdots & \vdots & \vdots & \vdots & \vdots & \vdots \\ \Psi_{1,2n-1} & \Psi_{2,2n-1} & \lambda_{2n-1} \Psi_{1,2n-1} & \cdots & \lambda_{2n-1}^{n-1} \Psi_{1,2n-1} & \lambda_{2n-1}^{n-1} \Psi_{2,2n-1} & \lambda_{2n-1}^n \Psi_{1,2n-1} \\ \Psi_{1,2n} & \Psi_{2,2n} & \lambda_{2n} \Psi_{1,2n} & \cdots & \lambda_{2n}^{n-1} \Psi_{1,2n} & \lambda_{2n}^{n-1} \Psi_{1,2n} & \lambda_{2n}^n \Psi_{1,2n} \end{vmatrix},$$

$(T_n)_{12}$

$$= \begin{vmatrix} 0 & 1 & 0 & \cdots & 0 & \lambda^{n-1} & 0 \\ \Psi_{1,1} & \Psi_{2,1} & \lambda_1 \Psi_{1,1} & \cdots & \lambda_1^{n-1} \Psi_{1,1} & \lambda_1^{n-1} \Psi_{2,1} & \lambda_1^n \Psi_{1,1} \\ \Psi_{1,2} & \Psi_{2,2} & \lambda_2 \Psi_{1,2} & \cdots & \lambda_2^{n-1} \Psi_{1,2} & \lambda_2^{n-1} \Psi_{2,2} & \lambda_2^n \Psi_{1,2} \\ \Psi_{1,3} & \Psi_{2,3} & \lambda_3 \Psi_{1,3} & \cdots & \lambda_3^{n-1} \Psi_{1,3} & \lambda_3^{n-1} \Psi_{2,3} & \lambda_3^n \Psi_{1,3} \\ \vdots & \vdots & \vdots & \vdots & \vdots & \vdots & \vdots \\ \Psi_{1,2n-1} & \Psi_{2,2n-1} & \lambda_{2n-1} \Psi_{1,2n-1} & \cdots & \lambda_{2n-1}^{n-1} \Psi_{1,2n-1} & \lambda_{2n-1}^{n-1} \Psi_{2,2n-1} & \lambda_{2n-1}^n \Psi_{1,2n-1} \\ \Psi_{1,2n} & \Psi_{2,2n} & \lambda_{2n} \Psi_{1,2n} & \cdots & \lambda_{2n}^{n-1} \Psi_{1,2n} & \lambda_{2n}^{n-1} \Psi_{1,2n} & \lambda_{2n}^n \Psi_{1,2n} \end{vmatrix},$$

$(T_n)_{21}$

$$
= \begin{vmatrix}
1 & 0 & \lambda & \cdots & \lambda^{n-1} & 0 & 0 \\
\Psi_{1,1} & \Psi_{2,1} & \lambda_1\Psi_{1,1} & \cdots & \lambda_1^{n-1}\Psi_{1,1} & \lambda_1^{n-1}\Psi_{2,1} & \lambda_1^n\Psi_{2,1} \\
\Psi_{1,2} & \Psi_{2,2} & \lambda_2\Psi_{1,2} & \cdots & \lambda_2^{n-1}\Psi_{1,2} & \lambda_2^{n-1}\Psi_{2,2} & \lambda_2^n\Psi_{2,2} \\
\Psi_{1,3} & \Psi_{2,3} & \lambda_3\Psi_{1,3} & \cdots & \lambda_3^{n-1}\Psi_{1,3} & \lambda_3^{n-1}\Psi_{2,3} & \lambda_3^n\Psi_{2,3} \\
\vdots & \vdots & \vdots & \vdots & \vdots & \vdots & \vdots \\
\Psi_{1,2n-1} & \Psi_{2,2n-1} & \lambda_{2n-1}\Psi_{1,2n-1} & \cdots & \lambda_{2n-1}^{n-1}\Psi_{1,2n-1} & \lambda_{2n-1}^{n-1}\Psi_{2,2n-1} & \lambda_{2n-1}^n\Psi_{2,2n-1} \\
\Psi_{1,2n} & \Psi_{2,2n} & \lambda_{2n}\Psi_{1,2n} & \cdots & \lambda_{2n}^{n-1}\Psi_{1,2n} & \lambda_{2n}^{n-1}\Psi_{1,2n} & \lambda_{2n}^n\Psi_{2,2n}
\end{vmatrix},
$$

$(T_n)_{22}$

$$
= \begin{vmatrix}
0 & 1 & 0 & \cdots & 0 & \lambda^{n-1} & \lambda^n \\
\Psi_{1,1} & \Psi_{2,1} & \lambda_1\Psi_{1,1} & \cdots & \lambda_1^{n-1}\Psi_{1,1} & \lambda_1^{n-1}\Psi_{2,1} & \lambda_1^n\Psi_{2,1} \\
\Psi_{1,2} & \Psi_{2,2} & \lambda_2\Psi_{1,2} & \cdots & \lambda_2^{n-1}\Psi_{1,2} & \lambda_2^{n-1}\Psi_{2,2} & \lambda_2^n\Psi_{2,2} \\
\Psi_{1,3} & \Psi_{2,3} & \lambda_3\Psi_{1,3} & \cdots & \lambda_3^{n-1}\Psi_{1,3} & \lambda_3^{n-1}\Psi_{2,3} & \lambda_3^n\Psi_{2,3} \\
\vdots & \vdots & \vdots & \vdots & \vdots & \vdots & \vdots \\
\Psi_{1,2n-1} & \Psi_{2,2n-1} & \lambda_{2n-1}\Psi_{1,2n-1} & \cdots & \lambda_{2n-1}^{n-1}\Psi_{1,2n-1} & \lambda_{2n-1}^{n-1}\Psi_{2,2n-1} & \lambda_{2n-1}^n\Psi_{2,2n-1} \\
\Psi_{1,2n} & \Psi_{2,2n} & \lambda_{2n}\Psi_{1,2n} & \cdots & \lambda_{2n}^{n-1}\Psi_{1,2n} & \lambda_{2n}^{n-1}\Psi_{1,2n} & \lambda_{2n}^n\Psi_{2,2n}
\end{vmatrix},
$$

T_n satisfies the next equations

$$
T_{nx} + T_n A = A^{[n]} T_n,
$$

$$
T_{nt} + T_n B = 2\lambda T_{ny} + B^{[n]} T_n.
$$

$$
A_0^{[n]} = A_0 + \left[\sigma_3, t_0^{[n]}\right], \quad B_{-1}^{[n]} = T_n\big|_{\lambda=-\mu} B_{-1} T_n^{-1}\big|_{\lambda=-\mu}.
$$

Then the solutions of the system (1)–(6) have the form

$$
q^{[n]} = q - 2i\frac{(T_n)_{12}}{\Delta_n},
$$

$$
v^{[n]} = v + 4i\left(\frac{(T_n)_{11}}{\Delta_n}\right)_y,
$$

$$
\eta^{[n]} = \frac{\left(|\omega+(T_n)_{11}|^2 - |(T_n)_{12}|^2\right)\eta + p(T_n)_{21}(\omega+(T_n)_{11}) - p^*(T_n)_{12}(\omega+(T_n)_{22})}{W},
$$

$$
p^{[n]} = \frac{p\left[(\omega+(T_n)_{11})^2 - p^*(T_n)_{12}^2\right] + 2\eta(T_n)_{12}\,(\omega+(T_n)_{11})}{W},
$$

$$p^{*[n]} = \frac{p^* \left((\omega + (T_n)_{22})^2 + p(T_n)_{21}^2 \right) - 2\eta (T_n)_{21} \left(\omega + (T_n)_{22} \right)}{W},$$

where

$$W = (T_2)_{11}(T_2)_{22} - (T_2)_{12}(T_2)_{21}.$$

We can find the transformation T_1 which has the following property

$$T_n(\lambda, \lambda_1, \lambda_2, \lambda_3, ..., \lambda_{2n})\big|_{\lambda = \lambda_i} \begin{pmatrix} \Psi_{1,i} \\ \Psi_{2,i} \end{pmatrix} = 0, \ i = 1, 2, 3, ..., 2n.$$

5 Soliton Solutions

To get the one-soliton solution we take the "seed" solution as $q = 0$, $v = 0$, $p = 0$, $\eta = 1$. Let $\lambda_1 = a + bi$. Then the corresponding associated linear system takes the form

$$\Psi_{1x} = -i\lambda \Psi_1,$$
$$\Psi_{2x} = i\lambda \Psi_2,$$
$$\Psi_{1t} = 2\lambda \Psi_{1y} + \frac{i}{\lambda + \omega} \Psi_1,$$
$$\Psi_{2t} = 2\lambda \Psi_{2y} - \frac{i}{\lambda + \omega} \Psi_2.$$

This system admits the following exact solutions

$$\Psi_1 = \Psi_{10} e^{-i\lambda_1 x + i\mu_1 y + i(2\lambda_1 \mu_1 + \frac{1}{\lambda_1 + \omega})t},$$
$$\Psi_2 = \Psi_{20} e^{i\lambda_1 x - i\mu_1 y - i(2\lambda_1 \mu_1 + \frac{1}{\lambda_1 + \omega})t},$$

or

$$\Psi_1 = e^{-i\lambda_1 x + i\mu_1 y + i(2\lambda_1 \mu_1 + \frac{1}{\lambda_1 + \omega})t + \delta_1 + i\delta_2},$$
$$\Psi_2 = e^{i\lambda_1 x - i\mu_1 y - i(2\lambda_1 \mu_1 + \frac{1}{\lambda_1 + \omega})t - \delta_1 - i\delta_2 + i\delta_0},$$

where $\mu_1 = c + id$, δ_i and c, d are real constants. Then the one-soliton solution of the (2+1)-dimensional SMBE is given by

$$q^{[1]} = \frac{2be^{L-L^*-i\delta_0}}{\cosh(L+L^*)},$$

$$v^{[1]} = -4i(a + bi\tanh(L+L^*))_y,$$

$$p^{[1]} = \frac{2bi[(w+a)\cosh(L+L^*) + bi\sinh(L+L^*)]e^{L-L^*-i\delta_0}}{((w+a)^2 + b^2)\cosh^2(L+L^*)},$$

where

$$L = -i\lambda_1 x + i\mu_1 y + i(2\lambda_1\mu_1 + \frac{1}{\lambda_1 + \omega})t + \delta_1 + i\delta_2;$$

$$L^* = i\lambda_1^* x - i\mu_1^* y - i(2\lambda_1^*\mu_1^* + \frac{1}{\lambda_1^* + \omega})t + \delta_1 - i\delta_2.$$

Using the above presented n-fold DT, similarly, we can construct the n-soliton solution of the (2+1)-dimensional SMBE. Below, we present the figures of one-soliton solutions (Figs. 1, 2 and 3).

Fig. 1 One-soliton solution $q^{[1]}$ when $t = 0$

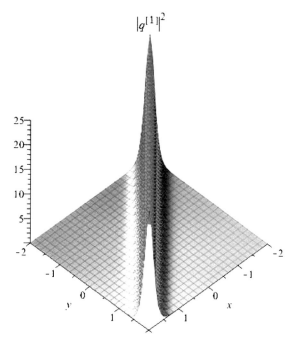

Fig. 2 One-soliton solution $v^{[1]}$ when $t = 0$

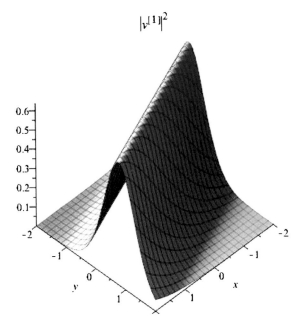

Fig. 3 One-soliton solution $p^{[1]}$ when $t = 0$

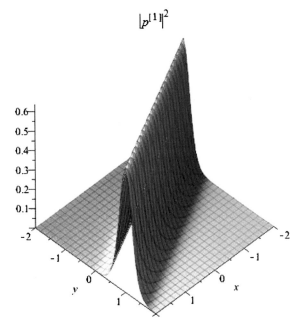

6 Conclusion

In this paper, we have constructed the DT for the (2+1)-dimensional SMBE. Using the derived DT, some exact solutions like, the one-soliton solution are obtained. The determinant representations are given for one-fold, two-fold and n-fold DT for the (2+1)-dimensional SMBE. Using the above presented results, one can also construct the n-solitons, breathers and rogue wave solutions of the (2+1)-dimensional SMBE. It is interesting to note that the rogue wave solutions of nonlinear equations are currently one of the hottest topics in nonlinear physics and mathematics. The application of the obtained solutions in physics is an interesting subject. In particular, we hope that the presented solutions may be used in experiments or optical fibre communication. Also we will study some important generalizations of the (2+1)-dimensional SMBE in future.

References

1. Zhunussova, Zh.Kh., Yesmakhanova, K.R., Tungushbaeva, D.I., Mamyrbekova, G.K., Nug-manova, G.N., Myrzakulov, R.: Integrable Heisenberg ferromagnet equations with self-consistent potentials, arXiv:1301.1649
2. Wang, L.H., Porsezian, K., He, J.S.: Breather and Rogue wave solutions of a generalized nonlinear Schrodinger equation. Phys. Rev. E **87**, 053202 (2013). arXiv:1304.8085
3. He, J., Xu, S., Porseizan, K.: N-order bright and dark rogue waves in a resonant erbium-doped fibre system. Phys. Rev. E **86**, 066603 (2012)
4. Xu, S., He, J.: The Rogue wave and breather solution of the Gerdjikov-Ivanov equation. arXiv:1109.3283
5. He, J., Xu, S., Cheng, Y.: The rational solutions of the mixed nonlinear Schrödinger equation. arXiv:1407.6917
6. Zhang, Y., Li, C., He, J.: Rogue waves in a resonant erbium-doped fiber system with higher-order effects. arXiv:1505.02237
7. Shan, S., Li, C., He, J.: On Rogue wave in the Kundu-DNLS equation. Commun. Nonlinear Sci. Numer. Simul. **18**(12), 3337–3349 (2013)
8. Senthilkumar, C., Lakshmanan, M., Grammaticos, B., Ramani, A.: Nonintegrability of image-dimensional continuum isotropic Heisenberg spin system: Painleve analysis. Phys. Lett. A **356**, 339–345 (2006)
9. Myrzakulov, R., Vijayalakshmi, S., Syzdykova, R., Lakshmanan, M.: On the simplest (2+1)-dimensional integrable spin systems and their equivalent nonlinear Schrödinger equations. J. Math. Phys. **39**, 2122–2139 (1998)
10. Myrzakulov, R., Lakshmanan, M., Vijayalakshmi, S., Danlybaeva, A.: Motion of curves and surfaces and nonlinear evolution equations in (2+1) dimensions. J. Math. Phys. **39**, 3765–3771 (1998)
11. Myrzakulov, R., Danlybaeva, A.K., Nugmanova, G.N.: Geometry and multidimensional soliton equations. Theor. Math. Phys. **118**(13), 441–451 (1999)
12. Myrzakulov, R., Rahimov, F.K., Myrzakul, K., Serikbaev, N.S.: On the geometry of stationary Heisenberg ferromagnets. In: Non-linear waves: classical and quantum aspects, pp. 543–549. Kluwer Academic Publishers, Dordrecht, Netherlands (2004)
13. Myrzakulov, R., Serikbaev, N.S., Myrzakul, K., Rahimov, F.K.: On continuous limits of some generalized compressible Heisenberg spin chains. J. NATO Sci. Series II Math. Phys. Chem. **153**, 535–542 (2004)

14. Myrzakulov, R.: Integrability of the Gauss-Codazzi-Mainardi equation in 2+1 dimensions. In: Mathematical problems of nonlinear dynamics. Proceedings of the international conference on progress in nonlinear sciences, vol. 1, pp. 314–319. Nizhny Novgorod, Russia, 2–6 July 2001

15. Zhao-Wen, Y., Min-Ru, C., Ke, W., Wei-Zhong, Z.: Integrable deformations of the (2+1)-dimensional Heisenberg Ferromagnetic model. Commun. Theor. Phys. **58**, 463–468 (2012)

16. He, J., Xu, S., Porseizan, K.: N-order bright and dark rogue waves in a resonant erbium-doped fibre system. Phys. Rev. E **86**, 066603 (2012). arXiv:1210.2522

17. He, J., Yi, C., Li, Y.-S.: The Darboux transformation for NLS-MB equations. Commun. Theor. Phys. **38**(4), 493–496 (2002)

18. Myrzakulov, R., Mamyrbekova, G.K., Nugmanova, G.N., Lakshmanan, M.: Integrable (2+1)-dimensional spin models with self-consistent potentials: relation to spin systems and soliton equations. Phys. Lett. A **378**, 2118–2123 (2014)

19. Shaikhova, G., Yesmakhanova, K., Zhussupbekov, K., Myrzakulov, R.: The (2+1)-dimensional Hirota-Maxwell-Bloch equation: Darboux transformation and soliton solutions. arXiv: 1404.5613

20. Chi, C., Zi-Xiang, Z.: Darboux tranformation and exact solutions of the Myrzakulov-I equations. Chin. Phys. Lett. **26**(8), 080504 (2009)

21. Li, C., He, J., Porsezian, K.: Rogue waves of the Hirota and the Maxwell-Bloch equations. Phys. Rev. E **87**, 012913 (2013). arXiv:1205.1191

22. Yang, J., Li, C., Li, T., Cheng, Z.: Darboux transformation and solutions of the two-component Hirota-Maxwell-Bloch system. Chin. Phys. Lett. **30**(10), 104201 (2013). arXiv:1310.0617

23. Li, C., He, J.: Darboux transformation and positons of the inhomogeneous Hirota and the Maxwell-Bloch equation. Commun. Theor. Phys. **38**, 493–496 (2002). arXiv:1210.2501

24. He, J., Zhang, L., Cheng, Y., Li, Y.: Determinant representation of Darboux transformation for the AKNS system. Sci. China Series A: Math. **49**(12), 1867–1878 (2006)

Numerical Solution of Nonlinear Klein-Gordon Equation Using Polynomial Wavelets

Jalil Rashidinia and Mahmood Jokar

Abstract The main aim of this paper is to apply the polynomial wavelets for the numerical solution of nonlinear Klein-Gordon equation. Polynomial scaling and wavelet functions are rarely used in the contexts of numerical computation. A numerical technique for the solution of nonlinear Klein-Gordon equation is presented. Our approach consists of finite difference formula combined with the collocation method, which uses the polynomial wavelets. Using the operational matrix of derivative, we reduce the problem to a set of algebraic equations by expanding the approximate solution in terms of polynomial wavelets with unknown coefficients. An estimation of error bound for this method is investigated. Some illustrative examples are included to demonstrate the validity and applicability of the approach.

1 Introduction

In this work, we are dealing with the numerical solutions of the following nonlinear partial differential equation, namely a Nonlinear Klein-Gordon equation:

$$u_{tt} + \alpha u_{xx} + \beta u + \gamma u^k = f(x, t), \qquad x \in \Omega = [-1, 1], \ t \in (0, T], \qquad (1)$$

subject to the initial conditions

$$\begin{cases} u(x, 0) = g_1(x), \ x \in \Omega, \\ u_t(x, 0) = g_2(x), \ x \in \Omega, \end{cases} \qquad (2)$$

J. Rashidinia · M. Jokar (✉)
School of Mathematics, Iran University of Science and Technology, Tehran, Iran
e-mail: jokar.mahmod@gmail.com

J. Rashidinia
e-mail: rashidinia@iust.ac.ir

© Springer International Publishing Switzerland 2016
G.A. Anastassiou and O. Duman (eds.), *Intelligent Mathematics II:*
Applied Mathematics and Approximation Theory, Advances in Intelligent Systems
and Computing 441, DOI 10.1007/978-3-319-30322-2_14

and Dirichlet boundary condition

$$u(x, t) = h(x, t), \; x \in \partial\Omega, \; t \in (0, T], \tag{3}$$

where α, β and γ are nonzero real constants and f is a known analytic function. Nonlinear phenomena occur in a wide variety of scientific applications such as plasma physics, solid state physics, fluid dynamics and chemical kinetics [1]. The nonlinear Klein-Gordon equation is one of the important models in quantum mechanics and mathematical physics. The equation has received considerable attention in studying solitons and condensed matter physics, in investigating the interaction of solitons in collisionless plasma, the recurrence of initial states, and in examining the nonlinear wave equations [2].

There are a lot of studies on the numerical solution of initial and initial-boundary problems of the linear or nonlinear Klein-Gordon equation. For instance, Chowdhury and Hashim [3] employed the homotopy-perturbation method to obtain approximate solutions of the Klein-Gordon and sine-Gordon equations. Four finite difference schemes for approximating the solution of nonlinear Klein-Gordon equation were discussed in [4]. Deeba and Khuri [5] presented a decomposition scheme for obtaining numerical solutions of the Eq. (1). In [6], a spline collocation approach for the numerical solution of a generalized nonlinear Klein-Gordon equation was investigated. Dehghan and Shokri [7] proposed a numerical scheme to solve the one-dimensional nonlinear Klein-Gordon equation with quadratic and cubic nonlinearity using the collocation points and approximating the solution by Thin Plate Splines radial basis functions. Authors in [8] considered a numerical method based on the cubic B-splines collocation technique on the uniform mesh points. Lakestani and Dehghan [9] presented two numerical techniques. The first one is Mixed Finite Difference in time and Collocation Methods using cubic B-spline functions in space (MFDCM) and the second one is fully Collocation Method (CM) which approximates the solution in both space and time variables using cubic B-spline functions. A fully implicit and discrete energy conserving finite difference scheme for the solution of an initial-boundary value problem of the nonlinear Klein-Gordon equation derived by Wong et al. [10]. A three-level spline-difference scheme to solve the one dimensional Klein-Gordon equation which is based on using the finite difference approximation for the time derivative and the spline approximation for the second-order spatial derivative was derived by authors in [11]. Most recently, authors in [12] proposed a spectral method using Legendre wavelets.

In this article we study the application of polynomial scaling functions and wavelets for computation of numerical solution of nonlinear Klein-Gordon equation. A numerical technique based on the finite difference and Collocation methods is presented. At the first stage, our method is based on the discretization of the time variable by means of the Crank-Nicolson method and freezing the coefficients of the resulting ordinary differential equation at each time step. At the second stage, we use the Wavelet-Collocation method on the yield linear ordinary differential equations at each time step resulting from the time semidiscretization. Considering this basis being wavelet functions, our method is essentially a spectral method.

Polynomial scaling and wavelet functions are rarely used in the contexts of numerical computation [13]. One of the advantages of using polynomial scaling function as expansion functions is the good representation of smooth functions by finite Chebyshev expansion. The Crank-Nicolson method is an unconditionally stable, implicit numerical scheme with second-order accuracy in both time and space. Our approach consists of reducing the Klein-Gordon equation to a set of algebraic equations by expanding the approximate solution in terms of wavelet functions with unknown coefficients. The operational matrix of derivative is presented. This matrix together with the Collocation method are then utilized to evaluate the unknown coefficients of the solution at each time step. Finally, the convergence analysis of the proposed method for the Eq. (1) is developed.

The organization of this paper is as follows. In Sect. 2, we describe the polynomial scaling and wavelet functions on $[-1, 1]$ and some basic properties. In Sect. 3, the proposed method is used to approximate the solution of the problem. As a result, a set of algebraic equations is formed and a solution of the considered problem is introduced at each time step. In Sect. 4, the error bounds of the method based on the Crank-Nicolson and polynomial wavelets are presented. In Sect. 5, we discuss the accuracy and efficiency of the employed method by applying to several test problems. A brief conclusion is given at the end of the paper in Sect. 6.

2 Preliminary

In this section, we shall give a brief introduction of the polynomial wavelets on $[-1, 1]$ and their basic properties [14]. Also the construction of the operational matrix of the derivative and some approximation results [15] are presented.

2.1 Polynomial Wavelets

Suppose that T_n and U_n be the following Chebyshev polynomials of the first and second kind respectively,

$$T_n(x) = \cos(n \arccos(x)) \quad \text{and} \quad U_{n-1}(x) = \frac{\sin(n \arccos(x))}{\sin(\arccos(x))},$$

here to introduce polynomial scaling function we need ω_j which is defined as:

$$\omega_j(x) = (1 - x^2)U_{2^j-1}(x) = \frac{(1 - x^2)}{2^j}T'_{2^j}(x), \quad j = 0, 1, 2, \ldots.$$

The zeros of $\omega_j(x)$ are $x_k = \cos(\frac{k\pi}{2^j})$ for $k = 0, 1, \ldots, 2^j$ and it should be pointed out that the zeros of ω_j are also zeros of ω_{j+1}.

Let

$$\varepsilon_{j,l} = \begin{cases} \frac{1}{2} & \text{for } l = 0 \text{ or } l = 2^j \\ 1 & \text{for } l = 1, 2, \ldots, 2^j - 1 \end{cases}$$

for any $j \in \mathbb{N}_0 = \mathbb{N} \cup \{0\}$, the polynomial scaling functions are defined as:

$$\phi_{j,l}(x) = \frac{\omega_j(x)}{2^j(-1)^{l+1}(x - x_l)} \varepsilon_{j,l}, \quad l = 0, 1, \ldots, 2^j. \tag{4}$$

Given $j \in \mathbb{N}_0$, the space of polynomial scaling functions on $[-1, 1]$ is defined by $V_j = \text{span}\{\phi_{j,l} : l = 0, 1, \ldots, 2^j\}$.

It is easy to see that the spaces $V_j = \Pi_{2^j}$ where Π_n denotes the set of all polynomials of degree at most n. The interpolatory property of this functions which helps to accelerate the computations is:

$$\phi_{j,l}(\cos(\frac{k\pi}{2^j})) = \delta_{k,l}, \quad k, l = 0, 1, \ldots, 2^j. \tag{5}$$

The wavelet spaces are defined by $W_j = \text{span}\{\psi_{j,l} : l = 0, 1, \ldots, 2^j - 1\}$, where

$$\psi_{j,l}(x) = \frac{T_{2^j}(x)}{2^j(x - \cos((2l+1)\pi/2^{j+1}))} \left(2\omega_j(x) - \omega_j\left(\cos\frac{(2l+1)\pi}{2^{j+1}}\right)\right).$$

The same interpolating property holds with the zeros of ω_{j+1} as:

$$\psi_{j,l}(\cos(\frac{(2k+1)\pi}{2^{j+1}})) = \delta_{k,l}, \quad k, l = 0, 1, \ldots, 2^j - 1.$$

We note that $\dim W_j = 2^j$ and $\dim V_j = 2^{j+1}$, also for all $j \in \mathbb{N}_0$ we have $V_{j+1} = V_j \oplus W_j$ and by denoting W_{-1} as V_0 we obtain

$$\Pi_{2^{j+1}} = V_{j+1} = \bigoplus_{k=-1}^{j} W_k. \tag{6}$$

2.2 Function Approximation

For any $j \in \mathbb{N}_0$, the operator L_j mapping any real-valued function $f(x)$ on $[-1, 1]$ into the space V_j by the Lagrange formula

$$L_j f(x) = \sum_{l=0}^{2^j} f(x_l)\phi_{j,l}(x) = U^T \Phi_j(x), \tag{7}$$

where U and Φ_j are vectors with $2^j + 1$ components as:

$$U = [f(x_0), f(x_1), ..., f(x_{2^j})]^T, \tag{8}$$

$$\Phi_j(x) = \left[\phi_{j,0}(x), \phi_{j,1}(x), ..., \phi_{j,2^j}(x)\right]^T. \tag{9}$$

Considering (6) it follows that

$$L_j f(x) = \sum_{k=0}^{1} a_k \phi_{0,k}(x) + \sum_{l=0}^{j-1} \sum_{i=0}^{2^l-1} b_{l,i} \psi_{l,i}(x) = C^T \Psi_{j-1}(x), \tag{10}$$

where $\phi_{0,k}(x)$ and $\psi_{l,i}(x)$ are scaling and wavelet functions, respectively, and C and Ψ_{j-1} are vectors with $2^j + 1$ components as:

$$C = \left[a_0, a_1, b_{0,0}, b_{1,0}, b_{1,1}, ..., b_{j-1,2^{j-1}-1}\right]^T, \tag{11}$$

$$\Psi_{j-1}(x) = \left[\phi_{0,0}, \phi_{0,1}, \psi_{0,0}, \psi_{1,0}, \psi_{1,1}, ..., \psi_{j-1,2^{j-1}-1}\right]^T. \tag{12}$$

The vector C can be obtained by considering,

$$\Psi_{j-1} = \mathbf{G}\Phi_j, \tag{13}$$

where $\mathbf{G}$ is a $(2^j + 1) \times (2^j + 1)$ matrix, which can be determined as follows. Using the two scale relations and decomposition between polynomial scaling and wavelet functions represented in [14, pp. 100], we have

$$\Phi_{j-1} = \lambda_{j-1}\Phi_j, \qquad \Psi_{j-1} = \mu_{j-1}\Phi_j, \tag{14}$$

where λ_{j-1} is a $(2^{j-1} + 1) \times (2^j + 1)$ matrix and μ_{j-1} is a $(2^{j-1}) \times (2^j + 1)$ matrix. Following [16] and by using Eqs. (13) and (14), we obtain

$$\mathbf{G} = \begin{bmatrix} \lambda_0 \times \lambda_1 \times \cdots \times \lambda_{j-1} \\ \mu_0 \times \lambda_1 \times \cdots \times \lambda_{j-1} \\ \vdots \\ \mu_{j-3} \times \mu_{j-2} \times \lambda_{j-1} \\ \mu_{j-2} \times \mu_{j-1} \\ \mu_{j-1} \end{bmatrix}.$$

By using Eqs. (13) and (7), we get

$$L_j f(x) = U^T \Phi_j - U^T \mathbf{G}^{-1} \Psi_{j-1},$$

so that we have $C^T = U^T \mathbf{G}^{-1}$.

Interpolation properties of polynomial scaling functions could help to obtaining the coefficients very fast, because it just needs to replacing the variable of function by the zeros of $\omega_j(x)$ and no need of integration. In the rest of the paper for simplicity and abbreviation we denote $\Phi_j(x)$ and $\Psi_{j-1}(x)$ by $\Phi(x)$ and $\Psi(x)$, respectively.

2.3 Operational Matrix of Derivative

Polynomial scaling functions operational matrix of derivative was derived in [15]. Here, we just list the theorem and a corollary as follows.

Theorem 1 *The differentiation of vector $\Phi(x)$ in (9) can be expressed as:*

$$\Phi'(x) = D_\phi \Phi(x), \tag{15}$$

where D_ϕ is $(2^j + 1) \times (2^j + 1)$ and the entries of operational matrix of derivative for polynomial scaling functions D_ϕ are:

$$d_{k,l} = \begin{cases} \displaystyle\sum_{i=0,\ i\neq k}^{2^j} \frac{1}{x_l - x_i}, & \text{if } l = k, \\[2ex] 2^{2^j - j - 1}(-1)^k \varepsilon_{j,k} \displaystyle\prod_{r=0,\ r\neq l,k}^{2^j} (x_l - x_r), & \text{if } l \neq k. \end{cases}$$

Proof See [15].

Corollary 2 *Using matrix $\mathbf{G}$ the operational matrix of derivative for polynomial wavelets can be represented as:*

$$D_\psi = \mathbf{G}^{-1} D_\phi \mathbf{G}. \tag{16}$$

3 The Polynomial Wavelet Method (PWM)

In this section, we solve nonlinear partial differential equation (1) on a bounded domain. For this end, we use finite difference method for one variable to reduce these equations to a system of ordinary differential equations, then we solve this system and find the solution of the given Klein-Gordon equation at the points $t_n = n\delta t$ for $\delta t = \frac{T-0}{N}, n = 0, 1, \ldots, N$.

In order to perform temporal discretization, we discretize (1) according to the following θ-weighted type scheme

$$\frac{u^{n+1} - 2u^n + u^{n-1}}{(\delta t)^2} + \theta \left(\alpha u_{xx}^{n+1} + \beta u^{n+1}\right) + (1-\theta)\left(\alpha u_{xx}^n + \beta u^n\right) + \gamma (u^n)^k = f(x, t_n),$$

$$\tag{17}$$

where δt is the time step size and u^{n+1} is used to show $u(x, t + \delta t)$. By choosing $\theta = \frac{1}{2}$ (Crank-Nicolson scheme) and rearranging Eq. (17) we obtain

$$u^{n+1} + \frac{\beta(\delta t)^2}{2}u^{n+1} + \frac{\alpha(\delta t)^2}{2}u_{xx}^{n+1} = \left(2 - \frac{\beta(\delta t)^2}{2}\right)u^n - \frac{\alpha(\delta t)^2}{2}u_{xx}^n$$
$$-\gamma(\delta t)^2(u^n)^k - u^{n-1} + (\delta t)^2 f(x, t_n). \quad (18)$$

Using Eq. (7), the approximate solution for $u^n(x)$ via scaling functions is represented by formula

$$L_j u^n(x) = \mathbf{U}_n^T \Phi(x), \quad (19)$$

where vectors $\mathbf{U}_n$ and $\Phi(x)$ are defined as (8) and (9) respectively.
For the derivatives of $u^n(x)$ by using (15) we can write the following relations,

$$L_j u_x^n(x) = \mathbf{U}_n^T \Phi'(x) = \mathbf{U}_n^T D_\phi \Phi(x), \quad (20)$$

$$L_j u_{xx}^n(x) = \mathbf{U}_n^T \Phi''(x) = \mathbf{U}_n^T D_\phi^2 \Phi(x). \quad (21)$$

Replacing Eqs. (19) and (21) in Eq. (18) we obtain

$$\mathbf{U}_{n+1}^T \Phi(x) + \frac{\beta(\delta t)^2}{2}\mathbf{U}_{n+1}^T \Phi(x) + \frac{\alpha(\delta t)^2}{2}\mathbf{U}_{n+1}^T D_\phi^2 \Phi(x)$$
$$= \left(2 - \frac{\beta(\delta t)^2}{2}\right)\mathbf{U}_n^T \Phi(x) - \frac{\alpha(\delta t)^2}{2}\mathbf{U}_n^T D_\phi^2 \Phi(x) - \gamma(\delta t)^2(\mathbf{U}_n^T \Phi(x))^k \quad (22)$$
$$-\mathbf{U}_{n-1}^T \Phi(x) + (\delta t)^2 f(x, t_n).$$

Substituting Eq. (13) into Eq. (22), we change current base to the polynomial wavelets bases

$$\mathbf{U}_{n+1}\left[I + \frac{\beta(\delta t)^2}{2}I + \frac{\alpha(\delta t)^2}{2}D_\phi^2\right]\mathbf{G}^{-1}\Psi(x)$$
$$= \left(2 - \frac{\beta(\delta t)^2}{2}\right)\mathbf{U}_n^T \mathbf{G}^{-1}\Psi(x) - \frac{\alpha(\delta t)^2}{2}\mathbf{U}_n^T D_\phi^2 \mathbf{G}^{-1}\Psi(x) \quad (23)$$
$$-\gamma(\delta t)^2(\mathbf{U}_n^T \mathbf{G}^{-1}\Psi(x))^k - \mathbf{U}_{n-1}^T \mathbf{G}^{-1}\Psi(x) + (\delta t)^2 f(x, t_n).$$

By collocating Eq. (23) in the points $x_k = \cos(\frac{k\pi}{2^j})$, $k = 0, 1, \ldots, 2^j$, we get,

$$\mathbf{U}_{n+1}\left[I + \frac{\beta(\delta t)^2}{2}I + \frac{\alpha(\delta t)^2}{2}D_\phi^2\right]\mathbf{G}^{-1}\Psi(x_k)$$
$$= \left(2 - \frac{\beta(\delta t)^2}{2}\right)\mathbf{U}_n^T \mathbf{G}^{-1}\Psi(x_k) - \frac{\alpha(\delta t)^2}{2}\mathbf{U}_n^T D_\phi^2 \mathbf{G}^{-1}\Psi(x_k) \quad (24)$$
$$-\gamma(\delta t)^2(\mathbf{U}_n^T \mathbf{G}^{-1}\Psi(x_k))^k - \mathbf{U}_{n-1}^T \mathbf{G}^{-1}\Psi(x_k) + (\delta t)^2 f(x_k, t_n),$$

which represents a system of $(2^j + 1) \times (2^j + 1)$ equations.

Using Eq. (19) in (3) we have

$$\mathbf{U}_{n+1}^T \mathbf{G}^{-1} \Psi(-1) = h(-1, t_{n+1}), \tag{25}$$

$$\mathbf{U}_{n+1}^T \mathbf{G}^{-1} \Psi(1) = h(1, t_{n+1}). \tag{26}$$

Because the rank of matrix D_ϕ is 2^j and the rank of D_ϕ^2 is $2^j - 1$ we replace Eqs. (25)–(26) instead of first and last equations of the system (24), so we finally obtain a following matrix form of equations,

$$\mathscr{A}_n \mathbf{U}_{n+1} = \mathscr{B}_n, \qquad n = 1, 2, \ldots \tag{27}$$

where $\mathscr{A}_n$ is a matrix with dimension $(2^j + 1) \times (2^j + 1)$ as,

$$\mathscr{A}_n = \begin{bmatrix} \Psi^T(-1)(\mathbf{G}^{-1})^T \\ \Psi^T(x_1)(\mathbf{G}^{-1})^T \left[I + \frac{\beta(\delta t)^2}{2} I + \frac{\alpha(\delta t)^2}{2} D_\phi^2 \right]^T \\ \vdots \\ \Psi^T(x_{2^j-1})(\mathbf{G}^{-1})^T \left[I + \frac{\beta(\delta t)^2}{2} I + \frac{\alpha(\delta t)^2}{2} D_\phi^2 \right]^T \\ \Psi^T(1)(\mathbf{G}^{-1})^T \end{bmatrix}$$

and

$$\mathscr{B}_n = \begin{bmatrix} h(-1, t_{n+1}) \\ \Gamma_1 \\ \vdots \\ \Gamma_{2^j-1} \\ h(1, t_{n+1}) \end{bmatrix}$$

with

$$\Gamma_i = \left(2 - \frac{\beta(\delta t)^2}{2} \right) \mathbf{U}_n^T \mathbf{G}^{-1} \Psi(x_i) - \frac{\alpha(\delta t)^2}{2} \mathbf{U}_n^T D_\phi^2 \mathbf{G}^{-1} \Psi(x_i) - \gamma(\delta t)^2 (\mathbf{U}_n^T \mathbf{G}^{-1} \Psi(x_i))^k$$

$$- \mathbf{U}_{n-1}^T \mathbf{G}^{-1} \Psi(x_i) + (\delta t)^2 f(x_i, t_n), \qquad i = 1, \ldots, 2^j - 1.$$

Using the first initial condition of Eq. (2), we have

$$\mathbf{U}_0^T \mathbf{G}^{-1} \Psi(x) = g_1(x), \tag{28}$$

By using the second initial condition of Eq. (2), one can get

$$\frac{u^1(x) - u^{-1}(x)}{2(\delta t)^2} = g_2(x), \qquad x \in \Omega. \tag{29}$$

Equation (29) can be rewritten as

$$\mathbf{U}_{-1}^T \mathbf{G}^{-1} \Psi(x) = \mathbf{U}_1^T \mathbf{G}^{-1} \Psi(x) - 2(\delta t)^2 g_2(x). \tag{30}$$

Equation (27) using Eqs. (28) and (30) as the starting points, gives the system of equations with $2^j + 1$ unknowns and equations, which can be solved to find $\mathbf{U}_{n+1}$ in any step $n = 1, 2, \ldots$. So the unknown functions $u(x, t_n)$ in any time $t = t_n, n = 0, 1, 2, \ldots$ can be found.

4 Error Bounds

Here we give the error analysis of the method presented in the previous section for the Nonlinear Klein-Gordon equation. Suppose that BV be the set of real valued functions $\mathscr{P} : \mathbb{R} \longrightarrow \mathbb{R}$ with bounded variation on $[-1, 1]$. The value $V(\mathscr{P}(x))$ is defined as total variation of $\mathscr{P}(x)$ on $[-1 \ 1]$. Let for the given weight function $w(x) = \frac{1}{\sqrt{1-x^2}}$, and for $2 \le p < \infty$, we define

$$\|\mathscr{P}\|_p := \left(\int_{-1}^{1} |\mathscr{P}(x)|^p w(x) dx \right)^{\frac{1}{p}}.$$

Here we need to recall two corollaries from [17].

Corollary 3 *Let $p \ge 2$, $\mathscr{P}^{(s)} \in BV$ and $0 \le s \le 2^j$ then,*

$$\| \mathscr{P} - L_j \mathscr{P} \|_p \le \xi 2^{-j(s+1/p)} V(\mathscr{P}^{(s)}). \tag{31}$$

Corollary 4 *Let $p \ge 2$, $0 \le l \le s$ and $\mathscr{P}^{(s)} \in BV$, then for the interpolatory polynomial based on the zeros of the Jacobi polynomial we have,*

$$\| (\mathscr{P} - L_j \mathscr{P})^{(l)} \|_p \le \xi 2^{-j(s+1/p-\max\{l, 2l-1/p\})} V(\mathscr{P}^{(s)}). \tag{32}$$

In the above corollaries ξ is constant depends on s.
We consider Eq. (18) as an operator equation in the form

$$\mathscr{H} u^{n+1} = \left((1 + \frac{\beta(\delta t)^2}{2}) \mathscr{I} + \frac{\alpha(\delta t)^2}{2} \mathscr{D} \right) u^{n+1} = F, \tag{33}$$

where $\mathscr{I}$ is an identity operator and

$$\mathscr{D} = \frac{d^2}{dx^2},$$

$$F = \left(2 - \frac{\beta(\delta t)^2}{2}\right)u^n - \frac{\alpha(\delta t)^2}{2}u_{xx}^n - \gamma(\delta t)^2(u^n)^k - u^{n-1} + (\delta t)^2 f(x, t_n).$$

For the operator equation (33) the approximate equation is

$$L_j(\mathscr{H})u_j^{n+1} = L_j\left((1 + \frac{\beta(\delta t)^2}{2})\mathscr{I} + \frac{\alpha(\delta t)^2}{2}\mathscr{D}\right)u_j^{n+1} = F_j. \qquad (34)$$

System (34) may be solved numerically to yield an approximate solution equation (1) at each level of time given by the expression $u_j^{n+1} = U_{n+1}^T \Phi(x)$. Next Lemma will give the approximation results of the differential operator, and then the total error bound for $||E_j^{n+1}||_p = ||u^{n+1} - u_j^{n+1}||_p$ will be presented.

Lemma 5 If $(u^{n+1})^{(s)} \in BV$, $s \geq 0$ then for the operator $\mathscr{D}$ we have

$$||\mathscr{D}u^{n+1} - L_j\mathscr{D}u^{n+1}||_p \leq C_1 2^{-j(s+2/p-4)}||V\left((u^{n+1})^{(s)}\right)||_p. \qquad (35)$$

Proof Using (32) by considering $l = 2$ we have

$$||\mathscr{D}u^{n+1} - L_j\mathscr{D}u^{n+1}||_p \leq ||u_{xx}^{n+1} - L_j u_{xx}^{n+1}||_p$$

$$\leq ||(u^{n+1} - L_j u^{n+1})_{xx}||_p$$

$$\leq C_1 2^{-j(s+2/p-4)} V\left((u^{n+1})^{(s)}\right).$$

Theorem 6 If u^{n+1} and u_j^{n+1} be the exact and approximate solution of (1) at each level of time $n+1$ respectively, also assume that the operator $\mathscr{H} = (1 + \frac{\beta(\delta t)^2}{2})\mathscr{I} + \frac{\alpha(\delta t)^2}{2}\mathscr{D}$ has bounded inverse and $(u^{n+1})^{(s)}$, $F^{(s)} \in BV$, $s \geq 0$, then

$$||E_j^{n+1}||_p \leq C\zeta ||(L_j\mathscr{H})^{-1}||_p 2^{-j(s+2/p-4)},$$

where

$$\zeta = \max\left\{V\left((u^{n+1})^{(s)}\right), V\left(F^{(s)}\right)\right\},$$

so for $s \geq 4$ we ensure the convergence when j goes to the infinity.

Proof Subtracting Eq. (34) form (1) yields

$$-L_j\mathscr{H}(u^{n+1} - u_j^{n+1}) = (\mathscr{H} - L_j\mathscr{H})u^{n+1} - (F - F_j),$$

provided that $\mathcal{H}^{-1}$ exists and bounded, we obtain the error bound

$$\|E_j^{n+1}\|_p = \|(L_j\mathcal{H})^{-1}\|_p \|(\mathcal{H} - L_j\mathcal{H})u^{n+1} - (F - F_j)\|_p, \tag{36}$$

where $\|(L_j\mathcal{H})^{-1}\|_p = \| \left((1 + \frac{\beta(\delta t)^2}{2})I + \frac{\alpha(\delta t)^2}{2}D_\phi^2 \right)^{-1} \|_p.$
Furthermore, by using Lemma 5 and relation (31), we have

$$\|(\mathcal{H} - L_j\mathcal{H})u^{n+1}\|_p \leq |1 + \frac{\beta(\delta t)^2}{2}|\|(\mathcal{I} - L_j\mathcal{I})u^{n+1}\|_p \tag{37}$$

$$+ |\frac{\alpha(\delta t)^2}{2}|\|(\mathcal{D} - L_j\mathcal{D})u^{n+1}\|_p \tag{38}$$

$$\leq C_1 |\frac{\alpha(\delta t)^2}{2}|2^{-j(s+2/p-4)}V\left((u^{n+1})^{(s)}\right) \tag{39}$$

$$+ C_2|1 + \frac{\beta(\delta t)^2}{2}|2^{-j(s+1/p)}V\left((u^{n+1})^{(s)}\right), \tag{40}$$

and

$$\|F - F_j\|_p \leq C_3 2^{-j(s+1/p)}V\left(F^{(s)}\right), \tag{41}$$

Substituting Eqs. (40)–(41) in (36), we have

$$\|u^{n+1} - u_j^{n+1}\|_p \leq \|(L_j\mathcal{H})^{-1}\|_p \left[C_1\frac{\alpha(\delta t)^2}{2}|2^{-j(s+2/p-4)}V\left((u^{n+1})^{(s)}\right) \right.$$
$$\left. + C_2|1 + \frac{\beta(\delta t)^2}{2}|2^{-j(s+1/p)}V\left((u^{n+1})^{(s)}\right) + C_3 2^{-j(s+1/p)}V\left(F^{(s)}\right) \right].$$

By choosing $C = max\left\{C_1|\frac{\alpha(\delta t)^2}{2}|, C_2|1 + \frac{\beta(\delta t)^2}{2}|, C_3\right\}$ finally we can obtain

$$\|u^{n+1} - u_j^{n+1}\|_p \leq C\zeta\|(L_j\mathcal{H})^{-1}\|_p 2^{-j(s+2/p-4)}.$$

Remark 7 We know that the order of accuracy for the Crank-Nicolson method is $O(\delta t^2)$. Therefore total error bound can be represented as

$$\|u - u_j^{n+1}\|_p \leq C\zeta\|(L_j\mathcal{H})^{-1}\|_p 2^{-j(s+2/p-4)} + O(\delta t^2).$$

5 Numerical Examples

In this section, we give some computational results of numerical experiments with method based on applying the technique discussed in Sect. 3 to find numerical solution of nonlinear Klein-Gordon equation and compare our results with exact solutions and those already available in literature [7, 12]. In order to test the accuracy of the presented method we use the error norms L_2, L_∞ and Root-Mean-Square (RMS) through the examples. The numerical computations have been done by the software Matlab.

Example 8 As the first test problem, we consider the nonlinear Klein-Gordon equation (1) with quadratic nonlinearity as

$$\frac{\partial^2 u}{\partial t^2}(x,t) + \alpha \frac{\partial^2 u}{\partial x^2}(x,t) + \beta u(x,t) + \gamma u^2(x,t) = -x\cos(t) + x^2\cos^2(t).$$

The provided parameters are $\alpha = -1$, $\beta = 0$ and $\gamma = 1$ in the interval $[-1, 1]$ and the initial conditions are given by

$$\begin{cases} u(x,0) = x, & x \in [-1, 1], \\ u_t(x,0) = 0, & x \in [-1, 1], \end{cases}$$

with the Dirichlet boundary condition

$$u(x,t) = h(x,t).$$

The analytical solution is given in [7] as $u(x,t) = x\cos(t)$. The prescribed Dirichlet boundary function $h(x,t)$ can be extracted from the exact solution. The L_2, L_∞ and RMS errors by applying method discussed in Sect. 3 for $j = 3$, in different times and $\delta t = 0.0001$ are presented in Table 1 and compared with the RBFs method proposed in [7]. As it can be shown from Table 1, our method (PWM) is more accurate than RBFs method while PWM uses much less number of grid points (9 grid points) in comparison with RBFs which uses 100 grid points. Figure 1, shows the graph of errors in the computed solution and approximate solution for $\delta t = 0.0001$ and $j = 3$. The graph of errors in the computed solution for different values of time and $\delta t = 0.0001$, $j = 3$ are plotted in Fig. 2.

Table 1 L_2, L_∞ and RMS errors for $j = 3$ and $\delta t = 0.0001$ compared with [7], Example 1

t	L_∞−error		L_2−error		RMS−error	
	PWM (9)	RBFs (100)	PWM (9)	RBFs (100)	PWM (9)	RBFs (100)
1.0	4.12×10^{-9}	1.25×10^{-5}	1.27×10^{-8}	6.54×10^{-5}	4.23×10^{-9}	6.50×10^{-6}
3.0	7.81×10^{-9}	1.55×10^{-5}	2.50×10^{-8}	1.17×10^{-4}	8.34×10^{-9}	1.16×10^{-5}
5.0	2.91×10^{-9}	3.37×10^{-5}	6.55×10^{-9}	2.20×10^{-4}	2.18×10^{-9}	2.19×10^{-5}
7.0	7.47×10^{-9}	3.77×10^{-5}	2.25×10^{-8}	2.58×10^{-4}	7.53×10^{-9}	2.57×10^{-5}
10	2.28×10^{-9}	1.30×10^{-5}	4.99×10^{-9}	7.98×10^{-5}	1.66×10^{-9}	7.94×10^{-6}

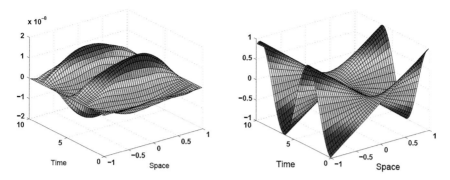

Fig. 1 Plot of errors (*left*) and approximate solution (*right*) with $\delta t = 0.0001$, $j = 3$, example 1

Fig. 2 Errors graph for Example 1, with $j = 3$, $\delta t = 0.0001$ and different times

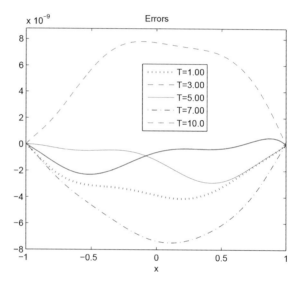

Example 9 This illustrated example presents the nonlinear Klein-Gordon equation (1) and cubic nonlinearity as

$$\frac{\partial^2 u}{\partial t^2}(x, t) + \alpha \frac{\partial^2 u}{\partial x^2}(x, t) + \beta u(x, t) + \gamma u^3(x, t)$$
$$= (x^2 - 2)\cosh(x + t) - 4x \sinh(x + t) + x^6 \cosh^3(x + t).$$

The provided parameters are $\alpha = -1$, $\beta = 1$ and $\gamma = 1$ in the interval $[-1, 1]$ with the initial conditions are given by

$$\begin{cases} u(x, 0) = x^2 \cosh(x + t), & x \in [-1, 1], \\ u_t(x, 0) = x^2 \sinh(x + t), & x \in [-1, 1], \end{cases}$$

Table 2 L_2, L_∞ and RMS errors for $j = 3$ and $\delta t = 0.0001$ compared with [12], Example 2

t	L_∞-error		L_2-error		RMS-error	
	PWM (9)	LWSCM (24)	PWM (9)	LWSCM (24)	PWM (9)	LWSCM (24)
1.0	6.38×10^{-5}	9.45×10^{-5}	1.47×10^{-4}	1.79×10^{-4}	4.90×10^{-5}	3.66×10^{-5}
2.0	1.19×10^{-4}	9.79×10^{-4}	3.55×10^{-4}	2.06×10^{-3}	1.18×10^{-4}	4.22×10^{-4}
3.0	1.52×10^{-4}	3.97×10^{-3}	3.91×10^{-4}	7.91×10^{-3}	1.30×10^{-4}	1.61×10^{-3}
4.0	2.20×10^{-4}	1.29×10^{-2}	4.34×10^{-4}	2.44×10^{-2}	1.44×10^{-4}	4.98×10^{-3}
5.0	3.40×10^{-4}	3.72×10^{-2}	4.49×10^{-4}	6.99×10^{-2}	2.16×10^{-4}	1.42×10^{-2}

Fig. 3 Errors graph for Example 2, with $j = 3$, $\delta t = 0.0001$ and different times

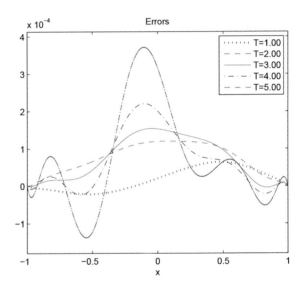

with the Dirichlet boundary condition

$$u(x, t) = h(x, t).$$

The analytical solution is given in [7, 12] as $u(x, t) = x^2 \cosh(x + t)$. We extract the boundary function $h(x, t)$ from the exact solution. The L_2, L_∞ and RMS errors by applying method discussed in Sect. 3 for $j = 3$, in different times and $\delta t = 0.0001$ are presented in Table 2 and compared with the Legendre wavelets spectral collocation method (LWSCM) [12]. As it can be seen from Table 2, our method is accurate than LWSCM while PWM uses 9 number of Polynomial wavelet basis functions in comparison with LWSCM which uses 24 number of Legendre wavelet basis. On the other hand, the accuracy of our results remains consistent when the time increases, which is the advantage of using PWM, but in the case of LWSCM the accuracy decreases fastly. Figure 3, shows the graph of errors in the computed solution for

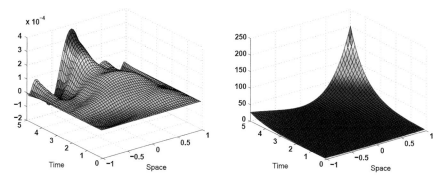

Fig. 4 Plot of errors (*left*) and approximate solution (*right*) with $\delta t = 0.0001$, $j = 3$, example 2

different values of time and $\delta t = 0.0001$, $j = 3$. The graph of errors in the computed solution and approximate solution for $\delta t = 0.0001$ and $j = 3$ are plotted in Fig. 4.

6 Conclusion

A numerical method was employed successfully for the Nonlinear Klein-Gordon equation. This approach is based on the Crank-Nicolson method for temporal discretization and the Wavelet-Collocation method in the spatial direction. After temporal discretization, the operational matrix of derivative along with a collocation method, is used to reduce the considered problem to the corresponding systems of algebraic equations at each time steps. One of the advantages of using polynomial wavelets is that the effort required to implement the method is very low, while the accuracy is high. The convergence analysis is developed. The method is computationally attractive and applications are demonstrated through illustrative examples.

References

1. Ablowitz, M.J., Clarkson, P.A.: Solitons, Nonlinear Evolution Equations and Inverse Scattering. Cambridge University Press, Cambridge (1990)
2. Dodd, R.K., Eilbeck, J.C., Gibbon, J.D., Morris, H.C.: Solitons and Nonlinear Wave Equations. Academic, London (1982)
3. Chowdhury, M.S.H., Hashim, I.: Application of homotopy-perturbation method to Klein-Gordon and sine-Gordon equations. Chaos, Solitons Fractals **39**(4), 1928–1935 (2009)
4. Jiminez, S., Vazquez, L.: Analysis of four numerical schemes for a nonlinear Klein-Gordon equation. Appl. Math. Comput. **35**, 61–94 (1990)
5. Deeba, E., Khuri, S.A.: A decomposition method for solving the nonlinear Klein-Gordon equation. J. Comput. Phys. **124**, 442–448 (1996)
6. Khuri, S.A., Sayfy, A.: A spline collocation approach for the numerical solution of a generalized nonlinear Klein-Gordon equation. Appl. Math. Comput. **216**, 1047–1056 (2010)

7. Dehghan, M., Shokri, A.: Numerical solution of the nonlinear Klein-Gordon equation using radial basis functions. J. Comput. Appl. Math. **230**, 400–410 (2009)
8. Rashidinia, J., Ghasemi, M., Jalilian, R.: Numerical solution of the nonlinear Klein-Gordon equation. J. Comput. Appl. Math. **233**, 1866–1878 (2010)
9. Lakestani, M., Dehghan, M.: Collocation and finite difference-collocation methods for the solution of nonlinear Klein-Gordon equation. Comput. Phys. Commun. **181**, 1392–1401 (2010)
10. Wong, Y.S., Chang, Q., Gong, L.: An initial-boundary value problem of a nonlinear Klein-Gordon equation. Appl. Math. Comput. **84**, 77–93 (1997)
11. Rashidinia, J., Mohammadi, R.: Tension spline approach for the numerical solution of nonlinear Klein-Gordon equation. Comput. Phys. Commun. **181**, 78–91 (2010)
12. Yin, F., Tian, T., Song, J., Zhu, M.: Spectral methods using Legendre wavelets for nonlinear Klein/Sine-Gordon equations. J. Comput. Appl. Math. **275**, 321–334 (2015)
13. Maleknejad, K., Khademi, A.: Filter matrix based on interpolation wavelets for solving Fredholm integral equations. Commun. Nonlinear. Sci. Numer. Simulat. **16**, 4197–4207 (2011)
14. Kilgore, T., Prestin, J.: Polynomial wavelets on the interval. Constr. Approx. **12**, 95–110 (1996)
15. Rashidinia, J., Jokar, M.: Application of polynomial scaling functions for numerical solution of telegraph equation. Appl. Anal. (2015). doi:10.1080/00036811.2014.998654
16. Lakestani, M., Razzaghi, M., Dehghan, M.: Semiorthogonal spline wavelets approximation for Fredholm integro-differential equations. Math. Probl. Eng. 1–12 (2006)
17. Prestin, J.: Mean convergence of Lagrange interpolation. Seminar Analysis, Operator Equation and Numerical Analysis 1988/89, pp. 75–86. Karl-Weierstraß-Institut für Mathematik, Berlin (1989)

A New Approach in Determining Lot Size in Supply Chain Using Game Theory

Maryam Esmaeili

Abstract Several seller-buyer supply chain models are suggested which emphasis simultaneously on production and market demand. In these models lot size is obtained based on different approaches. In this paper we present a novel approach to determine lot size in a seller-buyer supply chain. There is an interaction between the seller and the buyer, since the seller prefers large production lot sizes and the buyer likes small ones. Therefore for determining lot size, the seller and buyer's power is illustrated in the models. We consider two strategies for each situation (Seller-Stackelberg, Buyer-Stackelberg) whether the seller or the buyer as a leader, has more power. The leader can justify or enforce the strategy about the lot size to the follower, or let the follower determine their own lot size. Based on our findings we propose the optimal strategies for each situation. Each strategy's result will be compared by numerical examples presented. In addition, sensitivity analysis of some key parameters in the models and further research are presented.

1 Introduction

Seller-buyer supply chain models concern production and market demand decisions simultaneously [7, 17, 18]. The optimal solution of models are obtained by maximizing their profit under cooperative or non-cooperative seller and buyer's relation. In these models, the term retailer has been used correspond to the buyer. Similarly, the words manufacturer, vendor and supplier have been used interchangeably correspond to the seller. In this paper, words of the seller and the buyer have also been used according to the mentioned classification. The seller-buyer models can be observed from different views: certain or uncertain market demand and various coordination. We briefly summarize these models in order to compare the model with the new approach.

M. Esmaeili (✉)
Alzahra University, Tehran, Iran
e mail: esmaeili_m@alzahra.ac.ir

© Springer International Publishing Switzerland 2016   215
G.A. Anastassiou and O. Duman (eds.), *Intelligent Mathematics II:
Applied Mathematics and Approximation Theory*, Advances in Intelligent Systems
and Computing 441, DOI 10.1007/978-3-319-30322-2_15

Market demand plays an essential role in the seller and buyer's profit. In other words, if the structure of demand is changed the seller and buyer's optimal policy under constant, variable, stochastic or uncertain demand will also change. Chan and Kingsman [2], van den Heuvel et al. [16], Dai and Qi [15] present the seller-buyer models with a constant demand. The optimal order (production) cycles and lot size(order quantity) are determined by maximizing the whole supply chain profit in a cooperative structure. However, Yue et al. [20], Sajadieh and Jokar [13] consider some factors such as price or marketing expenditure which influence on market demand in their models. On the other hand, certain demand is avoided in some research. The optimal selling price and lot size are obtained in the models under market demand uncertainty [19].

Moreover, various types of coordination have been discussed in the literature on supply chain [10]. For instance sharing advertising cost is investigated in a supply chain includes a manufacturer and a retailer [14]. The demand is influenced by price and advertisement. Their model is developed by Giri and Sharma [8] with competing the retailers. The centralized and decentralized supply channel are considered by choosing different pricing strategies in the presence of consumers' reference-price effects [21]. However the multiple manufacturers and a common retailer are considered in a supply chain facing uncertain demand [9].

By reviewing recent publishes, the papers have covered all or some drawbacks of previous works. For example, Yugang et al. [18], introduce models in which the demand is a function of both price and marketing; also, not to assume that lot size is the same as the demand. However, the main question in the supply chain is who determines the lot size or order quantity which has been ignored by previous models. The optimal lot size in these models is determined based on their assumption which specifies who determines the lot size for the whole supply chain, the seller or the buyer, regardless to their power. Therefore, one of the participants determines the lot size while the other one has market power, which doesn't make sense. While the seller prefers large production lot sizes, the buyer likes small ones [11]. In this paper, we apply a novel approach to determine lot size in the seller-buyer model based on the seller or buyer's power. The seller as a manufacturer produces a product and wholesales it to the buyer, who then retails the product to the end consumer. The production rate of the seller is assumed to be linearly related to the market demand rate and demand is sensitive to selling price and marketing expenditure both charged by the buyer. The lot size is determined by either seller or the buyer based on their power. We consider two situations, when the seller dominates the buyer(Seller-Stackelberg), and also when the power has shifted from the seller to the buyer(Buyer-Stackelberg). In the Seller-Stackelberg(Buyer-Stackelberg) model, the seller(buyer) faces two strategies, either determines the lot size for the whole supply chain or lets the buyer(seller) determine their own lot size. The optimal solution of each strategy is obtained and it is shown that in each situation, which strategy for the seller and buyer is the best decision. Numerical examples presented in this paper, including sensitivity analysis of some key parameters, will compare the results between the models considered.

The remainder of this paper is organized as follows. The notation and assumptions are given in Sect. 2. In Sect. 3, the new approach in the seller and buyer's model is presented and compared all situations via optimal solution. In Sect. 4, some computational results, including a number of numerical examples and their sensitivity analysis that compares the results between different models are presented. Finally, the paper concludes in Sect. 5 with some suggestions for future work.

2 Notation and Problem Formulation

This section introduces the notation and formulation used in this paper.

- Decision variables

V The price charged by the seller to the buyer ($/unit)
Q_b Lot size (units) determined by the buyer
Q_s Lot size (units) determined by the seller
P Selling price charged by the buyer ($/unit)
M Marketing expenditure incurred by the buyer ($/unit)

- Input parameters

k Scaling constant for demand function ($k > 0$)
u Scaling constant for production function ($u \geq 1$)
i Percent inventory holding cost per unit per year
α Price elasticity of demand function ($\alpha > 1$)
β Marketing expenditure elasticity of demand ($0 < \beta < 1, \beta + 1 < \alpha$)
A_b Buyer's ordering cost ($/order)
A_s Seller's setup (ordering cost) ($/setup)
d Market demand rate ($units/day)
r Seller's production rate ($units/day)
C_s Seller's production cost including purchasing cost ($/unit)
$D(P, M)$ Annual demand; for notational simplicity we let $D \equiv D(P, M)$

The demand is assumed to be a function of P and M as follows:

$$D(P, M) = k P^{-\alpha} M^{\beta}. \tag{1}$$

- Assumptions

The proposed models are based on the following assumptions:

1. Planning horizon is infinite.
2. Parameters are deterministic and known in advance.
3. The annual demand depends on the selling price and marketing expenditure according to (1) (see [4]).
4. Shortages are not permitted.

3 Stackelberg Games by Considering Different Lot Sizing Approaches

In the literature, the relationships between the seller and buyer are modeled in the supply chain, by Stackelberg games where the seller and buyer take turn as a leader and follower [6]. In the previous models, the lot size is determined based on the assumption which specifies whether the seller or the buyer determines lot size. Therefore, the follower determines the lot size for the whole supply chain while the leader has market power. This doesn't make sense. In this section, we consider different approaches in seller-buyer models: first we consider the seller as the leader (Seller-Stackelberg). Regarding to the seller's role, the seller can justify or enforce the strategy about the lot size to the follower (the buyer), or let the follower determine their own lot size. We also consider the same role for the buyer when the buyer is a leader and the seller is a follower (Buyer-Stackelberg). The optimal policy of seller and buyer is obtained in each strategy in sequence as follows:

3.1 Seller-Stackelberg

The Seller-Stackelberg model has the seller as leader and the buyer as follower is widely reported in the literature as the conventional form [1, 3]. Therefore the seller can enforce the optimal lot size to the buyer, or can let the buyer determine their own lot size. We illustrate each strategy in order as follows:

3.1.1 Optimal Policy of Seller and Buyer Under Seller's Lot Size

Regarding the seller's power, for a given $Q_s = Q$ and $\mathscr{V}$ of the seller, according to the following model, the buyer obtains the best marketing expenditure M^* and selling price P^*, Eqs. (4) and (5) [4]:

$$\begin{aligned}
\Pi_b(P, M) &= PD - \mathscr{V}D - MD - A_b\frac{D}{Q} - 0.5i\mathscr{V}Q, \\
&= kP^{-\alpha+1}M^\beta - k\mathscr{V}P^{-\alpha}M^\beta - kP^{-\alpha}M^{\beta+1} \\
&\quad - A_bkP^{-\alpha}M^\beta Q^{-1} - 0.5i\mathscr{V}Q,
\end{aligned} \tag{2}$$

Then, the seller maximizes her profit $\Pi_{s_1}(\mathscr{V}, Q)$ based on the obtained P^* and M^* from the buyer's model. Thus, the problem reduces to

$$Max \ \Pi_{s_1}(\mathscr{V}, Q) = \mathscr{V}D - C_sD - A_s\frac{D}{Q} - 0.5iC_sQu^{-1}, \tag{3}$$

$$\text{subject to} \quad P^* = \frac{\alpha(\mathcal{V} + A_b Q^{-1})}{(\alpha - \beta - 1)}, \tag{4}$$

$$M^* = \frac{\beta(\mathcal{V} + A_b Q^{-1})}{(\alpha - \beta - 1)} \quad \beta + 1 < \alpha. \tag{5}$$

Substituting the constraints into the objective function, the problem transforms into an unconstrained nonlinear function of two variables $\mathcal{V}$ and Q, where the optimal solution can be found using a grid search.

3.1.2 Optimal Policy of Seller and Buyer Under Seller's Lot Size and Buyer's Lot Size

For a given $\mathcal{V}$ of the seller, according to the following model, the buyer obtains the best marketing expenditure M^*, selling price P^* and the buyer's lot size Q_b^*, Eqs. (8)–(11) [5].

$$\begin{aligned} \Pi_b(P, M, Q_b) &= PD - \mathcal{V}D - MD - A_b\frac{D}{Q_b} - 0.5i\mathcal{V}Q_b, \\ &= kP^{-\alpha+1}M^\beta - k\mathcal{V}P^{-\alpha}M^\beta - kP^{-\alpha}M^{\beta+1} \\ &\quad - A_b k P^{-\alpha}M^\beta Q_b^{-1} - 0.5i\mathcal{V}Q_b. \end{aligned} \tag{6}$$

The seller then maximizes their profit $\Pi_{s_2}(\mathcal{V}, Q)$; $Q = Q_s$ based on the pair P^*, M^* and Q_b^*. Thus, the problem reduces to

$$Max \; \Pi_{s_2}(\mathcal{V}, Q) = \mathcal{V}D - C_s D - A_s\frac{D}{Q} - 0.5iC_s Qu^{-1}, \tag{7}$$

$$\text{subject to} \quad P^* = \frac{\alpha(\mathcal{V} + A_b Q_b^{-1})}{(\alpha - \beta - 1)}, \tag{8}$$

$$M^* = \frac{\beta(\mathcal{V} + A_b Q_b^{-1})}{(\alpha - \beta - 1)} \quad \beta + 1 < \alpha, \tag{9}$$

$$Q_b^2(\mathcal{V} + A_b Q_b^{-1})^{\alpha-\beta} = \frac{2k A_b(\alpha - \beta - 1)^{\alpha-\beta}\beta^\beta}{i\mathcal{V}\alpha^\alpha}, \tag{10}$$

$$Q_b > \frac{A_b}{\mathcal{V}}\left(\frac{\alpha - \beta - 2}{2}\right). \tag{11}$$

By considering Eqs. (2), (8) and (9), (10) and (11) could be changed to:

$$Q_b^* = \sqrt{\frac{2A_b D}{i\mathcal{V}}}; \quad D = \frac{k\alpha^{-\alpha}\beta^\beta(\mathcal{V} + A_b Q_b^{-1})^{-\alpha+\beta}}{(\alpha - \beta - 1)^{-\alpha+\beta}}$$

Therefore, the seller's model would be:

$$Max \ \Pi_{s_2}(\mathcal{V}, Q) = \mathcal{V}D - C_sD - A_s\frac{D}{Q} - 0.5iC_sQu^{-1}, \tag{12}$$

$$subject \ to \quad Q_b^* = \sqrt{\frac{2A_bD}{i\mathcal{V}}}$$

By obtaining D from the constraint and substituting in the objective function the model transforms into an unconstrained nonlinear function of four variables $\mathcal{V}$, Q, P and M where the optimal solution can be found using a grid search.

3.2 Buyer-Stackelberg

In the last two decades, the buyer has increased their power relative to the seller's power [12]. Therefore, the buyer acts as the leader and the seller act as follower that is called Buyer-Stackelberg Model. For example, Wal-Mart effectively uses its power to get reduced prices from its sellers [20]. Since the buyer has more power, the lot size could be determined by the buyer or the buyer could let the seller determine their own lot size. Each situation is investigated in sequence as follows:

3.2.1 Optimal Policy of Seller and Buyer Under Buyer's Lot Size

For a given $Q_b = Q$, P and M of the buyer, according to the following model, the seller obtains the best price $\mathcal{V}^*$, Eq. (15) [4].

$$\begin{aligned} \Pi_s(\mathcal{V}) &= \mathcal{V}D - C_sD - A_s\frac{D}{Q} - 0.5iC_sQ\frac{d}{r}, \\ &= k\mathcal{V}P^{-\alpha}M^{\beta} - kC_sP^{-\alpha}M^{\beta} - A_skP^{-\alpha}M^{\beta}Q^{-1} \\ &\quad -0.5iC_sQu^{-1}. \end{aligned} \tag{13}$$

The buyer then maximizes their profit $\Pi_{b_3}(P, M, Q)$ based on $\mathcal{V}^*$. Thus, the problem reduces to

$$Max \ \Pi_{b_3}(P, M, Q) = PD - \mathcal{V}D - MD - A_b\frac{D}{Q} - 0.5i\mathcal{V}Q, \tag{14}$$

$$subject \ to \quad \mathcal{V}^* = F(C_s + A_sQ^{-1} + 0.5iC_sQ(uD)^{-1}). \tag{15}$$

Substituting the constraint into objective function, the above Buyer-Stackelberg problem reduces to optimizing an unconstrained nonlinear objective function. The optimal solution can again be found using a grid search.

3.2.2 Optimal Policy of Seller and Buyer Under Buyer's Lot Size and Seller's Lot Size

For a given P, M and $Q_b = Q$ of the buyer, according to the following model, the seller obtains the best price $\mathscr{V}^*$ and Q_s, Eqs. (18) and (19) [5].

$$
\begin{aligned}
\Pi_s(\mathscr{V}, Q_s) &= \mathscr{V}D - C_s D - A_s \frac{D}{Q_s} - 0.5 i C_s Q_s \frac{d}{r}, \\
&= k\mathscr{V}P^{-\alpha}M^{\beta} - kC_s P^{-\alpha}M^{\beta} - A_s k P^{-\alpha}M^{\beta}Q_s^{-1} \qquad (16) \\
&\quad -0.5 i C_s Q_s u^{-1}.
\end{aligned}
$$

The buyer then maximizes the profit $\Pi_{b_4}(P, M, Q_b)$ based on $\mathscr{V}^*$. Thus, the problem reduces to

$$
\text{Max } \Pi_{b_4}(P, M, Q) = PD - \mathscr{V}D - A_b \frac{D}{Q} - 0.5 i \mathscr{V} Q, \qquad (17)
$$

$$
\text{subject to} \quad \mathscr{V}^* = R(C_s + A_s Q_s^{-1} + 0.5 i C_s Q_s (uD)^{-1}), \qquad (18)
$$

$$
Q_s^* = \sqrt{\frac{2 A_s D}{i u^{-1} C_s}}. \qquad (19)
$$

Substituting the constraints into objective function, the above problem reduces to optimizing an unconstrained nonlinear objective function of four variables P, M, Q and $\mathscr{V}$. The optimal solution can again be found using a grid search.

4   Computational Results

In this section, we present numerical examples which are aimed at illustrating some significant features of the models established in previous sections. We will also perform sensitivity analysis of the main parameters of these models. We note that Examples 1, 2, 3 and 4 below illustrate the Seller-Stackelberg and Buyer-Stackelberg under different scenarios. In all these examples, we set $k = 36080$, $\beta = 0.15$, $\alpha = 1.7$, $i = 0.38$, $A_b = 38$, $A_S = 40$, $u = 1.1$ and $C_S = 1.5$.

4.1   Numerical Examples

Example 1 The Seller-Stackelberg model when the seller determines the lot size for the whole supply chain produces the following optimal values for our decision variables: $D^* = 364.6.1$, $P^* = 15.3$, $Q^* = 301.2$, $M^* = 1.3$ and $\mathscr{V}^* = 4.8$. The corresponding seller's and buyer's profits are $\Pi_s^* = 1071.8$ and $\Pi_b^* = 3010.1$ respectively.

Example 2 The Seller-Stackelberg model when the seller and buyer determine the lot size for themselves produces the following optimal values for our decision variables: $D^* = 371.3$, $P^* = 15.1$, $Q_b^* = 127.1$, $Q_s^* = 217.6 M^* = 1.3$ and $\mathscr{V}^* = 4.6$ The corresponding seller's and buyer's profits are $\Pi_s^* = 1014.7$ and $\Pi_b^* = 3196.7$ respectively. The second model, in contrast to the first one, utilizes less marketing expenditure, and has smaller seller's price and selling price charged by the buyer. Therefore, demand of the second model is higher than the first one. As expected from the theory in Sect. 3, the seller's profit here is less than in Example 1, although the demand is high. The reason would be the seller's role in determining a larger lot size for the whole supply chain. When the seller has more power the buyer's profit is less in the first model opposite to the seller's profit. In fact it makes sense that the buyer uses the freedom to choose lower lot size in the second model.

Example 3 The Buyer-Stackelberg model when the buyer determines the lot size for the whole supply chain, produces the following optimal values for the decision variables: $D^* = 1387.8$, $P^* = 6.5$, $Q^* = 393.1$, $M^* = 0.6$ and $\mathscr{V}^* = 2.1$. The seller's and buyer's profits are $\Pi_s^* = 586.5$ and $\Pi_b^* = 4960.7$ respectively. By comparing Examples 1, 2 and 3 we find out that the profit of the buyer in the buyer-Stackelberg is more than both models in the Seller-Stackelberg opposite to the seller's profit.

Example 4 The Buyer-Stackelberg model when the seller and buyer each one determines the lot size for themselves, produces the following optimal values for the decision variables: $D^* = 1387.9$, $P^* = 6.5$, $Q_s^* = 420.8$, $Q_b^* = 362.5 M^* = 0.6$ and $\mathscr{V}^* = 2.1$. The seller's and buyer's profits are $\Pi_s^* = 586.4$ and $\Pi_b^* = 4962.5$ respectively. As expected from the theorem in Sect. 3 the buyer's profit when the buyer enforces the optimal lot size to the seller is less than the buyer lets the seller determines their own lot size in the Buyer-Stackelberg. When the seller has more power, the seller would rather determine the lot size for the whole supply chain although the buyer would like determine the lot size. In addition, when the power shifts from the seller to the buyer, the buyer would like to let the seller determine the lot size, while the seller does not like such a strategy.

4.2 Sensitivity Analysis

We investigate the effects of parameters A_s, A_b, i, and C_s on P^*, $\mathscr{V}^*$, Q^*, M^*, D^*, Π_b^* and Π_s^* in scenarios of the Seller-Stackelberg and Buyer-Stackelberg models through a sensitivity analysis. We will fix $k = 36080$, $\alpha = 1.7$, $\beta = 0.15$, and, $u = 1.1$ as in the previous examples but allow A_b, A_S, C_s, and i to vary. Results of these sensitivity analysis are summarized in Tables 1, 2, 3, 4, 5, 6, 7, 8, 9, 10, 11, 12, 13, 14, 15 and 16.

Table 1 Sensitivity analysis of the model in Sect. 3.1.1 with respect to i

i	0.14	0.26	0.38	0.5	0.62
P^*	14.4	14.9	15.3	15.7	16.1
$\mathscr{V}^*$	4.6	4.7	4.8	4.9	5.03
Q^*	521.5	372.3	301.2	257.6	227.4
M^*	1.3	1.3	1.4	1.4	5.03
D^*	402.8	381.1	364.7	351.0	339.2
Π_s^*	1147.9	1104.8	1071.8	1044.3	1020.3
Π_b^*	3237.7	3110.3	3010.1	2924.7	1624.3

Table 2 Sensitivity analysis of the model in Sect. 3.1.2 with respect to i

i	0.14	0.26	0.38	0.5	0.62
P^*	14.3	14.7	15.1	15.5	15.8
$\mathscr{V}^*$	4.44	4.53	4.60	4.66	4.72
Q_s^*	375.7	268.6	217.7	186.4	164.7
Q_b^*	223.1	158.0	127.1	108.1	94.9
M^*	1.26	1.30	1.34	1.37	1.40
D^*	407.5	387.0	371.3	358.2	346.8
Π_b^*	3349.4	3263.7	3196.7	3139.9	3089.7
Π_s^*	1112.9	1057.4	1014.8	979.2	948.2

Table 3 Sensitivity analysis of the model in Sect. 3.2.1 with respect to i

i	0.14	0.26	0.38	0.5	0.62
P^*	6.2	6.3	6.5	6.6	6.7
$\mathscr{V}^*$	2.01	2.07	2.11	2.15	2.19
Q^*	677.5	484.8	393.1	336.9	297.9
M^*	0.54	0.56	0.57	0.58	0.59
D^*	1485.8	1430.3	1387.8	1352.2	1321.2
Π_S^*	598.7	591.9	586.5	582.0	577.9
Π_b^*	5213.8	5071.6	4960.7	4866.7	4783.8

The results in Tables 1, 2, 3, 4, 5, 6, 7 and 8 are also graphically displayed in Fig. 1. In each curve, the numbers 1, 2, 3 and 4 refer to the first, second scenarios in the Seller-Stackelberg and the third, forth in the Buyer-Stackelberg respectively.

Table 4 Sensitivity analysis of the model in Sect. 3.2.2 with respect to i

i	0.14	0.26	0.38	0.5	0.62
P^*	6.2	6.3	6.5	6.6	6.7
$\mathscr{V}^*$	2.01	2.07	2.11	2.15	2.19
Q_s^*	717.3	516.5	420.8	362.1	321.4
Q_b^*	632.8	449.6	362.5	309.1	272.2
M^*	0.55	0.56	0.57	0.58	0.59
D^*	1485.9	1430.5	1387.9	1352.3	1321.3
Π_s^*	598.7	591.8	586.4	581.8	577.7
Π_b^*	5214.5	5072.9	4962.5	4869.0	4786.6

Table 5 Sensitivity analysis of the model in Sect. 3.1.1 with respect to C_s

C_s	0.5	1.5	2.5	4.0	5.0
P^*	4.9	15.3	26.2	43.0	54.6
$\mathscr{V}^*$	1.6	4.8	8.2	13.5	17.0
Q^*	1263.6	301.2	154.0	82.8	61.6
M^*	0.4	1.4	2.3	3.8	4.8
D^*	2139.2	364.7	158.9	73.5	50.9
Π_s^*	2052.8	1071.8	788.0	591.5	515.5
Π_b^*	5784.5	3010.1	2206.7	1650.1	1434.7

Table 6 Sensitivity analysis of the model in Sect. 3.1.2 with respect to C_s

C_s	0.5	1.5	2.5	4.0	5.0
P^*	4.8	15.1	25.8	42.4	53.8
$\mathscr{V}^*$	1.5	4.6	7.8	12.6	15.9
Q_s^*	911.2	217.7	111.4	60.0	44.7
Q_b^*	538.1	127.1	64.6	34.5	25.6
M^*	0.4	1.3	2.3	3.7	4.7
D^*	2169.2	371.3	162.1	75.2	52.1
Π_b^*	6034.6	3196.7	2369.7	1794.3	1570.6
Π_s^*	1975.1	1014.8	738.7	548.4	475.1

Table 7 Sensitivity analysis of the model in Sect. 3.2.1 with respect to C_s

C_s	0.5	1.5	2.5	4.0	5.0
P^*	2.1	6.5	11.0	17.9	22.6
$\mathscr{V}^*$	0.7	2.1	3.6	5.9	7.4
Q^*	1643.9	393.1	201.4	108.5	80.9
M^*	0.18	0.57	0.97	1.58	1.99
D^*	7972.5	1387.8	612.2	287.2	200.2
Π_S^*	1088.0	586.5	439.3	336.4	296.2
Π_b^*	9384	4960.7	3672.9	2777.3	2429.4

Table 8 Sensitivity analysis of the model in Sect. 3.2.2 with respect to C_s

C_s	0.5	1.5	2.5	4.0	5.0
P^*	2.1	6.5	11.0	17.9	22.6
$\mathscr{V}^*$	0.68	2.11	3.59	5.85	7.39
Q_s^*	1746.9	420.8	216.5	117.2	87.5
Q_b^*	1528.8	362.5	184.8	99.1	73.6
M^*	0.18	0.57	0.97	1.58	1.99
D^*	7973.2	1387.9	612.3	287.2	200.2
Π_s^*	1087.9	586.4	439.2	336.2	296.1
Π_b^*	9385.9	4962.5	3674.6	2779.0	2431.1

Table 9 Sensitivity analysis of the model in Sect. 3.1.1 with respect to A_b

A_b	2	20	38	56	74
P^*	14.7	15.0	15.3	15.6	15.9
$\mathscr{V}^*$	4.74	4.78	4.83	4.88	4.93
Q^*	228.6	268.4	301.2	329.4	354.4
M^*	1.29	1.32	1.35	1.38	1.40
D^*	390.2	376.3	364.7	354.5	345.4
Π_b^*	3160.8	3079.6	3010.1	2948.3	2892.2
Π_s^*	1122.9	1095.2	1071.8	1051.3	1032.8

Table 10 Sensitivity analysis of the model in Sect. 3.1.2 with respect to A_b

A_b	2	20	38	56	74
P^*	14.74	15.00	15.14	15.26	15.36
γ^*	4.70	4.64	4.60	4.57	4.55
Q_s^*	222.3	219.3	217.7	216.4	215.3
Q_b^*	29.4	92.5	127.1	153.8	176.3
M^*	1.30	1.32	1.34	1.35	1.36
D^*	387.2	376.9	371.3	366.9	363.2
Π_s^*	1100.0	1044.6	1014.8	991.9	972.8
Π_b^*	3331.1	3244.1	3196.7	3160.2	3129.4

Table 11 Sensitivity analysis of the model in Sect. 3.2.1 with respect to A_b

A_b	2	20	38	56	74
P^*	6.30	6.39	6.47	6.54	6.61
γ^*	2.12	2.11	2.11	2.11	2.12
Q^*	310.3	355.0	393.1	426.6	456.5
M^*	0.56	0.56	0.57	0.58	0.58
D^*	1444.9	1414.3	1387.8	1364.1	1342.7
Π_b^*	5105.4	5028.1	4960.7	4900.3	4845.1
Π_s^*	612.7	598.0	586.5	577.0	568.7

Table 12 Sensitivity analysis of the model in Sect. 3.2.2 with respect to A_b

A_b	2	20	38	56	74
P^*	6.22	6.38	6.47	6.54	6.60
γ^*	2.11	2.11	2.11	2.11	2.12
Q_s^*	433.8	425.4	420.8	417.2	414.2
Q_b^*	85.9	266.0	362.5	436.1	497.4
M^*	0.55	0.56	0.57	0.58	0.58
D^*	1474.9	1418.5	1387.9	1364.4	1344.6
Π_s^*	621.1	598.6	586.4	577.0	569.2
Π_b^*	5190.0	5042.8	4962.5	4900.5	4848.2

Table 13 Sensitivity analysis of the model in Sect. 3.1.1 with respect to A_s

A_s	4	22	40	58	76
P^*	14.7	15.0	15.3	15.6	15.9
$\mathscr{V}^*$	4.58	4.72	4.83	4.93	5.03
Q^*	228.6	268.4	301.2	329.4	354.4
M^*	1.29	1.32	1.35	1.38	1.40
D^*	390.2	376.3	364.7	354.5	345.4
Π_b^*	3167.7	3083.0	3010.1	2944.9	2885.4
Π_s^*	1122.9	1095.2	1071.8	1051.3	1032.8

Table 14 Sensitivity analysis of the model in Sect. 3.1.2 with respect to A_s

A_s	4	22	40	58	76
P^*	13.97	14.68	15.14	15.52	15.86
$\mathscr{V}^*$	4.3	4.5	4.6	4.7	4.8
Q_s^*	73.3	165.3	217.7	257.1	289.5
Q_b^*	140.7	132.1	127.1	123.1	119.8
M^*	1.23	1.30	1.34	1.37	1.40
D^*	420.7	389.5	371.3	357.4	345.7
Π_s^*	1111.0	1050.4	1014.8	987.2	964.0
Π_b^*	3344.0	3252.4	3196.7	3152.8	3115.3

Table 15 Sensitivity analysis of the model in Sect. 3.2.1 with respect to A_s

A_s	4	22	40	58	76
P^*	6.25	6.37	6.47	6.56	6.64
$\mathscr{V}^*$	1.97	2.05	2.11	2.17	2.22
Q^*	284.5	344.6	393.1	434.3	470.5
M^*	0.55	0.56	0.57	0.58	0.59
D^*	1462.4	1421.5	1387.8	1358.6	1332.6
Π_b^*	5157.9	5050.6	4960.7	4881.8	4810.6
Π_s^*	575.8	582.8	586.5	588.9	590.4

Table 16 Sensitivity analysis of the model in Sect. 3.2.2 with respect to A_s

A_s	4	22	40	58	76
P^*	6.20	6.37	6.47	6.55	6.62
$\mathscr{V}^*$	1.95	2.05	2.11	2.16	2.21
Q_s^*	137.5	315.9	420.8	501.7	569.5
Q_b^*	390.0	372.6	362.5	354.7	348.1
M^*	0.55	0.56	0.57	0.58	0.58
D^*	1480.9	1422.6	1387.9	1360.9	1338.1
Π_s^*	576.9	583.0	586.4	589.0	591.1
Π_b^*	5203.1	5052.5	4962.5	4892.1	4832.7

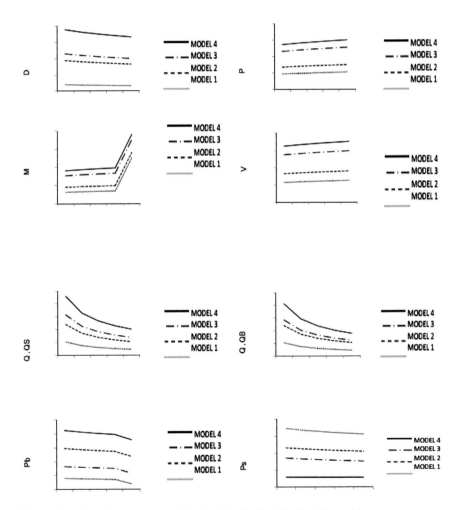

Fig. 1 The effect of i parameter on D^*, P^*, M^*, v^*, Q^*, Q_s^*, Q_b^*, Π_b^* and Π_s^*

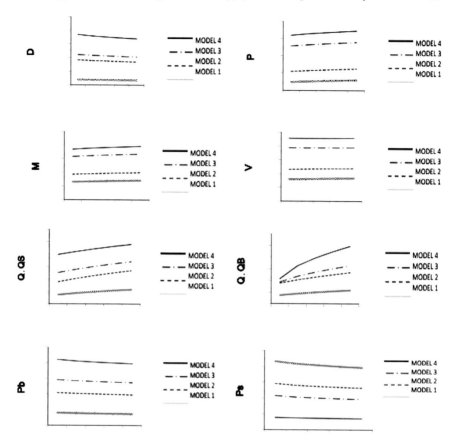

Fig. 2 The effect of A_b parameter on D^*, P^*, M^*, v^*, Q^*, Q_s^*, Q_b^*, Π_b^* and Π_s^*

Since i and C_s and also A_s and A_b have the same effect on P^*, $\mathcal{V}^*$, Q^*, M^*, D^*, Π_b^* and Π_s^*, we show only the effect of i and A_b in sequence on Figs. 1 and 2. As is seen in the figure, by increasing i and C_s, selling price, marketing expenditure and seller's price increase which cause the decrease of the demand, seller and the buyer's profits. Note due to increasing production cost and percent inventory holding cost, i and C_s, the holding cost will increase. Therefore, the seller and buyer would change their strategies to decrease lot size, which validates our model.

The effect of parameter A_s, A_b on P_i^*, M_i^*, Q_i^*, $\mathcal{V}_i^*$; $i = 1, 2, 3, 4$, i.e. for each scenario of the Seller-Stackelberg and the Buyer-Stackelberg are graphically displayed in Fig. 2. As A_b and A_s increase, P^*, M^* and $\mathcal{V}^*$ increase. By increasing A_b and A_s the seller and the buyer increase their lot size to decrease the ordering and the seller's setup cost respectively. As shown in Figs. 1 and 2 the results validate the models.

5 Conclusion

In this paper, the lot size problem in the seller-buyer supply chain is considered. The lot size in the supply chain is determined based on the seller and the buyer's power. The seller produces a product and wholesales it to the buyer, with the production rate linearly related to the market demand rate. The demand is sensitive to the selling price and the buyer's effort in marketing. We consider the seller-buyer relationship under Stackelberg games: Seller-Stackelberg, where the seller is the leader, and Buyer-Stackelberg, where the buyer is the leader. Regarding to the leader's role, the leader can justify or enforce the strategy about the lot size to the follower(the buyer), or let the follower determines their own lot size regardless to the leader's lot size. Optimal solution for each models is obtained. It is shown that when the seller has power, it would be better if the seller determines the lot size for the whole supply chain and when the buyer has power, it would be better if the buyer let the seller determines their own lot size. Numerical examples are presented which aim at illustrating the used approach. Through a sensitivity analysis, the effect of the main parameters of the model on the seller and the buyer's decisions are also investigated which illustrate our findings.

There is much to be extended the present work. For example, marketing expenditure could incorporate advertising expenditure and the seller may agree to share fraction of the advertising expenditure with the buyer by covering part of it. In this case, we could investigate cooperative games. Also, we have assumed in this paper that the production rate is greater than or equal to the demand rate, in order to avoid having to consider shortage cost. By not making this assumption, the extra cost could be incorporated into future models. Finally, even though the seller or buyer presumably knows their own costs and price charged to the consumers, it is unlikely that their opponent would be privy to such information. This would lead to incomplete knowledge on the part of the two participants and result in bargaining models with incomplete information along the line of models.

References

1. Abad, P., Jaggi, C.K.: Joint approach for setting unit price and the length of the credit period for a seller when end demand is price sensitive. Int. J. Prod. Econ. **83**, 115–122 (2003)
2. Chan, C.K., Kingsman, B.G.: Coordination in a single-vendor multi-buyer supply chain by synchronizing delivery and production cycles. Transp. Res. Part E **43**, 90–111 (2007)
3. Chen, M.-S., Chang, H.-J., Huang, C.-W., Liao, C.-N.: Channel coordination and transaction cost: a game-theoretic analysis. Ind. Mark. Manag. **35**, 178–190 (2006)
4. Esmaeili, M., Aryanezhad, M., Zeephongsekul, P.: A game theory approach in seller buyer supply chain. Eur. J. Oper. Res. **195**(2), 442–448 (2009)
5. Esmaeili, M., Abad, P., Aryanezhad, M.: Seller-buyer relationship when end demand is sensitive to price and promotion. Asia Pac. J. Oper. Res. **26**(5), 1–17 (2010)
6. Esmaeili, M., Gamchi, N.S., Asgharizadeh, E.: Three-level warranty service contract among manufacturer, agent and customer: a game-theoretical approach. Eur. J. Oper. Res. **239**, 177–186 (2014)

7. Esmaeili, M., Zeephongsekul, P.: Seller buyer models of supply chain management with an asymmetric information structure. Int. J. Prod. Econ. **239**, 146–154 (2014)
8. Giri, B., Sharma, S.: Manufacturer's pricing strategy in a two-level supply chain with competing retailers and advertising cost dependent demand. Econ. Model. **38**, 102–111 (2014)
9. Hsieh, C., Chang, Y., Wu, C.: Competitive pricing and ordering decisions in a multiple-channel supply chain. Int. J. Prod. Econ. **154**, 156–165 (2014)
10. Huang, Y., Huang, G., Newman, S.: Coordinating pricing and inventory decisions in a multi-level supply chain: a game-theoretic approach. Transp. Res. Part E **47**, 115–129 (2011)
11. Kelle, P., Faisal, A., Miller, P.A.: Partnership and negotiation support by joint optimal ordering-setup policies for JIT. Int. J. Prod. Econ. **81–82**, 431–441 (2003)
12. Li, S.X., Huang, Z., Zhu, J., Chau, P.Y.K.: Cooperative advertising, game theory and manufacturer retailer supply chains. Omega Int. J. Manag. Sci. **30**, 347–357 (2002)
13. Sajadieh, M.S., Jokar, M.R.: Optimizing shipment, ordering and pricing policies in a two-stage supply chain with price-sensitive demand. Transp. Res. Part E **45**, 564–571 (2009)
14. SeyedEsfahani, M., Biazaran, M., Gharakhani, M.: A game theoretic approach to coordinate pricing and vertical co-op advertising in manufacturer retailer supply chains. Eur. J. Oper. Res. **211**, 263–273 (2011)
15. Tinglong, D., Xiangtong, Q.: An acquisition policy for a multi-supplier system with a finite-time horizon. Comput. Oper. Res. **34**, 2758–2773 (2007)
16. van den Heuvel, W., Borm, P., Hamers, H.: Economic lot-sizing games. Eur. J. Oper. Res. **176**, 1117–1130 (2007)
17. Yang, S.-L., Zhou, Y.W.: Two-echelon supply chain models: considering duopolistic retailers' different competitive behaviors. Int. J. Prod. Econ. **103**, 104–116 (2006)
18. Yu, Y., Feng, C., Haoxun, C.: A Stackelberg game and its improvement in a VMI system with a manufacturing vendor. Eur. J. Oper. Res. **192**, 929–948 (2009)
19. Yue, D., Chao, X., Fang, S.-C., Nuttle, H.L.W.: Pricing in revenue management for multiple firms competing for customers. Int. J. Prod. Econ. **98**, 1–16 (2005)
20. Yue, J., Jill, A., Wang, M.C., Huang, Z.: Coordination of cooperative advertising in a two-level supply chain when manufacturer offers discount. Eur. J. Oper. Res. **168**(1), 65–85 (2006)
21. Zhang, J., Chiang, W., Liang, L.: Strategic pricing with reference effects in a competitive supply chain. Omega **211**, 1–10 (2015)

Voronovskaya Type Asymptotic Expansions for Multivariate Generalized Discrete Singular Operators

George A. Anastassiou and Merve Kester

Abstract Here we give asymptotic expansions including simultaneous ones, for the multivariate generalized discrete versions of unitary Picard, Gauss-Weierstrass, and Poisson-Cauchy type singular operators. These are Voronovskaya type expansions and they are connected to the approximation properties of these operators.

1 Introduction

In this article, we give multivariate Voronovskaya type asymptotic expansions for the multivariate generalized discrete singular Picard, Gauss-Weierstrass, and Poisson-Cauchy operators. We also demonstrate the simultaneous corresponding Voronovskaya asymptotic expansion for our operators. Our expansions show the rate of convergence of the operators mentioned above to unit operator.

2 Background

We are inspired by [4]. In [1], for $r \in \mathbb{N}$, $m \in \mathbb{Z}_+$, the author defined

$$
\alpha_{j,r}^{[m]} := \begin{cases} (-1)^{r-j} \dbinom{r}{j} j^{-m}, & \text{if } j = 1, 2, \ldots, r, \\ 1 - \sum\limits_{j=1}^{r} (-1)^{r-j} \dbinom{r}{j} j^{-m}, & \text{if } j = 0, \end{cases} \tag{1}
$$

G.A. Anastassiou · M. Kester (✉)
Department of Mathematical Sciences, University of Memphis, Memphis,
TN 38152, USA
e-mail: mkester@memphis.edu

G.A. Anastassiou
e-mail: ganastss@memphis.edu

© Springer International Publishing Switzerland 2016
G.A. Anastassiou and O. Duman (eds.), *Intelligent Mathematics II:
Applied Mathematics and Approximation Theory*, Advances in Intelligent Systems
and Computing 441, DOI 10.1007/978-3-319-30322-2_16

and

$$\delta_{k,r}^{[m]} := \sum_{j=1}^{r} \alpha_{j,r}^{[m]} j^k, \ k = 1, 2, \ldots, m \in \mathbb{N}. \tag{2}$$

See that

$$\sum_{j=0}^{r} \alpha_{j,r}^{[m]} = 1, \tag{3}$$

and

$$-\sum_{j=1}^{r} (-1)^{r-j} \binom{r}{j} = (-1)^r \binom{r}{0}. \tag{4}$$

Let μ_{ξ_n} be a probability Borel measure on $\mathbb{R}^N$, $N \geq 1$, $\xi_n > 0$, $n \in \mathbb{N}$. In [1], the author defined the multiple smooth singular integral operators as

$$\theta_{r,n}^{[m]} (f; x_1, \ldots, x_N) := \sum_{j=0}^{r} \alpha_{j,r}^{[m]} \int_{\mathbb{R}^N} f (x_1 + s_1 j, x_2 + s_2 j, \ldots, x_N + s_N j) \, d\mu_{\xi_n} (s),$$

$$\tag{5}$$

where $s := (s_1, \ldots, s_N)$, $x := (x_1, \ldots, x_N) \in \mathbb{R}^N$; $n, r \in \mathbb{N}, m \in \mathbb{Z}_+, f : \mathbb{R}^N \to \mathbb{R}$ is a Borel measurable function, and also $(\xi_n)_{n \in \mathbb{N}}$ is a bounded sequence of positive real numbers.

Above the $\theta_{r,n}^{[m]}$ are not in general positive operators and they preserve constants. In [1], they demonstrated

Theorem 1 *Let* $f \in C^m (\mathbb{R}^N)$, $m, N \in \mathbb{N}$,*with all* $\|f_\alpha\|_\infty \leq M$, $M > 0$, *all* $\alpha : |\alpha| = m$. *Let* $\xi_n > 0$, $(\xi_n)_{n \in \mathbb{N}}$ *bounded sequence,* μ_{ξ_n} *probability Borel measures on* $\mathbb{R}^N$.
    Call

$$c_{\alpha,n,\tilde{j}} = \int_{\mathbb{R}^N} \left(\prod_{i=1}^{N} s_i^{\alpha_i} \right) d\mu_{\xi_n} (s),$$

all $|\alpha| = \tilde{j} = 1, \ldots, m - 1$. *Assume*

$$\xi_n^{-m} \int_{\mathbb{R}^N} \left(\prod_{i=1}^{N} |s_i|^{\alpha_i} \right) d\mu_{\xi_n} (s) \leq \rho,$$

all $\alpha : |\alpha| = m$, $\rho > 0$, for any such $(\xi_n)_{n \in \mathbb{N}}$. Also $0 < \gamma \leq 1$, $x \in \mathbb{R}^N$. Then

$$\theta_{r,n}^{[m]}(f;x) - f(x) = \sum_{\widetilde{j}=1}^{m-1} \delta_{\widetilde{j},r}^{[m]} \left(\sum_{|\alpha|=\widetilde{j}} \frac{c_{\alpha,n,\widetilde{j}} f_\alpha(x)}{\left(\prod_{i=1}^N \alpha_i!\right)} \right) + o\left(\xi_n^{m-\gamma}\right). \qquad (6)$$

When $m = 1$ the sum collapses.

Above they assumed $\theta_{r,n}^{[m]}(f;x) \in \mathbb{R}$, $\forall\, x \in \mathbb{R}^N$.

In [1], they continued with

Theorem 2 *Let $f \in C^l(\mathbb{R}^N)$, $l, N \in \mathbb{N}$. Here μ_{ξ_n} is a Borel probability measure on $\mathbb{R}^N$, $\xi_n > 0$, $(\xi_n)_{n \in \mathbb{N}}$ a bounded sequence. Let $\beta := (\beta_1, \dots, \beta_N)$, $\beta_i \in \mathbb{Z}^+$, $i = 1, \dots, N$; $|\beta| := \sum_{i=1}^N \beta_i = l$. Here $f(x + sj)$, $x, s \in \mathbb{R}^N$, is μ_{ξ_n}-integrable wrt s, for $j = 1, \dots, r$. There exist μ_{ξ_n}-integrable functions*

$$h_{i_1,j}, h_{\beta_1,i_2,j}, h_{\beta_1,\beta_2,i_3,j}, \dots, h_{\beta_1,\beta_2,\dots,\beta_{N-1},i_N,j} \geq 0$$

($j = 1, \dots, r$) on $\mathbb{R}^N$ such that

$$\left| \frac{\partial^{i_1} f(x+sj)}{\partial x_1^{i_1}} \right| \leq h_{i_1,j}(s), \ i_1 = 1, \dots, \beta_1, \qquad (7)$$

$$\left| \frac{\partial^{\beta_1+i_2} f(x+sj)}{\partial x_2^{i_2} \partial x_1^{\beta_1}} \right| \leq h_{\beta_1,i_2,j}(s), \ i_2 = 1, \dots, \beta_2,$$

$$\vdots$$

$$\left| \frac{\partial^{\beta_1+\beta_2+\dots+\beta_{N-1}+i_N} f(x+sj)}{\partial x_N^{i_N} \partial x_{N-1}^{\beta_{N-1}} \dots \partial x_2^{\beta_2} \partial x_1^{\beta_1}} \right| \leq h_{\beta_1,\beta_2,\dots,\beta_{N-1},i_N,j}(s), \ i_N = 1, \dots, \beta_N,$$

$\forall\, x, s \in \mathbb{R}^N$. Then, both of the next exist and

$$\left(\theta_{r,n}^{[m]}(f;x) \right)_\beta = \theta_{r,n}^{[m]}(f_\beta;x). \qquad (8)$$

Finally, the author gave

Theorem 3 *Let $f \in C^{m+l}(\mathbb{R}^N)$, $m, l, N \in \mathbb{N}$. Assumptions of Theorem 2 are valid. Call $\gamma = 0$, β. Assume $\left\| f_{\gamma+\alpha} \right\|_\infty \leq M$, $M > 0$, for all $\alpha : |\alpha| = m$. Let $\xi_n > 0$, $(\xi_n)_{n \in \mathbb{N}}$ bounded sequence, μ_{ξ_n} probability Borel measures on $\mathbb{R}^N$. Call*

$$c_{\alpha,n,\widetilde{j}} - \int_{\mathbb{R}^N} \left(\prod_{i=1}^N s_i^{\alpha_i} \right) d\mu_{\xi_n}(s),$$

all $|\alpha| = \widetilde{j} = 1, \ldots, m - 1.$ *Assume*

$$\xi_n^{-m} \int_{\mathbb{R}^N} \left(\prod_{i=1}^N |s_i|^{\alpha_i} \right) d\mu_{\xi_n} (s) \leq \rho,$$

all $\alpha : |\alpha| = m,\ \rho > 0,$ *for any such* $(\xi_n)_{n \in \mathbb{N}}.$ *Also* $0 < \gamma \leq 1,\ x \in \mathbb{R}^N.$ *Then*

$$\left(\theta_{r,n}^{[m]} (f; x) \right)_\gamma - f_\gamma (x) = \sum_{\widetilde{j}=1}^{m-1} \delta_{\widetilde{j},r}^{[m]} \left(\sum_{|\alpha|=\widetilde{j}} \frac{c_{\alpha,n,\widetilde{j}} f_{\gamma+\alpha} (x)}{\left(\prod_{i=1}^N \alpha_i! \right)} \right) + o \left(\xi_n^{m-\gamma} \right). \quad (9)$$

When $m = 1$ *the sum collapses.*

Additionally, in [3], the authors defined:

Let μ_{ξ_n} be a Borel probability measure on $\mathbb{R}^N,\ N \geq 1,\ 0 < \xi_n \leq 1,\ n \in \mathbb{N}.$ Assume that $v := (v_1, \ldots, v_N),\ x := (x_{1,\ldots,} x_N) \in \mathbb{R}^N$ and $f : \mathbb{R}^N \to \mathbb{R}$ is a Borel measurable function.

(i) When

$$\mu_{\xi_n} (v) = \frac{e^{-\frac{\sum_{i=1}^N |v_i|}{\xi_n}}}{\sum_{v_1=-\infty}^\infty \cdots \sum_{v_N=-\infty}^\infty e^{-\frac{\sum_{i=1}^N |v_i|}{\xi_n}}}, \quad (10)$$

we define generalized multiple discrete Picard operators as:

$$P_{r,n}^{*\,[m]} (f; x_1, \ldots, x_N) \quad (11)$$

$$= \frac{\sum_{v_1=-\infty}^\infty \cdots \sum_{v_N=-\infty}^\infty \left(\sum_{j=0}^r \alpha_{j,r}^{[m]} f (x_1 + jv_1, \ldots, x_N + jv_N) \right) e^{-\frac{\sum_{i=1}^N |v_i|}{\xi_n}}}{\sum_{v_1=-\infty}^\infty \cdots \sum_{v_N=-\infty}^\infty e^{-\frac{\sum_{i=1}^N |v_i|}{\xi_n}}}.$$

(ii) When

$$\mu_{\xi_n} (v) = \frac{e^{-\frac{\sum_{i=1}^N v_i^2}{\xi_n}}}{\sum_{v_1=-\infty}^\infty \cdots \sum_{v_N=-\infty}^\infty e^{-\frac{\sum_{i=1}^N v_i^2}{\xi_n}}}, \quad (12)$$

we define generalized multiple discrete Gauss-Weierstrass operators as:

$$W_{r,n}^{*\,[m]} (f; x_1, \ldots, x_N) \quad (13)$$

$$= \frac{\sum_{v_1=-\infty}^\infty \cdots \sum_{v_N=-\infty}^\infty \left(\sum_{j=0}^r \alpha_{j,r}^{[m]} f (x_1 + jv_1, \ldots, x_N + jv_N) \right) e^{-\frac{\sum_{i=1}^N v_i^2}{\xi_n}}}{\sum_{v_1=-\infty}^\infty \cdots \sum_{v_N=-\infty}^\infty e^{-\frac{\sum_{i=1}^N v_i^2}{\xi_n}}}.$$

(iii) Let $\hat{\alpha} \in \mathbb{N}$ and $\beta > \frac{1}{\hat{\alpha}}$. When

$$\mu_{\xi_n}(v) = \frac{\prod_{i=1}^{N}\left(v_i^{2\hat{\alpha}} + \xi_n^{2\hat{\alpha}}\right)^{-\beta}}{\sum_{v_1=-\infty}^{\infty}\cdots\sum_{v_N=-\infty}^{\infty}\prod_{i=1}^{N}\left(v_i^{2\hat{\alpha}} + \xi_n^{2\hat{\alpha}}\right)^{-\beta}}, \tag{14}$$

we define the generalized multiple discrete Poisson-Cauchy operators as:

$$Q_{r,n}^{*\,[m]}(f; x_1, \ldots, x_N) \tag{15}$$
$$= \frac{\sum_{v_1=-\infty}^{\infty}\cdots\sum_{v_N=-\infty}^{\infty}\left(\sum_{j=0}^{r}\alpha_{j,r}^{[m]} f(x_1 + jv_1, \ldots, x_N + jv_N)\right)\prod_{i=1}^{N}\left(v_i^{2\hat{\alpha}} + \xi_n^{2\hat{\alpha}}\right)^{-\beta}}{\sum_{v_1=-\infty}^{\infty}\cdots\sum_{v_N=-\infty}^{\infty}\left(\prod_{i=1}^{N}\left(v_i^{2\hat{\alpha}} + \xi_n^{2\hat{\alpha}}\right)^{-\beta}\right)}.$$

Here the authors assumed that $0^0 = 1$. Finally, in [3], they observed

Lemma 4 (i) *For* $\widetilde{j} = 1, \ldots, m$, *and* $\alpha := (\alpha_1, \ldots, \alpha_N)$, $\alpha_i \in \mathbb{Z}^+$, $i = 1, \ldots, N$, $|\alpha| := \sum_{i=1}^{N}\alpha_i = \widetilde{j}$, *we have that*

$$c_{\alpha,n} := c_{\alpha,n,\widetilde{j}} := \frac{\sum_{v_1=-\infty}^{\infty}\cdots\sum_{v_N=-\infty}^{\infty}\left(\prod_{i=1}^{N}v_i^{\alpha_i}\right)e^{-\frac{\sum_{i=1}^{N}|v_i|}{\xi_n}}}{\sum_{v_1=-\infty}^{\infty}\cdots\sum_{v_N=-\infty}^{\infty}e^{-\frac{\sum_{i=1}^{N}|v_i|}{\xi_n}}} < \infty. \tag{16}$$

for all $\xi_n \in (0, 1]$. *Additionally, let* $\alpha_i \in \mathbb{N}$, *then as* $\xi_n \to 0$ *when* $n \to \infty$, *we get* $c_{\alpha,n,\widetilde{j}} \to 0$.

(ii) *For* $\widetilde{j} = 1, \ldots, m$, *and* $\alpha := (\alpha_1, \ldots, \alpha_N)$, $\alpha_i \in \mathbb{Z}^+$, $i = 1, \ldots, N$, $|\alpha| := \sum_{i=1}^{N}\alpha_i = \widetilde{j}$, *we have that*

$$p_{\alpha,n} := p_{\alpha,n,\widetilde{j}} := \frac{\sum_{v_1=-\infty}^{\infty}\cdots\sum_{v_N=-\infty}^{\infty}\left(\prod_{i=1}^{N}v_i^{\alpha_i}\right)e^{-\frac{\sum_{i=1}^{N}v_i^2}{\xi_n}}}{\sum_{v_1=-\infty}^{\infty}\cdots\sum_{v_N=-\infty}^{\infty}e^{-\frac{\sum_{i=1}^{N}v_i^2}{\xi_n}}} < \infty. \tag{17}$$

for all $\xi_n \in (0, 1]$. *Additionally, let* $\alpha_i \in \mathbb{N}$, *then as* $\xi_n \to 0$ *when* $n \to \infty$, *we get* $p_{\alpha,n,\widetilde{j}} \to 0$.

(iii) *For* $\widetilde{j} = 1, \ldots, m$, *and* $\alpha := (\alpha_1, \ldots, \alpha_N)$, $\alpha_i \in \mathbb{Z}^+$, $i = 1, \ldots, N$, $|\alpha| := \sum_{i=1}^{N}\alpha_i = \widetilde{j}$, *we have that*

$$q_{\alpha,n} := q_{\alpha,n,\widetilde{j}} := \frac{\sum_{v_1=-\infty}^{\infty}\cdots\sum_{v_N=-\infty}^{\infty}\left(\prod_{i=1}^{N}v_i^{\alpha_i}\left(v_i^{2\hat{\alpha}} + \xi_n^{2\hat{\alpha}}\right)^{-\beta}\right)}{\sum_{v_1=-\infty}^{\infty}\cdots\sum_{v_N=-\infty}^{\infty}\prod_{i=1}^{N}\left(v_i^{2\hat{\alpha}} + \xi_n^{2\hat{\alpha}}\right)^{-\beta}} < \infty. \tag{18}$$

for all $\xi_n \in (0, 1]$ *where* $\hat{\alpha} \in \mathbb{N}$ *and* $\beta > \frac{\alpha_i+r+1}{2\hat{\alpha}}$. *Additionally, let* $\alpha_i \in \mathbb{N}$, *then as* $\xi_n \to 0$ *when* $n \to \infty$, *we get* $q_{\alpha,n,\widetilde{j}} \to 0$.

3   Main Results

We start with

Proposition 5 (*i*) *Let* $\alpha := (\alpha_1, \ldots, \alpha_N)$, $\alpha_i \in \mathbb{Z}^+$, $i = 1, \ldots, N \in \mathbb{N}$, $|\alpha| := \sum_{i=1}^{N} \alpha_i = m \in \mathbb{N}$. *Then, there exist* $K_1 > 0$ *such that*

$$
S_{P^*, \xi_n} = \frac{\xi_n^{-m} \sum_{v_1=-\infty}^{\infty} \cdots \sum_{v_N=-\infty}^{\infty} \left(\prod_{i=1}^{N} |v_i|^{\alpha_i} \right) e^{-\frac{\sum_{i=1}^{N} |v_i|}{\xi_n}}}{\sum_{v_1=-\infty}^{\infty} \cdots \sum_{v_N=-\infty}^{\infty} e^{-\frac{\sum_{i=1}^{N} |v_i|}{\xi_n}}} \tag{19}
$$
$$
\leq K_1 < \infty,
$$

for all $\xi_n \in (0, 1]$ *where* $n \in \mathbb{N}$ *and* $v = (v_1, \ldots, v_N)$.

(*ii*) *Let* $\alpha := (\alpha_1, \ldots, \alpha_N)$, $\alpha_i \in \mathbb{Z}^+$, $i = 1, \ldots, N \in \mathbb{N}$, $|\alpha| := \sum_{i=1}^{N} \alpha_i = m \in \mathbb{N}$. *Then, there exist* $K_2 > 0$ *such that*

$$
S_{W^*, \xi_n} = \frac{\xi_n^{-m} \sum_{v_1=-\infty}^{\infty} \cdots \sum_{v_N=-\infty}^{\infty} \left(\prod_{i=1}^{N} |v_i|^{\alpha_i} \right) e^{-\frac{\sum_{i=1}^{N} v_i^2}{\xi_n}}}{\sum_{v_1=-\infty}^{\infty} \cdots \sum_{v_N=-\infty}^{\infty} e^{-\frac{\sum_{i=1}^{N} v_i^2}{\xi_n}}} \tag{20}
$$
$$
\leq K_2 < \infty,
$$

for all $\xi_n \in (0, 1]$ *where* $n \in \mathbb{N}$ *and* $v = (v_1, \ldots, v_N)$.

(*iii*) *There exist* $K_3 > 0$ *such that*

$$
S_{Q^*, \xi_n} = \frac{\xi_n^{-m} \sum_{v_1=-\infty}^{\infty} \cdots \sum_{v_N=-\infty}^{\infty} \left(\prod_{i=1}^{N} |v_i|^{\alpha_i} \right) \left(\prod_{i=1}^{N} \left(v_i^{2\hat{\alpha}} + \xi_n^{2\hat{\alpha}} \right)^{-\beta} \right)}{\sum_{v_1=-\infty}^{\infty} \cdots \sum_{v_N=-\infty}^{\infty} \prod_{i=1}^{N} \left(v_i^{2\hat{\alpha}} + \xi_n^{2\hat{\alpha}} \right)^{-\beta}}
$$
$$
\leq K_3 < \infty, \tag{21}
$$

for all $\xi_n \in (0, 1]$ *where* $n, \hat{\alpha} \in \mathbb{N}$, $\beta > \frac{m+1}{2\hat{\alpha}}$, *and* $v = (v_1, \ldots, v_N)$.

Proof We observe that

$$
m = \sum_{i=1}^{N} \alpha_i \geq \alpha_i, \text{ for all } i = 1, 2, \ldots, N \in \mathbb{N}. \tag{22}
$$

Thus, we get

$$
\prod_{i=1}^{N} |v_i|^{\alpha_i} \leq \prod_{i=1}^{N} |v_i|^{m}. \tag{23}
$$

Therefore

$$
\begin{aligned}
S_{P^*,\xi_n} &\leq \frac{\xi_n^{-m} \sum_{v_1=-\infty}^{\infty} \cdots \sum_{v_N=-\infty}^{\infty} \left(\prod_{i=1}^{N} |v_i|^m\right) e^{-\frac{\sum_{i=1}^{N} |v_i|}{\xi_n}}}{\sum_{v_1=-\infty}^{\infty} \cdots \sum_{v_N=-\infty}^{\infty} e^{-\frac{\sum_{i=1}^{N} |v_i|}{\xi_n}}} \\
&\leq \frac{\xi_n^{-mN} \sum_{v_1=-\infty}^{\infty} \cdots \sum_{v_N=-\infty}^{\infty} \left(\prod_{i=1}^{N} |v_i|^m\right) e^{-\frac{\sum_{i=1}^{N} |v_i|}{\xi_n}}}{\sum_{v_1=-\infty}^{\infty} \cdots \sum_{v_N=-\infty}^{\infty} e^{-\frac{\sum_{i=1}^{N} |v_i|}{\xi_n}}} \\
&= \frac{\sum_{v_1=-\infty}^{\infty} \cdots \sum_{v_N=-\infty}^{\infty} \left(\prod_{i=1}^{N} \left(\frac{|v_i|}{\xi_n}\right)^m e^{-\frac{|v_i|}{\xi_n}}\right)}{\sum_{v_1=-\infty}^{\infty} \cdots \sum_{v_N=-\infty}^{\infty} \left(\prod_{i=1}^{N} e^{-\frac{|v_i|}{\xi_n}}\right)} \\
&= \prod_{i=1}^{N} \left(\frac{\xi_n^{-m} \sum_{v_i=-\infty}^{\infty} \left(|v_i|^m e^{-\frac{|v_i|}{\xi_n}}\right)}{\sum_{v_i=-\infty}^{\infty} e^{-\frac{|v_i|}{\xi_n}}}\right).
\end{aligned}
\tag{24}
$$

In [2], the authors showed that there exist $M_1 > 0$ such that

$$
\frac{\xi_n^{-m} \sum_{v_i=-\infty}^{\infty} \left(|v_i|^m e^{-\frac{|v_i|}{\xi_n}}\right)}{\sum_{v_i=-\infty}^{\infty} e^{-\frac{|v_i|}{\xi_n}}} < M_1 < \infty,
\tag{25}
$$

for all $\xi_n \in (0, 1]$. Therefore, by (24) and (25), we obtain

$$
S_{P^*,\xi_n} \leq \prod_{i=1}^{N} M_1 = M_1^N := K_1 < \infty.
\tag{26}
$$

On the other hand, as in (24), we observe that

$$
S_{W^*,\xi_n} \leq \prod_{i=1}^{N} \left(\frac{\xi_n^{-m} \sum_{v_i=-\infty}^{\infty} \left(|v_i|^m e^{-\frac{v_i^2}{\xi_n}}\right)}{\sum_{v_i=-\infty}^{\infty} e^{-\frac{v_i^2}{\xi_n}}}\right).
\tag{27}
$$

In [2], the authors showed that there exist $M_2 > 0$ such that

$$
\frac{\xi_n^{-m} \sum_{v_i=-\infty}^{\infty} \left(|v_i|^m e^{-\frac{v_i^2}{\xi_n}}\right)}{\sum_{v_i=-\infty}^{\infty} e^{-\frac{v_i^2}{\xi_n}}} \leq M_2 < \infty,
$$

for all $\xi_n \in (0, 1]$. Hence, we have

$$S_{W^*, \xi_n} \le \prod_{i=1}^{N} M_2 = M_2^N := K_2 < \infty, \qquad (28)$$

for all $\xi_n \in (0, 1]$. Finally, similar to (24), we get the inequality

$$S_{Q^*, \xi_n} \le \prod_{i=1}^{N} \left(\frac{\xi_n^{-m} \sum_{\nu_i=-\infty}^{\infty} \left(|\nu_i|^m \left(\nu_i^{2\hat{\alpha}} + \xi_n^{2\hat{\alpha}} \right)^{-\beta} \right)}{\sum_{\nu_i=-\infty}^{\infty} \left(\nu_i^{2\hat{\alpha}} + \xi_n^{2\hat{\alpha}} \right)^{-\beta}} \right). \qquad (29)$$

In [2], the authors showed that there exist $M_3 > 0$ such that

$$\frac{\xi_n^{-m} \sum_{\nu_i=-\infty}^{\infty} \left(|\nu_i|^m \left(\nu_i^{2\hat{\alpha}} + \xi_n^{2\hat{\alpha}} \right)^{-\beta} \right)}{\sum_{\nu_i=-\infty}^{\infty} \left(\nu_i^{2\hat{\alpha}} + \xi_n^{2\hat{\alpha}} \right)^{-\beta}} < M_3 < \infty, \qquad (30)$$

for all $\xi_n \in (0, 1]$, $\hat{\alpha} \in \mathbb{N}$, and $\beta > \frac{m+1}{2\hat{\alpha}}$. Therefore, by (29) and (30), we obtain

$$S_{Q^*, \xi_n} \le \prod_{i=1}^{N} M_3 = M_3^N := K_3 < \infty. \qquad (31)$$

Hence, by (26), (28), and (31), the proof is done.

Next, we state

Theorem 6 *Let* $m, n \in \mathbb{N}$, $f \in C^m \left(\mathbb{R}^N \right)$, $N \ge 1$, *and* $\gamma, \xi_n \in (0, 1]$. *Assume* $\left\| \frac{\partial^m f(\cdot,\dots,\cdot)}{\partial x_1^{\alpha_1} \dots \partial x_N^{\alpha_N}} \right\|_{\infty} < \infty$, *for all* $\alpha_j \in \mathbb{Z}^+$, $j = 1, \dots, N : |\alpha| := \sum_{j=1}^{N} \alpha_j = m$. *Then, for all* $x \in \mathbb{R}^N$, *we have*

(i)

$$P_{r,n}^{*\,[m]} (f; x) - f(x) = \sum_{\tilde{j}=1}^{m-1} \delta_{\tilde{j},r}^{[m]} \left(\sum_{|\alpha|=\tilde{j}} \frac{c_{\alpha, n, \tilde{j}} f_\alpha(x)}{\left(\prod_{i=1}^{N} \alpha_i! \right)} \right) + o \left(\xi_n^{m-\gamma} \right), \qquad (32)$$

(ii)

$$W_{r,n}^{*\,[m]} (f; x) - f(x) = \sum_{\tilde{j}=1}^{m-1} \delta_{\tilde{j},r}^{[m]} \left(\sum_{|\alpha|=\tilde{j}} \frac{p_{\alpha, n, \tilde{j}} f_\alpha(x)}{\left(\prod_{i=1}^{N} \alpha_i! \right)} \right) + o \left(\xi_n^{m-\gamma} \right), \qquad (33)$$

and

(iii)

$$Q_{r,n}^{*\,[m]}(f;x) - f(x) = \sum_{\tilde{j}=1}^{m-1} \delta_{\tilde{j},r}^{[m]} \left(\sum_{|\alpha|=\tilde{j}} \frac{q_{\alpha,n,\tilde{j}} f_\alpha(x)}{\left(\prod_{i=1}^N \alpha_i!\right)} \right) + o\left(\xi_n^{m-\gamma}\right), \quad (34)$$

where $\hat{\alpha} \in \mathbb{N}$, and $\beta > \frac{m+1}{2\hat{\alpha}}$. Above when $m = 1$, the sums on R.H.S. collapse.

Proof By Theorem 1 and Proposition 5.

We have

Theorem 7 *Let $f \in C^l\left(\mathbb{R}^N\right)$, l, $N \in \mathbb{N}$. Here μ_{ξ_n} is a Borel probability measure on $\mathbb{R}^N$, $\xi_n > 0$, $(\xi_n)_{n \in \mathbb{N}}$ a bounded sequence. Let $\tilde{\beta} := \left(\tilde{\beta}_1, \ldots, \tilde{\beta}_N\right)$, $\tilde{\beta}_i \in \mathbb{Z}^+$, $i = 1, \ldots, N$; $\left|\tilde{\beta}\right| := \sum_{i=1}^N \tilde{\beta}_i = l$. Here $f(x + \nu j)$, $x \in \mathbb{R}^N$, $\nu \in \mathbb{Z}^N$, is μ_{ξ_n}-integrable with respect to ν, for $j = 1, \ldots, r$. There exist μ_{ξ_n}-integrable functions $h_{i_1,j}$, $h_{\tilde{\beta}_1,i_2,j}$, $h_{\tilde{\beta}_1,\tilde{\beta}_2,i_3,j}, \ldots, h_{\tilde{\beta}_1,\tilde{\beta}_2,\ldots,\tilde{\beta}_{N-1},i_N,j} \geq 0$ $(j = 1, \ldots, r)$ on $\mathbb{R}^N$ such that*

$$\left| \frac{\partial^{i_1} f(x + \nu j)}{\partial x_1^{i_1}} \right| \leq h_{i_1,j}(\nu), \quad i_1 = 1, \ldots, \tilde{\beta}_1, \quad (35)$$

$$\left| \frac{\partial^{\tilde{\beta}_1 + i_2} f(x + \nu j)}{\partial x_2^{i_2} \partial x_1^{\tilde{\beta}_1}} \right| \leq h_{\tilde{\beta}_1,i_2,j}(\nu), \quad i_2 = 1, \ldots, \tilde{\beta}_2,$$

$$\vdots$$

$$\left| \frac{\partial^{\tilde{\beta}_1 + \tilde{\beta}_2 + \cdots + \tilde{\beta}_{N-1} + i_N} f(x + \nu j)}{\partial x_N^{i_N} \partial x_{N-1}^{\tilde{\beta}_{N-1}} \ldots \partial x_2^{\tilde{\beta}_2} \partial x_1^{\tilde{\beta}_1}} \right| \leq h_{\tilde{\beta}_1,\tilde{\beta}_2,\ldots,\tilde{\beta}_{N-1},i_N,j}(\nu), \quad i_N = 1, \ldots, \tilde{\beta}_N,$$

$\forall x \in \mathbb{R}^N$, $\nu \in \mathbb{Z}^N$.

(i) When

$$\mu_{\xi_n}(\nu) = \frac{e^{-\frac{\sum_{i=1}^N |\nu_i|}{\xi_n}}}{\sum_{\nu_1=-\infty}^{\infty} \cdots \sum_{\nu_N=-\infty}^{\infty} e^{-\frac{\sum_{i=1}^N |\nu_i|}{\xi_n}}}, \quad (36)$$

then both of the next exist and

$$\left(P_{r,n}^{*\,[m]}(f;x)\right)_{\tilde{\beta}} = P_{r,n}^{*\,[m]}\left(f_{\tilde{\beta}};x\right). \quad (37)$$

(*ii*) When

$$\mu_{\xi_n}(v) = \frac{e^{-\frac{\sum_{i=1}^N v_i^2}{\xi_n}}}{\sum_{v_1=-\infty}^{\infty} \cdots \sum_{v_N=-\infty}^{\infty} e^{-\frac{\sum_{i=1}^N v_i^2}{\xi_n}}}, \tag{38}$$

then both of the next exist and

$$\left(W_{r,n}^{*\,[m]}(f;x)\right)_{\tilde{\beta}} = W_{r,n}^{*\,[m]}\left(f_{\tilde{\beta}};x\right). \tag{39}$$

(*iii*) Let $\hat{\alpha} \in \mathbb{N}$ and $\beta > \frac{1}{\hat{\alpha}}$. When

$$\mu_{\xi_n}(v) = \frac{\prod_{i=1}^N \left(v_i^{2\hat{\alpha}} + \xi_n^{2\hat{\alpha}}\right)^{-\beta}}{\sum_{v_1=-\infty}^{\infty} \cdots \sum_{v_N=-\infty}^{\infty} \prod_{i=1}^N \left(v_i^{2\hat{\alpha}} + \xi_n^{2\hat{\alpha}}\right)^{-\beta}}, \tag{40}$$

then both of the next exist and

$$\left(Q_{r,n}^{*\,[m]}(f;x)\right)_{\tilde{\beta}} = Q_{r,n}^{*\,[m]}\left(f_{\tilde{\beta}};x\right). \tag{41}$$

Proof By Theorem 2.

We give our final result as follows

Theorem 8 *Let* $f \in C^{m+l}\left(\mathbb{R}^N\right)$, $m, l, N \in \mathbb{N}$. *Assumptions of Theorem 7 are valid.* *Call* $\psi = 0$, $\tilde{\beta}$. *Assume* $\left\|f_{\psi+\alpha}\right\|_\infty \leq M$, $M > 0$, *for all* $\alpha : |\alpha| = m$. *Let* $1 \geq \xi_n > 0$ *and* μ_{ξ_n} *be probability Borel measures on* $\mathbb{R}^N$. *Also, let* $0 < \gamma \leq 1$ *and* $x \in \mathbb{R}^N$. *Then*

(*i*) *when*

$$\mu_{\xi_n}(v) = \frac{e^{-\frac{\sum_{i=1}^N |v_i|}{\xi_n}}}{\sum_{v_1=-\infty}^{\infty} \cdots \sum_{v_N=-\infty}^{\infty} e^{-\frac{\sum_{i=1}^N |v_i|}{\xi_n}}}, \tag{42}$$

we get

$$\left(P_{r,n}^{*\,[m]}(f;x)\right)_{\psi} - f_\psi(x) = \sum_{\tilde{j}=1}^{m-1} \delta_{\tilde{j},r}^{[m]}\left(\sum_{|\alpha|=\tilde{j}} \frac{c_{\alpha,n,\tilde{j}} f_{\psi+\alpha}(x)}{\left(\prod_{i=1}^N \alpha_i!\right)}\right) + o\left(\xi_n^{m-\gamma}\right). \tag{43}$$

When $m = 1$ *the sum on R.H.S. collapses.*

(*ii*) *When*

$$\mu_{\xi_n}(v) = \frac{e^{-\frac{\sum_{i=1}^N v_i^2}{\xi_n}}}{\sum_{v_1=-\infty}^{\infty} \cdots \sum_{v_N=-\infty}^{\infty} e^{\frac{\sum_{i=1}^N v_i^2}{\xi_n}}}, \tag{44}$$

we have

$$\left(W_{r,n}^{*\,[m]}\left(f;x\right)\right)_{\psi} - f_{\psi}\left(x\right) = \sum_{\tilde{j}=1}^{m-1} \delta_{\tilde{j},r}^{[m]}\left(\sum_{|\alpha|=\tilde{j}} \frac{p_{\alpha,n,\tilde{j}}\, f_{\psi+\alpha}\left(x\right)}{\left(\prod_{i=1}^{N} \alpha_i!\right)}\right) + o\left(\xi_n^{m-\gamma}\right).$$

(45)

When $m = 1$ the sum on R.H.S. collapses.

(iii) *Let $\hat{\alpha} \in \mathbb{N}$, and $\beta > \frac{m+1}{2\hat{\alpha}}$. When*

$$\mu_{\xi_n}\left(v\right) = \frac{\prod_{i=1}^{N}\left(v_i^{2\hat{\alpha}} + \xi_n^{2\hat{\alpha}}\right)^{-\beta}}{\sum_{v_1=-\infty}^{\infty}\cdots\sum_{v_N=-\infty}^{\infty}\prod_{i=1}^{N}\left(v_i^{2\hat{\alpha}} + \xi_n^{2\hat{\alpha}}\right)^{-\beta}},$$

(46)

we get

$$\left(Q_{r,n}^{*\,[m]}\left(f;x\right)\right)_{\psi} - f_{\psi}\left(x\right) = \sum_{\tilde{j}=1}^{m-1} \delta_{\tilde{j},r}^{[m]}\left(\sum_{|\alpha|=\tilde{j}} \frac{q_{\alpha,n,\tilde{j}}\, f_{\psi+\alpha}\left(x\right)}{\left(\prod_{i=1}^{N} \alpha_i!\right)}\right) + o\left(\xi_n^{m-\gamma}\right).$$

(47)

When $m = 1$ the sum on R.H.S. collapses.

Proof By Theorems 6, 7.

4 Applications

For $m = 1$, we have

Corollary 9 *Let $f \in C^1\left(\mathbb{R}^N\right)$, $N \geq 1$, with all $\left\|\frac{\partial f}{\partial x_i}\right\|_{\infty} \leq M$, $M > 0$, $i = 1, \ldots, N$. Let $1 \geq \xi_n > 0$, $(\xi_n)_{n \in \mathbb{N}}$ bounded sequence, μ_{ξ_n} probability Borel measures on $\mathbb{R}^N$ and $0 < \gamma \leq 1$. Then*

$$P_{r,n}^{*\,[1]}\left(f;x\right) - f\left(x\right) = o\left(\xi_n^{1-\gamma}\right),$$

$$W_{r,n}^{*\,[1]}\left(f;x\right) - f\left(x\right) = o\left(\xi_n^{1-\gamma}\right),$$

and for $\hat{\alpha} \in \mathbb{N}$, and $\beta > \frac{1}{\hat{\alpha}}$, we have

$$Q_{r,n}^{*\,[1]}\left(f;x\right) - f\left(x\right) = o\left(\xi_n^{1-\gamma}\right).$$

Proof By Theorem 6.

For $m = 2$, we get

Corollary 10 *Let* $f \in C^2(\mathbb{R}^2)$, *with all* $\left\|\frac{\partial^2 f}{\partial x_1^2}\right\|_\infty, \left\|\frac{\partial^2 f}{\partial x_2^2}\right\|_\infty, \left\|\frac{\partial^2 f}{\partial x_1 \partial x_2}\right\|_\infty \leq M$, $M > 0$. *Let* $1 \geq \xi_n > 0$, $(\xi_n)_{n\in\mathbb{N}}$ *bounded sequence*, μ_{ξ_n} *probability Borel measures on* $\mathbb{R}^2$. *Also* $0 < \gamma \leq 1$, $x \in \mathbb{R}^2$. *Then*

$$P_{r,n}^{*\,[2]}(f;x) - f(x) = \left(\sum_{j=1}^r \alpha_{j,r}^{[2]} j\right)\left(\sum_{|\alpha|=1} \frac{c_{\alpha,n,1} f_\alpha(x)}{\left(\prod_{i=1}^2 \alpha_i!\right)}\right) + o\left(\xi_n^{2-\gamma}\right),$$

$$W_{r,n}^{*\,[2]}(f;x) - f(x) = \left(\sum_{j=1}^r \alpha_{j,r}^{[2]} j\right)\left(\sum_{|\alpha|=1} \frac{p_{\alpha,n,1} f_\alpha(x)}{\left(\prod_{i=1}^2 \alpha_i!\right)}\right) + o\left(\xi_n^{2-\gamma}\right),$$

and for $\hat{\alpha} \in \mathbb{N}$, *and* $\beta > \frac{3}{2\hat{\alpha}}$, *we have*

$$Q_{r,n}^{*\,[2]}(f;x) - f(x) = \left(\sum_{j=1}^r \alpha_{j,r}^{[2]} j\right)\left(\sum_{|\alpha|=1} \frac{q_{\alpha,n,1} f_\alpha(x)}{\left(\prod_{i=1}^2 \alpha_i!\right)}\right) + o\left(\xi_n^{2-\gamma}\right).$$

Proof By *Theorem* 6.

References

1. Anastassiou, G.A.: Approximation by Multivariate Singular Integrals, Briefs in Mathematics. Springer, New York (2011)
2. Anastassiou, G.A., Kester, M.: Voronovskaya type asymptotic expansions for generalized discrete singular operators. J. Concr. Appl. Math. **13**(1–2), 108–119 (2015)
3. Anastassiou, G.A., Kester, M.: Uniform approximation with rates by multivariate generalized discrete singular operators. In: AMAT 2015 Proceedings. Springer, New York (2016), accepted for publication
4. Favard, J.: Sur les multiplicateurs d'interpolation. J. Math. Pures Appl. **23**(9), 219–247 (1944)

Variational Analysis of a Quasistatic Contact Problem

Mircea Sofonea

Abstract We start by proving an existence and uniqueness result for a new class of variational inequalities which arise in the study of quasistatic models of contact. The novelty lies in the special structure of these inequalities which involve history-dependent operators. The proof is based on arguments of monotonicity, convexity and fixed point. Then, we consider a mathematical model which describes the frictional contact between an elastic-viscoplastic body and a moving foundation. The mechanical process is assumed to be quasistatic, and the contact is modeled with a multivalued normal compliance condition with unilateral constraint and memory term, associated to a sliding version of Coulomb's law of dry friction. We prove that the model casts in the abstract setting of variational inequalities, with a convenient choice of spaces and operators. Further, we apply our abstract result to prove the unique weak solvability of the contact model.

1 Introduction

Contact phenomena involving deformable bodies abound in industry and everyday life. They lead to nonsmooth and nonlinear mathematical problems. Their analysis, including existence and uniqueness results, was carried out in a large number of works, see for instance [3, 4, 6, 9, 16, 17] and the references therein. The numerical analysis of the problems, including error estimation for discrete schemes and numerical simulations, can be found in [10, 11, 13, 14, 22]. The state of the art in the field, including applications in engineering, could be found in the recent special issue [15].

The study of both the qualitative and numerical analysis of various mathematical models of contact is made by using various mathematical tools, including the theory of variational inequalities. At the heart of this theory is the intrinsic inclusion of free boundaries in an elegant mathematical formulation. Existence and uniqueness

M. Sofonea (✉)
University of Perpignan Via Domitia, Perpignan, France
e-mail: sofonea@univ-perp.fr

© Springer International Publishing Switzerland 2016   245
G.A. Anastassiou and O. Duman (eds.), *Intelligent Mathematics II:*
Applied Mathematics and Approximation Theory, Advances in Intelligent Systems
and Computing 441, DOI 10.1007/978-3-319-30322-2_17

results in the study of variational inequalities can be found in [1, 2, 12, 16, 20], for instance. References concerning their numerical analysis of include [5, 11, 13].

The large variety of frictional or frictionless models of quasistatic contact led to different classes of time-dependent or evolutionary variational inequalities which, on occasion, have been studied in an abstract framework. Examples could be found in [7, 8, 19, 20]. Nevertheless, it was recently recognized that some models of contact lead to weak formulations expressed in terms of variational inequalities which are more general than those studied in the above-mentioned papers. Therefore, in order to prove the unique solvability of these models, there is a need to extend these results to a more general classes of inequalities.

The first aim of the present paper is to provide such extension. Thus, we provide here an abstract existence and uniqueness result in the study of a new class of history-dependent variational inequalities. Our second aim is to illustrate how this result is useful in the analysis of a new model of contact with viscoplastic materials.

The rest of the paper is structured as follows. In Sect. 2, we introduce some notation and preliminary material. Then, we state and prove our main abstract result, Theorem 2. In Sect. 3 and we describe the frictional contact problem, list the assumption on the data, derive its variational formulation and state its unique weak solvability, Theorem 3. The proof of Theorem 3, based on the abstract result provided by Theorem 2, is presented in Sect. 4.

2   An Abstract Existence and Uniqueness Result

Everywhere in this paper, we use the notation $\mathbb{N}$ for the set of positive integers and $\mathbb{R}_+$ will represent the set of nonnegative real numbers, i.e. $\mathbb{R}_+ = [0, +\infty)$. For a normed space $(X, \|\cdot\|_X)$ we use the notation $C(\mathbb{R}_+; X)$ for the space of continuously functions defined on $\mathbb{R}_+$ with values in X. For a subset $K \subset X$ we still use the symbol $C(\mathbb{R}_+; K)$ for the set of continuous functions defined on $\mathbb{R}_+$ with values on K. The following result, obtained in [18], will be used twice in this paper.

Theorem 1 *Let $(X, \|\cdot\|_X)$ be a real Banach space and let $\Lambda : C(\mathbb{R}_+; X) \to C(\mathbb{R}_+; X)$ be a nonlinear operator. Assume that for all $n \in \mathbb{N}$ there exist two constants $c_n \geq 0$ and $d_n \in [0, 1)$ such that*

$$\|\Lambda u(t) - \Lambda v(t)\|_X \leq c_n \int_0^t \|u(s) - v(s)\|_X \, ds + d_n \|u(t) - v(t)\|_X$$

for all $u, v \in C(\mathbb{R}_+; X)$ and all $t \in [0, n]$. Then the operator Λ has a unique fixed point $\eta^* \in C(\mathbb{R}_+; X)$.

The proof of Theorem 1 was carried out in several steps, based on the fact that the space $C(\mathbb{R}_+; X)$ can be organized as a Fréchet space with a convenient distance function.

We assume in what follows that X is real Hilbert space and Y is a real normed space. Let K be a subset of X, $A : K \subset X \to X$ and $\mathscr{S} : C(\mathbb{R}_+; X) \to C(\mathbb{R}_+; Y)$. Moreover, let $j : Y \times X \times K \to \mathbb{R}$ and $f : \mathbb{R}_+ \to X$. We consider the following assumptions.

$$K \text{ is a closed, convex, nonempty subset of } X. \tag{1}$$

$$\begin{cases} \text{(a) There exists } L > 0 \text{ such that} \\ \qquad \|Au_1 - Au_2\|_X \leq L\|u_1 - u_2\|_Y \quad \forall u_1, u_2 \in K. \\ \text{(b) There exists } m > 0 \text{ such that} \\ \qquad (Au_1 - Au_2, u_1 - u_2)_X \geq m\|u_1 - u_2\|_X^2 \quad \forall u_1, u_2 \in K. \end{cases} \tag{2}$$

$$\begin{cases} \text{(a) For all } y \in Y \text{ and } u \in X, \ j(y, u, \cdot) \text{ is convex and l.s.c on } K. \\ \text{(b) There exists } \alpha > 0 \text{ and } \beta > 0 \text{ such that} \\ \qquad j(y_1, u_1, v_2) - j(y_1, u_1, v_1) + j(y_2, u_2, v_1) - j(y_2, u_2, v_2) \\ \qquad \leq \alpha\|y_1 - y_2\|_Y \|v_1 - v_2\|_X + \beta\|u_1 - u_2\|_X\|v_1 - v_2\|_X \\ \qquad \forall y_1, y_2 \in Y, \ \forall u_1, u_2 \in X, \ \forall v_1, v_2 \in K. \end{cases} \tag{3}$$

$$\begin{cases} \text{For all } n \in \mathbb{N} \text{ there exists } s_n > 0 \text{ such that} \\ \|\mathscr{S}u_1(t) - \mathscr{S}u_2(t)\|_Y \leq s_n \int_0^t \|u_1(s) - u_2(s)\|_X \, ds \\ \forall u_1, u_2 \in C(\mathbb{R}_+; X), \ \forall t \in [0, n]. \end{cases} \tag{4}$$

$$f \in C(\mathbb{R}_+; X). \tag{5}$$

Concerning these assumptions we have the following comments. First, assumption (2) show that A is a Lipschitz continuous strongly monotone operator on K. Next, in (3) we use the abbreviation l.s.c. for a lower semicontinuous function. Finally, following the terminology introduced in [19] and used in various papers, condition (4) show that the operator $\mathscr{S}$ is a *history-dependent operator*. Example of operators which satisfies this condition could be find in [19, 20]. Variational inequalities involving history-dependent operators are also called *history-dependent variational inequalities*. In their study we have the following existence and uniqueness result.

Theorem 2 *Assume that (1)–(5) hold. Moreover, assume that*

$$m > \beta, \tag{6}$$

where m and β are the constants in (2) and (3), respectively. Then, there exists a unique function $u \in C(\mathbb{R}_+; K)$ such that, for all $t \in \mathbb{R}_+$, the following inequality holds:

$$\begin{aligned} u(t) \in K, \quad (Au(t), v - u(t))_X &+ j(\mathscr{S}u(t), u(t), v) \\ - j(\mathscr{S}u(t), u(t), u(t)) &\geq (f(t), v - u(t))_X \quad \forall v \in K. \end{aligned} \tag{7}$$

248   M. Sofonea

Proof The proof of Theorem 2 is based on argument similar to those presented in
[19] and, for this reason, we skip the details. The main step in the proof are the
followings.

(i) Let $\eta \in C(\mathbb{R}_+; X)$ be fixed and denote by $y_\eta \in C(\mathbb{R}_+; Y)$ the function given by

$$y_\eta(t) = \mathscr{S}\eta(t) \quad \forall t \in \mathbb{R}_+. \tag{8}$$

In the first step we use standard arguments on time-dependent elliptic variational
inequalities to prove that there exists a unique function $u_\eta \in C(\mathbb{R}_+; K)$ such that,
for all $t \in \mathbb{R}_+$, the following inequality holds:

$$u_\eta(t) \in K, \quad (Au_\eta(t), v - u_\eta(t))_X + j(y_\eta(t), \eta(t), v) \tag{9}$$
$$- j(z_\eta(t), \eta(t), u_\eta(t)) \geq (f(t), v - u_\eta(t))_X \quad \forall v \in K.$$

(ii) Next, in the second step, we consider the operator $\Lambda : C(\mathbb{R}_+; X) \to C(\mathbb{R}_+; K)$
$\subset C(\mathbb{R}_+; X)$ defined by equality

$$\Lambda\eta = u_\eta \quad \forall \eta \in C(\mathbb{R}_+; X) \tag{10}$$

and we prove that it has a unique fixed point $\eta^* \in C(\mathbb{R}_+; K)$. Indeed, let $\eta_1, \eta_2 \in$
$C(\mathbb{R}_+; X)$, and let y_i, be the functions defined by (8) for $\eta = \eta_i$, i.e. $y_i = y_{\eta_i}$, for
$i = 1, 2$. Also, denote by u_i the solution of the variational inequality (9) for $\eta = \eta_i$,
i.e. $u_i = u_{\eta_i}$, $i = 1, 2$. Let $n \in \mathbb{N}$ and $t \in [0, n]$. Then, using (9), (2) and (3) is easy
to see that

$$m\|u_1(t) - u_2(t)\|_X \leq \alpha\|y_1(t) - y_2(t)\|_Y + \beta\|\eta_1(t) - \eta_2(t)\|_X. \tag{11}$$

Moreover, by the assumptions (4) on the operator $\mathscr{S}$ one has

$$\|y_1(t) - y_2(t)\|_Y = \|\mathscr{S}\eta_1(t) - \mathscr{S}\eta_2(t)\|_Y \leq s_n \int_0^t \|\eta_1(s) - \eta_2(s)\|_X \, ds. \tag{12}$$

Thus, using (10)–(12) yields

$$\|\Lambda\eta_1(t) - \Lambda\eta_2(t)\|_X = \|u_1(t) - u_2(t)\|_X$$
$$\leq \frac{\alpha s_n}{m} \int_0^t \|\eta_1(s) - \eta_2(s)\|_X \, ds + \frac{\beta}{m}\|\eta_1(t) - \eta_2(t)\|_X$$

which, together with the smallness assumption (6) and Theorem 1, implies that the
operator Λ has a unique fixed point $\eta^* \in C(\mathbb{R}_+; X)$. Moreover, since Λ has values
on $C(\mathbb{R}_+; K)$, we deduce that $\eta^* \in C(\mathbb{R}_+; K)$.

(iii) Let $\eta^* \in C(\mathbb{R}_+; K)$ be the fixed point of the operator Λ. It follows from (8) and (10) that

$$y_{\eta^*}(t) = \mathscr{S}\eta^*(t), \qquad u_{\eta^*}(t) = \eta^*(t). \tag{13}$$

for all $t \in \mathbb{R}_+$. Now, letting $\eta = \eta^*$ in the inequality (9) and using (13) we conclude that $\eta^* \in C(\mathbb{R}_+; K)$ is a solution to the variational inequality (7). This proves the existence part in Theorem 2.

(iv) The uniqueness part is a consequence of the uniqueness of the fixed point of the operator Λ and can be proved as follows. Denote by $\eta^* \in C(\mathbb{R}_+; K)$ the solution of the variational inequality (7) obtained above, and let $\eta \in C(\mathbb{R}_+; K)$ be a different solution of this inequality, which implies that

$$(A\eta(t), v - \eta(t))_X + j(\mathscr{S}\eta(t), \eta(t), v) \tag{14}$$
$$- j(\mathscr{S}\eta(t), \eta(t), \eta(t)) \geq (f(t), v - \eta(t))_X \qquad \forall v \in K, \ t \in \mathbb{R}_+.$$

Letting $y_\eta = \mathscr{S}\eta \in C(\mathbb{R}_+; Y)$, inequality (14) implies that η is solution to the variational inequality (9). On the other hand, by step (i) this inequality has a unique solution u_η and, therefore,

$$\eta = u_\eta. \tag{15}$$

This shows that $\Lambda\eta = \eta$ where Λ is the operator defined by (10). Therefore, by Step (i) it follows that $\eta = \eta^*$, which concludes proof.

3   The Contact Model and Main Result

We turn now to an application of Theorem 2 in Contact Mechanics and, to this end, we start by presenting some notations and preliminaries. Let Ω a regular domain of $\mathbb{R}^d$ $(d = 2, 3)$ with surface Γ that is partitioned into three disjoint measurable parts Γ_1, Γ_2 and Γ_3, such that meas $(\Gamma_1) > 0$ and, in addition, Γ_3 is plane. We use the notation $\mathbf{x} = (x_i)$ for a typical point in Ω and $\boldsymbol{v} = (v_i)$ for the outward unit normal at Γ. In order to simplify the notation, we do not indicate explicitly the dependence of various functions on the spatial variable $\mathbf{x}$. Let $\mathbb{R}^d$ be d-dimensional real linear space and the let $\mathbb{S}^d$ denote the space of second order symmetric tensors on $\mathbb{R}^d$ or, equivalently, the space of symmetric matrices of order d. The canonical inner products and the corresponding norms on $\mathbb{R}^d$ and $\mathbb{S}^d$ are given by

$$\mathbf{u} \cdot \mathbf{v} = u_i v_i, \quad \|\mathbf{v}\| = (\mathbf{v} \cdot \mathbf{v})^{1/2} \qquad \forall \mathbf{u} = (u_i), \ \mathbf{v} = (v_i) \in \mathbb{R}^d,$$
$$\boldsymbol{\sigma} \cdot \boldsymbol{\tau} = \sigma_{ij} \tau_{ij}, \quad \|\boldsymbol{\tau}\| = (\boldsymbol{\tau} \cdot \boldsymbol{\tau})^{1/2} \qquad \forall \boldsymbol{\sigma} = (\sigma_{ij}), \ \boldsymbol{\tau} = (\tau_{ij}) \in \mathbb{S}^d,$$

respectively. Here and below the indices i, j, k, l run between 1 and d and, unless stated otherwise, the summation convention over repeated indices is used.

We use standard notation for the Lebesgue and the Sobolev spaces associated to Ω and Γ. Also, we introduce the spaces

$$V = \{ \mathbf{v} = (v_i) \in H^1(\Omega)^d : \mathbf{v} = \mathbf{0} \text{ a.e. on } \Gamma_1 \},$$
$$Q = \{ \boldsymbol{\tau} = (\tau_{ij}) \in L^2(\Omega)^{d \times d} : \tau_{ij} = \tau_{ji} \},$$
$$Q_1 = \{ \boldsymbol{\tau} = (\tau_{ij}) \in Q : \text{Div} \boldsymbol{\tau} \in L^2(\Omega)^d \}.$$

Here and below $\text{Div} \boldsymbol{\tau} = (\tau_{ij,j})$ denotes the divergence of the field $\boldsymbol{\tau}$, where the index that follows a coma indicates a partial derivative with the corresponding component of the spatial variable $\mathbf{x}$, i.e. $\tau_{ij,j} = \partial \tau_{ij} / \partial x_j$. The spaces Q and Q_1 are real Hilbert spaces with the canonical inner products given by

$$(\boldsymbol{\sigma}, \boldsymbol{\tau})_Q = \int_\Omega \boldsymbol{\sigma} \cdot \boldsymbol{\tau} \, dx \qquad \forall \boldsymbol{\sigma}, \boldsymbol{\tau} \in Q,$$

$$(\boldsymbol{\sigma}, \boldsymbol{\tau})_{Q_1} = \int_\Omega \boldsymbol{\sigma} \cdot \boldsymbol{\tau} \, dx + \int_\Omega \text{Div} \boldsymbol{\sigma} \cdot \text{Div} \boldsymbol{\tau} \, dx \qquad \forall \boldsymbol{\sigma}, \boldsymbol{\tau} \in Q_1.$$

In addition, since $\text{meas}(\Gamma_1) > 0$, it is well known that V is a real Hilbert space with the inner product

$$(\mathbf{u}, \mathbf{v})_V = \int_\Omega \boldsymbol{\varepsilon}(\mathbf{u}) \cdot \boldsymbol{\varepsilon}(\mathbf{v}) \, dx \qquad \forall \mathbf{u}, \mathbf{v} \in V$$

where $\boldsymbol{\varepsilon}$ is the deformation operator, i.e. $\boldsymbol{\varepsilon}(\mathbf{u}) = \varepsilon_{ij}(\mathbf{u})$, $\varepsilon_{ij}(\mathbf{u}) = \frac{1}{2}(u_{i,j} + u_{j,i})$, $u_{i,j} = \partial u_i / \partial x_j$. The associated norms on the spaces V, Q and Q_1 will be denoted by $\| \cdot \|_V$, $\| \cdot \|_Q$ and $\| \cdot \|_{Q_1}$, respectively.

For all $\mathbf{v} \in V$ we still write $\mathbf{v}$ for the trace of $\mathbf{v}$ to Γ. We recall that, by the Sobolev trace theorem, there exists a positive constant c_0 which depends on Ω, Γ_1 and Γ_3 such that

$$\|\mathbf{v}\|_{L^2(\Gamma_3)^d} \leq c_0 \|\mathbf{v}\|_V \qquad \forall \mathbf{v} \in V. \tag{16}$$

For $\mathbf{v} \in V$ we denote by v_ν and $\mathbf{v}_\tau$ the normal and tangential components of $\mathbf{v}$ on Γ, in the sense of traces, given by $v_\nu = \mathbf{v} \cdot \boldsymbol{\nu}$, $\mathbf{v}_\tau = \mathbf{v} - v_\nu \boldsymbol{\nu}$. Moreover, for $\boldsymbol{\sigma} \in Q_1$ we denote by $\sigma_\nu \in H^{-\frac{1}{2}}(\Gamma)$ its normal component, in the sense of traces. Let $R : H^{-\frac{1}{2}}(\Gamma) \to L^2(\Gamma)$ be a linear continuous operator. Then, there exists a positive constant $c_R > 0$ which depends on R, Ω and Γ_3 such that

$$\|R\sigma_\nu\|_{L^2(\Gamma_3)} \leq c_R \|\boldsymbol{\sigma}\|_{Q_1} \qquad \forall \boldsymbol{\sigma} \in Q_1. \tag{17}$$

Next, we recall that if $\boldsymbol{\sigma}$ is a regular function, then its normal and tangential components of the stress field $\boldsymbol{\sigma}$ on the boundary are defined by $\sigma_\nu = (\boldsymbol{\sigma}\boldsymbol{\nu}) \cdot \boldsymbol{\nu}$, $\boldsymbol{\sigma}_\tau = \boldsymbol{\sigma}\boldsymbol{\nu} - \sigma_\nu \boldsymbol{\nu}$ and the following Green's formula holds:

$$\int_\Omega \boldsymbol{\sigma} \cdot \boldsymbol{\varepsilon}(\mathbf{v})\, dx + \int_\Omega \mathrm{Div}\boldsymbol{\sigma} \cdot \mathbf{v}\, dx = \int_\Gamma \boldsymbol{\sigma} \boldsymbol{v} \cdot \mathbf{v}\, da \qquad \forall\, \mathbf{v} \in V. \qquad (18)$$

With these notation, we formulate the following problem.

Problem $\mathscr{P}$. *Find a displacement field* $\mathbf{u} = (u_i) : \Omega \times \mathbb{R}_+ \to \mathbb{R}^d$ *and a stress field* $\boldsymbol{\sigma} = (\sigma_{ij}) : \Omega \times \mathbb{R}_+ \to \mathbb{S}^d$ *such that*

$$\dot{\boldsymbol{\sigma}}(t) = \mathscr{E}\boldsymbol{\varepsilon}(\dot{\mathbf{u}}(t)) + \mathscr{G}(\boldsymbol{\sigma}(t), \boldsymbol{\varepsilon}(\mathbf{u}(t))) \quad \text{in } \Omega, \qquad (19)$$

$$\mathrm{Div}\boldsymbol{\sigma}(t) + \mathbf{f}_0 = \mathbf{0} \quad \text{in } \Omega, \qquad (20)$$

$$\mathbf{u}(t) = \mathbf{0} \quad \text{on } \Gamma_1, \qquad (21)$$

$$\boldsymbol{\sigma}(t)\boldsymbol{v} = \mathbf{f}_2(t) \quad \text{on } \Gamma_2, \qquad (22)$$

$$-\boldsymbol{\sigma}_\tau(t) = \mu|R\sigma_v(t)|\,\mathbf{n}^* \quad \text{on } \Gamma_3, \qquad (23)$$

for all $t \in \mathbb{R}_+$, *there exists* $\xi : \Gamma_3 \times \mathbb{R}_+ \to \mathbb{R}$ *which satisfies*

$$\begin{cases} u_v(t) \leq g, \qquad \sigma_v(t) + p(u_v(t)) + \xi(t) \leq 0, \\[4pt] (u_v(t) - g)\big(\sigma_v(t) + p(u_v(t)) + \xi(t)\big) = 0, \\[4pt] 0 \leq \xi(t) \leq F\Big(\displaystyle\int_0^t u_v^+(s)\, ds\Big), \\[4pt] \xi(t) = 0 \ \text{if} \ u_v(t) < 0, \\[4pt] \xi(t) = F\Big(\displaystyle\int_0^t u_v^+(s)\, ds\Big) \ \text{if} \ u_v(t) > 0 \end{cases} \quad \text{on } \Gamma_3, \qquad (24)$$

for all $t \in \mathbb{R}_+$ *and, moreover,*

$$\mathbf{u}(0) = \mathbf{u}_0, \qquad \boldsymbol{\sigma}(0) = \boldsymbol{\sigma}_0 \quad \text{in } \Omega. \qquad (25)$$

Problem $\mathscr{P}$ represents a mathematical model which describes the quasistatic process of contact between a viscoplastic body and a moving foundation. Here Ω represents the reference configuration of a the body and the dot above denotes the derivative with respect the time variable, i.e. $\dot{f} = \frac{\partial f}{\partial t}$. Equation (19) represents the viscoplastic constitutive law. Details and various mechanical interpretation concerning such kind of laws can be found in [9, 20], for instance. Equation (20) represents the equation of equilibrium in which $\mathbf{f}_0$ denotes the density of body forces, assumed to be time-independent. We use this equation since the process is quasistatic and, therefore, the inertial term in the equation of motion is neglected. Conditions (21) and (22) are the displacement and the traction boundary condition, respectively. They describe the fact that the body is fixed on Γ_1 and prescribed traction of density $\mathbf{f}_2$ act on Γ_2, during the contact process.

Conditions (23) and (24) represent a sliding version of Coulomb's law of dry friction and a normal compliance contact condition with unilateral constraint and memory term, respectively. Their are obtained from arguments presented in our recent paper [21] and, for this reason, we do not describe them with details. We just mention that μ denotes the coefficient of friction, $\mathbf{n}^*$ denotes a given unitary vector in the plane on Γ_3 and $v^* < 0$ is given. In addition, p and F are given function which describe the deformability and the memory effects of the foundation, $g > 0$ is a given depth and r^+ represent the positive part of r, i.e. $r^+ = \max\{r, 0\}$. Finally, conditions (25) represent the initial conditions for the displacement and the stress field, respectively.

In the study of Problem $\mathscr{P}$ we assume that the elasticity operator $\mathscr{E}$ and the nonlinear constitutive function $\mathscr{G}$ satisfy the following conditions.

$$
\left\{
\begin{aligned}
&\text{(a) } \mathscr{E} = (\mathscr{E}_{ijkl}) : \Omega \times \mathbb{S}^d \to \mathbb{S}^d. \\
&\text{(b) } \mathscr{E}_{ijkl} = \mathscr{E}_{klij} = \mathscr{E}_{jikl} \in L^\infty(\Omega),\ 1 \le i, j, k, l \le d. \\
&\text{(c) There exists } m_\mathscr{E} > 0 \text{ such that} \\
&\quad \mathscr{E}\boldsymbol{\tau} \cdot \boldsymbol{\tau} \ge m_\mathscr{E}\|\boldsymbol{\tau}\|^2 \text{ for all } \boldsymbol{\tau} \in \mathbb{S}^d, \text{ a.e. in } \Omega.
\end{aligned}
\right.
\tag{26}
$$

$$
\left\{
\begin{aligned}
&\text{(a) } \mathscr{G} : \Omega \times \mathbb{S}^d \times \mathbb{S}^d \to \mathbb{S}^d. \\
&\text{(b) There exists } L_\mathscr{G} > 0 \text{ such that} \\
&\quad \|\mathscr{G}(\mathbf{x}, \boldsymbol{\sigma}_1, \boldsymbol{\varepsilon}_1) - \mathscr{G}(\mathbf{x}, \boldsymbol{\sigma}_2, \boldsymbol{\varepsilon}_2)\| \le L_\mathscr{G}(\|\boldsymbol{\sigma}_1 - \boldsymbol{\sigma}_2\| + \|\boldsymbol{\varepsilon}_1 - \boldsymbol{\varepsilon}_2\|) \\
&\quad \text{for all } \boldsymbol{\sigma}_1, \boldsymbol{\sigma}_2, \boldsymbol{\varepsilon}_1, \boldsymbol{\varepsilon}_2 \in \mathbb{S}^d, \text{ a.e. } \mathbf{x} \in \Omega. \\
&\text{(c) The mapping } \mathbf{x} \mapsto \mathscr{G}(\mathbf{x}, \boldsymbol{\sigma}, \boldsymbol{\varepsilon}) \text{ is measurable on } \Omega, \\
&\quad \text{for all } \boldsymbol{\sigma}, \boldsymbol{\varepsilon} \in \mathbb{S}^d. \\
&\text{(d) The mapping } \mathbf{x} \mapsto \mathscr{G}(\mathbf{x}, \mathbf{0}, \mathbf{0}) \text{ belongs to } Q.
\end{aligned}
\right.
\tag{27}
$$

The densities of body forces and surface traction are such that

$$
\mathbf{f}_0 \in L^2(\Omega)^d, \quad \mathbf{f}_2 \in C(\mathbb{R}_+; L^2(\Gamma_2)^d).
\tag{28}
$$

The normal compliance function p and the surface yield function F satisfy

$$
\left\{
\begin{aligned}
&\text{(a) } p : \Gamma_3 \times \mathbb{R} \to \mathbb{R}_+. \\
&\text{(b) There exists } L_p > 0 \text{ such that} \\
&\quad |p(\mathbf{x}, r_1) - p(\mathbf{x}, r_2)| \le L_p |r_1 - r_2| \ \forall r_1, r_2 \in \mathbb{R}, \text{ a.e. } \mathbf{x} \in \Gamma_3. \\
&\text{(c) } (p(\mathbf{x}, r_1) - p(\mathbf{x}, r_2))(r_1 - r_2) \ge 0 \ \forall r_1, r_2 \in \mathbb{R}, \text{ a.e. } \mathbf{x} \in \Gamma_3. \\
&\text{(d) The mapping } \mathbf{x} \mapsto p(\mathbf{x}, r) \text{ is measurable on } \Gamma_3, \text{ for any } r \in \mathbb{R}. \\
&\text{(e) } p(\mathbf{x}, r) = 0 \text{ for all } r \le 0, \text{ a.e. } \mathbf{x} \in \Gamma_3.
\end{aligned}
\right.
\tag{29}
$$

$$
\left\{
\begin{aligned}
&\text{(a) } F : \Gamma_3 \times \mathbb{R} \to \mathbb{R}_+. \\
&\text{(b) There exists } L_F > 0 \text{ such that} \\
&\quad |F(\mathbf{x}, r_1) - F(\mathbf{x}, r_2)| \le L_F |r_1 - r_2| \ \forall r_1, r_2 \in \mathbb{R}, \text{ a.e. } \mathbf{x} \in \Gamma_3. \\
&\text{(c) The mapping } \mathbf{x} \mapsto F(\mathbf{x}, r) \text{ is measurable on } \Gamma_3, \text{ for any } r \in \mathbb{R}. \\
&\text{(d) } F(\mathbf{x}, 0) = 0 \text{ a.e. } \mathbf{x} \in \Gamma_3.
\end{aligned}
\right.
\tag{30}
$$

Also, the the coefficient of friction verifies

$$\mu \in L^\infty(\Gamma_3), \quad \mu(t, \mathbf{x}) \geq 0 \quad \text{a.e. } \mathbf{x} \in \Gamma_3, \tag{31}$$

and the initial data are such that

$$\mathbf{u}_0 \in V, \qquad \sigma_0 \in Q. \tag{32}$$

In what follows we consider the set of admissible displacements fields and the set of admissible stress fields defined by

$$U = \{ \mathbf{v} \in V \; : \; v_\nu \leq g \text{ on } \Gamma_3 \}, \tag{33}$$
$$\Sigma = \{ \tau \in Q \; : \; \text{Div}\tau + \mathbf{f}_0 = \mathbf{0} \text{ in } \Omega \}. \tag{34}$$

respectively. Note that assumptions $g > 0$ and $\mathbf{f}_0 \in L^2(\Omega)^d$ imply that U and Σ are closed, convex nonempty subsets of the spaces V and Q, respectively.

Assume in what follows that $(\mathbf{u}, \sigma)$ are sufficiently regular functions which satisfy (19)–(24) and let $\mathbf{v} \in U$ and $t > 0$ be given. First, we use the equilibrium equation (20) and the contact condition (23) to see that

$$\mathbf{u}(t) \in U, \qquad \sigma(t) \in \Sigma. \tag{35}$$

Then, we use Green's formula (18), the equilibrium equation (20) and the friction law (23) to obtain that

$$\int_\Omega \sigma(t) \cdot (\varepsilon(\mathbf{v}) - \varepsilon(\mathbf{u}(t))) \, dx \tag{36}$$

$$= \int_\Omega \mathbf{f}_0 \cdot (\mathbf{v} - \mathbf{u}(t)) \, dx + \int_{\Gamma_2} \mathbf{f}_2(t) \cdot (\mathbf{v} - \mathbf{u}(t)) \, da$$

$$+ \int_{\Gamma_3} \sigma_\nu(t)(v_\nu - u_\nu(t)) \, da - \int_{\Gamma_3} \mu |R\sigma_\nu(t)| \mathbf{n}^* \cdot (\mathbf{v}_\tau - \mathbf{u}_\tau(t)) \, da.$$

We now use the contact conditions (24) and the definition (33) of the set U to see that

$$\sigma_\nu(t)(v_\nu - u_\nu(t)) \geq -(p(u_\nu(t)) + \xi(t))(v_\nu - u_\nu(t)) \quad \text{on } \Gamma_3. \tag{37}$$

Next, we use (24), again, and the hypothesis (30)(a) on function F to deduce that

$$F\left(\int_0^t u_\nu^+(s)ds \right) (v_\nu^+ - u_\nu^+(t)) \geq \xi(t)(v_\nu - u_\nu(t)) \quad \text{on } \Gamma_3. \tag{38}$$

We now add the inequalities (37) and (38) and integrate the result on Γ_3 to find that

$$\int_{\Gamma_3} \sigma_\nu(t)(v_\nu - u_\nu(t))\, da \geq -\int_{\Gamma_3} p(u_\nu(t))(v_\nu - u_\nu(t))\, da \tag{39}$$

$$-\int_{\Gamma_3} F\left(\int_0^t u_\nu^+(s)ds\right)(v_\nu^+ - u_\nu^+(t))\, da.$$

Finally, we combine (36) and (39) to deduce that

$$\int_\Omega \sigma(t) \cdot (\varepsilon(\mathbf{v}) - \varepsilon(\mathbf{u}(t)))\, dx + \int_{\Gamma_3} p(u_\nu(t))(v_\nu - u_\nu(t))\, da \tag{40}$$

$$+\int_{\Gamma_3} F\left(\int_0^t u_\nu^+(s)ds\right)(v_\nu^+ - u_\nu^+(t))\, da + \int_{\Gamma_3} \mu|R\sigma_\nu(t)|\mathbf{n}^* \cdot (\mathbf{v}_\tau - \mathbf{u}_\tau(t))\, da$$

$$\geq \int_\Omega \mathbf{f}_0 \cdot (\mathbf{v} - \mathbf{u}(t))\, dx + \int_{\Gamma_2} \mathbf{f}_2(t).(\mathbf{v} - \mathbf{u}(t))\, da.$$

We now integrate the the constitutive law (19) with the initial conditions (25), then we gather the resulting equation with the regularity (35) and inequality (40) to obtain the following variational formulation of Problem $\mathscr{P}$.

Problem $\mathscr{P}_V$. *Find a displacement field* $\mathbf{u} : \mathbb{R}_+ \to U$ *and a stress field* $\sigma : \mathbb{R}_+ \to \Sigma$ *such that*

$$\sigma(t) = \mathscr{E}\varepsilon(\mathbf{u}(t)) + \int_0^t \mathscr{G}(\sigma(s), \varepsilon(\mathbf{u}(s))\, ds + \sigma_0 - \mathscr{E}\varepsilon(\mathbf{u}_0), \tag{41}$$

$$\int_\Omega \sigma(t) \cdot (\varepsilon(\mathbf{v}) - \varepsilon(\mathbf{u}(t)))\, dx + \int_{\Gamma_3} p(u_\nu(t))(v_\nu - u_\nu(t))\, da \tag{42}$$

$$+\int_{\Gamma_3} F\left(\int_0^t u_\nu^+(s)ds\right)(v_\nu^+ - u_\nu^+(t))\, da + \int_{\Gamma_3} \mu|R\sigma_\nu(t)|\mathbf{n}^* \cdot (\mathbf{v}_\tau - \mathbf{u}_\tau(t))\, da$$

$$\geq \int_\Omega \mathbf{f}_0 \cdot (\mathbf{v} - \mathbf{u}(t))\, dx + \int_{\Gamma_2} \mathbf{f}_2(t).(\mathbf{v} - \mathbf{u}(t))\, da$$

for all $t \in \mathbb{R}_+$.

Our main existence and uniqueness result in the study of the Problem $\mathscr{P}$, that we state here and prove in the next section is the following.

Theorem 3 *Assume that (26)–(32) hold. Then there exists a positive constant μ_0 which depends only on Ω, Γ_1, Γ_3, R and $\mathscr{E}$ such that Problem $\mathscr{P}_V$ has a unique solution, if*

$$\|\mu\|_{L^\infty(\Gamma_3)} < \mu_0. \tag{43}$$

Moreover, the solution satisfies $\mathbf{u} \in C(\mathbb{R}_+; U)$, $\boldsymbol{\sigma} \in C(\mathbb{R}_+; \Sigma)$.

Note that Theorem 3 provides the unique weak solvability of Problem $\mathscr{P}$, under the smallness assumption (43) on the coefficient of friction.

4 Proof of Theorem 3

The proof of the theorem will be carried out in several steps. To present it we assume in what follows that (26)–(32) hold. We start with the following existence and uniqueness result.

Lemma 4 *For each function $\mathbf{u} \in C(\mathbb{R}_+; V)$ there exists a unique function $\Theta\mathbf{u} \in C(\mathbb{R}_+; Q)$ such that*

$$\Theta\mathbf{u}(t) = \int_0^t \mathscr{G}(\Theta\mathbf{u}(s) + \mathscr{E}\boldsymbol{\varepsilon}(\mathbf{u}(s)), \boldsymbol{\varepsilon}(\mathbf{u}(s)))\,ds + \boldsymbol{\sigma}_0 - \mathscr{E}\boldsymbol{\varepsilon}(\mathbf{u}_0) \quad \forall t \in \mathbb{R}_+. \tag{44}$$

Moreover, the operator $\Theta : C(\mathbb{R}_+; V) \to C(\mathbb{R}_+; Q)$ is history-dependent, i.e. for all $n \in \mathbb{N}$ there exists $\theta_n > 0$ such that

$$\|\Theta\mathbf{u}_1(t) - \Theta\mathbf{u}_2(t)\|_Q \le \theta_n \int_0^t \|\mathbf{u}_1(s) - \mathbf{u}_2(s)\|_V\,ds \tag{45}$$

$$\forall \mathbf{u}_1, \mathbf{u}_2 \in C(\mathbb{R}_+; V), \ \forall t \in [0, n].$$

Proof Let $\mathbf{u} \in C(\mathbb{R}_+; V)$ and consider the operator $\Lambda : C(\mathbb{R}_+; Q) \to C(\mathbb{R}_+; Q)$ defined by

$$\Lambda\boldsymbol{\tau}(t) = \int_0^t \mathscr{G}(\boldsymbol{\tau}(s) + \mathscr{E}\boldsymbol{\varepsilon}(\mathbf{u}(s)), \boldsymbol{\varepsilon}(\mathbf{u}(s)))ds + \boldsymbol{\sigma}_0 - \mathscr{E}\boldsymbol{\varepsilon}(\mathbf{u}_0) \tag{46}$$

$$\forall \boldsymbol{\tau} \in C(\mathbb{R}_+; Q), \ t \in \mathbb{R}_+.$$

The operator Λ depends on $\mathbf{u}$ but, for the sake of simplicity, we do not indicate it explicitly. Let $\boldsymbol{\tau}_1, \boldsymbol{\tau}_2 \in C(\mathbb{R}_+; Q)$ and let $t \subset \mathbb{R}_+$. Then, using (46) and (27) we have

$$\|\Lambda \tau_1(t) - \Lambda \tau_2(t)\|_Q \le L_{\mathscr{G}} \int_0^t \|\tau_1(s) - \tau_2(s)\|_Q \, ds.$$

This inequality combined with Theorem 1 shows that the operator Λ has a unique fixed point in $C(\mathbb{R}_+; Q)$. We denote by $\Theta \mathbf{u}$ the fixed point of Λ and we combine (46) with the equality $\Lambda(\Theta \mathbf{u}) = \Theta \mathbf{u}$ to see that (44) holds.

To proceed, let $n \in \mathbb{N}$, $t \in [0, n]$ and let $\mathbf{u}_1, \mathbf{u}_2 \in C(\mathbb{R}_+; V)$. Then, using (44) and taking into account (27), (26) we write

$$\|\Theta \mathbf{u}_1(t) - \Theta \mathbf{u}_2(t)\|_Q$$

$$= L_0 \left(\int_0^t \|\Theta \mathbf{u}_1(s) - \Theta \mathbf{u}_2(s)\|_Q \, ds + \int_0^t \|\mathbf{u}_1(s) - \mathbf{u}_2(s)\|_V \, ds \right),$$

where L_0 is a positive constant which depends on $\mathscr{G}$ and $\mathscr{E}$. Using now a Gronwall argument we deduce that

$$\|\Theta \mathbf{u}_1(t) - \Theta \mathbf{u}_2(t)\|_Q \le L_0 \, e^{L_0 n} \int_0^t \|\mathbf{u}_1(s) - \mathbf{u}_2(s)\|_V \, ds.$$

This inequality shows that (45) holds with $\theta_n = L_0 \, e^{L_0 n}$.

Next, we consider the operators $A : V \to V$ and $\mathscr{R} : V \times Q \to L^2(\Gamma_3)$ defined by

$$(A\mathbf{u}, \mathbf{v})_V = (\mathscr{E}\boldsymbol{\varepsilon}(\mathbf{u}), \boldsymbol{\varepsilon}(\mathbf{v}))_Q + \int_{\Gamma_3} p(u_\nu) v_\nu \, da \qquad \forall \mathbf{u}, \mathbf{v} \in V, \tag{47}$$

$$\mathscr{R}(\mathbf{u}, \mathbf{z}) = |R(P_\Sigma(\mathscr{E}\boldsymbol{\varepsilon}(\mathbf{u}) + \mathbf{z}))_\nu| \qquad \forall \mathbf{u} \in V, \ \mathbf{z} \in Q, \tag{48}$$

where $P_\Sigma : Q \to \Sigma$ represents the projection operator. Note that, since $\Sigma \subset Q_1$, the operator $\mathscr{R}$ is well defined. Denote $Y = Q \times L^2(\Gamma_3) \times Q$ where, here and below, $X_1 \times \ldots \times X_m$ represents the product of the Hilbert spaces $X_1, \ldots, X_m$ ($m = 2, 3$), endowed with its canonical inner product. Besides the operator $\Theta : C(\mathbb{R}_+; V) \to C(\mathbb{R}_+; Q)$ defined in Lemma 4, let $\Phi : C(\mathbb{R}_+; V) \to C(\mathbb{R}_+; L^2(\Gamma_3))$ and $\mathscr{S} : C(\mathbb{R}_+; V) \to C(\mathbb{R}_+; Y)$ be the operators given by

$$(\Phi \mathbf{v})(t) = F\left(\int_0^t v_\nu^+(s) ds \right), \tag{49}$$

$$\mathscr{S}\mathbf{v}(t) = (\Theta \mathbf{v}(t), \Phi \mathbf{v}(t), \Theta \mathbf{v}(t)) \tag{50}$$

for all $\mathbf{v} \in C(\mathbb{R}_+; V)$, $t \in \mathbb{R}_+$. Finally, let $j : Y \times V \times V \to \mathbb{R}$ and $\mathbf{f} : \mathbb{R}_+ \to V$ denote the functions defined by

$$j(\mathbf{w}, \mathbf{u}, \mathbf{v}) = (\mathbf{x}, \boldsymbol{\varepsilon}(\mathbf{v}))_Q + (y, v_\nu^+)_{L^2(\Gamma_3)} + (\mu \mathscr{R}(\mathbf{u}, \mathbf{z})\mathbf{n}^*, \mathbf{v}_\tau)_{L^2(\Gamma_3)^d} \quad (51)$$
$$\forall \mathbf{w} = (\mathbf{x}, y, \mathbf{z}) \in Y, \ \mathbf{u}, \mathbf{v} \in V,$$

$$(\mathbf{f}(t), \mathbf{v})_V = \int_\Omega \mathbf{f}_0 \cdot \mathbf{v}\, dx + \int_{\Gamma_2} \mathbf{f}_2(t) \cdot \mathbf{v}\, da \quad \forall \mathbf{v} \in V, \ t \in \mathbb{R}_+. \quad (52)$$

We have the following equivalence result.

Lemma 5 *Assume that $\mathbf{u} \in C(\mathbb{R}_+; U)$ and $\boldsymbol{\sigma} \in C(\mathbb{R}_+; \Sigma)$. Then, the couple $(\mathbf{u}, \boldsymbol{\sigma})$ is a solution of Problem $\mathscr{P}_V$ if and only if*

$$\boldsymbol{\sigma}(t) = \mathscr{E}\boldsymbol{\varepsilon}(\mathbf{u}(t)) + \Theta\mathbf{u}(t), \quad (53)$$
$$(A\mathbf{u}(t), \mathbf{v} - \mathbf{u}(t))_V + j(\mathscr{S}\mathbf{u}(t), \mathbf{u}(t), \mathbf{v}) \quad (54)$$
$$-j(\mathscr{S}\mathbf{u}(t), \mathbf{u}(t), \mathbf{u}(t)) \geq (\mathbf{f}(t), \mathbf{v} - \mathbf{u}(t))_V \quad \forall \mathbf{v} \in U$$

for all $t \in \mathbb{R}_+$.

Proof Let $(\mathbf{u}, \boldsymbol{\sigma}) \in C(\mathbb{R}_+; U \times \Sigma)$, be a solution of Problem $\mathscr{P}_V$ and let $t \in \mathbb{R}_+$. By (41) we have

$$\boldsymbol{\sigma}(t) - \mathscr{E}\boldsymbol{\varepsilon}(\mathbf{u}(t)) = \int_0^t \mathscr{G}(\boldsymbol{\sigma}(s) - \mathscr{E}\boldsymbol{\varepsilon}(\mathbf{u}(s)) + \mathscr{E}\boldsymbol{\varepsilon}(\mathbf{u}(s)), \boldsymbol{\varepsilon}(\mathbf{u}(s)))\, ds + \boldsymbol{\sigma}_0 - \mathscr{E}\boldsymbol{\varepsilon}(\mathbf{u}_0),$$

and, using the definition (44) of the operator Θ, we obtain (53). Moreover, we substitute (41) in (42), then we use (49) and equality $P_\Sigma \boldsymbol{\sigma}(t) = \boldsymbol{\sigma}(t)$. As a result, we deduce that

$$\int_\Omega \mathscr{E}\boldsymbol{\varepsilon}(\mathbf{u}(t)) \cdot (\boldsymbol{\varepsilon}(\mathbf{v}) - \boldsymbol{\varepsilon}(\mathbf{u}(t)))\, dx + \int_\Omega \Theta\mathbf{u}(t) \cdot (\boldsymbol{\varepsilon}(\mathbf{v}) - \boldsymbol{\varepsilon}(\mathbf{u}(t)))\, dx \quad (55)$$

$$+ \int_{\Gamma_3} p(u_\nu(t))(v_\nu - u_\nu(t))\, da + \int_{\Gamma_3} \Phi\mathbf{u}(t)\, (v_\nu^+ - u_\nu^+(t))\, da$$

$$+ \int_{\Gamma_3} \mu |R\big(P_\Sigma(\mathscr{E}\boldsymbol{\varepsilon}(\mathbf{u}(t)) + \Theta\mathbf{u}(t))\big)_\nu| \mathbf{n}^* \cdot (\mathbf{v}_\tau - \mathbf{u}_\tau(t))\, da$$

$$> \int_\Omega \mathbf{f}_0 \cdot (\mathbf{v} - \mathbf{u}(t))\, dx + \int_{\Gamma_2} \mathbf{f}_2(t) \cdot (\mathbf{v} - \mathbf{u}(t))\, da \quad \forall \mathbf{v} \in U.$$

Using now the definitions (47), (48) and (52) yields

$$(A\mathbf{u}(t), \mathbf{v} - \mathbf{u}(t))_V + (\Theta\mathbf{u}(t) \cdot (\boldsymbol{\varepsilon}(\mathbf{v}) - \boldsymbol{\varepsilon}(\mathbf{u}(t)))_Q$$
$$+ (\Phi\mathbf{u}(t)) \, (v_\nu^+ - u_\nu^+(t)))_{L^2(\Gamma_3)} + (\mu\mathscr{R}(\mathbf{u}(t), \Theta(\mathbf{u}(t))\mathbf{n}^*, \mathbf{v}_\tau - \mathbf{u}_\tau(t))_{L^2(\Gamma_3)^d}$$
$$\geq (\mathbf{f}(t), \mathbf{v} - \mathbf{u}(t))_V \quad \forall \mathbf{v} \in U.$$

This inequality combined with the definitions (50) and (51) shows that the variational inequality (54) holds.

Conversely, assume that $(\mathbf{u}, \boldsymbol{\sigma}) \in C(\mathbb{R}_+; U \times \Sigma)$ is a couple of functions which satisfies (53) and (54) and let $t \in \mathbb{R}_+$. Then, using the definitions (47), (48), (50)–(52) it follows that (55) holds. Moreover, recall that the regularity $\boldsymbol{\sigma} \in C(\mathbb{R}_+; \Sigma)$ implies that $P_\Sigma\boldsymbol{\sigma}(t) = \boldsymbol{\sigma}(t)$ and, in addition, (53) yields $\boldsymbol{\sigma}(t) = \mathscr{E}\boldsymbol{\varepsilon}(\mathbf{u}(t)) + \Theta\mathbf{u}(t)$. Substituting these equalities in (55) and using (49) we see that (42) holds. Finally, to conclude, we note that (41) is a direct consequence of (53) and the definition of the operator Θ in Lemma 5.

The interest in Lemma 5 arrises in the fact that it decouples the unknowns $\mathbf{u}$ and $\boldsymbol{\sigma}$ in the system (41)–(42). The next step is to provide the unique solvability of the variational inequality (54) in which the unknown is the displacement field. To this end we need the following intermediate result on the operator $\mathscr{R}$.

Lemma 6 *There exists $L_{\mathscr{R}} > 0$ which depends only on Ω, Γ_3 and R, such that*

$$\|\mathscr{R}(\mathbf{u}_1, \mathbf{z}_1) - \mathscr{R}(\mathbf{u}_2, \mathbf{z}_2)\|_{L^2(\Gamma_3)} \leq L_{\mathscr{R}} \left(\|\mathbf{u}_1 - \mathbf{u}_2\|_V + \|\mathbf{z}_1 - \mathbf{z}_2\|_Q\right) \quad (56)$$
$$\forall \mathbf{u}_1, \mathbf{u}_2 \in V, \; \mathbf{z}_1, \mathbf{z}_2 \in Q.$$

Proof Let $\mathbf{u}_1, \mathbf{u}_2 \in V$, $\mathbf{z}_1, \mathbf{z}_2 \in Q$. Then, by the definition (48) of the operator $\mathscr{R}$ combined with inequality (17) we have

$$\|\mathscr{R}(\mathbf{u}_1, \mathbf{z}_1) - \mathscr{R}(\mathbf{u}_2, \mathbf{z}_2)\|_{L^2(\Gamma_3)} \quad (57)$$
$$\leq c_R \|P_\Sigma(\mathscr{E}\boldsymbol{\varepsilon}(\mathbf{u}_1) + \mathbf{z}_1) - P_\Sigma(\mathscr{E}\boldsymbol{\varepsilon}(\mathbf{u}_2) + \mathbf{z}_2)\|_{Q_1}.$$

On the other hand, the definition of the set Σ and the nonexpansivity of the operator P_Σ yields

$$\|P_\Sigma(\mathscr{E}\boldsymbol{\varepsilon}(\mathbf{u}_1) + \mathbf{z}_1) - P_\Sigma(\mathscr{E}\boldsymbol{\varepsilon}(\mathbf{u}_2) + \mathbf{z}_2)\|_{Q_1} \quad (58)$$
$$\leq \|\mathscr{E}\boldsymbol{\varepsilon}(\mathbf{u}_1) - \mathscr{E}\boldsymbol{\varepsilon}(\mathbf{u}_2) + \mathbf{z}_1 - \mathbf{z}_2\|_Q$$

We now combine inequalities (57) and (58) to see that

$$\|\mathscr{R}(\mathbf{u}_1, \mathbf{z}_1) - \mathscr{R}(\mathbf{u}_2, \mathbf{z}_2)\|_{L^2(\Gamma_3)} \leq c_R \left(\|\mathscr{E}\boldsymbol{\varepsilon}(\mathbf{u}_1) - \mathscr{E}\boldsymbol{\varepsilon}(\mathbf{u}_2) + \mathbf{z}_1 - \mathbf{z}_2\|_Q\right) \quad (59)$$

Lemma 6 is now a consequence of inequality (59) and assumption (26).

We proceed with the following existence and uniqueness result.

Lemma 7 *The variational inequality (54) has a unique solution with regularity* $\mathbf{u} \in C(\mathbb{R}_+, U)$.

Proof It is straightforward to see that inequality (54) represents a variational inequality of the form (7) in which $X = V$, $K = U$ and $Y = Q \times L^2(\Gamma_3) \times Q$. Therefore, in order to prove its unique solvability, we check in what follows the assumptions of Theorem 2.

First, we note that assumption (1) is obviously satisfied. Next, we use the definition (47), assumptions (26), (29)(b) and inequality (16) to obtain that

$$\|A\mathbf{u} - A\mathbf{v}\|_V \le (L_{\mathscr{E}} + c_0^2 L_p)\|\mathbf{u} - \mathbf{v}\|_V \qquad \forall \mathbf{u}, \mathbf{v} \in V, \tag{60}$$

where $L_{\mathscr{E}}$ is a positive constant which depends on the elasticity operator $\mathscr{E}$. On the other hand, from (26)(c) and (29)(c) the we deduce that

$$(A\mathbf{u} - A\mathbf{v}, \mathbf{u} - \mathbf{v})_V \ge m_{\mathscr{E}}\|\mathbf{u} - \mathbf{v}\|_V^2. \tag{61}$$

We conclude from above that the operator A satisfies condition (2) with $L = L_{\mathscr{E}} + c_0^2 L_p$ and $m = m_{\mathscr{E}}$.

Let $\mathbf{w} = (\mathbf{x}, \mathbf{y}, \mathbf{z}) \in Y$ and $\mathbf{u} \in V$ be fixed. Then, using the properties of the traces it is easy to see that the function $\mathbf{v} \mapsto j(\mathbf{w}, \mathbf{u}, \mathbf{v})$ is convex and continuous and, therefore, it satisfies condition (3)(a). We now consider the elements $\mathbf{w}_1 = (\mathbf{x}_1, y_1, \mathbf{z}_1)$, $\mathbf{w}_2 = (\mathbf{x}_2, y_2, \mathbf{z}_2) \in Y$, $\mathbf{u}_1, \mathbf{u}_2, \mathbf{v}_1, \mathbf{v}_2 \in V$. Then, using inequality (56), assumption (31) and inequality (16) we find that

$$j(\mathbf{w}_1, \mathbf{u}_1, \mathbf{v}_2) - j(\mathbf{w}_1, \mathbf{u}_1, \mathbf{v}_1) + j(\mathbf{w}_2, \mathbf{u}_2, \mathbf{v}_1) - j(\mathbf{w}_2, \mathbf{u}_2, \mathbf{v}_2)$$
$$\le \|\mathbf{x}_1 - \mathbf{x}_2\|_Q \|\mathbf{v}_1 - \mathbf{v}_2\|_V + c_0 \|y_1 - y_2\|_{L^2(\Gamma_3)} \|\mathbf{v}_1 - \mathbf{v}_2\|_V$$
$$+ c_0 L_{\mathscr{R}} \|\mu\|_{L^\infty(\Gamma_3)} \left(\|\mathbf{u}_1 - \mathbf{u}_2\|_V + \|\mathbf{z}_1 - \mathbf{z}_2\|_Q \right) \|\mathbf{v}_1 - \mathbf{v}_2\|_V$$
$$\le \alpha \|\mathbf{w}_1 - \mathbf{w}_2\|_Z \|\mathbf{v}_1 - \mathbf{v}_2\|_V + \beta \|\mathbf{u}_1 - \mathbf{u}_2\|_V \|\mathbf{v}_1 - \mathbf{v}_2\|_V$$

where $\alpha = 2 \max \{1, c_0, c_0 L_{\mathscr{R}}\|\mu\|_{L^\infty(\Gamma_3)}\}$ and $\beta = c_0 L_{\mathscr{R}}\|\mu\|_{L^\infty(\Gamma_3)}$. It follows from here that j satisfies condition (3)(b). Let

$$\mu_0 = \frac{m_{\mathscr{E}}}{c_0 L_{\mathscr{R}}}, \tag{62}$$

which, clearly, depends only on Ω, Γ_1, Γ_3, R and $\mathscr{E}$. Then, it is easy to see that if the smallness assumption $\|\mu\|_{L^\infty(\Gamma_3)} < \mu_0$ is satisfied we have $\beta < m$ and, therefore, condition (6) holds.

Next, let $\mathbf{u}, \mathbf{v} \in C(\mathbb{R}_+; V), n \in \mathbb{N}$ and let $t \in [0, n]$. Then, using (49) and taking into account (30)(b) and (16) we obtain that

$$
\|\boldsymbol{\Phi}\mathbf{u}(t) - \boldsymbol{\Phi}\mathbf{v}(t)\|_{L^2(\Gamma_3)} = \left\| F\left(\int_0^t u_\nu^+(s)ds\right) - F\left(\int_0^t v_\nu^+(s)ds\right)\right\|_{L^2(\Gamma_3)}
$$

$$
\leq L_F \left\|\int_0^t (u_\nu^+(s) - v_\nu^+(s))ds\right\|_{L^2(\Gamma_3)} \leq c_0 L_F \int_0^t \|\mathbf{u}(s) - \mathbf{v}(s)\|_V \, ds.
$$

Therefore, using this the definition (50) of the operator $\mathscr{S}$ and (45) we have

$$
\|\mathscr{S}\mathbf{u}(t) - \mathscr{S}\mathbf{v}(t)\|_{Q \times L^2(\Gamma_3) \times Q} \leq (2\theta_n + c_0 L_F) \int_0^t \|\mathbf{u}(s) - \mathbf{v}(s)\|_V \, ds.
$$

It follows from here that the operator $\mathscr{S}$ satisfies condition (4). Finally, we note that assumption (28) on the body forces and traction and definition (52) imply that $\mathbf{f} \in C(\mathbb{R}_+; V)$.

We conclude from above that all the assumptions of Theorem 2 are satisfied. Therefore, we deduce that inequality (54) has a unique solution $\mathbf{u} \in C(\mathbb{R}_+; U)$ which concludes the proof.

We now have all the ingredients to provide the proof of Theorem 3.

Proof (Proof of Theorem 3) Let $\mathbf{u} \in C(\mathbb{R}_+; U)$ be the unique solution of inequality (54) obtained in Lemma 7 and let $\boldsymbol{\sigma}$ the function defined by (53). Then, using assumption (26) it follows that $\boldsymbol{\sigma} \in C(\mathbb{R}_+; Q)$. Let $t \in \mathbb{R}_+$ be given. Arguments similar to those used in the proof of Lemma 7 show that

$$
\int_\Omega \boldsymbol{\sigma}(t) \cdot (\boldsymbol{\varepsilon}(\mathbf{v}) - \boldsymbol{\varepsilon}(\mathbf{u}(t))) \, dx + \int_{\Gamma_3} p(u_\nu(t))(v_\nu - u_\nu(t)) \, da
$$

$$
+ \int_{\Gamma_3} F\left(\int_0^t u_\nu^+(s)ds\right) (v_\nu^+ - u_\nu^+(t)) \, da
$$

$$
+ \int_{\Gamma_3} \mu |R(P_\Sigma(\sigma_\nu(t))| \mathbf{n}^* \cdot (\mathbf{v}_\tau - \mathbf{u}_\tau(t)) \, da
$$

$$
\geq \int_\Omega \mathbf{f}_0 \cdot (\mathbf{v} - \mathbf{u}(t)) \, dx + \int_{\Gamma_2} \mathbf{f}_2(t) \cdot (\mathbf{v} - \mathbf{u}(t)) \, da \qquad \forall \mathbf{v} \in U.
$$

Let $\varphi \in C_0^\infty(\Omega)^d$. We test in this inequality with $\mathbf{v} = \mathbf{u}(t) \pm \varphi$ to deduce that

$$\int_\Omega \sigma(t) \cdot \varepsilon(\varphi) \, dx = \int_\Omega \mathbf{f}_0 \cdot \varphi \, dx$$

which implies that $\mathrm{Div}\,\sigma(t) + \mathbf{f}_0 = \mathbf{0}$ in Ω. It follows from here that $\sigma(t) \in \Sigma$ and, moreover, $\sigma \in C(\mathbb{R}_+; \Sigma)$.

We conclude from above that $(\mathbf{u}, \sigma)$ represents a couple of functions which satisfies (53)–(54) and, in addition, it has the regularity $(\mathbf{u}, \sigma) \in C(\mathbb{R}_+; U \times \Sigma)$. The existence part in Theorem 3 is now a direct consequence of Lemma 5. The uniqueness part follows from the uniqueness of the solution of the variational inequality (54), guaranteed by Lemma 7.

References

1. Baiocchi, C., Capelo, A.: Variational and Quasivariational Inequalities: Applications to Free-Boundary Problems. John Wiley, Chichester (1984)
2. Brezis, H.: Equations et inéquations non linéaires dans les espaces vectoriels en dualité. Ann. Inst. Fourier **18**, 115–175 (1968)
3. Capatina, A.: Variational Inequalities and Frictional Contact Problems, Advances in Mechanics and Mathematics, vol. 31. Springer, New York (2014)
4. Duvaut, G., Lions, J.L.: Inequalities in Mechanics and Physics. Springer-Verlag, Berlin (1976)
5. Glowinski, R.: Numerical Methods for Nonlinear Variational Problems. Springer, New York (1984)
6. Eck, C., Jarušek, J., Krbeč, M.: Unilateral Contact Problems: Variational Methods and Existence Theorems, Pure and Applied Mathematics, vol. 270. Chapman/CRC Press, New York (2005)
7. Han, W., Sofonea, M.: Time-dependent variational inequalities for viscoelastic contact problems. J. Comput. Appl. Math. **136**, 369–387 (2001)
8. Han, W., Sofonea, M.: Evolutionary variational inequalities arising in viscoelastic contact problems. SIAM J. Numer. Anal. **38**, 556–579 (2000)
9. Han, W., Sofonea, M.: Quasistatic Contact Problems in Viscoelasticity and Viscoplasticity, Studies in Advanced Mathematics, vol. 30. American Mathematical Society, Providence, RI-International Press, Somerville, MA (2002)
10. Haslinger, J., Hlavácek, I., Nečas, J.: Numerical methods for unilateral problems in solid mechanics. In: Lions, J.-L., Ciarlet, P. (eds.) Handbook of Numerical Analysis, vol. IV, pp. 313–485. North-Holland, Amsterdam (1996)
11. Hlaváček, I., Haslinger, J., Nečas, J., Lovíšek, J.: Solution of Variational Inequalities in Mechanics. Springer, New York (1988)
12. Kikuchi, N., Oden, J.T.: Theory of variational inequalities with applications to problems of flow through porous media. Int. J. Eng. Sci. **18**, 1173–1284 (1980)
13. Kikuchi, N., Oden, J.T.: Contact Problems in Elasticity: A Study of Variational Inequalities and Finite Element Methods. SIAM, Philadelphia (1988)
14. Laursen, T.: Computational Contact and Impact Mechanics. Springer, Berlin (2002)
15. Migórski, S., Shillor, M., Sofonea, M. (Guest Editors): Special section on Contact Mechanics, Nonlinear Analysis: Real World Applications **22**, 435–680 (2015)
16. Panagiotopoulos, P.D.: Inequality Problems in Mechanics and Applications. Birkhäuser, Boston (1985)
17. Shillor, M., Sofonea, M., Telega, J.: Models and Variational Analysis of Quasistatic Contact. Lecture Notes in Physics, vol. 655. Springer, Heidelberg (2004)

18. Sofonea, M., Avramescu, C., Matei, A.: A fixed point result with applications in the study of viscoplastic frictionless contact problems. Commun. Pure Appl. Anal. **7**, 645–658 (2008)
19. Sofonea, M., Matei, A.: History-dependent quasivariational inequalities arising in contact mechanics. Eur. J. Appl. Math. **22**, 471–491 (2011)
20. Sofonea, M., Matei, A.: Mathematical Models in Contact Mechanics, London Mathematical Society Lecture Note Series, vol. 398. Cambridge University Press, Cambridge (2012)
21. Sofonea, M., Souleiman, Y.: A viscoelastic sliding contact problem with normal compliance, unilateral constraint and memory term. Mediterr. J. Math. (2015). doi:10.1007/s00009-015-0661-9 (to appear)
22. Wriggers, P.: Computational Contact Mechanics. Wiley, Chichester (2002)

Efficient Lower Bounds for Packing Problems in Heterogeneous Bins with Conflicts Constraint

Mohamed Maiza, Mohammed Said Radjef and Lakhdar Sais

Abstract In this paper we discuss a version of the classical one dimensional bin-packing problem, where the objective is to minimize the total cost of heterogeneous bins needed to store given items, each with some space requirements. In this version, some of the items are incompatible with each other, and cannot be packed together. This problem with various real world applications generalizes both the Variable Sized Bin Packing Problem and the Vertex Coloring Problem. We propose two lower bounds for this problem based on both the relaxation of the integrity constraints and the computation of the large clique in the conflicts graph.

1 Introduction and Problem Positioning

The purpose of this paper, is to investigate the problem of determining the minimal possible cost generated by the use of heterogeneous warehouses required in a supply chain to store given items, where some of the items are in conflicting with each other, and cannot be stored together in the same warehouse. We say with compatible items the items that can be stored simultaneously in the same warehouse, otherwise they are called incompatible items. This problem occur in a number of industrial and transportation contexts within a supply chain where generally we prospect to ask the best transfer cost of diverse available products. For example, a manufacturing company has to ship n customer orders from its factory to a distribution center. All shipments

M. Maiza (✉)
Laboratoire de Mathématiques Appliquées-Ecole Militaire Polytechnique,
Bordj El Bahri-Alger, Algiers, Algeria
e-mail: maiza_mohamed@yahoo.fr

M. Said Radjef
Laboratoire de Modélisation et Optimisation des Systémes, Université de Béjaia, Béjaia, Algeria
e-mail: msradjef@yahoo.fr

L. Sais
Centre de Recherche en Informatique de Lens CNRS-UMR8188, Université d'Artois,
F-62300 Lens, France
e-mail: sais@cril.fr

© Springer International Publishing Switzerland 2016
G.A. Anastassiou and O. Duman (eds.), *Intelligent Mathematics II:
Applied Mathematics and Approximation Theory*, Advances in Intelligent Systems
and Computing 441, DOI 10.1007/978-3-319-30322-2_18

are carried out by leased trucks and several truck sizes are available with different leasing costs. Besides, some customer orders cannot be stored simultaneously in the same truck for reasons of deterioration or to avoid fire danger. The problem is to determine the least-cost fleet mix and size for shipping all the orders with the respect of compatibility constraint. It is convenient to associate with the problem a conflict graph $G = (V; E)$, where an edge $(i; i') \in E$ exists if and only if items i and i' are the incompatible items. This problem which we called in what follows the Variable Sized Bin-Packing Problem with Conflicts (VSBPPC) is strongly NP-hard since it generalizes both the Variable Sized Bin Packing Problem (VSBPP) in which all items are mutually compatible, and the Vertex Coloring Problem (VCP) in which all items weights take value 0. Also, it is clear that the particular case where we consider only one type of warehouses (i.e. all warehouses are identical) turns out to be the studied one-dimensional Bin Packing Problem with Conflicts (BPPC).

Although the VSBPPC is NP-hard with important real world application, few works which it investigate are outlined. Indeed, only the online version of this problem has been addressed by [1] and [2] whose the goal was to pack all items into a set of bins with a minimum total size. In this version, a set of unpacked items is defined at the beginning of the assignment and heterogeneous bins arrive one by one. Before arriving of such a bin, authors proposed and analyzed the asymptotic competitive ratio of algorithms need to decide on which items to pack into the existing bin. Whereas, in this paper we address the offline version of the VSBPPC which can be showed as a common problem with many other combinatorial problems such as the BPPC, VSBPP and VCP where we considered them as a particular cases of described problem. So, these problems has been extensively investigated in numerous papers and therefore we restrict our reviewing to the papers which are used heuristics approaches that we will base to solve a problem in the remainder of this paper.

For solving the VSBPP, it is worth to note that [3] proposed and analyzed two variants of the well-known first-fit decreasing and best-fit decreasing algorithms. They show that these algorithms give a solution which is less than $\frac{3}{2}c_{opt} + 1$ for the general case where c_{opt} refers to the value of an optimal cost solution. Haouari and Serairi [4] proposed four constructive greedy heuristics based on iteratively solving a subset-sum problem, together with a set-covering based heuristic and a genetic based algorithm. Performances of their proposition are analyzed on a large set of randomly generated test instances with up to seven bin-type and 2000 items. Among the proposed greedy heuristics, the best compromise between solution quality and running time is obtained by the named SSP3 heuristic. In [5], the same authors proposed six different lower bounds for the VSBPP where the number of bins for each bin-type is limited, as well as an exact algorithm. They show that the particular case with all size items larger than a third the largest bin capacity can be restated and solved in polynomial-time as a maximum-weight matching problem in a non-bipartite graph. Introducing the conflict constraints, it is noteworthy to studies the propriety of the conflicts graph, but most problem are also NP-hard for the arbitrary conflicts graph, particularly the VCP and the maximal clique search problem. These last problems are extensively discussed in the literature. Johnson [6] described an efficient polynomial greedy heuristic to determine the largest clique set. This heuristic method initializes the clique with the vertex of maximum degree in the graph and then adds successive vertices each of which has edges with all the vertices already included in the clique. At each iteration, vertices are considered according to a decreasing degree order. Thereafter, this greedy heuristic was widely applied to solve numerous related problems such as the BPPC where the main heuristic approaches resolution available in the literature are those given in [7, 8], in which authors surveyed previous results in the literature, presented new lower and upper bounds, and introduced benchmark instances.

Based on these cited works, this paper aimed at providing new lower and upper bounds for the VSBPPC. So, it is organized as follows: in the next section (Sect. 2) we provide the description and the mathematical formulation of our problem. Then, we present in the followed section (Sect. 3) our new lower bounds based on both the computation of the maximal clique set and the relaxation scheme of the mathematical formulation. Finally, we close this paper in Sect. 4 with concluding remarks.

2   Problem Description and Formulation

The VSBPPC is formally defined as follows; Given a set V of n items of sizes $w_1, w_2, \ldots, w_n$ with

$$w_1 \geqslant w_2 \geqslant \cdots \geqslant w_n,$$

where we it associate an undirected graph named as a conflict graph $G = (V; E)$ such that $(i, i') \in E$ when the two items i and i' of V are incompatible items. Likewise, given m different bin-types (Categories of warehouses), each bin-type j $(j = 1, \ldots, m)$ includes an infinite number of identical bins (bins are warehouses), each having a capacity W_j and a fixed cost c_j with

$$W_1 \leqslant W_2 \leqslant \cdots \leqslant W_m$$

and

$$c_1 \leqslant c_2 \leqslant \cdots \leqslant c_m.$$

The VSBPPC problem consists in assigning each items from V into one bin while ensuring that no bin contains incompatible items and the cumulated size in each bin does not exceed its capacity and the total cost of the needed bins is minimized. Without loss of generality, we assume that the data is deterministic and presented in off-line mode and that the largest item can be assigned at least to one type of bins ($w_1 \leqslant W_m$). Note that if $c_j > c_{j'}$ for $1 \leqslant j < j' \leqslant m$ then bin-type j is dominated and should therefore not be considered.

The VSBPPC can be formulated as an integer linear program (ILP). Let x_{ij_k} be a binary variable equal to 1 if and only if item i is assigned to bin k of bin-type j and

0 otherwise, and let y_{j_k} be a binary variable equal to 1 if and only if bin k of bin-type j is used, 0 otherwise. The formulation is then

$$\min \sum_{j=1}^{m} \sum_{k=1}^{n} c_j y_{j_k}, \tag{1}$$

$$\text{s.t. } \sum_{j=1}^{m} \sum_{k=1}^{n} x_{ij_k} = 1, i = 1, \ldots, n, \tag{2}$$

$$\sum_{i=1}^{n} w_i x_{ij_k} \leq W_j y_{j_k}, j = 1, \ldots, m, \ k = 1, \ldots, n, \tag{3}$$

$$x_{ij_k} + x_{i'j_k} \leqslant 1, (i, i') \in E \quad j = 1, \ldots, m, \ k = 1, \ldots, n, \tag{4}$$

$$x_{ij_k} \in \{0, 1\}, i = 1, \ldots, n \quad j = 1, \ldots, m, \ k = 1, \ldots, n, \tag{5}$$

$$y_{j_k} \in \{0, 1\}, j = 1, \ldots, m, k = 1, \ldots, n. \tag{6}$$

The objective function to be minimized (1) represents the cost of the bins used to pack all items. Constraint (2) ensures that each item i has to be packed. Constraint (3) indicates that the amount packed in each bin does not exceed the capacity of this bin, whereas, constraint (4) make sure that each incompatible pair (i, i') of items does not assigned to the same bin. Finally, constraints (5) and (6) restrict decision variables to be binary. Note that the above ILP without constraint (4) is the well formulation of the VSBPP without conflict.

In the remaining of the paper we will use the definition of extended conflict graph $G'(V, E')$ where

$$E' = E \cup \left\{ (i, i') : i, i' \in V \text{ and } w_i + w_{i'} > W_m \right\}.$$

3   Lower Bounds

We have developed two lower bounds L_1 and L_2 for the VSBPPC in order to provide the means to measure the solution quality of the various procedure and also to be used as performance criterion of the heuristics developed in the next section. Obtained through the resolution of the relaxed mathematical formulation, these lower bounds are based on the principle of relaxing the constraint (5) which ensures that each item should be entirely—not in fractional way—assigned to such a bin. The similarly principle has been adopted on splitting-based lower bound of [5] for the VSBPP without conflicts—denoted in what follows L_0-. This last bound is based on the resolution of the relaxation problem (7)–(9) where authors consider the set S of large items in which each item should be fit alone, only in the largest bin. Therefore one bin is initialized for each item in this set, and only the remaining items in $V \setminus S$ are assigned in a fractional way to the residual capacity of initialized bins and possibly to other new bins.

$$L_0 = \min \sum_{j=1}^{m} c_j x_j + q c_m \tag{7}$$

s.t.

$$\sum_{j=1}^{m} W_j x_j \geqslant \sum_{i=q+1}^{m} w_i - \min \left(q W_m - \sum_{i=1}^{q} w_i, \sum_{i=p}^{m} w_i \right) \tag{8}$$

$$x_j \in \mathbb{N}; \quad j = 1, \ldots, m, \tag{9}$$

x_j, q and p are respectively, the number of bins of type j, the largest index of item such that $w_q > \max \left(W_{m-1}, \frac{W_m}{2} \right)$, and the smallest index of item such that $w_q + w_p \leqslant W_m$.

The VSBPPC can be shown as a particular case of the VSBPP in which the corresponding conflicts graph is the discrete graph D_n of order n ($G(V, E)$ with $E = \emptyset$; Graph without edges in which all the items are mutually compatible). Therefore, any lower bound for the VSBPP without conflicts is also a valid lower bound for the problem with conflicts. Indeed, if the total cost of packing of such an instance without a conflicts constraint is defined then it is trivial to constant that by addition of restrictions on the packing process, this initial cost increases or, in the worst cases, remains unchanged. Then we have

$$L_0 = L(D_n) \leqslant L(G), \tag{10}$$

where $L(G)$ is any lower bound for the VSBPPC with conflicts graph G.

Now, consider the case where the conflicts graph is a complete graph (K_n) of n items in which each item should be loaded into a distinct bin (A complete graph is defined here as a graph in which every pair of items is mutually incompatible). Under the assumption imposed on this paper ($W_1 < W_2 < \cdots < W_m$ and $c_1 < c_2 < \cdots < c_m$), each item should be loaded in the smallest possible (i.e. cheapest) bin that fits. Then the problem becomes to finding for each item the smallest index of bin-type that which can fit and therefore the optimal cost solution $c_{opt}(K_n)$ that is also the VSBPPC lower bound for a complete conflicts graph $L(K_n)$ can be given in $O(n)$ *time* as follows.

Define $q_j (j = 1, \ldots, m)$ be the largest index of item i such as $W_{j-1} < w_i \leqslant W_j$ with $W_0 = 0$.

$$L(K_n) = c_{opt}(K_n) = \sum_{j=1}^{m} c_j \left(q_j - q_{j-1} \right) \text{ with } q_0 = 0. \tag{11}$$

By definition, the value $(q_j - q_{j-1})$ is the number of items with size w_i in $[W_{j-1}, W_j]$. The cost obtained by formula (11) represents an optimal solution for the problem (1)–(6) where the constraints (4) is instantiated for each items pair. Obviously, this optimal cost solution is an upper bound for the VSBPPC in which we cannot be increases when we have to relaxing or to remove some instances of this constraint, which is the case of the VSBPPC with no complete conflicts graph. Then from formulas (10) and (11), we have

$$L_0 = L(D_n) \leqslant L(G) \leqslant L(K_n) = c_{opt}(K_n). \tag{12}$$

Here again, we derive our lower bounds for the VSBPPC for the general conflicts graph. The main principle of our proposition can be summarized in the three following steps:

1. Find the maximal clique set V_K from the conflicts graph;
2. Initialize one bin for each item of V_K;
3. Assign the items of $V \setminus V_K$ into the initialized bins in fractional way and possibly into new bins without considering the conflicts.

The two last steps will be carried out by solving the relaxed mathematical formulation and both lower bounds L_1 and L_2 execute the same procedure to solve this mathematical model. The main difference between these two propositions lies in how to obtain the maximal clique set (Step 1). Indeed, finding the maximal clique set V_K is also an NP-hard problem. In L_1 this set is determined by application of Johnson's [6] greedy heuristic directly on the extended conflicts graph $G'(V, E')$. Whilst in L_2, it is obtained through Muritiba's improvement computation [8] where authors compute the large clique set on the initial conflicts graph $G(V; E)$ using Johnson's greedy heuristic and remove this set from the graph G, then from the obtained partial graph, they updating the set E to E' and enlarge the initial clique by added items from this new partial extended conflicts graph using constantly Johnson's greedy heuristic. In [8], authors indicate that this strategy leads to larger cliques set than those obtained by applying the Johnson's algorithm directly on the extended conflicts graph.

Now, consider the following notation:

- V_K: a set of a maximal clique from the extended conflict graph $G'(V, E')$;
- $\overline{V}_K$: a complement set of V_K in V ($\overline{V}_K = V \setminus V_K$);
- $\overline{s} = \sum_{i \in \overline{V}_K} w_i$ be the total items size of the set $\overline{V}_K$.

Obviously, each item from V_K should be assigned into a distinct bin that can fits and consequently we have to initialize $|V_K|$ bins at least. Let j^* be the smallest index of bin-type that can receive the smallest item of the clique set ($W_{j^*} \geqslant \min_{i \in V_K}(w_i)$), and consider the partition of V_K into the subsets V_h as follows:

$$V_h = \{i \in V_K : W_{h-1} < w_i \leqslant W_h\} \text{ for } h = j^*, \ldots, m.$$

Clearly, each item i from V_h ($h = j^*, \ldots, m$.) should be assigned alone into one bin from a selected bin-type j $j \in [h, m]$. Let y_{hj} be the number of bins of type j ($j = h, \ldots, m$) that are used for packing the items of subset V_h. Then, we have

$$\sum_{j=h}^{m} y_{hj} = |V_h| \text{ for } h = j^*, \ldots, m \tag{13}$$

Also, the assignment of any item from V_h into one bin of a given bin-type j ($j \in [h, m]$) generates a residual capacity $\widetilde{W}_{hj}$ that it is at most equal to $W_j - min_{i \in V_h} w_i$. In the calculation of our lower bound, the residual capacity will be occupied with the items of $\overline{V}_K$ without considering the conflicts in this last subset. Hence, for every bin-type j and for a given subset V_h, the maximum value of $\widetilde{W}_{hj}$ can be obtained by solving the following subset sum problem (SSP)

$$\widetilde{W}_{hj} = \text{Max} \sum_{i \in \overline{V}_K} w_i z_i \tag{14}$$

s.t.

$$\sum_{i \in \overline{V}_K} w_i z_i \leqslant W_j - \min_{i \in V_h} w_i \tag{15}$$

$$z_i \in \{0, 1\}, i \in \overline{V}_K \tag{16}$$

Both lower bounds consist in the resolution of the VSBPPC by relaxing the integrality constraints in which we assume that only items from $\overline{V}_K$ might be split. For that, define an integer variable $\overline{y}_j$ as the number of added bins of type j used for complete the packing of the items of $\overline{V}_K$ set. Then, we have to solve the following mathematical programming formulation

$$LB = \text{Min} \left(\sum_{j=1}^{m} c_j \overline{y}_j + \sum_{h=j^*}^{m} \sum_{j=h}^{m} c_j y_{hj} \right) \tag{17}$$

s.t.

$$\sum_{j=h}^{m} y_{hj} = |V_h| \, h = j^*, \ldots, m \tag{18}$$

$$\sum_{j=1}^{m} W_j \overline{y}_j + \sum_{h=j^*}^{m} \sum_{j=h}^{m} W_j y_{hj} \geqslant \sum_{i \in V} w_i \tag{19}$$

$$\sum_{j=1}^{m} W_j \overline{y}_j + \sum_{h=j^*}^{m} \sum_{j=h}^{m} \widetilde{W}_{hj} y_{hj} \geqslant \sum_{i \in \overline{V}_K} w_i \tag{20}$$

$$\overline{y}_j \in \mathbb{N}, \, j = 1, \dots, m \tag{21}$$

$$y_{hj} \in \mathbb{N}, \, h = j^*, \dots, m \, \, j = h, \dots, m. \tag{22}$$

The objective function (17) is to minimize the total cost of required bins (both the additional bins that are added for complete the packing of items from $\overline{V}_K$ and the bins used for the packing of items from the maximal clique set V_K). Constraint (18) is derived from formula (13) ensure that each item of the maximal clique set V_K is packed into a distinct bin. Constraint (19) ensures that the total capacity of used bins enables to pack all the items. Constraint (20) informs that the residual capacities generated by the packing of the maximal clique items together with the capacity of additional bins are enable to pack the set of items from $\overline{V}_K$. Whilst, formulas (21) and (22) represent the integrality constraints.

4 Conclusion

The variable sized bin-packing problem with conflicts is an NP-hard combinatorial problem often encountered in the practical field; it generalizes both the bin-packing problem with heterogeneous bins and the vertex coloring problem, both well known and notoriously difficult. In this paper, we have addressed the offline version of this problem, where we have proposed two variant of lower bound based on the combination of the clique computation and the constraint relaxing formulation.

Finally, considering the practical relevance of discussed problem and the fact that the offline version of this problem has not been excessively addressed in the literature, this work can be considered as a comparison basis for the future works.

References

1. Zhang, G.: A new version of on-line variable-sized bin packing. Discrete Appl. Math. **72**(3), 193–197 (1997)
2. Epstein, L., Favrholdt, L.L.M., Levin, A.: Online variable-sized bin packing with conflicts. Discrete Optim. **8**, 333–343 (2011)
3. Kang, J., Park, S.: Algorithms for the variable sized bin-packing problem. Eur. J. Oper. Res. **147**, 365–372 (2003)
4. Haouari, M., Serairi, M.: Heuristics for the variable sized bin-packing problem. Comput. Oper. Res. **36**(10), 2877–2884 (2009)
5. Haouari, M., Serairi, M.: Relaxations and exact solution of the variable sized bin packing problem. Comput. Optim. Appl. **48**, 345–368 (2011)
6. Johnson, D.S.: Approximation algorithms for combinatorial problems. J. Comput. Syst. Sci. **9**, 256–278 (1974)
7. Gendreau, M., Laporte, G., Semet, F.: Heuristics and lower bounds for the bin packing problem with conflicts. Comput. Oper. Res. **31**, 347–358 (2004)
8. Muritiba, A.E.F., Iori, M., Malaguti, E., Toth, P.: Algorithms for the bin packing problem with conflicts. INFORMS J. Comput. **22**, 401–415 (2010)

Global Existence, Uniqueness and Asymptotic Behavior for a Nonlinear Parabolic System

Naima Aïssa and H. Tsamda

Abstract This paper is dealing with a cancer invasion model proposed by Chaplain (J Theor Biol 241:564–589, 2006). We are interested by the case of small initial data and when matrix-degrading enzymes initial data is bounded by below by a strictly positive constant. We provide global existence and uniqueness of weak solutions. We prove that the global solution converges with an exponential decay to a study state which is a solution of the corresponding stationary equation.

1 The Model Equation

The model is described by the following system [2]

$$
\begin{cases}
\partial_t n = D\Delta n - \rho\nabla\cdot(n\nabla f), (t,x) \in \mathbb{R}^+ \times \Omega, \\
\partial_t f = -\gamma m f, (t,x) \in \mathbb{R}^+ \times \Omega, \\
\partial_t m = \varepsilon\Delta m + \alpha n - \nu m, (t,x) \in \mathbb{R}^+ \times \Omega, \\
n(0) = n_0, \ f(0) = f_0, \ m(0) = m_0, \\
\partial_\nu n = \partial_\nu m = 0, \quad \text{on} \ \ \mathbb{R}^+ \times \partial\Omega,
\end{cases}
\tag{1}
$$

where n is the cancer cell density, m is the concentration of the matrix degrading enzyme, f is the extracellular matrix density, $D, \rho, \gamma, \varepsilon, \nu$ are nonnegative constant and Ω is a bounded regular domain of $\mathbb{R}^2$.

One can summarize briefly the invasion-degradation process as follow. The extra-cellular matrix is invaded by the migration of cancer cells following the haptotactic gradient $\nabla\cdot(n\nabla f)$. The migrated cells produce degrading matrix enzyme m which degrades the surrounded extracellular matrix leading to the cancer growth. Local

N. Aïssa (✉) · H. Tsamda
USTHB, Laboratoire AMNEDP, Algiers, Algeria
e-mail: aissa.naima@gmail.com

H. Tsamda
e-mail: hocinemaths@gmail.com

© Springer International Publishing Switzerland 2016
G.A. Anastassiou and O. Duman (eds.), *Intelligent Mathematics II:*
Applied Mathematics and Approximation Theory, Advances in Intelligent Systems
and Computing 441, DOI 10.1007/978-3-319-30322-2_19

271

existence and uniqueness of regular solutions for such a model has been proved by
[4] for classical initial data. The most feature of such equations is the possibility of
blowing-up in finite time for arbitrary initial conditions. However, global existence
can be obtained for small initial data.

In the sequel, $L^2(\Omega)$ will denote the Hilbert space of measurable and square
integrable functions endowed with the norm $\|.\|$ while the Banach space $L^p(\Omega)$
will be endowed with the usual norm denoted by $\|.\|_{0,p}$. The weak solutions are
constructed the Sobolev spaces $H^m(\Omega)$ endowed with the classical norm $\|.\|_{H^m}$.

The plan of the manuscript is the following. In Sect. 2 we will prove existence
and uniqueness of a local in time solution while in Sect. 3, we will prove that the
previous solution is global in time. Finally, the Sect. 4 is devoted to the asymptotic
behavior of the solution. More precisely, we will prove that the solution converges
with an exponential decay to the average of the initial data.

2   Local in Time Existence and Uniqueness of a Solution

Definition 1 The triplet (n, f, m) with

$$n \in L^\infty(0, T; \ L^\infty(\Omega)) \cap L^2(0, T; \ H^1(\Omega)), \quad \partial_t n \in L^2(0, T; (H^1(\Omega))^\star),$$

$$f \in \mathscr{C}(0, T; \ H^1(\Omega)) \cap L^\infty(0, T; \ H^2(\Omega)),$$

$$m \in L^\infty(0, T; \ L^2(0, T)) \cap L^2(0, T; \ H^1(\Omega)); \ \partial_t m \in L^2(Q_T),$$

is called a weak solution of (1) if f satisfies the ODE a.e; and if for all
$\varphi \in L^2(0, T; \ H^1(\Omega))$

$$\int_0^T \langle \partial_t n, \varphi \rangle + \int_0^T \int_\Omega \{D\nabla n - \rho n \nabla f\} \cdot \nabla \varphi \, dx dt = 0,$$

$$\int_{Q_T} \partial_t m \varphi + \varepsilon \nabla m \cdot \nabla \varphi - (\nu m - \alpha n)\varphi \, dx dt = 0.$$

The aim of this section is to prove the following Theorem

Theorem 2 *Assume that*

$$n_0 \in L^\infty(\Omega), f_0 \in W^{2,p}(\Omega), m_0 \in W^{1,p}(\Omega), \quad p > 2$$

$$n_0 \geq 0, \quad f_0 \geq 0, \quad m_0 \geq \gamma_\star > 0, \quad \partial_\nu f_0 = 0,$$

for some positive constant $\gamma_\star$. Then, there exists a unique local nonnegative weak solution

$$(n, f, m) \in L^2(0, T^\star, H^1(\Omega)) \times L^2(0, T^\star, H^2(\Omega)) \times L^2(0, T^\star, H^1(\Omega))$$

satisfying $m \geq \gamma_\star$.

The proof will be done in several steps.

2.1   The Equation Satisfied by *m* for Given *n*

Suppose that the initial condition satisfies $m_0 \geq \gamma_\star$, and let $n \geq 0$,

$$n \in L^2(0, T; H^1(\Omega))$$

be given. Our aim is to construct a solution satisfying $m \geq \gamma_\star$ to the problem

$$\begin{cases} \partial_t m = \varepsilon \Delta m + \alpha n - \nu m, \ (t, x) \in \mathbb{R}^+ \times \Omega. \\ m(0, .) = m_0, \quad \partial_\nu m = 0. \end{cases} \tag{2}$$

To do so, we will consider the following intermediate problem

$$\begin{cases} \partial_t m = \varepsilon \Delta m + \alpha n - \nu m 1_{\{m \geq \gamma_\star\}}, \ (t, x) \in \mathbb{R}^+ \times \Omega, \\ m(0, .) = m_0, \quad \partial_\nu m = 0. \end{cases} \tag{3}$$

Solution for such a problem can be constructed by an iteration argument. Indeed, let us consider the sequence (m^k), defined for given m^1, by

$$\begin{cases} \partial_t m^{k+1} = \varepsilon \Delta m^{k+1} + \alpha n - \nu m^{k+1} 1_{\{m^k \geq \gamma_\star\}}, \ (t, x) \in \mathbb{R}^+ \times \Omega, \\ m^{k+1}(0, .) = m_0, \quad \partial_\nu m^{k+1} = 0. \end{cases} \tag{4}$$

We check easily by standard calculus that for fixed $T > 0$, there exists a constant $C_{T,n}$ independent of k such that

$$\left\| m^{k+1} \right\|^2_{L^\infty(0,T;L^2(\Omega))} + \left\| m^{k+1} \right\|^2_{L^2(0,T;\,H^1(\Omega))} + \left\| \partial_t m^{k+1} \right\|^2_{L^2(\Omega_T)} \leq C_{T,n}.$$

Then, for a subsequence, $m^k \to m$ strongly in $L^2(\Omega_T)$, where m is a solution to (3). Then multiplying (3) by $(m - \gamma_\star)^-$ and integrating by parts, we get

$$-\partial_t \left\| (m - \gamma_\star)^- \right\|^2 = \left\| \nabla (m - \gamma_\star)^- \right\|^2 + \int_\Omega n(m - \gamma_\star)^- \, dx \geq 0.$$

As $m_0 \geq \gamma_\star$ then $\|(m - \gamma_\star)^-\|^2 = 0$ hence $m \geq \gamma_\star$. Consequently the solution m to (3) is also a solution to (2). Then we proved that if $m_0 \geq \gamma_\star$, $n \geq 0$ and $n \in L^2(0, T; H^1(\Omega))$, the solution to (2) satisfies $m \geq \gamma^\star$. Moreover, by standard computations on parabolic equations, there exists a constant $C > 0$ depending on the initial data such that

$$\|m\|^2_{L^\infty(0,T; L^2(\Omega))} + \|m\|^2_{L^2(0,T; H^k(\Omega))} \leq C \left(\|n\|^2_{L^2(0,T; H^1(\Omega))} + 1 \right), \tag{5}$$

$k = 1, 2, 3$.

2.2 The Equation Satisfied by the Extracellular Matrix Density f for Given m

Let m be given by the previous subsection and f be the solution of the ordinary differential equation,

$$\partial_t f = -\gamma m f, \quad f(0) = f_0.$$

We prove easily that

$$\|f\|^2_{H^k(\Omega)}(t) \leq C_{f_0} \left(1 + t \|m\|^2_{L^2(0,T; H^k(\Omega))} \right) e^{-2\gamma\gamma_\star t}, \quad k = 1, 2, 3. \tag{6}$$

$$\left\| \Delta f - f_0 \left| \gamma \int_0^t \nabla m(\tau) d\tau \right|^2 \right\|^2_{L^2(\Omega)} (t) \leq C_{f_0} \left(1 + t \|n\|^2_{L^2(Q_T)} \right) e^{-2\gamma\gamma_\star t}, \tag{7}$$

where

$$C_{f_0} = \|\Delta f_0\|^2_{L^2(\Omega)} + 2\gamma \|\nabla f_0\|^2_{L^2(\Omega)} + \gamma \|f_0\|^2_{L^\infty(\Omega)}. \tag{8}$$

2.3 The Parabolic Equation Satisfied by Cell Density for Given ECM Density f

For f satisfying properties of the previous subsection, we consider $n^\star$ the weak solution of the linear parabolic problem

$$\begin{cases} \partial_t n^\star = D \Delta n^\star - \rho \nabla \cdot (n^\star \nabla f), & (t, x) \in (0, T) \times \Omega, \\ n(0) = n_0, \quad \partial_\nu n^\star = 0. \end{cases} \tag{9}$$

From classical results on parabolic equations, there exists a unique solution $n^\star$

$$n^\star \in L^\infty(0, T; L^2(\Omega)) \cap L^2(0, T; H^1(\Omega)), \quad \partial_t n^\star \in L^2(0, T, (H^1(\Omega)^\star)),$$

where $(H^1(\Omega))^\star$ is the dual space of $H^1(\Omega)$.

2.4 Fixed Point Procedure

For fixed $T > 0$, $R_i > 0$ $i = 1, 2, 3$, we define V_T by

$$V_T = \left\{ \begin{array}{l} n \in L^\infty(0, T; \ L^2(\Omega)) \cap L^2(0, T; H^1(\Omega)), \ n \geq 0, \ n(0, x) = n_0 \\ \|n\|_{L^\infty(0,T;L^2(\Omega))} \leq R_1, \ \|\nabla n\|_{L^2(0,T;L^2(\Omega))} \leq R_2, \ \|\partial_t n\|_{L^2(0,T;(H^1(\Omega)^\star)} \leq R_3 \end{array} \right\}$$

Define $\mathscr{L}$ on V_T as follows. Let $n \in K$ and m be the corresponding solution of

$$\partial_t m = \Delta m + \alpha n - \nu m, \quad m(0) = m_0, \quad \partial_\nu m = 0.$$

Then consider f the solution of the ODE

$$\partial_t f = -\gamma m f, \quad f(0) = f_0$$

Then $\mathscr{L}(n)$ is defined as the weak solution

$$\partial_t n^\star = D \Delta n^\star - \rho \nabla \cdot (n^\star \nabla f), \quad n^\star(0) = n_0, \quad \partial_\nu n^\star = 0.$$

It is clear that V_T is a compact and convex subset of $L^2(Q_T)$ and we use standard arguments to prove that there exists R_i, $i = 1, 2, 3$ such that for T small enough, the operator $\mathscr{L}$ sends V_T into itself and that $\mathscr{L}$ is continuous. Indeed, Multiplying the parabolic equation by $n^\star$ and integrating by parts we get

$$\frac{d}{2dt} \left\| n^\star \right\|^2 + D \left\| \nabla n^\star \right\|^2 = \rho \int_\Omega n^\star \nabla f \cdot \nabla n^\star = \frac{\rho}{2} \int_\Omega \nabla ((n^\star)^2) \cdot \nabla f$$

$$= -\frac{\rho}{2} \int_\Omega ((n^\star)^2) \cdot \Delta f.$$

Consequently

$$\frac{d}{2dt} \left\| n^\star \right\|^2 + D \left\| \nabla n^\star \right\|^2 \leq \frac{\rho}{2} \|\Delta f\| \left\| (n^\star)^2 \right\| \leq \frac{\rho}{2} \|\Delta f\| \left\| n^\star \right\|_{0,3} \left\| n^\star \right\|_{0,6}.$$

Then using the Gagliardo-Nirenberg inequality

$$\|n\|_{0,3} \leq C \|n\|_{H^1(\Omega)}^{\frac{1}{2}} \|n\|_{0,2}^{\frac{1}{2}},$$

and the fact that $H^1(\Omega)$ is continuously embedded into $L^6(\Omega)$, we obtain

$$\frac{d}{2dt}\left\|n^\star\right\|^2 + D\left\|\nabla n^\star\right\|^2 \le C\left\|\Delta f\right\|\left\|n^\star\right\|^{\frac{3}{2}}_{H^1(\Omega)}\left\|n^\star\right\|^{\frac{1}{2}}.$$

Furthermore, by Young's inequality ($p=\frac{4}{3}, q=4$)

$$\frac{d}{2dt}\|n^\star\|^2 + D\|\nabla n^\star\|^2 \le C\alpha\|n^\star\|^2_{H^1(\Omega)} + C_\alpha\|\Delta f\|^4\left\|n^\star\right\|^2.$$

Then, choosing α small enough, we get

$$\frac{d}{2dt}\|n^\star\|^2 + \frac{D}{2}\|\nabla n^\star\|^2 \le (C_\alpha\|\Delta f\|^4 + C\alpha)\|n^\star\|^2.$$

Combining (6), (5) and the fact that $n \in V_T$, we deduce that

$$\frac{d}{2dt}\|n^\star\|^2 + \frac{D}{2}\|\nabla n^\star\|^2 \le C(R_1, R_2, \alpha, T)\|n^\star\|^2,$$

for some positive constant depending on R_1, R_2, α, T. Using Gronwall's lemma, we get

$$\|n^\star\|(t) \le e^{\int_0^t C(R_1,R_2,\alpha,T)d\tau}\|n_0\|^2,$$

$$\|\nabla n^\star\|^2_{L^2(Q_T)} \le \left(1 + \int_0^T e^{\int_0^t C(R_1,R_2,\alpha,T)d\tau}\right)\|n_0\|^2. \tag{10}$$

To provide time derivative estimate, we test the parabolic equation satisfied by $n^\star$ by $\varphi \in L^2(0, T; H^1(\Omega))$

$$\langle \partial_t n^\star; \varphi \rangle = -D\int_\Omega \nabla n^\star \cdot \nabla\varphi dx + \rho\int_\Omega n^\star\nabla f \cdot \nabla\varphi dx.$$

Then

$$|\langle \partial_t n^\star; \varphi \rangle| \le D\|\nabla n^\star\|\|\nabla\varphi\| + \rho\|n^\star\nabla f\|\|\nabla\varphi\|$$
$$\le D\|\nabla n^\star\|\|\nabla\varphi\| + \rho\|n^\star\|_{0,4}\|\nabla f\|_{0,4}\|\nabla\varphi\|.$$

As $H^1(\Omega)$ is continuously embedded in $L^4(\Omega)$, we deduce that

$$|\langle \partial_t n^\star; \varphi \rangle| \le D\|\nabla n^\star\|\|\nabla\varphi\| + \rho\|n^\star\|_{H^1(\Omega)}\|\nabla f\|_{H^1(\Omega)}\|\nabla\varphi\|,$$

consequently

$$\int_0^t |\langle \partial_t n^\star; \varphi \rangle| dt$$

$$\leq \left(D\|\nabla n^\star\|_{L^2(0,T,L^2(\Omega))} + \rho\|n^\star\|_{L^2(0,T,H^1(\Omega))}\|f\|_{L^\infty(0,T,H^2(\Omega))} \right) \|\varphi\|_{L^2(0,T;H^1(\Omega))}.$$

This means

$$\|\partial_t n^\star\|_{L^2(0,T;(H^1)^\star)}$$
$$\leq D\|\nabla n^\star\|_{L^2(0,T,L^2(\Omega))} + \rho\|f\|_{L^\infty(0,T,H^2(\Omega))}\|n^\star\|_{L^2(0,T,H^1(\Omega))} \qquad (11)$$
$$\leq DR_2 + \rho(R_1 + R_2) + C(R_1, R_2, T).$$

Then, recalling (10) and (11), we check that for $R_1^2 = 2\|n_0\|^2$, $R_2^2 = \|n_0\|^2 + 1$ and $R_3 = DR_2 + \rho(R_1 + R_2) + C(R_1, R_2, T)$, the operator $\mathcal{L}$ sends V_T into itself for small time T.

Next, using again (10) and (11) and the uniqueness of the solution of (9) we prove that the operator $\mathcal{L}$ is continuous on $L^2(0, T, L^2(\Omega))$.

Finally, the existence of a solution (n, f, m) to (1) is a consequence of Schauder's fixed point theorem.

However, the uniqueness cannot be obtained by standard arguments because the solution n is not smooth enough. We will use the idea developed by Gajewski-Zakarias [3]. We set

$$g(n) = n(\log(n) - 1)$$

Let (n_i, f_i, m_i) two solutions of (1) with the same initial data. Then setting

$$h(n_1, n_2) = g(n_1) + g(n_2) - 2g\left(\frac{n_1 + n_2}{2}\right),$$

we check that

$$\partial_t(h(n_1, n_2)) = (\partial_t n_1) \log\left(\frac{2n_1}{n_1 + n_2}\right) + (\partial_t n_2) \log\left(\frac{2n_2}{n_1 + n_2}\right).$$

Then replacing $\partial_t n_i$ by $D\Delta n_i - \rho\nabla \cdot (n_i \nabla f_i)$ and integrating by parts, we get

$$\int_\Omega h(n_1, n_2)dx \leq C\|n_1 + n_2\|_{L^\infty(Q_T)} \int_0^t \|\nabla f_1 - \nabla f_2\|^2 ds. \qquad (12)$$

Using the equations satisfied by $f_1 - f_2$ and $m_1 - m_2$ respectively, we prove easily that

$$\int_0^t \|\nabla f_1 - \nabla f_2\|^2 ds \leq C \left(\int_0^t \|\nabla m_1 - \nabla m_2\|^2 + \|m_1 - m_2\|^2 ds \right), \qquad (13)$$

and

$$\|\nabla m_1 - \nabla m_2\|^2(t) + \|m_1 - m_2\|^2(t)$$
$$\leq C \int_0^t \|n_1 - n_2\|^2 ds \qquad (14)$$
$$\leq C \|n_1 + n_2\|_{\infty, Q_T} \int_0^t \|\sqrt{n_1} - \sqrt{n_2}\|^2 ds.$$

Then combining (12), (13) and (14), we get

$$\int_\Omega h(n_1, n_2) dx + \|m_1 - m_2\|_{H^1(\Omega)}^2(t)$$

$$\leq C \|n_1 + n_2\|_{\infty, Q_T} \left\{ \int_0^t \|\sqrt{n_1} - \sqrt{n_2}\|^2 + \|m_1 - m_2\|_{H^1(\Omega)}^2(s) ds \right\}$$

Finally uniqueness follows from the following inequality ([3, Lemma 6.4; p. 112])

$$h(n_1, n_2) \geq \frac{1}{4}(\sqrt{n_1} - \sqrt{n_2})^2,$$

and Gronwall's Lemma.

3   Global Existence of a Solution

We prove that the previous local solution is global in time by adapting some ideas of Yagi [7].

Theorem 3 *Denoting by C_Ω the constant appearing in the interpolation inequality*

$$\|a\|_{L^8(\Omega)} \leq C_\Omega \|a\|_{L^2(\Omega)}^{\frac{1}{4}} \|a\|_{H^1(\Omega)}^{\frac{3}{4}}, \quad \forall a \in L^8(\Omega) \cap H^1(\Omega),$$

and by $p(x) = C_\Omega x^2(x+1)$. Assuming that the initial data n_0, f_0 satisfy

$$4D - \frac{1}{2}\left(1 + \frac{C_{f_0}}{\gamma^2 \gamma_\star^2} p^2(\|n_0\|_{L^1(\Omega)}^{\frac{1}{6}})\right) > 0, \qquad (15)$$

where D is the diffusion coefficient of cancer cells and C_{f_0} the constant

$$C_{f_0} = \|\Delta f_0\|_{L^2(\Omega)}^2 + 2\gamma \|\nabla f_0\|_{L^2(\Omega)}^2 + \gamma \|f_0\|_{L^\infty(\Omega)}^2. \qquad (16)$$

Then there exists a constant $C > 0$ independent of time, depending only on the initial data such that

$$\left\|\nabla\sqrt{(n+1)}\right\|^2_{L^2(Q_T)} \leq C, \quad \forall T > 0. \tag{17}$$

Consequently, there exists $C > 0$ depending only on the initial data such that

$$\|n\|_{L^2(0,t,H^2(\Omega))} + \|m\|_{L^2(0,t,H^2(\Omega))} + \|f\|_{L^2(0,t,H^2(\Omega))} \leq C(1+t), \quad t \in (0,T) \tag{18}$$

which leads to the global existence of a solution to (1).

Proof The proof relies on interpolation inequalities and mass conservation leading to

$$\|n\|_{L^2(\Omega)}(t)$$
$$\leq p\left(\|n_0\|^{\frac{1}{6}}_{L^1(\Omega)}\right)\left\|\nabla\sqrt{(n+1)}\right\|(t) + (\|n_0\|_{L^1(\Omega)} + 1)^{\frac{1}{2}}p(\|n_0\|^{\frac{1}{6}}_{L^1(\Omega)}), \tag{19}$$

for $0 \leq t \leq T$ (p is the polynomial defined above) and the following entropy inequality obtained by multiplying the first equation of (1) by the entropy term $ln(n+1)$ and integrating by parts

$$\varphi(t) + (4D - \tfrac{1}{2}(1 + \tfrac{C_{f_0}}{(\gamma\gamma_\star)^2})p^2(\|n_0\|^{\frac{1}{6}}_{L^1(\Omega)}))\left\|\nabla\sqrt{(n+1)}\right\|^2_{L^2((0,T)\times\Omega)} \leq C + \varphi(0), \tag{20}$$

where

$$\varphi(t) = \int_\Omega (n+1)(ln(n+1) - n)\,dx$$

and C is a positive constant which is independent of T, depending only on the initial data and γ, $\gamma_\star$. Finally, using (20) and (15) we deduce that

$$\left\|\nabla\sqrt{(n+1)}\right\|^2_{L^2((0,T)\times\Omega)}$$

is bounded by a constant which is independent of T then (17) is proved.

Next, combining (19) and (17) we deduce that there exists C, independent of T, depending only on the data such that

$$\|n\|_{L^2(0,T); L^2(\Omega)} \leq C. \tag{21}$$

Consequently (18) follows from classical results on parabolic equations. To end this section, it remains to prove that $n \in L^\infty(0, T, L^\infty(\Omega))$. This is done by controlling L^p norms and using the theorem of Alikakos [1] based on the iterations theorem of Moser. See also [3].

4 Asymptotic Behavior of the Solution

We will need more regularity on the initial data to provide the asymptotic behavior of the solution. We will prove that $n \to \overline{n_0}$, $m \to \frac{\alpha}{\nu}\overline{n_0}$, $f \to 0$ in $L^\infty(\Omega)$ with exponential decay, where

$$\overline{n_0} = \frac{1}{|\Omega|} \int_\Omega n_0(x)dx.$$

Theorem 4 *Assume that the initial data satisfy (15) and*

$$(n_0, m_0, f_0) \in H^2(\Omega) \times H^2(\Omega) \times H^3(\Omega).$$

Then for arbitrary $0 < \beta < \gamma\gamma_\star$

$$\lim_{t \to \infty} e^{\beta t} \|f(t, .)\|_{L^\infty(\Omega)} = 0. \tag{22}$$

Moreover, there exists a constant $k > 0$ depending on D and $|\Omega|$ such

$$\lim_{t \to \infty} e^{kt} \|n(t, .) - \overline{n_0}\|_{L^\infty(\Omega)} = 0. \tag{23}$$

Finally for $0 < \delta < \min(2\nu, \kappa)$

$$\lim_{t \to \infty} e^{\delta t} \|m(t, .) - \frac{\alpha}{\nu}\overline{n_0}\|_{L^\infty(\Omega)} = 0. \tag{24}$$

Proof As

$$f(t, x) = f_0(x)e^{-\int_0^t m(\tau, x)d\tau}, \quad m \geq \gamma_\star, \quad \text{a.e.,}$$

then $\|f(t, .)\|_{L^\infty(\Omega)} \leq \|f_0\|_{L^\infty(\Omega)}e^{-\gamma\gamma_\star t}$ then (22) holds. Denoting by

$$P_1(u) = \partial_t n - D\Delta n + \rho\nabla \cdot (n\nabla f)$$

and testing $\nabla P_1(u)$ by $\nabla\Delta n$ and integrating by parts, we get

$$\frac{d}{2dt}\|\Delta n\|_{L^2(\Omega)}^2 + D\|\nabla\Delta n\|_{L^2(\Omega)}^2 \leq \frac{D}{2}\|\nabla\Delta n\|_{L^2(\Omega)}^2 + C\|\nabla \cdot (n\nabla f)\|_{H^1(\Omega)}^2.$$

Hence

$$\frac{d}{2dt}\|\Delta n\|_{L^2(\Omega)}^2 + \frac{D}{2}\|\nabla\Delta n\|_{L^2(\Omega)}^2 \leq C\|n\nabla f\|_{H^2(\Omega)}^2 \leq \|n\|_{H^2(\Omega)}^2\|f\|_{H^3(\Omega)}^2.$$

Consequently, using (6), (5) and (18)

$$\frac{d}{2dt}\|\Delta n\|_{L^2(\Omega)}^2 + \frac{D}{2}\|\nabla\Delta n\|_{L^2(\Omega)}^2 \le C(1+t)e^{-2\gamma\gamma_2 t} \le Ce^{-\gamma\gamma_2 t}. \tag{25}$$

Moreover, testing $P_1(u)$ by Δu and integrating by parts

$$\frac{d}{2dt}\|\nabla n\|_{L^2(\Omega)}^2 + D\|\Delta n\|_{L^2(\Omega)}^2 \le \frac{D}{2}\|\Delta n\|_{L^2(\Omega)}^2 + C\|\nabla \cdot (n\nabla f)\|_{L^2(\Omega)}^2.$$

By the previous arguments we get

$$\frac{d}{2dt}\|\nabla n\|_{L^2(\Omega)}^2 + \frac{D}{2}\|\Delta n\|_{L^2(\Omega)}^2 \le C(1+t)e^{-2\gamma\gamma_2 t} \le Ce^{-\gamma\gamma_2 t}. \tag{26}$$

Next, we will need the following Lemma whose proof is based on Poincaré-Wirtinger inequality

Lemma 5 *Let Ω be a bounded and regular domain of $\mathbb{R}^N$, $N \le 3$. Then there exists a constant $C_\Omega > 0$ depending only on Ω such that for every $uv \in H^2(\Omega)$ satisfying $\partial_\nu u = 0$*

$$\|\nabla u\|_{L^2(\Omega)} \le C_\Omega \|\Delta u\|_{L^2(\Omega)} \tag{27}$$

$$\|u - \overline{u}\|_{L^\infty(\Omega)} \le C_\Omega \|\Delta u\|_{L^2(\Omega)}. \tag{28}$$

Using (27) we get $\|\nabla\Delta n\|^2 \ge \frac{1}{C_\Omega^2}\|\nabla n\|^2$ and adding (25) and (26)

$$\frac{d}{dt}(\|\nabla n\|^2 + \|\Delta n\|^2) + \kappa(\|\nabla n\|^2 + \|\Delta n\|^2) \le Ce^{-\gamma\gamma_* t} \tag{29}$$

for $0 < \kappa \le \max(\frac{D}{2}, \frac{C_\Omega^2 D}{2})$. Consequently

$$\|\nabla n\|^2(t) + \|\Delta n\|^2(t) \le e^{-\kappa t}(c_{n_0} + \int_0^t e^{(\kappa - \gamma\gamma_*)t}ds), \tag{30}$$

then choosing κ such that $\kappa - \gamma\gamma_* \le 0$, we get

$$\|\Delta n\|^2 \le c_{n_0}(1+t)e^{-\kappa t}.$$

Finally, as $\overline{n} = \overline{n_0}$, we deduce from (28) and the previous estimate that

$$\|n - \overline{n}_0\|_{L^\infty(\Omega)}^2 \le c_{n_0}(1+t)e^{-\kappa t},$$

then (23) follows. We will use the same arguments for m. Denoting by $P_2(m) = \partial_t m - \alpha n - \varepsilon\Delta m + \nu m$ and testing $\nabla P_2(m)$ by $\nabla\Delta m$ we get

$$\frac{d}{2dt}\|\Delta m\|^2 + \frac{\varepsilon}{2}\|\nabla \Delta m\| + \nu\|\Delta m\|^2 \le c\|\nabla n\|^2.$$

Hence using (30),

$$\frac{d}{2dt}\|\Delta m\|^2 + \nu\|\Delta m\|^2 \le c(1+t)e^{-\kappa t}.$$

Consequently

$$\|\Delta m\|^2 \le e^{-\nu t}c(1 + \int_0^t (1+s)e^{(\nu-\kappa)s}ds). \tag{31}$$

Moreover, simple calculus yield

$$\overline{m} = e^{-\nu t}(\overline{m}_0 - \frac{\alpha}{\nu}\overline{n}_0) + \frac{\alpha}{\nu}\overline{n}_0, \tag{32}$$

So that

$$\|\overline{m} - \frac{\alpha}{\nu}\overline{n}_0\|_{L^\infty} \le C e^{\nu t} \tag{33}$$

Then combining (31), (28), (32) and (33) we get (24) and the proof of the theorem is achieved.

Remark 6 We proved existence and uniqueness of a global strong solution to (1) for small data in the two dimensional case. In the papers of [5] and [6], we can find existence and uniqueness of a global classical solution without restriction on the initial data, to a similar model with a logistic source term $n(1 - n - f)$ in the cancer cell equation, which is known to prevent blow up. Comparing to those papers, blow up is not excluded in our case, so condition (15) is justified.

References

1. Alikakos, N.D.: L^p bounds of solutions of reaction-diffusion equations. Commun. Partial Diff. Eqn. **4**(8), 827–868 (1979)
2. Chaplain, M.A.J., et al.: Mathematical modeling of dynamic adaptive tumor angiogenesis. J. Theor. Biol. **241**, 564–589 (2006)
3. Gajewski, H., Zakarias, K.: Global behavior of a reaction-diffusion system modelling chemotaxis. Math. Nachr. **195**, 77–114 (1998)
4. Morales-Rorrigo, C.: Local existence and uniqueness of regular solutions in a model of tissue invasion by solid tumours. Math. Comput. Modell. **47**, 604–613 (2008)
5. Tao, Y.: Global existence for a haptotactic model of cancer invasion with tissue remodeling. Nonlinear Anal.: Real World Appl. **12**, 418–435 (2011)
6. Tello, J.I., Winkler, M.: Chemotaxis system with logistic source. Commun. Partial Diff. Eqn. **32**, 849–877 (2007)
7. Yagi, A.: Abstract Parabolic Evolution Equations. Springer (2000)

Mathematical Analysis of a Continuous Crystallization Process

Amira Rachah and Dominikus Noll

Abstract In this paper we discuss a mathematical model of crystallization of KCl in continuous operational mode with fines dissolution and classified product removal. We prove the global existence and the uniqueness of solutions of the model under realistic hypotheses.

1 Introduction

Crystallization is the unitary operation of formation of solid crystals from a solution. It is a key technology for a wide range of pharmaceutical, food and chemical industries where it is used to produce solid particles with desirable characteristics. The crystallization process is initiated when a solution becomes supersaturated, either by cooling, evaporation of solvent, addition of anti-solvent, or by chemical reaction. The principal processes in crystallization include nucleation or crystal birth, crystal growth, breakage, attrition and possibly agglomeration, in tandem with external processes like heating and cooling, product removal, fines dissolution [1–3], and much else.

Crystallizers can be operated in batch, semi-batch or continuous mode. In the continuous operational mode, solution is continuously fed to the crystallizer and product is continuously removed [4–6] in order to maintain a steady state. Mathematical models of crystallization processes are based on population, molar and energy balances. The population balance is described by a first order hyperbolic partial differential equation. The molar and energy balances are described by integro-differential equations.

A. Rachah (✉) · D. Noll
Institut de Mathématiques de Toulouse, Université Paul Sabatier,
Toulouse Cedex 9, 31062 Toulouse, France
e-mail: amira.rachah@math.univ-toulouse.fr

D. Noll
e-mail: dominikus.noll@math.univ-toulouse.fr

© Springer International Publishing Switzerland 2016
G.A. Anastassiou and O. Duman (eds.), *Intelligent Mathematics II:*
Applied Mathematics and Approximation Theory, Advances in Intelligent Systems
and Computing 441, DOI 10.1007/978-3-319-30322-2_20

In this work we discuss a mathematical model of crystallization of potassium chloride (KCl) in continuous operational mode with fines dissolution and classified product removal. Crystalline KCl is used in medicine and food processing, and as a sodium-free substitute for table salt. The process operates as follows: liquid solution is fed to the crystallizer. The supersaturation is generated by cooling. Due to supersaturation, crystals are formed from the solution and grow. Solution and crystals are continuously removed from the crystallizer by the product outlet (see several applications [4, 7, 8]). To justify our model we prove the existence and the uniqueness of solutions of the model under realistic hypotheses [9–14].

The structure of the paper is as follows. In first section we present the mathematical model of continuous crystallization of KCl. In the second part we justify the model by proving global existence and the uniqueness of solutions under realistic hypotheses.

2 Modelling and Dynamics of Process

In this section we present the population and mass balance equations which describe the dynamic model of continuous crystallization of KCl.

2.1 Population Balance Equation

The population balance equation describes a first interaction between the population of solid crystals, classified by their size, the characteristic length L, and a ageless populations of solute molecules of the constituent in liquid phase. The population balance equation models birth, growth and death of crystals due to breakage, washout, removal and is given by

$$
\frac{\partial \left(V(t)n(L,t) \right)}{\partial t} + G(c(t)) \frac{\partial \left(V(t)n(L,t) \right)}{\partial L}
$$
$$
= -q h_{f,p}(L) \, n(L,t) - a(L) \, V(t)n(L,t) \tag{1}
$$
$$
+ \int_{L}^{\infty} a(L')b(L' \to L) \, V(t)n(L',t) \, dL'.
$$

The boundary value is given by

$$
n(0,t) = \frac{B(c_t(t))}{G(c_t(t))}, \qquad t \geq 0 \tag{2}
$$

and the initial condition is given by

$$n(L, 0) = n_0(L), \qquad L \in [0, \infty). \tag{3}$$

The second term on the left of (1) describes the growth of the population of crystals of size L, while the terms on the right describe external effects like fines dissolution, product removal, flow into and out of the crystallizer, breakage and attrition. Extended modeling could also account for agglomeration of crystals [1, 2]. The term $h_{fp}(L) = h_f(L) + h_p(L)$ regroups the classification and the dissolution functions $h_f(L)$ and $h_p(L)$, depends on the volume $V(t)$. The classification function $h_p(L)$ describes the profile of the product removal filter, which removes large particles with a certain probability according to size. In the ideal case, assumed e.g. in [15], one has

$$h_p(L) = \begin{cases} R_p, & \text{if } L \geq L_p \\ 0, & \text{if } L < L_p, \end{cases} \tag{4}$$

where L_p is the product removal size and R_p the product removal rate. This corresponds to an ideal high-pass filter. Fines removal is characterized by the classification function h_f, which ideally is a low-pass filter of the form

$$h_f(L) = \begin{cases} 0, & \text{if } L > L_f \\ R_f, & \text{if } L \leq L_f, \end{cases} \tag{5}$$

where R_f is the fines removal rate, and L_f is the fine size.

The growth rate $G(c(t))$ in (1) is dependent of crystal size L and depends on the concentration of solute $c(t)$. One often assumes a phenomenological formula

$$G(c(t)) = k_g \, (c(t) - c_s)^g , \tag{6}$$

where growth coefficient k_g and growth exponent g depend on the constituent, and where c_s is the saturation concentration, [4, 6, 15]. For theory it suffices to assume that G is locally Lipschitz with $G(c) > 0$ for supersaturation $c > c_s$, and $G(c) < 0$ for $c < c_s$, in which case crystals shrink.

The breakage integral on the right of (1) can be explained as follows. The breakage rate $a(L)$ represents the probability that a particle of size L and volume $k_v L^3$ undergoes breakage. The daughter distribution $b(L, L')$ represents the conditional probability that a particle of size L, when broken, produces a particle of size $L' < L$. Equation (1) goes along with initial and boundary conditions. The initial crystal distribution $n_0(L)$ is called a *seed*. The boundary condition $n(0, t) = B(c(t))/G(c(t))$ models birth of crystals at size $L = 0$ and is governed by the ratio B/G of birth rate $B(c)$ over growth rate $G(c)$. Again it is customary to assume a phenomenological law of the form

$$B(c(t)) = k_b \left((c(t) - c_s)_+\right)^b \tag{7}$$

for the birth rate, where k_b is the nucleation or birth coefficient, b the birth exponent, and $q_+ = \max\{0, q\}$. For theory it is enough to assume that B is locally Lipschitz with $B > 0$ for $c > c_s$ and $B = 0$ for $c \leq c_s$, meaning that nucleation only takes place in a supersaturated suspension.

2.2 Mole Balance Equation

We now derive a second equation which models the influence of various internal and external effects on the second population, the concentration $c(t)$ of solute molecules in the liquid. In order to derive the so-called mole balance equation, we start by investigating the mass balance within the crystallizer. The total mass m of the suspension in the crystallizer is given by

$$m = m_{\text{liquid}} + m_{\text{solid}} = m_{\text{solvent}} + m_{\text{solute}} + m_{\text{solid}}. \tag{8}$$

In this study we consider non-solvated crystallization, where solute molecules transit directly into solid state without integrating (or capturing) solvent molecules. We therefore have

$$\frac{dm_{\text{solvent}}}{dt} = \dot{m}_{\text{solvent}}^{\pm}.$$

In continuous or semi-batch mode this equation has to be completed by external sources and sinks, takes the form

$$\frac{dm}{dt} = \frac{dm_{\text{solute}}}{dt} + \frac{dm_{\text{solid}}}{dt} + \frac{dm_{\text{solvent}}}{dt} \tag{9}$$

$$= \dot{m}_{\text{solute}}^{+} - \dot{m}_{\text{solute}}^{-} + \dot{m}_{\text{solid}}^{+} - \dot{m}_{\text{solid}}^{-} + \dot{m}_{\text{solvent}}^{+} - \dot{m}_{\text{solvent}}^{-}.$$

We will now have to relate this equation to the population balance equation (1). In analogy with (8) we decompose the total volume V of the suspension as

$$V = V_{\text{liquid}} + V_{\text{solid}} = V_{\text{solute}} + V_{\text{solvent}} + V_{\text{solid}}.$$

Since the crystallization is non-solvated, we have $\frac{dm_{\text{solute}}}{dt} = -\frac{dm_{\text{solid}}}{dt}$. The mass of solute is given by

$$m_{\text{solute}} = VcM. \tag{10}$$

Here M is the molar mass of the constituent, V the total volume of the suspension, and V_{solute} is the volume of solute in the suspension. The unit of c is [mol/ℓ]. In this study we will consider the total volume of the suspension as time-varying, which leads to the formula

$$\frac{dm_{\text{solute}}}{dt} = \frac{d(VcM)}{dt} = \frac{dV}{dt}cM + V\frac{dc}{dt}M. \tag{11}$$

Let us now get back to (9). We start by developing the expressions on the right hand side. Decomposing

$$\dot{m}_{\text{solid}}^{\pm} = \dot{m}_{\text{fines}}^{\pm} + \dot{m}_{\text{product}}^{\pm} + \dot{m}_{\text{general}}^{\pm},$$

we have

$$\dot{m}_{\text{fines}}^{+} + \dot{m}_{\text{product}}^{+} + \dot{m}_{\text{general}}^{+} = 0,$$

meaning that we do not add crystals during the process. For crystal removal we have

$$\dot{m}_{\text{product}}^{-} = qk_v\rho \int_0^\infty h_p(L)n(L,t)L^3 dL,$$

which means that crystals of size L are filtered with a certain probability governed by the classification function h_p. Similarly, fines are removed according to

$$\dot{m}_{\text{fines}}^{-} = qk_v\rho \int_0^\infty h_f(L)n(L,t)L^3 dL,$$

where h_f is the fines removal filter profile. Here $\rho := \rho_{\text{solute}} = \rho_{\text{solid}}$ is the density of solute, and also the crystal density.

The term

$$\dot{m}_{\text{general}}^{-} = qk_v\rho \int_0^\infty n(L,t)L^3 dL$$

corresponds to a size indifferent removal of particles caused by the flow with rate q. The external terms for solute include $\dot{m}_{\text{solute}}^{-} = \frac{q}{V}m_{\text{solute}}$, meaning that due to the flow with rate q a portion of the solute mass is lost. In the input we have $\dot{m}_{\text{solute}}^{+} = qc_f M + \dot{m}_{\text{fines}}^{-}$, where $qc_f M$ means solute feed and is a control input. The second term $\dot{m}_{\text{fines}}^{-}$ means that the mass which is subtracted from m_{solid} in the dissolution phase is recycled and added to m_{solute}. Altogether fines removal does not alter the mass balance. We now have related the dotted expressions to quantities used in the population balance equation.

Our next step is to relate the internal dynamics of the mass balance to the population balance equation. We start by noting that

$$m_{\text{solid}}(t) = k_v\rho V(t) \int_0^\infty n(L,t)L^3 dL.$$

Therefore, differentiating with respect to time and substituting the integrated right hand side of the population balance equation $\int_0^\infty \{\ldots\} L^3 dL$ in (1) gives

$$
\begin{aligned}
\frac{dm_{\text{solid}}(t)}{dt} &= k_v \rho \int_0^\infty \frac{\partial\,(V(t)n(L,t))}{\partial t} L^3 dL \\
&= -k_v \rho G(c(t)) \int_0^\infty \frac{\partial\,(V(t)n(L,t))}{\partial L} L^3 dL \\
&\quad - k_v \rho q \int_0^\infty \big(1 + h_f(L) + h_p(L)\big) n(L,t) L^3\, dL \\
&= 3k_v V(t) \rho G(c(t)) \int_0^\infty n(L,t) L^2\, dL \\
&\quad - k_v \rho q \int_0^\infty \big(1 + h_f(L) + h_p(L)\big) n(L,t) L^3\, dL,
\end{aligned}
$$

where the third line uses integration by parts and also the fact that breakage conserves mass at all times

$$
\int_0^\infty \big(\mathscr{D}_{\text{break}}^+(L,t) - \mathscr{D}_{\text{break}}^-(L,t)\big) L^3 dL = 0,
$$

so that terms related to breakage cancel. Altogether, Eq. (9) becomes

$$
\begin{aligned}
\frac{d(VcM)}{dt} &+ 3k_v \rho V(t) G(c(t)) \int_0^\infty n(L,t) L^2\, dL \\
&- k_v \rho q \int_0^\infty \big(1 + h_f(L) + h_p(L)\big) n(L,t) L^3\, dL \\
&= -\frac{q}{V} VcM + qc_f M - qk_v \rho \int_0^\infty (1 + h_p(L)) n(L,t) L^3 dL.
\end{aligned}
$$

Canceling the term $qk_v \rho \int_0^\infty (1 + h_p(L)) n(L,t) L^3 dL$ on both sides gives

$$
\frac{d(VcM)}{dt} = qc_f M - qcM - 3V k_v \rho G(c(t)) \int_0^\infty n(L,t) L^2 dL \tag{12}
$$

$$+qk_v\rho \int_0^\infty h_f(L)n(L,t)L^3\,dL,$$

where c_f is the feed concentration.

Finally, using (11) and (6), Eq. (12) is transformed to

$$\frac{dc(t)}{dt} = -\left(\frac{V'(t)}{V(t)} + \frac{q}{V(t)} + \frac{3k_v\rho\mu(t)}{M}\right)c(t) + \frac{qc_f}{V(t)} \qquad (13)$$

$$+3\frac{k_v\rho c_s\mu(t)}{M} + \frac{qk_v\rho}{V(t)M}v(t)$$

where

$$v(t) = \int_0^\infty h_f(L)n(L,t)L^3\,dL.$$

and

$$\mu(t) = \int_0^\infty n(L,t)L^2\,dL.$$

3   Introducing Characteristics

In this part we present a transformation of the model using moments and characteristic curves. Let us recall the complete model described by the population and molar balances. The population balance equation is given by

$$\frac{\partial (V(t)n(L,t))}{\partial t} + G(c(t))\frac{\partial V(t)n(L,t)}{\partial L} = -h(L)V(t)n(L,t) + w(L,t) \quad (14)$$

where

$$w(L,t) = \int_L^\infty a(L')b(L' \to L)\left(V(t)n(L',t)\right)\,dL'$$

is the source term due to breakage and attrition.

$$h(L) = \tilde{h}(L) + a(L) > 0 \qquad (15)$$

where $\tilde{h} = qh_{f+p}/V$ since h_{f+p} depends on the volume V. The boundary value and the initial condition are given respectively by (2) and (3).

Then, we can define

$$w_0 = w(L,0) = \int\limits_{L}^{\infty} a(L')b(L' \to L)V(t)n_0(L')\,dL' \tag{16}$$

Lemma 1 *Suppose the non-negative function $c(t)$ with $c(0) = c_0$ satisfies the mole balance equation (13). Then*

$$c(t) = e^{\int\limits_0^t -\left(\frac{V'(\tau)}{V(\tau)} + \frac{q}{V(\tau)} + \frac{3k_v\rho\mu(\tau)}{M}\right)d\tau}$$
$$\times \left(c_0 + \int\limits_0^t \left[\frac{qc_f}{V(\tau)} + \frac{3k_v\rho c_s\mu(\tau)}{M} + \frac{qk_v\rho v(\tau)}{V(\tau)M}\right]\right.$$
$$\left. \times e^{\int\limits_0^\tau \left(\frac{V'(s)}{V(s)} + \frac{q}{V(s)} + \frac{3k_v\rho\mu(s)}{M}\right)ds}\,d\tau.\right) \tag{17}$$

Proof By solving (13) with respect to $c(t)$ using variation of the constant, we obtain (17).

Next we introduce characteristic curves. For t_0 and L_0 fixed we let ϕ_{t_0,L_0} be the solution of the initial value problem

$$\frac{d\phi(t)}{dt} = G(c(t)), \quad \phi(t_0) = L_0.$$

Since the right hand side does not depend on L, we have explicitly

$$\phi_{t_0,L_0}(t) = L_0 + \int\limits_{t_0}^{t} G(c(\tau))d\tau. \tag{18}$$

We write specifically $z(t) := \phi_{0,0}(t)$. Now we introduce a family of functions $N_{t,L}$ which we use later to define $n(L,t)$ via

$$N_{t_0,L_0}(t) := V(t)n(\phi_{t_0,L_0}(t),t).$$

We let $L = \phi_{t_0,L_0}(t)$, then N_{t_0,L_0} satisfies

$$N'_{t_0,L_0}(t) = \frac{\partial(V(t)n(L,t))}{\partial L}\phi'_{t_0,L_0}(t) + \frac{\partial(V(t)n(L,t))}{\partial t}$$
$$= \frac{\partial(V(t)n(L,t))}{\partial L}G(c(t)) + \frac{\partial(V(t)n(L,t))}{\partial t}.$$

Therefore (14) transforms into

$$N'_{t_0,L_0}(t) = -h(\phi_{t_0,L_0}(t))N_{t_0,L_0}(t) + w(\phi_{t_0,L_0}(t),t), \tag{19}$$

and we consequently use these ODEs to define the functions $N_{t,L}$. Integration of (19) gives

$$N_{t_0,L_0}(t) = \left(N_{t_0,L_0}(t_0) + \int_{t_0}^{t} w(\phi_{t_0,L_0}(\tau),\tau) \exp\left\{ \int_{t_0}^{\tau} h(\phi_{t_0,L_0}(\sigma))d\sigma \right\} d\tau \right)$$

$$\times \exp\left\{ -\int_{t_0}^{t} h(\phi_{t_0,L_0}(\tau))d\tau \right\}. \tag{20}$$

We can exploit this for two possible situations, where $N_{t_0,L_0}(t_0)$ can be given an appropriate value.

Before putting this to work, we will need two auxiliary functions τ and ξ, which are easily defined using the characteristics. First we define $\tau = \tau(t, L)$ implicitly by

$$\phi_{\tau,0}(t) = L, \text{ or equivalently } \phi_{t,L}(\tau) = 0, \tag{21}$$

or again,

$$\int_{\tau(t,L)}^{t} G(c(t))\, d\sigma = L. \tag{22}$$

Then we define $\xi = \xi(t, L) = \phi_{t,L}(0)$, which gives $\xi = L + \int_{t}^{0} G(c(t))\, d\tau$. Using $N_{t,L}$, respectively (20), we can now define

$$V(t_0)n(L_0,t_0) \tag{23}$$

$$= \begin{cases} \left(V(t_0)\frac{B(c(t))}{G(c(t))} + \int_{\tau_0}^{t_0} w(\phi_{\tau_0,0}(s),s) \exp\left\{ \int_{\tau_0}^{s} h(\phi_{\tau_0,0}(\sigma))d\sigma \right\} ds \right) \\ \qquad \times \exp\left(-\int_{\tau_0}^{t_0} h(\phi_{\tau_0,0}(s))\, ds \right), \quad \text{if } L_0 < z(t_0) \\[2em] \left(V(t_0)n_0(\phi_{t_0,L_0}(0)) + \int_{0}^{t_0} w(\phi_{t_0,L_0}(s),s) \exp\left\{ \int_{0}^{s} h(\phi_{t_0,L_0}(\sigma))d\sigma \right\} ds \right) \\ \qquad \times \exp\left(-\int_{0}^{t_0} h(\phi_{t_0,L_0}(s))\, ds \right), \quad \text{if } L_0 \geq z(t_0) \end{cases}$$

where $\tau_0 = \tau(t_0, L_0)$ and $t_0 \in [0, t_f]$. The formula is justified as follows. Let t_0, L_0 be such that $L_0 < z(t_0) = \phi_{0,0}(t_0)$. This is the case where $\tau_0 = \tau(t_0, L_0) > 0$. Here we consider Eq. (19) for $N_{\tau_0,0}$ with initial value

$$N_{\tau_0,0}(\tau_0) = V(\tau_0)n(\phi_{\tau_0,0}(\tau_0), \tau_0) = V(\tau_0)n(0, \tau_0)$$
$$= V(\tau_0)B(c(\tau_0))/G(c(\tau_0)).$$

This uses the fact that $\phi_{\tau_0,0}(\tau_0) = 0$ according to the definition of $\phi_{\tau,0}$. Integration gives the upper branch of (23).

Next consider t_0, L_0 such that $L_0 \geq z(t_0)$. Then $\tau_0 < 0$, so that we do not want to use it as initial value. We therefore apply (19), (20) to N_{t_0,L_0}, now with initial time 0. Then we get

$$N_{t_0,L_0}(t_0) = \left(N_{t_0,L_0}(0) + \int_0^{t_0} w(\phi_{t_0,L_0}(s), s) \exp\left\{ \int_0^{s} h(\phi_{t_0,L_0}(\sigma))d\sigma \right\} ds \right)$$

$$\times \exp\left(-\int_0^{t_0} h(\phi_{t_0,L_0}(s)) ds \right). \tag{24}$$

Here $N_{t_0,L_0}(0) = V(0)n(\phi_{t_0,L_0}(0), 0) = V(0)n_0(\phi_{t_0,L_0}(0))$, so we get the lower branch of (23) all right. This justifies the formula and completes the definition of the characteristic curves and the representation of the $V(t)n(L, t)$ via the characteristics.

4 Apriori Estimates

In this section we discuss an hypothesis on the control input q, which lead to apriori estimates, under which later on global existence of a solution will be shown. This conditions is motivated by the physics of the process and lead to bounds on mass and volume. Note that we do not get prior estimates on surface, length and number of solids, which as we shall see presents a difficulty when proving global existence of solutions. In this section we focus on the total volume and the feed rate. The total mass of slurry is given by (9), then $m(0) = m_{solvent}(0) + m_{solute}(0) + m_{solid}(0)$. The total volume of slurry V is

$$V(t) = m_{solvent}(t) + m_{solute}(t) + m_{solid}(t) \tag{25}$$

$$= \frac{m_{solvent}(t)}{\rho_{H_2O}} + \frac{m_{solute}(t)}{\rho} + \frac{m_{solid}(t)}{\rho}.$$

where ρ_{H_2O} is the density of the solvent and $\rho = \rho_{solute} = \rho_{solid}$. Now suppose we are allowed a maximum volume V_{max} of slurry in the crystallizer. We then have to steer the process such that $V(t) \leq V_{max}$ at all times t. Naturally, this can only be arranged by a suitable control of the feed rate. Let us recall that the crystallization is not solvated (means $\frac{dm_{solute}}{dt} = -\frac{dm_{solid}}{dt}$). Then, the control of the feed rate is linked to m_{solvent} by the relation

$$\frac{dm_{\text{solvent}}}{dt} = q(t)\rho_{H_2O}. \tag{26}$$

Lemma 2 *Suppose the feed rate q satisfies the constraint*

$$(H_1) \int_0^t q(\tau)d\tau \leq \left[\frac{V_{max} - V(0)}{\rho_{H_2O}\left(\rho_{H_2O}^{-1} + 2\rho^{-1}\right)} \right] \tag{27}$$

at all times $t \geq 0$. Then the total volume of slurry $V(t)$ satisfies $V(t) \leq V_{\max}$.

Proof From (9) we obviously have

$$m'(t) = m'_{\text{solvent}}(t) + m'_{\text{solute}}(t) + m'_{\text{solid}}(t). \tag{28}$$

Using the fact that the crystallization is not solvated and (26) in (28), we obtain

$$m'(t) = q(t)\rho_{H_2O} \tag{29}$$

which on integration gives

$$\int_0^t q(\tau)d\tau = \frac{m(t) - m(0)}{\rho_{H_2O}}$$

for every $t > 0$. Hence, using the hypothesis (H_1) and (29), we obtain

$$m(t) \leq \frac{V_{max}}{\rho_{H_2O}^{-1} + 2\rho^{-1}}. \tag{30}$$

Using (9) and (25), this implies $V(t) \leq V_{\max}$.

In a practical process, q could be steered by feedback to avoid overflow of the crystallizer.

As we have seen, bounding the feed rate via (H_1) gives a bound on the total volume of slurry, and also on the total mass, namely

$$m(t) \leq m_{\max} = \frac{V_{max}}{\rho_{H_2O}^{-1} + 2\rho^{-1}}. \tag{31}$$

5   Existence and Uniqueness of Solutions

In this section we will assemble the results from the previous sections and prove global existence and uniqueness. We use the method of characteristics with a fixed-point argument for an operator $\mathscr{D}$ which we now define.

5.1   Setting up the Operator

For the function w we introduce the Banach space

$$\mathbf{E} = C_\infty(\mathbb{R}^+ \times [0, t_f]) \cap \mathscr{L}_u^1(\mathbb{R}^+ \times [0, t_f], h(L)dL),$$

where we set $h(L) = \max\{1, L^3\}$ and introduce the norm on $\mathbf{E}$ as

$$\|w\| = \|w\|_\infty + \sup_{0 \le t \le t_f} \int_0^\infty |w(L, t)| h(L) dL.$$

Let us introduce the moments of the CSD as

$$\mu_i(t) = \int_0^\infty V(t) n(L, t) L^i \, dL, \, i = 0, 1, \ldots, \tag{32}$$

This initial data are then

$$\mu_{i,0} = \int_0^\infty V(t) n_0(L) L^i \, dL, \, i = 0, 1, \ldots. \tag{33}$$

We consider $\mathbf{x} = (c, v, \mu_2, \mu_1, \mu_0, w)$ an element of the space

$$\mathbf{F} = C[0, t_f]^5 \times \mathbf{E},$$

where each copy of $C[0, t_f]$ is equipped with the supremum norm, so that the norm on $\mathbf{F}$ is

$$|\mathbf{x}| = \|c\|_\infty + \|v\|_\infty + \|\mu_2\|_\infty + \|\mu_1\|_\infty + \|\mu_0\|_\infty + \|w\|.$$

We proceed to define the action of the operator $\mathscr{D}$ on element $\mathbf{x} \in \mathbf{F}$, writing

$$\mathscr{D}\mathbf{x} = \tilde{\mathbf{x}} = (\tilde{c}, \tilde{v}, \tilde{\mu}_2, \tilde{\mu}_1, \tilde{\mu}_0, \tilde{w}).$$

The initial values at $t = 0$ are

$$\mathbf{x}_0 = (c_0, v_0, \mu_{2,0}, \mu_{1,0}, \mu_{0,0}, w_0),$$

In order to define the elements $\widetilde{w}$ and $\widetilde{\mu}_2$, we first have to introduce the characteristics $\phi_{t,L}$ via formula (18). Then we define the functions $N_{t,L}$ via formula (20). We are now ready to define $\widetilde{\mu}_2$ as

$$\widetilde{\mu}_2(t)$$

$$= \rho k_v \int_0^\infty V(t) n(L, t) L^2 dL$$

$$= \rho k_v \int_0^{z(t)} \left(\frac{V(\tau) B(c(\tau))}{G(c(\tau))} + \int_\tau^t w(\phi_{\tau,0}(s), s) \exp\left\{ \int_\tau^s a(\phi_{\tau,0}(\sigma)) d\sigma \right\} ds \right)$$

$$\times \exp\left(-\int_\tau^t a(\phi_{\tau,0}(s)) \, ds \right) L^2 dL \qquad (34)$$

$$+ \rho k_v \int_{z(t)}^\infty \left(V(0) n_0(\phi_{t,L}(0)) + \int_0^t w(\phi_{t,L}(s), s) \exp\left\{ \int_0^s a(\phi_{t,L}(\sigma)) d\sigma \right\} ds \right)$$

$$\times \exp\left(-\int_0^t a(\phi_{t,L}(s)) \, ds \right) L^2 dL.$$

In the first integral $\int_0^{z(t)}$ we use the change of variables $L \to \tau = \tau(t, L)$. Then

$$[0, z(t)] \ni L \mapsto \tau(t, L) \in [0, t], \qquad dL = G(c(t)) d\tau.$$

In the second integral $\int_{z(t)}^\infty$ we use the change of variables $L \to \xi(t, L) := \phi_{t,L}(0)$. Then

$$[z(t), \infty) \ni L \mapsto \xi \in [0, \infty), \qquad dL = d\xi.$$

The inverse relation is

$$L = \xi + \int_0^t G(c(\sigma)) d\sigma = \xi + z(t).$$

From (34) we obtain

$$
\tilde{\mu}_2(t)
$$

$$
= \rho k_v \int_0^t \left(V(\tau)B(c(\tau)) + \int_\tau^t w(\phi_{\tau,0}(s), s) \exp\left\{ \int_\tau^s a(\phi_{\tau,0}(\sigma))d\sigma \right\} ds \right)
$$

$$
\times \exp\left(-\int_\tau^t a(\phi_{\tau,0}(s))\, ds \right) L(\tau)^2 \, d\tau \tag{35}
$$

$$
+ \rho k_v \int_0^\infty \left(V(0)n_0(\xi) + \int_0^t w(\phi_{0,\xi}(s), s) \exp\left\{ \int_0^s a(\phi_{0,\xi}(\sigma))d\sigma \right\} ds \right)
$$

$$
\times \exp\left(-\int_0^t a(\phi_{0,\xi}(s))\, ds \right) L(\xi)^2 d\xi,
$$

where $L(\tau) = \int_\tau^t G(c(\sigma))\, d\sigma$ and $L(\xi) = \xi + \int_0^t G(c(\sigma))\, d\sigma$, and where we use $\phi_{t,L}(s) = \phi_{0,\xi}(s)$ in the second integral. For fixed t the functions $L \mapsto \tau(t, L)$ and $\tau \mapsto L(\tau)$ are inverses of each other, and similarly, $L \leftrightarrow \xi$ is one-to-one via the formula $L = \xi - \int_0^t G(c(\sigma))\, d\sigma$. A Similar formula is obtained for $\tilde{v}$:

$$
\tilde{v}(t)
$$

$$
= \rho k_v \int_0^t \left(V(\tau)B(c(\tau)) + \int_\tau^t w(\phi_{\tau,0}(s), s) \exp\left\{ \int_\tau^s a(\phi_{\tau,0}(\sigma))d\sigma \right\} ds \right)
$$

$$
\times \exp\left(-\int_\tau^t a(\phi_{\tau,0}(s))\, ds \right) h_f(\tau)L(\tau)^3 \, d\tau
$$

$$
+ \rho k_v \int_0^\infty \left(V(0)n_0(\xi) + \int_0^t w(\phi_{0,\xi}(s), s) \exp\left\{ \int_0^s a(\phi_{0,\xi}(\sigma))d\sigma \right\} ds \right)
$$

$$
\times \exp\left(-\int_0^t a(\phi_{0,\xi}(s))\, ds \right) h_f(\xi)L(\xi)^3 d\xi,
$$

Similarly, continuing to define the operator $\mathscr{D}$, we define the moments $\tilde{\mu}_i$, $i = 0, 1$ within $\tilde{\mathbf{x}}$ via $\tilde{\mu}_i(t) = \int_0^\infty V(t)n(L, t)L^i \, dL$, where we express the right hand side via characteristics and the elements of $\mathbf{x}$ in much the same way as done for $\tilde{\mu}_2$.

We also need to get back to the function $\widetilde{w}(L, t)$. We introduce

$$\beta(L, L') = \begin{cases} a(L)b(L \to L'), & \text{if } L \geq L' \\ 0, & \text{else} \end{cases}$$

then we can write

$$\widetilde{w}(L', t) = \int_{L'}^{\infty} a(L)b(L \to L')V(t)n(L, t)dL = \int_{0}^{\infty} \beta(L, L')V(t)n(L, t)dL,$$

which is essentially like the moment integral (35), the function $L \mapsto L^i$ being replaced by $\beta(L, L')$. Applying the same technique as in the case of (35), we obtain

$$\widetilde{w}(L', t)$$

$$= \int_{0}^{t} \left(V(\tau)B(c(\tau)) + \int_{\tau}^{t} w(\phi_{\tau,0}(s), s) \exp\left\{ \int_{\tau}^{s} a(\phi_{\tau,0}(\sigma))d\sigma \right\} ds \right)$$

$$\times \exp\left(-\int_{\tau}^{t} a(\phi_{\tau,0}(s))\, ds \right) \beta(L(\tau), L')\, d\tau$$

$$+ \int_{0}^{\infty} \left(V(0)n_0(\xi) + \int_{0}^{t} w(\phi_{t,L}(s), s) \exp\left\{ \int_{0}^{s} a(\phi_{t,L}(\sigma))d\sigma \right\} ds \right)$$

$$\times \exp\left(-\int_{0}^{t} a(\phi_{t,L}(s))\, ds \right) \beta(L(\xi), L')d\xi,$$

which expresses $\widetilde{w}$ in terms of w and the characteristics, hence by elements of $\mathbf{x}$. Here we use $w \in \mathbf{E}$, and it is routine to check that $\widetilde{w} \in \mathbf{E}$, so that $\widetilde{\mathbf{x}} \in \mathbf{F}$. This completes the definition of $\mathcal{Q}$.

We will also need the following hypotheses on the breakage terms a and b:

$$(H_2) \qquad \|a\|_\infty := \max_{0 \leq L < \infty} a(L) < +\infty,$$

$$\|a\|_L = \sup_{0 \leq L < L'} \left| \frac{a(L) - a(L')}{L - L'} \right| < \infty,$$

$$(H_3) \qquad \|b\|_\infty := \max_{0 \le L \le L'} b(L' \to L) < +\infty,$$

$$\|b\|_L := \sup_{L \ge 0} \sup_{L \le L' < L''} \left| \frac{b(L \to L') - b(L \to L'')}{L' - L''} \right| < \infty.$$

5.2 Setting up the Space

We now have to define a closed subset **X** of **F** on which the operator $\mathscr{Q}$ acts as a contraction with respect to the metric induced by the norm of **F**. In other words, we need to assure $\mathscr{Q}(\mathbf{X}) \subset \mathbf{X}$ and

$$\left| \mathscr{Q}\mathbf{x}^{(1)} - \mathscr{Q}\mathbf{x}^{(2)} \right| \le \gamma \left| \mathbf{x}^{(1)} - \mathbf{x}^{(2)} \right|$$

for a constant $0 < \gamma < 1$ and all $\mathbf{x}^{(1)}, \mathbf{x}^{(2)} \in \mathbf{X}$. This will be achieved by choosing t_f *small enough*, as usual, but in order to prove global existence, we will have to make a careful quantification of γ in terms of t_f and the initial values $\mathbf{x}_0$.

We distinguish between globally bounded states and those for which no prior bound can be put forward. Due to our hypothesis (H_1), c is globally bounded thanks to the formulas (31). We write $\mathbf{y} = c$ for the globally bounded states. On the other hand, it is not clear whether the moments v, μ_2, μ_1, μ_0 nor w are globally bounded, and whether such a bound can be obtained from the physical constraints. This is due to particle breakage, which may lead to an exceedingly large number of small particles, or *fines*. We write $\mathbf{z} = (\mu_0, \mu_1, \mu_2, v)$ and $w \in \mathbf{E}$ for the unbounded state. Altogether $\mathbf{x} = (\mathbf{y}, \mathbf{z}, w) \in \mathbf{F}$.

The axiom (H_1) on the controls q will then assure that on every interval of existence of a solution $\mathbf{x}(t)$, the bounded states $\mathbf{y}(t)$ satisfy the same global bound. We split the initial conditions into $\mathbf{x}_0 = (\mathbf{y}_0, \mathbf{z}_0, w_0)$, where $\mathbf{y}_0$ is the bounded part, $\mathbf{z}_0, w_0$ that part which does not have a prior bound. The state $\mathbf{x}(t)$ is split accordingly as $\mathbf{x}(t) = (\mathbf{y}(t), \mathbf{z}(t), w(t))$. Now define the moments of $w(L, t)$ as $\mu_{w,i}(t) = \int_0^\infty w(L,t) L^i \, dL$ and fix $K > |\mathbf{z}_0| + |\mu_{w,0}(0)| + |\mu_{w,1}(0)| + |\mu_{w,2}(0)| + |\mu_{w,3}(0)|$. Then we define the space **X** as

$$\mathbf{X} = \left\{ \mathbf{x} \in \mathbf{F} : \mathbf{x}(0) = \mathbf{x}_0, \mathbf{x} = (\mathbf{y}, \mathbf{z}, w), |\mathbf{z}(t)| + \sum_{i=0}^3 |\mu_{w,i}(t)| \le K \text{ for all } t \in [0, t_f] \right\}.$$

Lemma 3 *There exists a constant $s > 0$ depending only on the global volume bound $V_{\max}$ such that for every $\mathbf{y}_0$, every $\mathbf{z}_0, w_0$, and every $K > |\mathbf{z}_0| + \sum_{i=0}^3 |\mu_{w,i}(0)|$ the following is true: Suppose $t_f > 0$ satisfies*

$$0 < t_f < \left(K - |\mathbf{z}_0| - \sum_{i=0}^3 |\mu_{w,i}(0)| \right) / s. \qquad (36)$$

Then $\mathcal{Q}(\mathbf{X}) \subset \mathbf{X}$.

Proof Estimating the states over an interval $[0, t_f]$ leads to estimates of the form

$$\widetilde{\mathbf{z}}(t) = \mathbf{z}_0 + \int_0^t f(\mathbf{y}(\tau), \mathbf{z}(\tau), w(\tau)) \, d\tau \le |\mathbf{z}_0| + (s + K) t_f, \text{ for all } 0 \le t \le t_f$$

where a global bound s for $\mathbf{y}(t)$ is used on $[0, t_f]$, while the bounds $|\mathbf{z}(t)| \le K$, $|\mu_{w,i}(t)| \le K$ are used for the states $\mathbf{z}$, w. Then in order to assure $|\widetilde{\mathbf{z}}(t)| \le K$, we have but to force the condition

$$|\mathbf{z}_0| + s(1 + K) t_f \le K,$$

which gives $t_f \le \frac{K - |\mathbf{z}_0|}{s(1+K)} \le \frac{K - |\mathbf{z}_0|}{s}$. A similar argument applies to the moments $\mu_{w,i}$ of w, and combining the two gives (36).

5.3 Proving $\gamma \in (0, 1)$

For the globally bounded states gathered in $\mathbf{y}$ we obtain an estimate of the form

$$\|\widetilde{\mathbf{y}}^{(1)} - \widetilde{\mathbf{y}}^{(2)}\| \le s_1 (1 + |\mathbf{z}_0| + K) t_f \left| \mathbf{x}^{(1)} - \mathbf{x}^{(2)} \right|$$

for a global constant s_1 depending on $V_{\max}$. For the state $\mathbf{z} = (\mu_0, \mu_1, \mu_2, v)$ on the other hand we obtain an estimation of the form

$$\|\widetilde{\mathbf{z}}^{(1)} - \widetilde{\mathbf{z}}^{(2)}\| \le s_2 \left(1 + |\mathbf{z}_0| + \sum_{i=0}^{3} |\mu_{w,i}(0)| + K \right) t_f \left| \mathbf{x}^{(1)} - \mathbf{x}^{(2)} \right|,$$

where s_2 is another global constant. The Lipschitz constant now depends on the initial condition $\mathbf{z}_0$, $\mu_{w,i}(0)$, for which no global bound is available. A similar estimate

$$\|\widetilde{w}^{(1)} - \widetilde{w}^{(2)}\| \le s_3 \left(1 + \|\mathbf{z}_0\| + \sum_{i=0}^{3} |\mu_{w,i}(0)| + K \right) t_f \left| \mathbf{x}^{(1)} - \mathbf{x}^{(2)} \right|$$

is obtained for the terms involving w. Altogether, we have

Lemma 4 *There exists a constant s depending only on $V_{\max}$ such that for every initial condition $\mathbf{y}_0$, every initial $\mathbf{z}_0$, w_0, and every K with $\|\mathbf{z}_0\| + \sum_{i=0}^{3} |\mu_{w,i}(0)| < K$, the operator $\mathcal{Q}$ satisfies the following Lipschitz estimate on $\mathbf{X}$:*

$$\left| \mathcal{Q}\mathbf{x}^{(1)} - \mathcal{Q}\mathbf{x}^{(2)} \right| \leq s \left(1 + |\mathbf{z}_0| + \sum_{i=0}^{3} |\mu_{w,i}(0)| + K \right) t_f \left| \mathbf{x}^{(1)} - \mathbf{x}^{(2)} \right|. \quad (37)$$

Condition (37) in tandem with (36) allows us now to apply the Banach contraction theorem to the operator $\mathcal{Q}$. Using this, we can prove the following

Theorem 5 *Suppose hypothesis* (H_1) *assuring global bounds on masses and volume are satisfied, and that the initial condition of the crystallizer respects the global volume bound* $V_{\max}$. *Assume further that the breakage kernel satisfies the axioms* $(H_2) - (H_3)$. *Then the crystallizer system (14), (13) has a unique global solution on* $[0, \infty)$.

Proof (1) Suppose we choose $K > |\mathbf{z}_0| + \sum_{i=0}^{3} |\mu_{w,i}(0)|$, then

$$t_f = \min \left\{ \frac{K - |\mathbf{z}_0| - \sum_{i=0}^{3} |\mu_{w,i}(0)|}{s + K}, \frac{1}{2s(1 + |\mathbf{z}_0| + \sum_{i=0}^{3} |\mu_{w,i}(0)| + K)} \right\} \quad (38)$$

assures that $\mathcal{Q}$ is a self-map and a contraction on $\mathbf{X}$, so that a unique solution with the initial condition $\mathbf{x}_0$ exists on $[0, t_f]$.

(2) We will now have to iterate the process, and for that we shall have to change our notation. We write $\mathbf{w}(t) = |\mathbf{z}(t)| + \sum_{i=0}^{3} |\mu_{w,i}(t)|$. We start the method at $t_0 = 0$. Putting $\mathbf{w}_0 = \mathbf{w}(0)$, we chose $K = \mathbf{w}_0 + 1 > \mathbf{w}(0)$ in part (1) above, so that condition (38) re-written at $t_f = t_1$ becomes

$$t_1 = \min \left\{ \frac{1}{s + \mathbf{w}_0 + 1}, \frac{1}{4s(1 + \mathbf{w}_0)} \right\}.$$

This gives a unique solution on $[0, t_f] = [t_0, t_1]$.

(3) Since by construction of $\mathbf{X}$ on $[t_0, t_1] = [0, t_f]$ we have $\mathbf{w}(t) \leq K$ for every $t \in [t_0, t_1]$, we get $\mathbf{w}_1 \leq K = \mathbf{w}_0 + 1$. In addition, the initial condition respect $m_{\text{solvent}}(0) + m_{\text{solute}}(0) + m_{\text{solid}}(0) < m_{\max}$ and therefore the global bound related to $V_{\max}$, so that we will be able to continue to use the same global constant s in the next iteration. We now use $\mathbf{x}(t_1)$ as the new initial condition at the new initial time t_1 and repeat the same argument to the right of t_1. That requires choosing a new constant $K > |\mathbf{z}(t_1)| + \sum_{i=0}^{3} |\mu_{w,i}(t_1)| = \mathbf{w}(t_1) = \mathbf{w}_1$. We choose again $K = \mathbf{w}_1 + 1$. Then the final time corresponding to t_f, which is now called $t_2 - t_1$, has to satisfy (38), which reads

$$t_2 - t_1 = \min \left\{ \frac{1}{s + \mathbf{w}_1 + 1}, \frac{1}{2s(1 + \mathbf{w}_1 + K)} \right\}$$

$$\geq \min \left\{ \frac{1}{s + \mathbf{w}_0 + 2}, \frac{1}{4s(2 + \mathbf{w}_0)} \right\}.$$

By recursion we find that

$$t_n - t_{n-1} \geq \min \left\{ \frac{1}{s + \mathbf{w}_0 + n}, \frac{1}{4s(n + \mathbf{w}_0)} \right\},$$

so that for some constant s' depending only on $V_{\max}$ and $\mathbf{w}_0 = |\mathbf{z}_0| + \sum_{i=0}^{3} |\mu_{w,i}(0)|$,

$$t_N = \sum_{n=1}^{N} t_n - t_{n-1} \geq s' \sum_{n=1}^{N} \frac{1}{n} \to \infty \; (N \to \infty).$$

Since the solution can be continues from 0 to any t_N, this proves global existence and uniqueness.

References

1. Mersmann, A.: Crystallization Technology Handbook. Marcel Dekker (2001)
2. Jones, A.G.: Crystallization Process Systems. Butterworth-Heinemann (2002)
3. Mullin, J.W., Nyvlt, J.: Programmed cooling of batch crystallizers. Chem. Eng. Sci. **26**, 369–377 (1971)
4. Randolph, A.D., Larson, M.A.: Theory of Particulate Processes, 2nd edn. Academic Press, San Diego (1988)
5. Pathath, P.K., Kienle, A.: A Numerical Bifurcation Analysis of Nonlinear Oscillations in Crystallization Process, Chemical Engineering Science (2002)
6. Tavare, N.N.: Industrial Cystallization: Process Simulation, Analysis and Design. New York and London (1995)
7. Gerstlauer, A., Gahn, C., Zhou, H., Rauls, M., Schreiber, M.: Application of population balances in the chemical industry-current status and future needs. Chem. Eng. Sc. **61**, 205–217 (2006)
8. Griffin, D.W., Mellichamp, D.A., Doherty, M.F.: Reducing the mean size of API crystals by continuous manufacturing with product classification and recycle. Chem. Eng. Sci. **65**, 5770–5780 (2010)
9. Walker, C., Simonett, G.: On the solvability of a mathematical model for prion proliferation. J. Math. Anal. Appl. **324**, 580–603 (2006)
10. Calsina, A., Farkas, J.Z.: Steady states in a structured epidemic model with Wentzell boundary condition. J. Evol. Eqn. **12**, 495–512 (2012)
11. Cushing, J.M.: A competition model for size-structured species. SIAM J. Appl. Math. **49**, 838–858 (1989)
12. Gurtin, M.E., Maccamy, R.C.: Non-linear age-dependent population dynamics. Arch. Ration. Mech. Anal. **54**, 281–300 (1974)
13. Calsina, A., Saldaña, J.: A model of physiologically structured population dynamics with a nonlinear individual growth rate. J. Math. Biol. **33**, 335–364 (1995)
14. Smith, H.L.: Existence and uniqueness of global solutions for a size-structured model of an insect population with variable instal duration. Rocky Mt. J. Math. **24**, 311–334 (1994)
15. Vollmer, U., Raisch, J.: H_∞-control of a continuous crystallizer. Control Eng. Pract. **9**, 837–845 (2001)

Wave Velocity Estimation in Heterogeneous Media

Sharefa Asiri and Taous-Meriem Laleg-Kirati

Abstract In this paper, modulating functions-based method is proposed for estimating space-time dependent unknown velocity in the wave equation. The proposed method simplifies the identification problem into a system of linear algebraic equations. Numerical simulations on noise-free and noisy cases are provided in order to show the effectiveness of the proposed method.

1 Introduction

Inverse velocity problem for the wave equation is crucial in many applications such as medical imaging and oil exploration [2, 11]. This inverse problem consists in estimating the velocity from available measurements which are practically limited. In heterogeneous media, estimating the velocity helps in characterizing the media and detecting its discontinuities. Different methods have been proposed to solve this problem such as optimization-based techniques [3, 4] or observer-based approaches [1, 6, 10]. However, these methods are heavy computationally and not very efficient when the data is corrupted with noise.

In this work, we propose a new technique based on modulating functions. Modulating functions-based method has been introduced in the early fifties [13, 14] and has been used in parameters identification for ordinary differential equations. In 1966, Perdreauville and Goodson [9] extended the method to identify parameters for partial differential equations using distributed measurements for all space and for all the time. However, it is not practically possible to use a large number of sensors.

S. Asiri (✉) · T.-M. Laleg-Kirati
Computer, Electrical and Mathematical Sciences and Engineering Division,
King Abdullah University of Science and Technology (KAUST),
Thuwal, Kingdom of Saudi Arabia
e-mail: sharefa.asiri@kaust.edu.sa

T.-M. Laleg-Kirati
e-mail: taousmeriem.laleg@kaust.edu.sa

© Springer International Publishing Switzerland 2016
G.A. Anastassiou and O. Duman (eds.), *Intelligent Mathematics II:*
Applied Mathematics and Approximation Theory, Advances in Intelligent Systems
and Computing 441, DOI 10.1007/978-3-319-30322-2_21

The objective of this paper is to extend the method to estimate the wave velocity in heterogeneous media using a small number of measurement points. We will investigate the properties of this method when solving the inverse problem.

The major advantages of this approach are significant. First, it does not require solving the direct problem by which the computation is simplified. Moreover, it converts the identification problem into a system of linear algebraic equations. In addition, it does not require knowing the initial values which are usually unknown in real applications. Also, it is robust against noise.

The paper is organized as follows: Sect. 2 presents the problem. In Sect. 3, modulating functions are defined then the method is applied to one-dimensional wave equation to estimate the velocity. Section 4 provides numerical simulations in order to show the effectiveness of the proposed method. A discussion and concluding remarks are presented in Sects. 5 and 6; respectively.

2 Problem Statement

Consider the one-dimensional wave equation in the domain $\Omega := (0, L) \times (0, T]$:

$$\begin{cases} u_{tt}(x, t) - c(x, t)u_{xx}(x, t) = f(x, t), \\ u(0, t) = g_1(t), \ u(L, t) = g_2(t), & t \in (0, T) \\ u(x, 0) = r_1(x), \ u_t(x, 0) = r_2(x), & x \in (0, L). \end{cases} \quad (1)$$

where x is the space coordinate, t is the time coordinate, L is the final point in the spatial interval, and T is the final time. $g_1(t)$, $g_2(t)$ and $r_1(x)$, $r_2(x)$ are the boundary conditions and the initial conditions; respectively. The source function is denoted by $f(x, t)$. $c(x, t)$ is the square of the velocity of wave propagation at point x and time t, and it is assumed to be positive and bounded. All the functions, including the velocity, are assumed to be sufficiently regular. The notations u_a and u_{aa} refer to the first and second derivatives of u with respect to a, respectively.

The objective of this paper is to estimate the velocity using the following measurements: $u(x^*, t)$ and $u_{xx}(x^*, t)$; where x^* refers to specific points in the spatial interval $[0, L]$ such that $x^* = \{x_1^*, x_2^*, \ldots, x_n^*\}$. Moreover, the initial conditions and the boundary conditions are not necessary to be known.

3 Velocity Estimation

In this section, first we recall the definition of modulating functions. Then, modulating functions-based method is applied on (1) to estimate $c(x, t)$.

Definition 1 [5] A function $\phi(t)$ is called a modulating function of order N in the interval $[0, T]$ if it satisfies:

$$\begin{cases} \phi(t) \in C^N([0, T]) \\ \phi_m^{(p)}(0) = \phi_m^{(p)}(T) = 0, \quad \forall p = 1, 2, \ldots, N-1, \end{cases} \qquad (2)$$

where p is the derivative order.

The wave velocity estimation in heterogeneous media using the proposed method is illustrated in the following proposition:

Proposition 2 *Let $c(x^*, t) = \sum_{i=1}^{I} \gamma_i \xi_i(t)$ be a basis expansion of the unknown coefficient $c(x^*, t)$ in (1) where x^* is a specific point in the spatial interval; $\xi_i(t)$ and γ_i are basis functions and basis coefficients, respectively. Let $\{\phi_m(t)\}_{m=1}^{m=M}$ be a class of modulating functions with $M \geq I$. Then, the unknown coefficients γ_i can be estimated by solving the system:*

$$\mathscr{A}\Gamma = K, \qquad (3)$$

where the components of the $M \times I$ matrix $\mathscr{A}$ have the form:

$$a_{mi} = \int_0^T \phi_m(t)\xi_i(t)u_{xx}(x^*, t)\, dt, \qquad (4)$$

the components of the vector K, which has length M, are:

$$k_m = \int_0^T \phi_m''(t)u(x^*, t)\, dt - \int_0^T \phi_m(t)f(x^*, t)\, dt, \qquad (5)$$

and Γ is the vector of the unknowns γ_i.

Proof First, at fixed spatial point x^*, we multiply Eq. (1) by $\phi_m(t)$ and integrate over the time interval, so we obtain:

$$\int_0^T \phi_m(t)c(x^*, t)u_{xx}(x^*, t)\, dt = \int_0^T \phi_m(t)u_{tt}(x^*, t)\, dt - \int_0^T \phi_m(t)f(x^*, t)\, dt. \qquad (6)$$

Then by integrating the first term in the right-hand side by parts twice, one can obtain:

$$\int_0^T \phi_m(t)c(x^*, t)u_{xx}(x^*, t)\, dt = \int_0^T \phi_m''(t)u(x^*, t)\, dt - \phi_m(t)f(x^*, t)\, dt; \qquad (7)$$

where the initial conditions are eliminated thanks to the properties of the modulating functions. Finally, by writing $c(x^*, t)$ in its basis expansion, system (3) is obtained with components as in (4) and (5).

Remark 3 Note that matrix $\mathscr{A}$ is full-column rank; hence, system (3) has a unique solution.

Remark 4 Proposition 2 can be applied directly to estimate time-dependent velocity $c(t)$ or space-time-dependent velocity $c(x, t)$. For constant velocity case, the method becomes simpler and more accurate as the basis expansion is not required. In this case, the estimated velocity is:

$$\hat{c} = \frac{\mathscr{A}^t K}{\mathscr{A}^t \mathscr{A}}. \tag{8}$$

where the components of the vectors $\mathscr{A}$ and K have the form

$$a_m = \int_0^T \phi_m(t) u_{xx}(x^*, t) \, dt, \tag{9}$$

and

$$k_m = \int_0^T \phi_m''(t) u(x^*, t) \, dt - \int_0^T \phi_m(t) f(x^*, t) \, dt; \tag{10}$$

respectively.

4 Numerical Simulations

The system has been discretized using finite difference scheme. A set of synthetic data has been generated using the following parameters: $L = 4$, $T = 1$, $N_x = N_t = 801$, and a source $f(x, t) = \sin(x)t^2$; N_x and N_t refer to the number of grid points in space and time; respectively. For the velocity, three cases have been studied: $c = 0.5$,

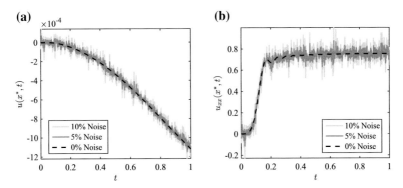

Fig. 1 *Black dashed lines* represent the exact measurements while the *gray solid lines* represent the noisy measurements with 5 and 10 % of noise. The sub-figures **a** is for the displacement $u(x, t)$ and **b** for $u_{xx}(x, t)$; both at fixed point x^*

$c(t) = t^2$, and $c(x, t) = (xt)^2$. The method has been implemented in Matlab and applied in: noise-free and noisy data cases. In the noise-corrupted case, 1, 3, 5 and 10 % white Gaussian random noises with zero means have been added to the data (see Fig. 1).

For the modulating functions, one can note that there are many functions which satisfy (2), see e.g. [7, 8, 12, 15]. In this paper, we propose to use the following polynomial modulating functions [5]:

$$\phi_m(t) = (T - t)^{5+m} t^{5+M+1-m}, \tag{11}$$

where $m = 1, 2, \ldots, M$ with $M = 4, 14, 15$ for the first, second, and third case of the velocity; respectively.

The exact constant velocity c and the estimated one $\hat{c}$ versus different noise levels are shown in Fig. 2.

Fig. 2 Constant case: the exact speed $c = 0.5$ is given by the *horizontal red line*, the bars represent the estimated velocity for different noise levels: 0, 1, 3, 5 and 10 %

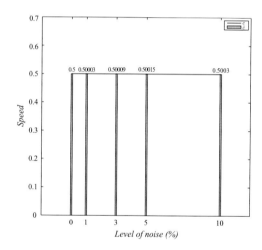

Fig. 3 Time-dependent case: the exact velocity $c(t)$ (*dashed blue*) and the estimated one $\hat{c}(t)$ (*solid red*) using modulating function-based method, in noise-free case

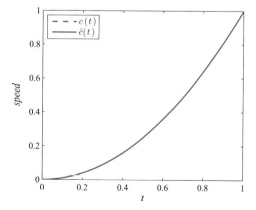

Figures 3 and 4 show the estimated time-dependent velocity $\hat{c}(t)$ in noise-free and noise-corrupted cases; respectively. The corresponding relative error values for these two figures are shown in Table 1. The errors, $c(t) - \hat{c}(t)$, in the noisy case are exhibited in Fig. 5 which shows that the maximum error is significantly small. In these two cases, constant velocity and time-dependent velocity, it is sufficient to have the measurements at only one point, x^*, in the spatial interval.

For $c(x, t)$, first the velocity has been estimated at three fixed spatial points x_1^*, x_2^* and x_3^*. Then, the estimated velocities $c(x_1^*, t)$, $c(x_2^*, t)$ and $c(x_3^*, t)$ have been interpolated to find $c(x, t)$ for all $x \in [0, L]$.

Figure 6 shows the estimated velocity at these different points where the level of noise is 5 %; and the estimated velocity $\hat{c}(x, t)$ after interpolation is presented in Fig. 7. Table 2 presents the relative errors of estimating $c(x, t)$ versus different noise levels. Certainly, as the number of measurements $\bar{n}$ increases, the error decreases.

The presented figures and tables show that the estimated unknown is in quite good agreement with the exact one; therefore, it proves the efficiency and the robustness

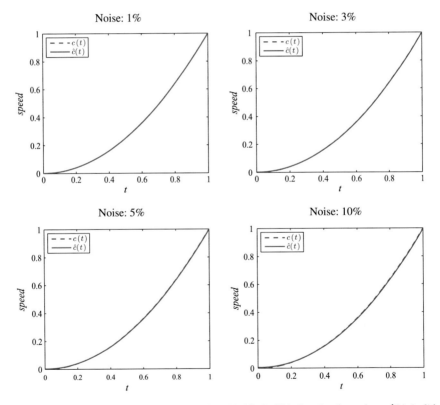

Fig. 4 Time-dependent case: the exact velocity $c(t)$ (*dashed blue*) and estimated one $\hat{c}(t)$ (*solid red*) corresponding to different noise levels

Table 1 Relative errors of $\hat{c}(t)$ versus different levels of noise

Noise level (%)	Relative error (%)
0	3.1×10^{-6}
1	0.080162
3	0.24016
5	0.39974
10	0.79681

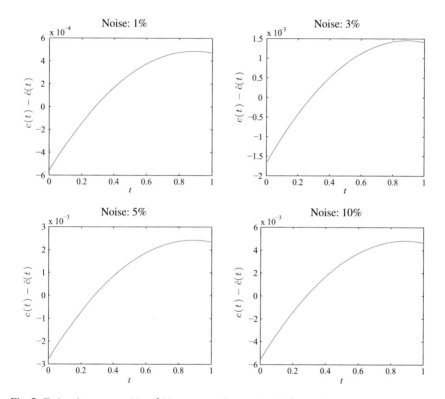

Fig. 5 Estimation errors, $c(t) - \hat{c}(t)$, corresponding to the sub-figures in Fig. 4

of the proposed method for solving velocity inverse problem for wave equation in heterogeneous media.

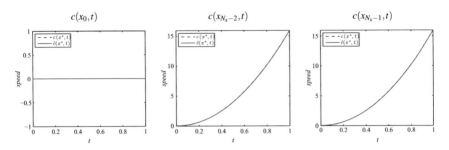

Fig. 6 Exact velocity $c(x^*, t)$ (*dashed blue*) and estimated one $\hat{c}(x^*, t)$ (*solid red*) at three different points in the spatial interval; the level of noise $= 5\%$

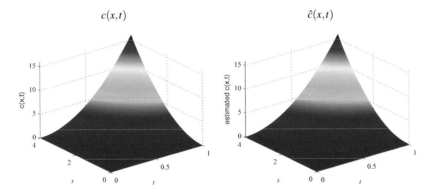

Fig. 7 Space-time-dependent case: exact velocity $c(x, t)$ (*left*) and estimated one $\hat{c}(x, t)$ after applying modulating functions-based method and doing interpolation; noise level $= 5\%$

Table 2 Relative errors of $\hat{c}(x, t)$ versus different levels of noise	Noise level (%)	Relative error (%)
	0	0.0002
	1	0.3763
	3	1.1305
	5	1.8860
	10	3.7812

5 Discussion

The obtained results in Sect. 4 prove the effectiveness of the proposed method and its robustness even with high level of noise and few number of measurements.

The effect of the number of modulating functions M on the estimation is shown in Fig. 8; the figure exhibits the relative error versus the number of modulating functions for the three cases and with 5% noise level. Although it shows that there exist an optimum number of modulating functions, M^*; it also shows that the estimation is generally good within a relatively large interval for M. Moreover, it proves that this

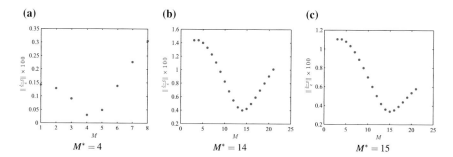

Fig. 8 Relative error (in percent) versus different number of modulating functions: **a** constant velocity c, **b** time-dependent velocity $c(t)$, **c** space-time-dependent velocity $c(x^*, t)$ at fixed point x^*. M^* refers to the optimum number of modulating functions. The level of noise is 5 % in this figure

optimum number depends on the considered problem; $M^* = 4, 14, 15$ for the first, second, and third cases; respectively.

In this paper, we illustrated the method on velocities of low-order polynomial types. Thus, it was sufficient to use polynomial basis with $I = 3$. However, if the unknown is non polynomial, then the use of another type of basis functions may be required.

In case of applications that depend on final time measurements, one can apply space-dependent modulating function, $\phi(x)$, instead of time-dependent function. In addition, if full measurements are available, N-dimensional modulating functions can be applied as in [9].

6   Conclusion

In this paper, modulating functions-based method for solving inverse velocity problem for one-dimensional wave equation was proposed. By applying this method, the problem was converted into a system of linear algebraic equations which can be solved using least square. Numerical simulations in both noise-free and noise-corrupted cases showed the effectiveness of this method.

As a future work, more investigations on the number of modulating functions M will be carried on in order to find an efficient and systematic approach for choosing this number.

Acknowledgments Research reported in this publication was supported by the King Abdullah University of Science and Technology (KAUST).

References

1. Asiri, S., Zayane-Aissa, C., Laleg-Kirati, T.-M.: An adaptive observer-based algorithm for solving inverse source problem for the wave equation. Math. Probl. Eng. **2015** (2015)
2. Fear, E., Stuchly, M.: Microwave detection of breast cancer. IEEE Trans. Microwave Theory Tech. **48**(11), 1854–1863 (2000)
3. Fomel, S., Guitton, A.: Regularizing seismic inverse problems by model reparameterization using plane-wave construction. Geophysics **71**(5), A43–A47 (2006)
4. Fu, H., Han, B., Gai, G.: A wavelet multiscale-homotopy method for the inverse problem of two-dimensional acoustic wave equation. Appl. Math. Comput. **190**(1), 576–582 (2007)
5. Liu, D.-Y., Laleg-Kirati, T.-M., Gibaru, O., Perruquetti, W.: Identification of fractional order systems using modulating functions method. In American Control Conference (ACC), 2013, IEEE, pp. 1679–1684 (2013)
6. Moireau, P., Chapelle, D., Le Tallec, P.: Joint state and parameter estimation for distributed mechanical systems. Comput. Methods Appl. Mech. Eng. **197**(6), 659–677 (2008)
7. Patra, A., Unbehauen, H.: Identification of a class of nonlinear continuous-time systems using hartley modulating functions. Int. J. Control **62**(6), 1431–1451 (1995)
8. Pearson, A., Lee, F.: Parameter identification of linear differential systems via Fourier based modulating functions. Control-Theory Adv. Technol. **1**, 239–266 (1985)
9. Perdreauville, F.J., Goodson, R.: Identification of systems described by partial differential equations. J. Fluids Eng. **88**(2), 463–468 (1966)
10. Ramdani, K., Tucsnak, M., Weiss, G.: Recovering the initial state of an infinite-dimensional system using observers. Automatica **46**, 1616–1625 (2010)
11. Robinson, E.A.: Predictive decomposition of time series with application to seismic exploration. Geophysics **32**(3), 418–484 (1967)
12. Saha, D.C., Rao, B.P., Rao, G.P.: Structure and parameter identification in linear continuous lumped systems: the poisson moment functional approach. Int. J. Control **36**(3), 477–491 (1982)
13. Shinbrot, M.: On the analysis of linear and nonlinear dynamical systems from transient-response data. In: National Advisory Committee for Aeronautics NACA (1954)
14. Shinbrot, M.: On the analysis of linear and nonlinear systems. Trans. ASME **79**(3), 547–552 (1957)
15. Takaya, K.: The use of hermite functions for system identification. IEEE Trans. Autom. Control **13**(4), 446–447 (1968)

Asymptotic Rate for Weak Convergence of the Distribution of Renewal-Reward Process with a Generalized Reflecting Barrier

Tahir Khaniyev, Başak Gever and Zulfiye Hanalioglu

Abstract In this study, a renewal-reward process $(X(t))$ with a generalized reflecting barrier is constructed mathematically and under some weak conditions, the ergodicity of the process is proved. The explicit form of the ergodic distribution is found and after standardization, it is shown that the ergodic distribution converges to the limit distribution $R(x)$, when $\lambda \to \infty$, i.e.,

$$Q_X(\lambda x) \equiv \lim_{t \to \infty} P\{X(t) \le \lambda x\} \to R(x)$$

$$\equiv \frac{2}{m_2} \int_0^x \int_v^\infty [1 - F(u)]dudv.$$

Here, $F(x)$ is the distribution function of the initial random variables $\{\eta_n\}$, $n = 1, 2, \ldots$, which express the amount of rewards and $m_2 \equiv E(\eta_1^2)$. Finally, to evaluate asymptotic rate of the weak convergence, the following inequality is obtained:

$$|Q_X(\lambda x) - R(x)| \le \frac{2}{\lambda} |\pi_0(x) - R(x)|.$$

T. Khaniyev (✉) · B. Gever
TOBB University of Economics and Technology, Ankara, Turkey
e-mail: tahirkhaniyev@etu.edu.tr

B. Gever
e-mail: bgever@etu.edu.tr

T. Khaniyev
Institute of Control Systems, Azerbaijan National Academy of Sciences,
Baku, Azerbaijan

Z. Hanalioglu
Karabuk University, Karabuk, Turkey
e-mail: zulfiyyamammadova@karabuk.edu.tr

© Springer International Publishing Switzerland 2016
G.A. Anastassiou and O. Duman (eds.), *Intelligent Mathematics II:*
Applied Mathematics and Approximation Theory, Advances in Intelligent Systems
and Computing 441, DOI 10.1007/978-3-319-30322-2_22

313

Here,

$$\pi_0(x) = \left(\frac{1}{m_1}\right) \int_0^x (1 - F(u))du$$

is the limit distribution of residual waiting time generated by $\{\eta_n\}$, $n = 1, 2, \ldots$, and $m_1 = E(\eta_1)$.

1 Introduction

A number of very interesting problems of queuing, reliability, stock control theory, stochastic finance, mathematical insurance, physics and biology are expressed by means of random walks and renewal-reward processes. Because of the theoretical and practical importance, random walk and renewal-reward processes are investigated very well in literature (e.g., Feller [5], Gihman and Skorohod [7], Borovkov [3], Brown and Solomon [4], Lotov [15], Kastenbaum [9], Weesakul [19], Zang [20], Khaniyev et al. [10–12], Nasirova [16], Aliyev et al. [1] Patch et al. [17]). Moreover, many modifications of renewal—reward processes can be used for solutions of some problems in these fields, as well. These modifications are mostly given with various types of barriers such as absorbing, delaying, reflecting and elastic barriers or some kinds of discrete interference of chance. For instance, a renewal-reward process with triangular distributed interference of chance is dealt with in the studies of [13, 14], to apply a stochastic inventory problem.

Note that the stochastic processes with reflecting barriers are more complex than the processes with other types of barriers (e.g., [11]). Because, when barriers are reflecting, the dependency between interferences causes a complexity on the probabilistic calculations. We confronted with this difficulty while we were proving the ergodicity of our process.

Recently, the stochastic processes with reflecting barrier are begun to apply to some real-life problems (e.g., motion of the particle with high energy in a diluted environment). To examine these problems, it is necessary to investigate the stochastic processes with reflecting barrier. To fill the lack of information about reflecting barriers, researchers get to obtain some results on this subject (e.g., Feller [5], Gihman and Skorohod [7], Borovkov [3]). However, in these studies, authors generally have obtained analytical results which consist of highly complex mathematical structures, unfortunately. This caused to have some difficulties to implement them to real-world applications. For this reason, even though these results are approximate, nowadays, researchers tend to get asymptotic results for the applications (e.g., Aliyev et al. [2], Janssen and Leeuwaarden [8], Khaniyev et al. [10]). Therefore, in this study, we also aim to obtain asymptotic results for the ergodic distribution of the considered process and its rate of the convergence.

In this study, the following model is considered.

The Model. Suppose that there is a depot and the initial level of which is $\lambda z : (\lambda > 0,$ $z > 0)$ at the beginning of the time. There are demands $(\{\eta_n\} : n = 1, 2, \dots)$ which consecutively come over to depot with an inter-arrival times denoted by $\{\xi_n\}$, $n = 1, 2, \dots$. The amounts of demand and inter-arrival times are both random variables. According to the demand quantities, the level of depot begins to decrease until it drops to a negative value. It means that depot becomes empty and it gets into debt $(\{\zeta_n\} : n = 1, 2, \dots)$. When the time debt come up, the new initial value is determined as λ times of this debt quantity $(\lambda \zeta_1)$. Since the demands arrive successively, the inventory of depot continues to reduce until the depot gets into debt. Then, the process proceeds likewise. Here, these debts $\{\zeta_n\}$ are interpreted as reflections and the zero-axis is reflecting barrier of the process.

A process serves like that is called as a renewal-reward process with a generalized reflecting barrier.

Let us construct the process mathematically.

2   Mathematical Construction of the Process $X(t)$

Let $\{(\xi_n, \eta_n)\}$, $n = 1, 2, 3, \dots$, be a sequence of independent and identically distributed random pairs defined on a probability space $(\Omega, \mathscr{F}, P)$, such that the random variables ξ_n and η_n are also mutually independent and take only positive values. Suppose that the distribution functions of ξ_n and η_n are given and these are denoted by $\Phi(t)$ and $F(x)$, respectively, i.e.,

$$\Phi(t) = P\{\xi_n \leq t\}, \quad F(x) = P\{\eta_n \leq x\}; \quad t \geq 0 \quad x \geq 0, \quad n = 1, 2, \dots$$

Define the renewal sequences $\{T_n\}$ and $\{S_n\}$ as follows:

$$T_0 = S_0 = 0, \quad T_n = \sum_{i=1}^{n} \xi_i, \quad S_n = \sum_{i=1}^{n} \eta_i, \quad n = 1, 2, \dots,$$

and construct sequences of random variables $\{N_n\}$ and $\{\zeta_n\}$, $n = 0, 1, 2, \dots$, as follows:

$$N_0 = 0; \quad \zeta_0 = z \geq 0; \quad N_1 = N_1(\lambda z) = \inf\{k \geq 1 : \lambda z - S_k < 0\};$$
$$\zeta_1 = \zeta_1(\lambda z) = |\lambda z - S_{N_1}|;$$
$$N_n \equiv N_n(\lambda \zeta_{n-1}) = \inf\{k \geq N_{n-1} + 1 : \lambda \zeta_{n-1} - (S_k - S_{N_{n-1}}) < 0\};$$
$$\zeta_n \equiv \zeta_n(\lambda \zeta_{n-1}) = |\lambda \zeta_{n-1} - (S_{N_n} - S_{N_{n-1}})|, \quad n = 2, 3, \dots$$

Here, $\lambda > 0$ is an arbitrary positive constant.

Using $\{N_n : n = 0, 1, 2, \dots\}$, define the following sequence $\{\tau_n : n = 0, 1, 2, \dots\}$:

$$\tau_0 = 0; \quad \tau_1 \equiv \tau_1(\lambda z) = \sum_{i=1}^{N_1} \xi_i, \dots, \quad \tau_n = \sum_{i=1}^{N_n} \xi_i, \quad n = 1, 2, \dots$$

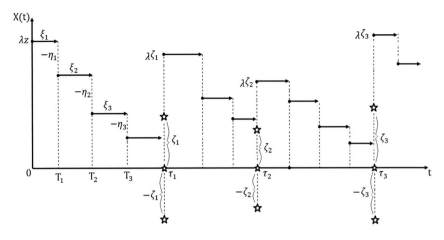

Fig. 1 A trajectory of the process $X(t)$

Moreover, let $v(t) = \max\{n \geq 1 : T_n \leq t\}, t > 0$.

We can now construct the desired stochastic process $X(t)$ as follows:

$$X(t) = \lambda \zeta_n - \left(S_{v(t)} - S_{N_n}\right)$$

for all $\tau_n \leq t < \tau_{n+1}, \quad n \geq 0$.

The process $X(t)$ can be also rewritten as follows:

$$X(t) = \sum_{n=0}^{\infty} \left\{\lambda \zeta_n - \left(S_{v(t)} - S_{N_n}\right)\right\} I_{[\tau_n; \tau_{n+1})}(t).$$

Here $I_A(t)$ represents the indicator function of the set A, such that

$$I_A(t) = \begin{cases} 1, & t \in A \\ 0, & t \notin A \end{cases}$$

A trajectory of the process $X(t)$ is given as in Fig. 1.

The process $X(t)$ is called renewal-reward process with a generalized reflecting barrier. In the case $\lambda = 1$, the process $X(t)$ is known as renewal-reward process with a reflecting barrier.

3   The Ergodicity of the Process $X(t)$

To investigate the stationary characteristics of the process considered, it is necessary to prove that $X(t)$ is ergodic under some assumptions. Before giving this property as a theorem, let us define some notations.

Put $\pi_\lambda(z) = \lim_{n \to \infty} P\{\zeta_n \leq z\}$, i.e., $\pi_\lambda(z)$ is the ergodic distribution of the Markov chain $\{\zeta_n\}$, $n = 1, 2, 3 \ldots$, which represents the reflections of the process. Let us define a new random variable $\hat{\zeta}_\lambda$ as follows:

$$\hat{\zeta}_\lambda : P\{\hat{\zeta}_\lambda \leq z\} = \pi_\lambda(z),$$

i.e., $\hat{\zeta}_\lambda$ is a random variable, the distribution function of which is $\pi_\lambda(z)$. Furthermore, $U_\eta(x)$ is a renewal function generated by the sequence $\{\eta_n\}$, $n = 1, 2, \ldots$ and $m_n \equiv E\left(\eta_1^n\right)$, $n = 1, 2, \ldots$.

Now, state the following Lemma.

Lemma 1 *Suppose that $m_3 \equiv E(\eta_1^3) < \infty$ is satisfied. Then, the following asymptotic expansion can be given, when $\lambda \to \infty$:*

$$E\left(U_\eta\left(\lambda\hat{\zeta}_\lambda\right)\right) = \frac{m_2}{2m_1^2}\lambda + \frac{m_2}{2m_1^2} + o\,(1)\,. \tag{1}$$

Proof By definition,

$$E\left(U_\eta\left(\lambda\hat{\zeta}_\lambda\right)\right) = \int_0^\infty U_\eta(\lambda z)d\pi_\lambda(z).$$

According to refined renewal theorem, when $m_2 = E(\eta_1^2) < \infty$ and $\lambda \to \infty$, for each $z > 0$ the following expansion can be written (Feller: [5, p. 366]):

$$U_\eta(\lambda z) = \frac{\lambda z}{m_1} + \frac{m_2}{2m_1^2} + g(\lambda z). \tag{2}$$

In Eq. (2), $g(\lambda z)$ is a bounded function such that $\lim_{\lambda \to \infty} g(\lambda z) = 0$. Therefore,

$$\int_0^\infty U_\eta(\lambda z)d\pi_\lambda(z) = \int_0^\infty \left[\frac{\lambda z}{m_1} + \frac{m_2}{2m_1^2} + g(\lambda z)\right] d\pi_\lambda(z)$$

$$= \frac{\lambda}{m_1}\int_0^\infty z d\pi_\lambda(z) + \frac{m_2}{2m_1^2} + \int_0^\infty g(\lambda z)d\pi_\lambda(z).$$

$$= \frac{\lambda}{m_1}E(\hat{\zeta}_\lambda) + \frac{m_2}{2m_1^2} + \int_0^\infty g(\lambda z)d\pi_\lambda(z).$$

Here, $\int_0^\infty g(\lambda z)d\pi_\lambda(z) \to 0$, when $\lambda \to \infty$ (Gever[6]). Additionally, it is known that (Feller[5] and Rogozin[18]), when $\lambda \to \infty$, $\pi_\lambda(x)$ converges to a limit distribution $\pi_0(x)$, i.e.,

$$\pi_\lambda(x) \to \pi_0(x) \equiv \frac{1}{m_1} \int_0^x (1 - F(v))dv. \tag{3}$$

Note that, $\pi_0(x)$ is the limit distribution of a residual waiting time generated by the random variables η_n, $n = 1, 2, \ldots$, and $m_1 = E(\eta_1)$. According to Eq. (3) and property of the convergence of moments (Feller: [5, p. 251]), the following convergence also satisfied, when $\lambda \to \infty$:

$$E(\hat\zeta_\lambda) \equiv \int_0^\infty z d\pi_\lambda(z) \to \int_0^\infty z d\pi_0(z) = \frac{m_2}{2m_1}.$$

Additionally, $m_3 = E(\eta_1^3) < \infty$ is hold, the following asymptotic expansion can be written, when $\lambda \to \infty$:

$$E(\hat\zeta_\lambda) = \frac{m_2}{2m_1} + o\left(\frac{1}{\lambda}\right). \tag{4}$$

Using the asymptotic expansion in (4), the following calculations can be done:

$$\int_0^\infty U_\eta(\lambda z)d\pi_\lambda(z) = \frac{\lambda}{m_1}E(\hat\zeta_\lambda) + \frac{m_2}{2m_1^2} + o(1)$$

$$= \frac{m_2}{2m_1^2}\lambda + \frac{m_2}{2m_1^2} + o(1).$$

Thus, the Lemma 1 is proved.

Now, we can express the theorem on the ergodicity of the process $X(t)$ which is the main aim of this section as follows.

Theorem 2 *Let the initial sequence of the random pairs $\{(\xi_n, \eta_n)\}$, $n = 1, 2\ldots$, satisfy the following supplementary conditions:*

(i) $0 < E(\xi_1) < \infty$;
(ii) $E(\eta_1) > 0$;
(iii) $E(\eta_1^3) < \infty$;
(iv) η_1 *is a non-arithmetic random variable.*

Then, the process $X(t)$ is ergodic and the following relation is correct with probability 1 for each measurable bounded function $f(x)$ ($f : [0, +\infty) \to R$):

$$\lim_{T \to \infty} \frac{1}{T} \int_0^T f(X(u))du = S_f \qquad (5)$$

$$\equiv \frac{1}{E(\bar{\tau}_1)} \int_0^\infty \int_0^\infty \int_0^\infty f(v) P_{\lambda z}\{\tau_1 > t : X(t) \in dv\}dt d\pi_\lambda(z).$$

Here $\pi_\lambda(z)$ is the ergodic distribution of the Markov chain $\{\zeta_n\}$, $n = 0, 1, 2, \ldots$ and

$$E(\bar{\tau}_1) \equiv \int_0^\infty E(\tau_1(\lambda z))d\pi_\lambda(z).$$

Proof The process $X(t)$ belongs to a wide class of processes which is known in the literature as "the class of semi-Markov processes with a discrete interference of chance". Furthermore, the general ergodic theorem of type Smith's "key renewal theorem" exists in the literature (Gihman and Skorohod [7]) for these processes. Let us show the conditions of general ergodic theorem are satisfied under the conditions of Theorem 2. According to the general ergodic theorem, to make sure the process $X(t)$ is ergodic, the following assumptions must be provided.

Assumption 1 There exist a monotone increasing random times

$$0 < \gamma_1 < \gamma_2 < \cdots < \gamma_{n-1} < \gamma_n < \cdots$$

such that the values of the process $X(t)$ at these times, i.e., $\kappa_n = X(\gamma_n)$, form an embedded ergodic Markov chain.

For satisfying Assumption 1, it is sufficient to choose the stopping times $\{\tau_n\}$, $n = 0, 1, 2, \ldots$ which is defined in Sect. 2 instead of γ_n. Due to definition of stopping times τ_n, it is obvious that $X(\tau_n) = \lambda \zeta_n$, $n = 0, 1, 2, \ldots$. Here, ζ_n expresses the quantity of nth reflection which is defined in Sect. 2. Furthermore, the reflection quantities ζ_n form a Markov chain with a stationary distribution $\pi_\lambda(z)$. Therefore, the sequence $\{\lambda \zeta_n\}$ can be considered as an embedded Markov chain with a stationary distribution function $\pi_\lambda(z/\lambda)$. Here, the distribution function $\pi_\lambda(x)$ can be obtained from the following integral equation:

$$\pi_\lambda(x) = \int_0^\infty G_{\lambda z}(x)d\pi_\lambda(z),$$

where $G_{\lambda z}(x) \equiv P\{\zeta_n \leq x \mid \zeta_{n-1} = z\}$. In the study Rogozin [18], this probability is represented as follows:

$$G_{\lambda z}(x) = P_{\lambda z}\{\zeta_1 \le x\} = -\int_0^x [1 - F(t)]\, d_t U_\eta(\lambda z + x - t).$$

Here, $U_\eta(t)$ is a renewal function generated by the random variables η_n, i.e., $U_\eta(t) = \sum_{n=0}^\infty F^{*n}(t)$. Thus, Assumption 1 is satisfied.

Assumption 2 The expected values of the differences $\gamma_n - \gamma_{n-1}, n = 1, 2, \ldots$ must be finite, so $E(\gamma_n - \gamma_{n-1}) < \infty$.

Let us show that the second assumption is satisfied under the conditions of Theorem 2. For this purpose, it is sufficient to prove that $E(\tau_1)$ and $E(\tau_2 - \tau_1)$ are finite. Due to the conditions of Theorem 2, $E(\xi_1) < \infty$ is hold. On the other hand, for each $0 < \lambda < \infty$ and $0 < z < \infty$, the renewal function $U_\eta(\lambda z) < \infty$ (see, Feller: [5, p. 185]). Therefore,

$$E(\tau_1) = E(\tau_1(\lambda z)) = E(\xi_1) U_\eta(\lambda) < \infty.$$

On the other hand,

$$E(\tau_2 - \tau_1) = \int_0^\infty E(\tau_1(\lambda z)) d\pi_\lambda(z) = E(\xi_1) \int_0^\infty E(N_1(\lambda z))\, d\pi_\lambda(z)$$

$$= E(\xi_1) \int_0^\infty U_\eta(\lambda z) d\pi_\lambda(z).$$

According to Lemma 1,

$$\int_0^\infty U_\eta(\lambda z) d\pi_\lambda(z) = \frac{m_2}{2m_1^2}(\lambda + 1 + o(1)).$$

Thus, for the finite values of λ, the following inequality holds:

$$\int_0^\infty U_\eta(\lambda z) d\pi_\lambda(z) < \infty.$$

Therefore, $E(\tau_2 - \tau_1) < \infty$ is hold. Hence, the Assumption 2 is proved. Consequently, under the conditions of the Theorem 2, the process $X(t)$ is ergodic.

Moreover, under the conditions of the Theorem 2, time average of the process $X(t)$ converges with probability 1, to space average, when $t \to \infty$. Hence, the Eq. (5) is true (see, Gihman and Skorohod: [7, p. 243]).

We can derive many important conclusion from Theorem 2, one of which is given in Theorem 3.

Theorem 3 *Assume that the conditions of Theorem 2 are satisfied. Then, for each $x > 0$, the ergodic distribution function $(Q_X(x))$ of the process $X(t)$ can be written as follows:*

$$Q_X(x) = 1 - \frac{E(U_\eta(\lambda\hat{\zeta}_\lambda - x))}{E(U_\eta(\lambda\hat{\zeta}_\lambda))}. \tag{6}$$

Here, $Q_X(x) = \lim_{t\to\infty} P\{X(t) \le x\}$ is the ergodic distribution function of the process $X(t)$; $U_\eta(z)$ is a renewal function generated by the sequence $\{\eta_n\}$, $n = 1, 2, \dots$. Furthermore, for each bounded function $M(x)$, $E\left(M\left(\lambda\hat{\zeta}_\lambda\right)\right)$ is given by the following integral:

$$E\left(M\left(\lambda\hat{\zeta}_\lambda\right)\right) \equiv \int_0^\infty M(\lambda z)d\pi_\lambda(z).$$

Proof Define the indicator function as follows $(x > 0)$:

$$I_{(0,x]}(v) = \begin{cases} 1, & v \le x \\ 0, & v > x. \end{cases}$$

Then, substituting the indicator function $I_{(0,x]}(v)$ instead of $f(v)$ in Eq. (5), the following equation is hold:

$$S_f \equiv \frac{1}{E(\bar{\tau}_1)} \int_{z=0}^\infty \int_{v=0}^\infty \int_{t=0}^\infty f(v)P_{\lambda z}\{\tau_1 > t; X(t) \in dv\}dtd\pi_\lambda(z)$$

$$= \frac{1}{E(\bar{\tau}_1)} \int_{z=0}^\infty \int_{v=0}^\infty P_{\lambda z}\{\tau_1 > t; X(t) \le x\}dtd\pi_\lambda(z).$$

Therefore, for each $x > 0$, the function $Q_X(x)$ is obtained as follows:

$$Q_X(x) = \frac{1}{E(\bar{\tau}_1)} \int_{t=0}^\infty \int_{z=0}^\infty P_{\lambda z}\{\tau_1 > t; X(t) \le x\}dtd\pi_\lambda(z).$$

For shortness $G(t, x, \lambda z) \equiv P_{\lambda z}\{\tau_1 > t; X(t) \le x\}$. First, calculate the function $G(t, x, \lambda z)$:

$$G(t, x, \lambda z) = \sum_{n=0}^\infty P_{\lambda z}\{v(t) = n; \tau_1 > t; X(t) \le x\}.$$

Here, $\nu(t) = \max\{n \geq 0; T_n \leq t\}$. Using the renewal sequences T_n and S_n, $G(t, x, \lambda z)$ is calculated as follows:

$$G(t, x, \lambda z) = \sum_{n=0}^{\infty} P_{\lambda z} \{T_n \leq t \leq T_{n+1}; \tau_1 > t; X(t) \leq x\} \tag{7}$$

$$= \sum_{n=0}^{\infty} P\{T_n \leq t \leq T_{n+1}\} P\{\lambda z - S_n > 0; \lambda z - S_n \leq x\}$$

$$= \sum_{n=0}^{\infty} (\Phi_n(t) - \Phi_{n+1}(t)) (F_n(\lambda z) - F_{n+1}(\lambda z - x)).$$

Here, $\Phi_n(t) = P\{T_n \leq t\}$ and $F_n(z) = P\{S_n \leq z\}$. Applying Laplace transform with respect to parameter t to the Eq. (7),

$$\tilde{G}(s, x, \lambda z) \equiv \int_0^{\infty} e^{-st} G(t, x, \lambda z) dt$$

$$= \sum_{n=0}^{\infty} (F_n(\lambda z) - F_{n+1}(\lambda z - x)) \frac{\varphi^n(s)(1 - \varphi(s))}{s}, \quad s > 0,$$

is obtained. Here

$$\varphi(s) = E(\exp\{-s\xi_1\}) \equiv \int_0^{\infty} e^{-st} d\Phi(t).$$

Since $E(\xi_1) < \infty$, the following relation is hold (Feller [5]):

$$\lim_{s \to 0} \frac{1 - \varphi(s)}{s} = E(\xi_1).$$

We can write the Laplace transform of $\tilde{G}(0, x, \lambda z)$ as follows:

$$\tilde{G}(0, x, \lambda z) \equiv \lim_{s \searrow 0} \tilde{G}(s, x, \lambda z) = E(\xi_1) \sum_{n=0}^{\infty} (F_n(\lambda z) - F_{n+1}(\lambda z - x))$$

$$= E(\xi_1) (U_\eta(\lambda z) - U_\eta(\lambda z - x)).$$

Therefore,

$$Q_X(x) = \frac{1}{E(\bar{\tau}_1)} \int_{z=0}^{\infty} \int_{t=0}^{\infty} P_{\lambda z} \{\tau_1 \geq t; X(t) \leq x\} dt d\pi_\lambda(z)$$

$$= \frac{1}{E(\bar{\tau}_1)} \left\{ E(\xi_1) \int_{z=0}^{\infty} [U_\eta(\lambda z) - U_\eta(\lambda z - x)] d\pi_\lambda(z) \right\}$$

is hold. According to Wald identity, we can obtained the expected value of τ_1:

$$E(\bar{\tau}_1) \equiv \int_0^\infty E(\tau_1(\lambda z)) d\pi_\lambda(z) = E(\xi_1) \int_0^\infty E(N_1(\lambda z)) d\pi_\lambda(z)$$

$$= E(\xi_1) \int_0^\infty U_\eta(\lambda z) d\pi_\lambda(z) = E(\xi_1) E(U_\eta(\lambda \hat{\zeta}_\lambda)).$$

As a result, $Q_X(x)$ is calculated as follows:

$$Q_X(x) = \frac{E(\xi_1) \left[E(U_\eta(\lambda \hat{\zeta}_\lambda)) - E(U_\eta(\lambda \hat{\zeta}_\lambda - x)) \right]}{E(\xi_1) E\left(U_\eta(\lambda \hat{\zeta}_\lambda) \right)}$$

$$= 1 - \frac{E\left(U_\eta(\lambda \hat{\zeta}_\lambda - x) \right)}{E\left(U_\eta(\lambda \hat{\zeta}_\lambda) \right)}.$$

Here

$$E\left(M\left(\lambda \hat{\zeta}_\lambda \right) \right) \equiv \int_0^\infty M(\lambda z) d\pi_\lambda(z).$$

Thus, the proof of Theorem 3 is hold.

4 Asymptotic Rate of the Weak Convergence for the Ergodic Distribution of the Process

To prove the weak convergence theorem for the ergodic distribution of the process, we need to standardize the process $X(t)$ as follows:

$$Y_\lambda(t) \equiv \frac{X(t)}{\lambda}, \quad \lambda > 0.$$

The aim is to calculate the limit form for the ergodic distribution of the process $Y_\lambda(t)$, when $\lambda \to \infty$. Therefore, first of all, give the following Lemmas.

Lemma 4 *The following relation can be written for the distribution $\pi_\lambda(x)$, when $\lambda \to \infty$:*

$$\pi_\lambda(x) \equiv \lim_{n \to \infty} P\{\zeta_n \le x\} = \pi_0(x) + \frac{1}{\lambda} g_\lambda(x). \tag{8}$$

Here, $\pi_0(x) = \frac{1}{m_1} \int_0^x (1 - F(u)) \, du$; $m_1 = E(\eta_1)$ and $g_\lambda(x)$ is a measurable bounded function such that $\lim_{\lambda \to \infty} g_\lambda(x) = 0$.

Proof Since ζ_1 is a residual waiting time generated by $\{\eta_n\}$, $n = 1, 2, \ldots$ at the first cycle, the exact expression for its distribution function can be given as follows (Feller [5, p. 369]):

$$H(\lambda z; x) \equiv P\{\zeta_1(\lambda z) \leq x\} = \int_0^\infty [F(\lambda z - v + x) - F(\lambda z - v)] \, dU_\eta(v). \quad (9)$$

Applying Laplace transform to Eq. (9) and using Tauber-Abel Theorems, the distribution function $H(\lambda z; x)$ can be presented as follows:

$$H(\lambda z; x) \equiv P\{\zeta_1(\lambda z) \leq x\} = \pi_{\lambda 1}(\lambda z; x) = \pi_0(x) + \frac{1}{\lambda} g_1(\lambda z; x).$$

Here, for all $z, x > 0$, $g_1(\lambda z; x)$ is a measurable and bounded function such that $\lim_{\lambda \to \infty} g_1(\lambda z; x) = 0$.

Now, let us obtain the distribution function of ζ_2 as follows:

$$\pi_{2\lambda}(\lambda z; x) \equiv P\{\zeta_2(\lambda \zeta_1) \leq x\} = \int_{v=+0}^\infty P\{\zeta_1 \in dv; \zeta_2(\lambda \zeta_1) \leq x\}$$

$$= \int_{v=+0}^\infty P\{\zeta_1 \in dv\} P\{\zeta_1(\lambda v) \leq x\}$$

$$= \int_{v=+0}^\infty H(\lambda v; x) d_v H(\lambda z; v). \quad (10)$$

Here, when $\lambda \to \infty$, $H(\lambda v; x) = \pi_0(x) + \frac{1}{\lambda} g_1(\lambda v; x)$. Substituting $H(\lambda v; x)$ in Eq. (10):

$$\pi_{2\lambda}(\lambda z; x) = \int_{v=+0}^\infty \left\{ \pi_0(x) + \frac{1}{\lambda} g_1(\lambda v; x) \right\} d_v H(\lambda z; v)$$

$$= \pi_0(x) \int_{v=+0}^\infty H(\lambda z; dv) + \frac{1}{\lambda} \int_{v=+0}^\infty g_1(\lambda v; x) d_v H(\lambda z; v)$$

$$= \pi_0(x) + \frac{1}{\lambda} g_2(\lambda z; x).$$

is obtained. Here, $g_2(\lambda z; x) = \int_{v=+0}^{\infty} g_1(\lambda v; x) d_v H(\lambda z; v)$ and it is possible to show that $g_2(\lambda z; x)$ is a measurable and bounded function such that $\lim_{\lambda \to \infty} g_2(\lambda z; x) = 0$. Hence, when $\lambda \to \infty$, for all $v, x > 0$, the following equality can be written:

$$\pi_{2\lambda}(\lambda z; x) = \pi_0(x) + \frac{1}{\lambda} g_2(\lambda z; x).$$

Now, let us obtain the distribution function of ζ_3 as follows:

$$\pi_{3\lambda}(\lambda z; x) \equiv P\{\zeta_3(\lambda \zeta_2) \le x\} = \int_{v=+0}^{\infty} P\{\zeta_2 \in dv; \zeta_3(\lambda \zeta_2) \le x\}$$

$$= \int_{v=+0}^{\infty} P\{\zeta_2 \in dv\} P\{\zeta_3(\lambda \zeta_2) \le x | \zeta_2 = v\}$$

$$= \int_{v=+0}^{\infty} H(\lambda v; x) \pi_{2\lambda}(\lambda z; dv)$$

$$= \int_{v=+0}^{\infty} \left\{ \pi_0(x) + \frac{1}{\lambda} g_1(\lambda v; x) \right\} \pi_{2\lambda}(\lambda z; dv)$$

$$= \pi_0(x) + \frac{1}{\lambda} \int_{v=+0}^{\infty} g_1(\lambda v; x) \pi_{2\lambda}(\lambda z; dv).$$

Here let us denote the integral $\int_{v=+0}^{\infty} g_1(\lambda v; x) \pi_{2\lambda}(\lambda z; dv)$ with $g_3(\lambda z; x)$, i.e.,

$$g_3(\lambda z; x) = \int_{v=+0}^{\infty} g_1(\lambda v; x) \pi_{2\lambda}(\lambda z; dv)$$

and it is possible to show that $g_3(\lambda z; x)$ is a measurable and bounded function such that $\lim_{\lambda \to \infty} g_3(\lambda z; x) = 0$. Hence, the distribution function $\pi_{3\lambda}(\lambda z; x)$ can be written as follows:

$$\pi_{3\lambda}(\lambda z; x) = \pi_0(x) + \frac{1}{\lambda} g_3(\lambda z; x).$$

Similarly, when $\lambda \to \infty$, it is finally shown that the following asymptotic expansion can be written by induction:

$$\pi_{n\lambda}(\lambda z; x) = \pi_0(x) + \frac{1}{\lambda} g_n(\lambda z; x). \tag{11}$$

Here, $\pi_0(x) = (1/m_1) \int_0^x (1 - F(v)) \, dv$ and $g_n(\lambda z; x)$ is a measurable and bounded function such that $\lim_{\lambda \to \infty} g_n(\lambda z; x) = 0$.

Taking limit in Eq. (11), when $n \to \infty$, the following relation is hold:

$$\pi_\lambda(x) = \lim_{n \to \infty} \pi_{n\lambda}(\lambda z; x) = \pi_0(x) + \frac{1}{\lambda} \lim_{n \to \infty} g_n(\lambda z; x). \tag{12}$$

In the relation (12), $\lim_{n \to \infty} g_n(\lambda z; x)$ can be denoted by $g_\lambda(x)$. Therefore,

$$\pi_\lambda(x) = \pi_0(x) + \frac{1}{\lambda} g_\lambda(x).$$

is obtained. Thus, Lemma 4 is proved.

Lemma 5 *Under the conditions of Theorem 2, for each $x > 0$, the following asymptotic expansion can be written, when $\lambda \to \infty$:*

$$E \left(U_\eta \left(\lambda \left(\hat{\xi}_\lambda - x \right) \right) \right) = \frac{\lambda}{m_1} \int_x^\infty (z - x) \, d\pi_\lambda(z) + \frac{m_2}{2m_2^2} (1 - \pi_\lambda(x)) + o(1). \tag{13}$$

Here, $m_n = E(\eta_1^n)$, $n = 1, 2, \ldots$.

Proof It is easy to conclude that it can be proved with the same method on the proof of Lemma 1. Briefly, the following equation can be written by definition:

$$E \left(U_\eta \left(\lambda \left(\hat{\xi}_\lambda - x \right) \right) \right) = \int_0^\infty U_\eta (\lambda (z - x)) \, d\pi_\lambda(z). \tag{14}$$

From Eq. (14), the following asymptotic expansion can be hold:

$$E \left(U_\eta \left(\lambda \left(\hat{\xi}_\lambda - x \right) \right) \right) = \frac{\lambda}{m_1} \int_x^\infty (z - x) \, d\pi_\lambda(z) + \frac{m_2}{2m_2^2} (1 - \pi_\lambda(x)) + o(1).$$

Thus, it proves the Lemma 5.

Using these lemmas, the weak convergence theorem for the standardized process $Y_\lambda(t)$ can be stated as follows:

Theorem 6 *Under the conditions of Theorem 2, $Y_\lambda(t)$ is ergodic and the ergodic distribution of this process $(Q_Y(x))$ convergences to the following limit distribution $R(x)$, for each $x > 0$, when $\lambda \to \infty$:*

$$\lim_{\lambda \to \infty} Q_Y(x) = R(x) \equiv \frac{2}{m_2} \int_0^x \left\{ \int_z^\infty (1 - F(v)) \, dv \right\} dz \tag{15}$$

Here, $m_2 = E(\eta_1^2)$.

Proof According to Theorem 3, Lemmas 1 and 5, the following calculations can be done:

$$Q_Y(x) = Q_X(\lambda x) = 1 - \frac{E\left(U_\eta\left(\lambda\hat{\zeta}_\lambda - \lambda x\right)\right)}{E\left(U_\eta\left(\lambda\hat{\zeta}_\lambda\right)\right)}$$

$$= 1 - \left[\frac{\lambda}{m_1}\int_x^\infty (z-x)\,d\pi_\lambda(z) + \frac{m_2}{2m_2^2}(1-\pi_\lambda(x)) + o(1)\right]\left[\frac{m_2}{2m_1^2}\lambda + \frac{m_2}{2m_1^2} + o(1)\right]^{-1}$$

$$= 1 - \frac{1}{\hat{m}_1}\int_x^\infty [1-\pi_\lambda(z)]\,dz + o(1) = \frac{1}{\hat{m}_1}\int_0^x [1-\pi_\lambda(z)]\,dz + o(1).$$

Here $\hat{m}_1 \equiv E(\hat{\zeta}_\lambda) = \int_0^\infty (1-\pi_\lambda(z))\,dz$, according to the property of convergence of moments (Feller [5]), $\hat{m}_1 = \frac{m_2}{2m_1} + o(1)$ can be written. Then, the following expansion can be given:

$$Q_Y(x) = \frac{2m_1}{m_2}\int_0^x \left\{\frac{1}{m_1}\int_z^\infty (1-F(v))\,dv\right\}dz + o(1)$$

$$= \frac{2}{m_2}\int_0^x \left\{\int_z^\infty (1-F(v))\,dv\right\}dz + o(1) = R(x) + o(1).$$

When $\lambda \to \infty$, the following relation is obtained:

$$\lim_{\lambda\to\infty} Q_Y(x) = R(x) \equiv \frac{2}{m_2}\int_0^x \left\{\int_z^\infty (1-F(v))\,dv\right\}dz.$$

Thus, Theorem 6 is proved.

Theorem 7 *Under the conditions of Theorem 6, for the sufficiently large values of* λ, *the following inequality can be written:*

$$|Q_Y(x) - R(x)| \le \frac{2}{\lambda}|\pi_0(x) - R(x)|. \tag{16}$$

Here

$$\pi_0(x) = \frac{1}{m_1}\int_0^x [1-F(v)]\,dv$$

is the limit distribution of the residual waiting time generated by $\{\eta_n\}$ *and*

$$R(x) = \frac{2}{m_2} \int\limits_0^x \left\{ \int\limits_z^\infty (1 - F(v)) \, dv \right\} dz$$

is the limit distribution of the process $Y_\lambda(t)$, where $m_n = E(\eta_1^n)$, $n = 1, 2 \dots$

Proof We can rewrite the Eq. (6) as follows:

$$Q_Y(x) = \frac{E\left(U_\eta\left(\lambda\left(\hat{\xi}_\lambda - x\right)\right)\right) - E\left(U_\eta\left(\lambda\hat{\xi}_\lambda\right)\right)}{E\left(U_\eta\left(\lambda\hat{\xi}_\lambda\right)\right)}.$$

Denote that $C_F \equiv \frac{m_2}{2m_1^2}$ and using Lemmas 1 and 5, the following equality can be written:

$$\Delta \equiv E\left(U_\eta\left(\lambda\left(\hat{\xi}_\lambda - x\right)\right)\right) - E\left(U_\eta\left(\lambda\hat{\xi}_\lambda\right)\right)$$
$$= \frac{\lambda}{m_1} D(x) + C_F \pi_\lambda(x) + o(1).$$

Here, $D(x) = \frac{1}{m_1} \int_0^x (1 - \pi_\lambda(z)) \, dz$. According to Lemma 1, the following expansion can be written:

$$E\left(U_\eta\left(\lambda\hat{\xi}_\lambda\right)\right) = C_F\lambda\left(1 + \frac{1}{\lambda} + o\left(\frac{1}{\lambda}\right)\right)$$

Hence, we obtain that

$$Q_Y(x) = \frac{\Delta}{E\left(U_\eta\left(\lambda\hat{\xi}_\lambda\right)\right)} = \frac{\Delta}{\left(C_F\lambda\left(1 + \frac{1}{\lambda} + o\left(\frac{1}{\lambda}\right)\right)\right)}.$$

Using Taylor expansion, the following calculations can be done:

$$Q_Y(x) = \frac{\Delta}{C_F\lambda}\left(1 - \frac{1}{\lambda} + o\left(\frac{1}{\lambda}\right)\right)$$
$$= \left\{\frac{D(x)}{C_F} + \frac{\pi_\lambda(x)}{\lambda} + o\left(\frac{1}{\lambda}\right)\right\}$$
$$= \frac{2m_1}{m_2} \int\limits_0^x (1 - \pi_\lambda(z)) \, dz$$
$$+ \frac{1}{\lambda}\left[\pi_\lambda(x) + \frac{2m_1}{m_2} \int\limits_0^x (1 - \pi_\lambda(z)) \, dz\right] + o\left(\frac{1}{\lambda}\right). \qquad (17)$$

Recall that according to Lemma 4:

$$\pi_\lambda(z) \equiv P\left\{\hat{\xi}_\lambda \le z\right\} = \pi_0(z) + o\left(\frac{1}{\lambda}\right). \tag{18}$$

Substituting the Eq. (18) into Eq. (17), we obtain that

$$
\begin{aligned}
Q_Y(x) &= \frac{2m_1}{m_2} \int_0^x (1 - \pi_0(z))\, dz \\
&\quad + \frac{1}{\lambda}\left[\pi_0(x) + \frac{2m_1}{m_2}\int_0^x (1 - \pi_0(z))\, dz\right] + o\left(\frac{1}{\lambda}\right) \\
&= \frac{2m_1}{m_2}\int_0^x \left\{\frac{1}{m_1}\int_z^\infty (1 - F(v))\, dv\right\} dz \\
&\quad + \frac{1}{\lambda}\left[\frac{1}{m_1}\int_0^x (1 - F(v))\, dv - \frac{2m_1}{m_2}\int_0^x \left\{\frac{1}{m_1}\int_z^\infty (1 - F(v))\, dv\right\} dz\right] \\
&\quad + o\left(\frac{1}{\lambda}\right).
\end{aligned}
$$

According to Theorem 6,

$$Q_Y(x) = R(x) + \frac{1}{\lambda}[\pi_0(x) - R(x)] + o\left(\frac{1}{\lambda}\right)$$

To obtain the asymptotic rate,

$$|Q_Y(x) - R(x)| \le \frac{1}{\lambda}|\pi_0(x) - R(x)| + \left|o\left(\frac{1}{\lambda}\right)\right|.$$

When λ is large enough, then the following inequality can be written:

$$\left|o\left(\frac{1}{\lambda}\right)\right| \le \frac{1}{\lambda}|\pi_0(x) - R(x)|.$$

Hence, it can be written as follows:

$$|Q_Y(x) - R(x)| \le \frac{2}{\lambda}|\pi_0(x) - R(x)|.$$

Thus, Theorem 7 is proved.

5 Conclusion

In this study, the weak convergence theorem for the ergodic distribution of renewal-reward process with a generalized reflecting barrier $(X(t))$ is proved. As a result, the asymptotic relation for the ergodic distribution function $Q_Y(x)$ is hold as in the Eq. (15). It can be inferred from this result that the limit distribution $R(x)$ is a distribution of a residual waiting time which is generated by another residual waiting times. That is a very interesting interpretation in the aspect of stochastic process theory. According to this result, the rate is expressed by the distribution functions of $\pi_0(x)$ and $R(x)$. Since they are distribution functions, they can only take values in the interval $[0, 1]$. Thus, the absolute difference between $\pi_0(x)$ and $R(x)$ can't be greater than 1. On the other hand, when $x \to \infty$, this absolute difference goes to zero. Therefore, in that case, the rate can be $\frac{2}{\lambda}$ at most and it gets smaller as well, when $\lambda \to \infty$.

References

1. Aliyev, R., Khaniyev, T., Kesemen, T.: Asymptotic expansions for the moments of a semi-Markovian random walk with gamma distributed interference of chance. Commun. Statis. Theory Meth. **39**(1), 130–143 (2010)
2. Aliyev, R.T., Kucuk, Z., Khaniyev, T.: A three-term asymptotic expansions for the moments of the random walk with triangular distributed interference of chance. Appl. Math. Modell. **34**(11), 3599–3607 (2010)
3. Borovkov, A.A.: Stochastic Processes in Queuing Theory. Spinger, New York (1976)
4. Brown, M., Solomon, H.: A second—order approximation for the variance of a renewal-reward process. Stochas. Process.Appl. **34**(11), 3599–3607 (2010)
5. Feller, W.: Introduction to Probability Theory and Its Applications II. Wiley, New York (1971)
6. Gever, B.: Genelleştirilmiş Yansıtan Bariyerli Öd üllü Yenileme Sürecinin Asimtotik Yöntemlerle İ ncelenmesi, Yüksek Lisans Tezi, TOBB ETÜ, Ankara (in Turkish)
7. Gihman, I.I., Skorohod, A.V.: Theory of Stochastic Processes II. Springer, Berlin (1975)
8. Janssen, A.J.E.M., Leeuwaarden, J.S.H.: On Lerch's transcendent and the Gaussian random walk. Annals Appl. Prob. **17**(2), 421–439 (2007)
9. Kastenbaum, M.A.: A dialysis system with one absorbing and one semi-reflecting state. J. Appl. Probab. **3**, 363–371 (1996)
10. Khaniyev, T., Gever, B., Mammadova, Z.: Approximation formulas for the ergodic moments of Gaussian random walk with a reflecting barrier. Advances in Applied Mathematics and Approximation Theory: Contributions from AMAT. Springer Proceedings in Mathematics and Statistics, vol. 13, pp. 219–233 (2012)
11. Khaniev, T.A., Unver, I., Maden, S.: On the semi-Markovian random walk with two reflecting barriers. Stochas. Anal. Appl. **19**(5), 799–819 (2001)
12. Khaniyev, T.A., Mammadova, Z.: On the stationary characteristics of the extended model of type (s, S) with Gaussian distribution of summands. J. Statis. Comput. Simul. **76**(10), 861–874 (2006)
13. Khaniev, T., Atalay, K.: On the weak convergence of the ergodic distribution in an inventory model of type (s, S),. Hacettepe J. Math. Statis. **39**(4), 599–611 (2010)
14. Khaniyev, T., Kokangul, A., Aliyev, R.: An asymptotic approach for a semi-Markovian inventory model of type (s, S). Appl. Stochas. Models Bus. Indus. **29**, 439–453 (2013)
15. Lotov, V.I.: On some boundary crossing problems for Gaussian random walks. Annals Prob. **24**(4), 2154–2171 (1996)

16. Nasirova, T.I.: Processes of the semi—Markovian Random Walks. Elm, Baku (1984)
17. Patch, B., Nazarathy, Y., Taimre, T.: A correction term for the covariance of multivariate renewal-reward processes. Statis. Prob. Lett. **102**, 1–7 (2014)
18. Rogozin, B.A.: On the distribution of the first jump. Theory Prob. Appl. **9**(3), 450–465 (1964)
19. Weesakul, B.: The random walk between a reflecting and an absorbing barrier. Annals Math. Statis. **32**, 765–773 (1961)
20. Zang, Y.L.: Some problems on a one dimensional correlated random walk with various types of barrier. J. Appl. Prob. **29**, 196–201 (1992)

Bias Study of the Naive Estimator in a Longitudinal Linear Mixed-Effects Model with Measurement Error and Misclassification in Covariates

Jia Li, Ernest Dankwa and Taraneh Abarin

Abstract This research presents a generalized likelihood approach to estimate the parameters in a longitudinal linear mixed-effects model. In our model, we consider measurement error and misclassification in the covariates. Through simulation studies, we observe the impact of each parameter of the model on the bias of the naive estimation that ignores the errors in the covariates.

1 Introduction

Longitudinal studies are research studies that involve repeated observations of the same factors over a period of time. As observational studies, they are applied in the health science to uncover predictors of certain outcome. Frequently, existing studies examine such predicting variables under the assumption that they are measured accurately. However, unobserved or error-prone variables are unavoidable. It is now well-known that measurement and/or misclassification error can influence the results of a study [1]. The impact of ignoring these errors varies from bias and large variability in estimators to low power in testing hypothesis [2, 3]. In order to improve the accuracy and precision in assessing the variables, one needs to take into account of the imprecise data of the studies.

In this research, we consider a longitudinal linear mixed-effects model. The model contains a repeatedly measured response, y_{it}, continuous predictor(s), X_{it}, subject to measurement error, and a classified predictor, G_i, subject to misclassification. We consider the model error term, ε_{it}, with an autoregressive model with lag one, AR(1),

J. Li · E. Dankwa · T. Abarin (✉)
Department of Mathematics and Statistics, Memorial University,
St John's, Canada
e-mail: tabarin@mun.ca

J. Li
e-mail: jial7642@gmail.com

E. Dankwa
e-mail: ed6780@mun.ca

© Springer International Publishing Switzerland 2016
G.A. Anastassiou and O. Duman (eds.), *Intelligent Mathematics II:*
Applied Mathematics and Approximation Theory, Advances in Intelligent Systems
and Computing 441, DOI 10.1007/978-3-319-30322-2_23

to account for the correlation between the time points. Autoregressive models with lag one are widely applied in science [4–6]. We also consider a time-independent random effect, γ_i, for a specific individual. As both X and G are random variables, we calculated the marginal moments of the response in order to obtain a closed-form for the parameter estimates using Generalized Least Square (GLS) estimation [7].

This paper is organized as follows: In Sect. 2, we describe the models in the research. We begin with a longitudinal linear mixed effect model without any errors in the covariates; then we introduce measurement error and misclassification models as well as a final model with both measurement error and misclassification in covariates. Section 3 is a discussion of simulations studies for the final model where there are both measurement error and misclassification involved. The last section draws conclusion on the whole research.

2 The Models

2.1 Error-Free Model

Without considering any errors in covariates, we define the longitudinal linear mixed-effects model as follows:

$$y_{it} = X'_{it}\beta + \gamma_i + G_i\alpha + \varepsilon_{it}, \quad i = 1, \ldots, k, \quad t = 1, \ldots, T, \qquad (1)$$

where $y_{it} \in \mathbb{R}$ is the response for the ith individual, at time point t, $t = 1, \ldots, T$, γ_i is the individual random effect with mean zero and variance σ_γ^2. Moreover, $X_{it} \in \mathbb{R}^p$ is the p dimensional continuous covariate with coefficient vector β, and independent of $G_i \in \mathbb{R}$, which is the categorical time-invariant predictor. In model (1) also, $\alpha \in \mathbb{R}$ is the coefficient of the categorical predictor, ε_{it} is an error term that follows an AR(1), such that $\varepsilon_{it} = \rho_1 \varepsilon_{i,t-1} + a_{it}$ and $|\rho_1| < 1$. In this autoregressive model, a_{it} is a random error term with mean zero and variance σ_ε^2, independent of γ_i. The model error term ε_{it} can, therefore, be expressed as $\varepsilon_{it} = \sum_{t=0}^{\infty} \rho_1^t a_{it}$. The Generalized Least Square (GLS) estimate of $\theta = (\beta', \alpha)'$, the vector of coefficient parameters, based on the marginal (in terms of the random effects) moments of the response, has a closed-form given as follows:

$$\hat{\theta} = \begin{pmatrix} \hat{\beta} \\ \hat{\alpha} \end{pmatrix} = \left[\sum_{i=1}^{n} \begin{pmatrix} X_i \\ \cdot\cdot \\ \mathbf{1}'_T G_i \end{pmatrix} \Sigma_i^{-1} (X'_i : \mathbf{1}_T G_i) \right]^{-1} \left[\sum_{i=1}^{n} \begin{pmatrix} X_i \\ \cdot\cdot \\ \mathbf{1}'_T G_i \end{pmatrix} \Sigma_i^{-1} y_i \right]$$

where $X_i = (X'_{i1}, \ldots, X'_{ip})'$, $X'_{ip} = (X_{i1p}, \ldots, X_{iTp})'$, $\mathbf{1}_T$ is a T-dimensional column vector of ones, y_i is $(y_{i1}, \ldots, y_{iT})'$, and Σ_i is the covariance matrix of y_i which satisfies:

1. $var(y_{it}|X_{it}, G_i) = \sigma_\gamma^2 + \frac{\sigma_a^2}{1-\rho_1^2}$,

2. for $t \neq u$, $cov(y_{it}, y_{iu}|X_{it}, G_i) = \sigma_\gamma^2 + \frac{\sigma_a^2 \rho_1^{|t-u|}}{1-\rho_1^2}$.

The GLS estimators of the model coefficient parameters may be computed, iteratively, based on other model parameters.

2.2 Model with Measurement Error

In this model, the true continuous predictor X_{it} is not observed, instead another variable W_{it} is observed subject to measurement error. We consider an additive measurement error model as follows.

$$W_{it} = X_{it} + U_{it}, \quad i = 1, \ldots, n, \quad t = 1 \ldots T \tag{2}$$

Here, U_{it} is the p-dimensional measurement error, independent of X_{it} with mean vector $\mathbf{0}$ and variances $\sigma_{u1}^2 \ldots \sigma_{up}^2$. We assume measurement error for any two covariates are independent irrespective of their occurrence times. Furthermore, measurement errors for measuring the same covariate at different time points are likely to be correlated.

2.3 Model with Misclassification

First, we examined the case where the categorical predictor is a binary variable with values 0 or 1; where 0 is failure and 1 is success. Thus, instead of the true categorical predictor G, a binary variable, G^*, is observed subject to misclassification. The conditional probabilities of G^* given G are as follow.

$$\theta_{ij} = P(G^* = i|G = j), \quad i = 0, 1, \quad j = 0, 1. \tag{3}$$

$$\sum_{i=0}^{1} \theta_{ij} = 1, \quad j = 0, 1.$$

In literature, $\theta_{11} = P(G^* = 1|G = 1)$, or the probability of correct classification of success, is known as *sensitivity* while $\theta_{00} = P(G^* = 0|G = 0)$, or the probability of correct classification of failure, is known as *specificity*. We also extended the study to a more complicated case where the categorical covariate has three categories with values 0, 1 or 2 (not presented in this paper). For detailed discussions on the case where the categorical variable has three categories, please refer to Chap. 5 of [8].

2.4   Model with Measurement Error and Misclassification

In this model, error-prone variables, W_{it} and G_i^*, are observed instead of the true covariates, X_{it} and G_i, respectively. Using the model assumptions and law of iterative expectations, we obtained marginal mean, variance and covariance of the response as follows. Detailed steps on calculating these moments, can be found in Chap. 5 of [8].

1. $E(y_{it}|W_{it}, G_i^*) = E(X_{it}'|W_{it})\beta + \alpha E(G_i|G_i^*)$
2. $var(y_{it}|W_{it}, G_i^*) = \beta' var(X_{it}'|W_{it})\beta + \alpha^2 var(G_i|G_i^*) + \sigma_\gamma^2 + \frac{\sigma_a^2}{1-\rho_1^2}$,
3. for $t \neq u$, $cov(y_{it}, y_{iu}|W_{it}, W_{iu}, G_i^*) = \beta' cov((X_{it}', X_{iu}')|W_{it}, W_{iu})\beta + var(G_i| G_i^*)\alpha^2 + \sigma_\gamma^2 + \frac{\sigma_a^2 \rho_1^{|t-u|}}{1-\rho_1^2}$.

The *naive* GLS estimate of the model coefficient parameters based on the observed W and G^* rather than X and G, is expressed as follows.

$$\hat{\theta}_n = \left[\sum_{i=1}^n \begin{pmatrix} W_i \\ \cdot \cdot \\ \mathbf{1}_T' G_i^* \end{pmatrix} \Phi_i^{*-1} (W_i' : \mathbf{1}_T G_i^*) \right]^{-1} \left[\sum_{i=1}^n \begin{pmatrix} W_i \\ \cdot \cdot \\ \mathbf{1}_T' G_i^* \end{pmatrix} \Phi_i^{*-1} y_i \right], \quad (4)$$

where $\hat{\theta}_n = \begin{pmatrix} \hat{\beta}_n \\ \hat{\alpha}_n \end{pmatrix}$ and Φ_i^* is the covariance matrix of the response based on W_i and G^*, which satisfy expressions (2) and (3) given above. The naive estimator is generally biased for θ, as the marginal moments of the response based on the observed covariates, W and G^*, differ from the ones based on the true covariates, X and G.

The covariance matrix of $\hat{\theta}_n$ conditioned on W_i and G_i^*, can be expressed as

$$Cov(\hat{\theta}_n|W_i, G_i^*) = \left[\sum_{i=1}^n \begin{pmatrix} W_i \\ \cdot \cdot \\ \mathbf{1}_T' G_i^* \end{pmatrix} \Phi_i^{*-1} (W_i' : \mathbf{1}_T G_i^*) \right]^{-1}$$

3   Simulation Results

We now present the common set-ups for all the scenarios. For each of the continuous covariates (in here, $p = 2$), and $T = 4$ time points, we generated independent time-variant continuous predictors from uniform distributions $U(0, 1)$. We used autoregressive models to account for such correlations in the covariates. Therefore, the measurement error for covariate one, (say η_{1it}), follows an AR(1), such that $\eta_{1it} = \rho_2 \eta_{1i,t-1} + u_{1it}$ and $|\rho_2| < 1$. In this autoregressive model, u_{1it} is a random error term with mean zero and variance $\sigma_{u_1}^2$. The measurement error for covariate two, (say η_{2it}), follows an AR(1), such that $\eta_{2it} = \rho_3 \eta_{2i,t-1} + u_{2it}$ and $|\rho_3| < 1$. Similarly, u_{2it} is a random error term with mean zero and variance $\sigma_{u_2}^2$. ρ_2 and ρ_3 were both set for 0.8, except in the scenarios that they changed. The categorical time-invariant, G, was generated from a binary distribution with probability of success $\pi = 0.4$. The regression model parameters for most scenarios were set to be $\alpha = 0.2$ and $\beta = (1, -0.5)'$. However, in order to investigate the impact of the signs (negative/positive) of the coefficients, we also considered $\alpha = -0.2$, and other possible signs of each βs. The model error term, ε_{it}, follows a first order auto-regressive model, such that $\varepsilon_{it} = \rho_1 \varepsilon_{i,t-1} + a_{it}$ and $|\rho_1| < 1$. We generated a_{it} from a normal distribution with mean zero. Except when they changed, we set ρ_1 and σ_ε^2 to be 0.8 and 1, respectively.

For the misclassification, the categorical time-invariant, G^*, was generated based on G, in order to guarantee specific values of sensitivity and specificity. A sample size of 200 units were selected for all the scenarios. For each of the sample sizes, 1000 Monte Carlo replicates were simulated and the Monte-Carlo mean estimates and standard errors of the estimators were computed. In our simulation studies, we observed the bias in the naive estimator as we vary the parameters of the model one after the other with all the other parameters remaining the same. We present here some results from the simulations.

The results of changing α from -3 to 3 is shown in Fig. 1a. The bias in the naive estimate of α increases sharply as α changes. As it could be expected, when $\alpha = 0$, the bias in the naive estimator of α is approximately zero. The bias of the estimators of β_1 and β_2 were not affected by the changes in α.

Figure 1b shows the bias in the naive estimator as π, the probability of the of success of the categorical covariate, varies from 0 to 1. It is interesting to note that when the parameter α is set as a negative value, the bias in the naive estimator of α increases, however when α is set positive (not shown in here) the bias in the naive estimator decreases.

As ρ_1 increases from -1 to 1, Fig. 1c indicates that the bias in the naive estimator of β_1 initially decreases to a point a point where $\rho_1 = -0.25$ and then rises again giving it a $\cup$ shape. The bias in the naive estimator of α increases to the same point for ρ_1 and declines afterwards giving it an $\cap$ shape. However, the bias in the naive estimator of β_2 decreases to a point a point where $\rho_1 = 0$ and increases slowly, thereafter.

From the graph in Fig. 1d, we observe that as the variance of the measurement error of covariate X_1 increases, bias in the naive estimator of β_1 increases, as was expected. Since X_1 and X_2 were generated independently, the bias in the naive estimator of β_2 remains unchanged while the bias of the naive estimator of α declines. Although, the main goal of our research was to study the bias in the naive estimates, we also investigated the variabilities in them. From Table 1, we observe that the increase in measurement error of X_1 decreases the variabilities in the naive estimators of the three parameters, making the naive estimator of β_1 even worse.

In Fig. 2, we take a critical look at the magnitude of bias in the naive estimators as we vary β_1 from -3 to 3. For graphs (a) and (b), we set β_2 to be positive (0.5).

Table 1 Bias and standard error for the different values of $\sigma_{u_1}^2$

$\sigma_{u_1}^2$	$\hat{\alpha}$		$\hat{\beta}_1$		$\hat{\beta}_2$	
	Bias	SE	Bias	SE	Bias	SE
0.2	0.5050	0.0395	−0.6946	0.0060	−0.6105	0.0015
0.4	0.3774	0.0355	−0.4345	0.0033	−0.6194	0.0014
0.6	0.2788	0.0332	−0.2328	0.0022	−0.6260	0.0013
0.8	0.2041	0.0316	−0.0798	0.0016	−0.6308	0.0012
1.0	0.1456	0.0305	−0.0401	0.0012	−0.6345	0.0012
1.2	0.0983	0.0296	0.1371	0.0010	−0.6375	0.0011
1.4	0.0592	0.0289	0.2174	0.0008	−0.6399	0.0011
1.6	0.0261	0.0284	0.2851	0.0007	−0.6419	0.0011
1.8	−0.022	0.0279	0.3433	0.0006	−0.6436	0.0010
2.0	−0.0021	0.0275	0.3436	0.0023	−0.6456	0.0009

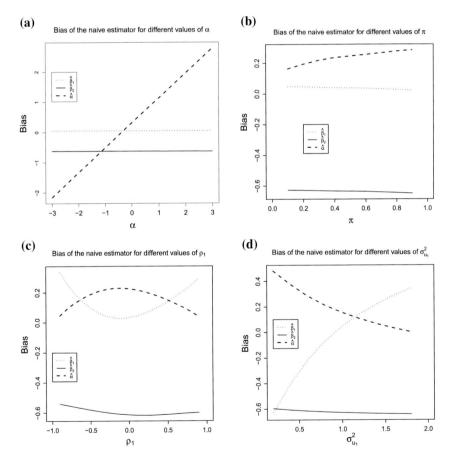

Fig. 1 Graphs for scenarios where the categorical predictor has two categories

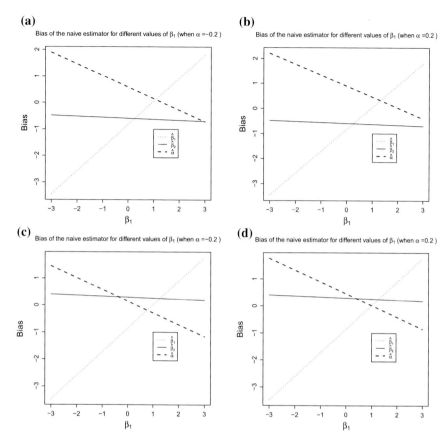

Fig. 2 Graphs for the study of bias in naive estimator as β_1 changes. **a** $\beta_2 = 0.5$. **b** $\beta_2 = 0.5$. **c** $\beta_2 = -0.5$. **d** $\beta_2 = -0.5$

The results indicate that as β_1 changes, in both graphs, the bias in the naive estimator of α declines. However, the magnitude of bias is larger in the scenario in which α is set positive. Similar observations were made in graphs (c) and (d), where we set β_2 to be negative (-0.5). In addition to the above results, we observe that although in each graph, the bias in the naive estimator of β_2 remains unchanged, the naive estimator tends to underestimate β_1, when it is actually positive and over estimate it when it is negative. Comparing graphs (a) and (b) with (c) and (d) also reveals that the magnitude of the bias in $\hat{\alpha}$ is higher for positive β_2 as comparing to the negative value.

3.1 More Interesting Results Not Shown in Graphs or Table

- By increasing either *sensitivity* or *specificity*, when α is set as a negative value, the bias in the naive estimator of α increases but when α is set as a positive value the bias in the naive estimator of α decreases.
- We also observe the behaviour of the bias in the naive estimates, as sample size changes from 100 to 1000. It is interesting that increasing the sample size does not improve the naive estimators of the coefficient parameters. More interestingly, as the sample size increases, the variabilities of all the estimates decline, leaving the naive estimates to perform very poorly.
- Increasing either σ_γ^2 or σ_ε^2 (from zero to one) had almost the same effects on the bias of naive estimator of the parameters. While the bias of naive estimator of α increases, that of β_1 and β_2 decrease slightly.

4 Conclusions

- Even though assessing bias through the closed-form naive estimates is quite challenging, one may assess the bias (and variabilities) of naive estimator by varying model parameters one at a time while keeping the others constant. These results provide useful insights on the effect of errors in covariates in the linear mixed effect models.
- Bias in the naive estimator tend to be high when the parameter have high contributions in the model. A typical case was where the magnitude of bias for the naive estimator of α was high when the parameter α was set as a positive value. So when the categorical covariate has a positive contribution in the model, we expect measurement error and misclassification to have worse influence on the estimate of α, the coefficient of the categorical predictor.
- Bias, accompanied by low variabilities, indicate how poor estimates are when we have error-prone covariates in the model. This has been a very consistent observation in our study of the effects of errors in the covariates of our model. This implies that one cannot do any better inference without accounting for the errors in the covariates.
- Changes in some of the parameters of the model such as the coefficients of the predictor variables, α, β_1 and β_2, lag correlations coefficients, ρ_1 and ρ_2, and *sensitivity*, were found to have much influence on the magnitude and direction of bias of the estimators of the parameters as compared to the other parameters.

References

1. Fuller, W.A.: Measurement Error Models, vol. 305. John Wiley and Sons (2009)
2. Carroll, R.J., David, R., Stefanski, L.A., Ciprian, M.C.: Measurement Error in Nonlinear Models: A Modern Perspective. CRC Press (2012)
3. Bounaccorsi, J.P.: Measurement Error: Models, Methods and Applications. CRC Press (2010)
4. Dakos, V., Carpenter, S.R., Brock, W.A., Ellison, A.M., Guttal, V., Ives, A.R., Scheffer, M. et al.: Methods for detecting early warnings of critical transitions in time series illustrated using simulated ecological data. PLoS One **7**(7), e41010 (2012)
5. Zhang, X., Zhang, T., Young, A.A., Li, X.: Applications and comparisons of four time series models in epidemiological surveillance data. PloS One **9**(2), e88075 (2014)
6. Nelder, J., Wedderburn, R.: Generalized linear models. J. R. Stat. Soc. Ser. A (General) (Blackwell Publishing) **135**(3), (1972)
7. McCulloch, C.E., Neuhaus, J.M.: Generalized Linear Mixed Models. John Wiley and Sons, Ltd (2001)
8. Li, J.: A Study of Bias in the Naive Estimator in Longitudinal Linear Mixed-Effects Models with Measurement Error and Misclassification in Covariates, Master's Dissertation. Memorial University (2014)

Transformations of Data in Deterministic Modelling of Biological Networks

Melih Ağraz and Vilda Purutçuoğlu

Abstract The Gaussian graphical model (GGM) is a probabilistic modelling approach used in the system biology to represent the relationship between genes with an undirected graph. In graphical models, the genes and their interactions are denoted by nodes and the edges between nodes. Hereby, in this model, it is assumed that the structure of the system can be described by the inverse of the covariance matrix, Θ, which is also called as the precision, when the observations are formulated via a lasso regression under the multivariate normality assumption of states. There are several approaches to estimate Θ in GGM. The most well-known ones are the neighborhood selection algorithm and the graphical lasso (glasso) approach. On the other hand, the multivariate adaptive regression splines (MARS) is a non-parametric regression technique to model nonlinear and highly dependent data successfully. From previous simulation studies, it has been found that MARS can be a strong alternative of GGM if the model is constructed similar to a lasso model and the interaction terms in the optimal model are ignored to get comparable results with respect to the GGM findings. Moreover, it has been detected that the major challenge in both modelling approaches is the high sparsity of Θ due to the possible non-linear interactions between genes, in particular, when the dimensions of the networks are realistically large. In this study, as the novelty, we suggest the Bernstein operators, namely, Bernstein and Szasz polynomials, in the raw data before any lasso type of modelling and associated inference approaches. Because from the findings via GGM with small and moderately large systems, we have observed that the Bernstein polynomials can increase the accuracy of the estimates. Hence, in this work, we perform these operators firstly into the most well-known inference approaches used in GGM under realistically large networks. Then, we investigate the assessment of these transformations for the MARS modelling as the alternative of GGM again under the same large complexity. By this way, we aim to propose these transformation techniques for all sorts of modellings under the steady-state condition of the protein-protein interaction networks in order

M. Ağraz · V. Purutçuoğlu (✉)
Middle East Technical University, Ankara, Turkey
e-mail: vpurutcu@metu.edu.tr

M. Ağraz
e-mail: agraz@metu.edu.tr

© Springer International Publishing Switzerland 2016
G.A. Anastassiou and O. Duman (eds.), *Intelligent Mathematics II:*
Applied Mathematics and Approximation Theory, Advances in Intelligent Systems
and Computing 441, DOI 10.1007/978-3-319-30322-2_24

343

to get more accurate estimates without any computational cost. In the evaluation of the results, we compare the precision and F-measures of the simulated datasets.

1 Introduction

The description of the biochemical activations via networks and the mathematical modelling are very powerful approaches to understand the actual behaviour of the bio-logical process and present the structure of the complex systems. There are different levels to present biochemical events. The protein-protein interaction and metabolic networks are the two well-known representations. Here we deal with the former type of networks which aims to explain the functional/physical interactions between proteins. This biological network can be modelled with different techniques. One of the very well known models to describe and to visualize biological networks is the Gaussian Graphical Model (GGM). GGM is extensively used in many fields, includ-ing image process [1], economy [2], gene regularity network [3] and the process of sampling [4]. Basically, it is a parametric method and applies the inverse of the covariance matrix, i.e. precision Θ, to explain relationships between genes. Here every gene is regressed by the other genes in such a way that the coefficients of the regression indicate conditionally dependent structure [5]. Thus the entries of the precision matrix are the model parameters of interest. But the estimation of these parameters are challenging, especially, under high dimensional systems with sparse observations. Because it leads to the ill-posed problem. There are numerous works to overcome this problem [6–8]. One of the common solutions is the neighbor-hood selection method [9]. This method is a subproblem of the covariance selection methods. It uses the neighborhoods of the nodes and assumes that two nodes in the graph are conditionally independent to each other given the remaining nodes. Finally the given neighbors of the variable are estimated by fitting a lasso regression. The other common approach for the estimation is the L_1-penalized likelihood, i.e. glasso method [10]. This approach simply performs the lasso regression by putting the L_1-penalty into the precision matrix.

MARS is an innovative nonparametric and nonlinear regression modelling tech-nique which is introduced by statistician Friedman in 1991 [11]. It efficiently models the variables' nonlinearities and interactions by using the linear terms and does not make any assumption about the functional relationship between the response and the exploratory variables [12]. When constructing the model, it partitions the input variables into regions where each has its own regression equation. This method is implemented in various fields such as financial data clustering [13], data mining [14], time series [15] and the engineering analyses [16].

In this study, we suggest to work on the Bernstein and Szasz-Mirakyan Polinamials in inference with GGM and MARS to accurately estimate realistically complex biological networks and to compare their results. By this way, we aim to detect whether these operators can be suggested in advance of all lasso types of models for protein interaction networks. In general the Bernstein operators are applied to prove polynomials ability on the approximation of any function over an interval and smoothing the datasets by transforming them on the [0, 1] interval. Moreover, the Bernstein operators are not only used for smoothing the functions, but also, for smoothing statistics [17], solving numerical analysis [18] and drawing computer graphs via the bezier curves [19]. Finally, they can be also performed in generalized Fourier series in order to proximate curves and surfaces [20].

Hence, in the rest of the paper, we present the following organization. In Sect. 2, GGM is explained briefly and its estimation technique via two major methods are introduced. In Sect. 3, the MARS method and Bernstein operators are shortly presented. In order to evaluate the results, the simulation studies are applied in Sect. 4. Then the findings are compared and future works are presented in Sect. 5.

2 Graphical Models

A graph is a representation of pairs of nodes and edges where the nodes are the basic components of biological networks and the edges show the interactions between nodes. In biological networks, the graphical models are one of the useful ways to extract the meaning of networks from a dataset. The graphical models can be divided into the two groups, known as directed and undirected graphical structures. As it is seen in Fig. 1, the directed graph shows the direction and interactions between nodes, but the undirected graph only presents the interactions between nodes without their directional information.

2.1 Gaussian Graphical Models

The Gaussian graphical models (GGM) are one of the well-known modelling approaches, which show the undirected graphical interactions over a set of random variables under the multivariate normal distribution. GGM have been firstly used in the literature under the name of the covariance selection models by Dempster [21]. But the graphical representation of these models is firstly introduced by Whittaker [22]. In GGM, the nodes can be formulated as $Y = (Y^1, Y^2, \ldots, Y^p)$ which is multivariate normally distributed via

Fig. 1 Basic representation of the **a** directed and **b** the undirected graph between three nodes

(a)   **(b)**

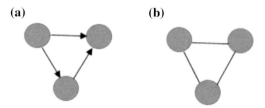

$$Y \sim N(\mu, \Sigma), \tag{1}$$

where μ is a p-dimensional vector with $\mu = (\mu_1, \mu_2, \ldots, \mu_p)$ and Σ is a $(p \times p)$-dimensional covariance matrix. So the probability distribution function of Y can be presented by

$$f(y) = \frac{1}{(2\pi)^{n/2}|\Sigma|^{1/2}} e^{-\frac{1}{2}(y-\mu)'\Sigma^{-1}(y-\mu)} \tag{2}$$

in which Y describes a multivariate normally distributed variable, μ refers to a mean vector and Σ is the variance-covariance matrix as stated beforehand. Finally, $|.|$ denotes the determinant of the given matrix. The main idea of GGM is based on the conditionally independent structure. For three nodes A, B and C, we say that A and B are conditionally independent with C if the structure of A and B is separated by C. In gene networks, if two nodes, i.e. genes, are conditionally independent, there is no edge between two nodes and is represented with a zero entry in the precision matrix. On the other hand, the precision is the inverse of the covariance matrix Σ, denoted by Θ with a $(p \times p)$-dimensional matrix. Thus, the pairwise dependency between two nodes i and j can be shown by θ_{ij} as follows.

$$\Theta = \Sigma^{-1} = \theta_{ij}. \tag{3}$$

As the precision matrices consist of partial covariances, its diagonal entries are obtained from $\theta_{ii} = 1/var(Y^{(i)}|Y^1, Y^2, \ldots, Y^{i-1}, Y^i, \ldots, Y^p)$ and the off-diagonal entries are found from

$$\phi_{ij} = \frac{-\theta_{ij}}{\sqrt{\theta_{ii}\theta_{ij}}}, \tag{4}$$

where θ_{ij} shows the partial correlation between Y^i and Y^j and $var(.)$ denotes the variance term in the given random variable. In gene networks, the number of nodes p is much more than the number of observations n (i.e. $p \gg n$) that leads to the singularity problem. In other words, the estimated sample covariance matrix S is not invertible. There are many methods to infer this partial covariance matrix. One of them is to apply the lasso regression method. Let's assume that Y is a vector and all the observed networks are contained by Y. So a regression model is constructed between a response variable Y^p and the explanatory variables Y^{-p}. Hereby, the model is described as

$$Y^p = Y^{-p}\beta + \varepsilon. \tag{5}$$

In this expression, ε is the error term which has a normal distribution with zero mean. Thus, the mean vector μ and the variance-covariance matrix Σ of the model in Eq. (5) can be shown by

$$\mu = \begin{pmatrix} \mu_{-p} \\ \mu_p \end{pmatrix} \quad \text{and} \quad \Sigma = \begin{pmatrix} \Sigma_{-p,p} & \sigma_{-p,p} \\ \sigma_{-p,p} & \sigma_{p,p} \end{pmatrix}, \tag{6}$$

respectively. Here, μ_{-p} indicates the mean vector of all nodes except the pth node, $\Sigma_{-p,p}$ is the $(p-1) \times (p-1)$ -dimensional variance-covariance matrix of all nodes except the pth node, $\sigma_{-p,p}$ refers to a $(p-1)$-dimensional vector and $\sigma_{p,p}$ is the covariance value of the pth node. In Eq. (5), β stands for the regression coefficient. Interestingly, there is a relation between β and the precision matrix Θ which is formalized by

$$\beta = -\frac{\Theta_{-p,p}}{\Theta_{p,p}}. \tag{7}$$

In Eq. (7), $\Theta_{-p,p}$ describes the precision of nodes except the pth entry and $\Theta_{p,p}$ represents the $(p \times p)$-dimensional full-rank precision matrix. Therefore, the entries of Θ are used to infer the strength of the interactions between nodes.

There are some approaches to estimate the model parameters of GGM. When the number of observations is greater than the number of dimensions (i.e. $n > p$), one can estimate θ by the maximum likelihood estimation (MLE) easily. On the other hand if $n < p$, then the singularity problem can occur. In order to overcome this challenge, the L_1-penalized method and the neighborhood selection with the lasso approach are the two well-known approaches. The mathematical details of both techniques are presented as below.

2.1.1   Graphical Lasso (L_1-Penalized Method)

One of the efficient ways to estimate a sparse and symmetric matrix Θ is the graphical lasso approach (glasso) which is introduced by Friedman et al. [7]. According to the Lagrangian dual form, the problem is the maximization of the loglikelihood function with respect to the nonnegative matrix as follows.

$$\max_{\Theta} \left(log(|\Theta|) - \text{Trace}(S\Theta) \right), \tag{8}$$

where $S = XX'/n$ is an estimate of the covariance matrix. Yuan and Lin [6] show that instead of maximizing Eq. (8), the penalized loglikelihood function can be maximized via

$$\max_{\Theta} \left\{ log(|\Theta|) - \text{Trace}(S\Theta) - \lambda ||\Theta||_1 \right\} \tag{9}$$

in which Trace(.) denotes the trace matrix as used before. $||\Theta||_1$ is the L_1-norm which is the summation of the absolute values of the elements of the precision matrix. According to the Karush-Kuhn-Ticker condition to maximize Θ, Eq. (9)

must provide the following equation.

$$\Theta^{-1} - S - \lambda\Gamma(\Theta) = 0, \tag{10}$$

where $\Gamma(x)$ shows the subgradient of $|x|$. That means, if $\Theta_{ij} > 0$, $\Gamma(\Theta_{i,j})$ equals to 1. If $\Theta_{ij} < 0$, $\Gamma(\Theta_{i,j})$ sets to -1 and $\Theta_{ij} = 0$.

A sufficient condition for the solution of the graphical lasso is to block the diagonal matrix with blocks if the inequality $S_{ii'} < \lambda$ is satisfied for all $i \in C_k$, $i' \in C_{k'}$ and $k \neq k'$, where $C_1, C_2, \ldots, C_k$ represent a partition of p features. $\widehat{\Theta}$ is a block diagonal matrix with k blocks by

$$\widehat{\Theta} = \begin{bmatrix} \Theta_1 & & & \\ & \Theta_2 & & \\ & & \ddots & \\ & & & \Theta_k \end{bmatrix}.$$

Here the kth block of $\widehat{\Theta}$ satisfies Eq. (8) and $\widetilde{\Theta}$ is estimated. From the findings [8], it is seen that the blocking idea is computationally efficient in inference.

2.1.2   Neighborhood Selection with the Lasso Approach

A popular alternative way to overcome the underlying singularity of the variance-covariance matrix is to apply the neighborhood selection with the lasso approach [9]. This method is computationally attractive for sparse and high dimensional graphes.

The neighborhood selection method is a sub-problem of the covariance selection. If Φ is a set of nodes, the neighborhood of ne_p of the node $p \in \Phi$ is the smallest subset of $\Phi \setminus \{p\}$, which denotes the set of nodes except the pth node. So all variables Y_{ne_p} in the neighborhood, Y_p is conditionally independent on all remaining variables. The neighborhoods of the node p consist of the node $b \in \Phi \setminus \{p\}$ so that $(p, b) \in E$ when E denotes the set of edges.

This method can be converted as a standard regression problem and can be solved efficiently with the lasso approach [10]. Hereby, the lasso estimate of $\widehat{\theta}$ for the pth node and under the penalty constant λ is given by

$$\widehat{\Theta}^{p,\lambda} = argmin\left(||Y_p - Y\Theta||_2^2 + \lambda_p||\Theta||_1\right) \tag{11}$$

where $||\Theta||_1 = \sum\limits_{b \in \Phi(n)} |\Theta_b|$ is the L_1-norm of the coefficient vector and $||.||_2^2$ shows the L_2-norm. But the solution of Eq. (11) is not unique. Because each choice of the penalty parameter λ indicates an estimate of the neighborhood ne_p for the node $p \in \Phi(n)$.

3 Multivariate Adaptive Regression Splines and Bernstein Polynomials

3.1 Multivariate Adaptive Regression Splines (MARS)

The Multivariate Adaptive Regression Splines (MARS) is a nonparametric regression technique that makes no assumption about the functional relationship between dependent and independent variables and it has an increasing number of applications in many areas of the science over the last few years. Because it builds a flexible model for the high-dimensional non-linear data by introducing piecewise linear regressions. The classical nonparametric model has the following structure.

$$y_i = f(\beta, x_i') + \varepsilon, \tag{12}$$

where β is the unknown parameter, n represents the sample size and x stands for the independent variable. Moreover, f describes an unknown functional form and ε denotes the random error term. Hence, the MARS method affords to proximate the nonlinear functions of f by using piecewise linear basis elements, known as basis functions BFs [11]. The form of BFs can be shown as $(x - t)_+$ and $(t - x)_+$ in which x is an input variable on the positive side "+". So

$$(x - t)_+ = \begin{cases} x - t & \text{if } x > t \\ 0 & \text{otherwise} \end{cases}, (t - x)_+ = \begin{cases} t - x & \text{if } x < t \\ 0 & \text{otherwise} \end{cases}. \tag{13}$$

In Eq. (13), t is a univariate knot obtained from the dataset simply shown in Fig. 2 too.

From Fig. 2, it is seen that each function is piecewise linear with a knot at the value t. These two functions are called the reflected pairs. The aim of such applications is to construct the reflected pairs for each input variable X_j with knots at each observed value x_{ij} of that input. Therefore, the collection of BFs under $(i = 1, 2, \ldots, N; j = 1, 2, \ldots, p)$ is defined as

$$C = \left\{ (X_j - t)_+, (t - X_j)_+ | t \in \{x_{1,j}, x_{2,j}, \ldots, x_{N,j}, j \in \{1, 2, \ldots, p\} \right\}, \tag{14}$$

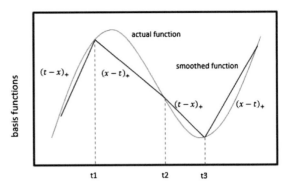

Fig. 2 Simple representation of the smoothing method for the curvature structure via BFs of MARS with t_1, t_2, t_3 knots

where N is the number of observations and p shows the dimension of the input space. If all of the input values are distinct, we can construct $2Np$ basis functions altogether.

The general method to produce spline fitting in higher dimensions is to employ basis functions that are tensor products of univariate spline functions. Hence, the multivariate spline BFs which take the following form is performed as the mth BF that are tensor products of the univariate spline functions.

$$B_m(x) = \prod_{k=1}^{K_m}[s_{km}(x_{v(km)} - t_{km})]_+ \tag{15}$$

in which K_m is the total number of truncated linear functions in the mth BF and $x_{v(km)}$ describes the input variable corresponding to the kth truncated linear function in the mth basis function. Moreover, t_{km} refers to the corresponding knot value and s_{km} takes the value ∓ 1 and indicates the (right/left) sense of the combined step function. The $v(km)$ identifies the predictor variable and t_{km} substitutes for values on the corresponding variable. Finally, $[.]_+$ indicates the partial function as described in Eq. (13). Accordingly, the construction of the modelling strategy is similar to a forward stepwise linear regression. But different from this model, the functions from the set C are allowed to be used in MARS, instead of the original inputs. Therefore, the MARS model is represented by

$$f(x) = c_0 + \sum_{m=1}^{M} c_m B_m(X) + \varepsilon, \tag{16}$$

where $B_m(x)$ is a function of C as shown in Eq. (15), $X = (X_1, X_2, \ldots, X_p)'$, c_0 presents the intercepts, c_m's are the regression coefficient for each basis function and it is estimated by minimizing the residual sum of squares in the linear regression model. Furthermore, M denotes the number of basis functions and finally, ε corresponds to the uncorrelated random error term with a zero mean and an unknown constant variance.

MARS performs both forward and backward methods. At the end of the forward stage, it generates the best fitted and the largest model similar to Eq. (16). So we need to reduce the complexity. Friedman [11] suggests to perform a modified form of the generalized cross validation criterion (GCV) as denoted in Eq. (17) in order to choose the best model. GCV produces an estimated best fitted model $\hat{f}_\lambda$ of each size of λ produced at the end of the backward process.

$$GCV(\lambda) = \frac{\sum_{i=1}^{N}(y_i - \hat{f}_\lambda(x_i))^2}{(1 - M(\lambda)/N)^2}, \tag{17}$$

where N represents the number of observations and $M(\lambda)$ is the effective number of parameters. In this equation, $M(\lambda)$ is found via $M(\lambda) = r + cK$ in which r refers to the number of linearly independent basis functions and K describes the number of selected knots during the forward stage. Additionally, c is the cost in the optimization of BF and the smoothing parameter of the model that is generally taken as $c = 3$ [23]. Finally, y and $\hat{f}_\lambda$ show the response variable and the estimated f with data y, respectively.

3.2   Bernstein Polynomials

The Bernstein polynomials are based on the theorem of the Weierstrass approximation. Assuming that f is a function over the range $C[a, b]$, f can be uniformly approximated by polynomials. Hereby, the Bernstein polynomials are one of the most well-known polynomials with a real-valued function f bounded on the interval $[0, 1]$. These polynomials are defined by

$$B_n(f; x) = \sum_{k=0}^{n} f\left(\frac{k}{n}\right)\binom{n}{k}b_{k,n}(x) \tag{18}$$

in which n is the degree of the Bernstein polynomials. $f\left(\frac{k}{n}\right)$ is equivalent to the approximation of the values for the function f at points k/n $(k = 0, \ldots, n)$ in the domain of f implying that any interval $[a, b]$ can be transformed into the interval $[0, 1]$. Finally, $b_{k,n}(x)$ is the Bernstein basis with the degree n on the parameter $x \in [0, 1]$ via

$$b_{k,n}(x) = \binom{n}{k}(1 - x)^{n-k}x^k. \tag{19}$$

In Eq. (19), $\binom{n}{k}$ is a binomial coefficient that can be obtained from the Pascals triangle where $k = 0, 1, \ldots, n$. For instance, some of the first Bernstein basis polynomials can be listed as below.

$b_{0,0} = 1,$

$b_{0,1} = 1 - x, \ b_{1,1} = x,$

$b_{0,2} = (1 - x)^2, \ b_{1,2} = 2x(1 - x), \ b_{2,2} = x^2,$

$b_{0,3} = (1 - x)^3, \ b_{1,3} = 3x(1 - x)^2, \ b_{2,3} = 3x^2(1 - x), \ b_{3,3} = x^3.$

Furthermore, the Bernstein polynomial approximates to a given function $f(x)$ is always at least as smooth as $f(x)$ is allocated uniformly in [0, 1] for a continuous $f(x)$ on the range [0, 1] as shown in Eq. (18).

$$\lim_{n \to \infty} B_n(f; x) = f(x).$$

This expression satisfies the fundamental property of the Bernstein polynomials. On the other hand, the Szasz-Mirakyan operators are the generalizations of the Bernstein polynomials [24, 25] still keeping the properties of these polynomials. These operators are defined by

$$S_n(f; x) = e^{-nx} \sum_{k=0}^{\infty} f\left(\frac{k}{n}\right) \frac{(nx)^k}{k!},$$

where $x \in [0, 1]$ and the function f is presented in an infinite interval $R^+ = [0, \infty)$.

4 Application

In the application, we show the comparison of the MARS and GGM approaches via different estimation techniques together with the Bernstein and Szasz polynomials. For the analyses of both models, we generate 500, 900 and 1000 dimensional datasets in which each gene has 20 observations. In the data generation, we arbitrarily set the off-diagonal of the precision matrix Θ to 0.9 so that the interactions between genes can be clearly observed and we generate scale-free networks [26] under the given Θ by running the huge package in the R programme. Accordingly, in the calculation based on the 1000 Monte Carlo simulations, we initially produce a network structure for the true network and generate sample datasets from this true network. Then we transform these data by the Bernstein and Szasz operators and finally, use them for modelling and inferring Θ.

In modelling via MARS, every single node is implemented as a response and the remaining nodes are taken as covariates similar to the lasso regression. Hereby, we consider only main effects and eliminate all interaction terms. Then, we take into account the significant β parameters in Eq. (5) to estimate Θ. These steps are repeated until every gene i explained by the remaining other genes as the lasso regression applies. Furthermore, the forward and backward steps are performed for constructing the optimal model and the GCV criterion is calculated to eliminate overfitted coefficients. Finally, we convert the estimated Θ to the binary form. To obtain a symmetric Θ, the AND rule is performed. Hereby, if the covariate j for the lasso model with the response i is significant as well as the covariate j for the lasso model with the response i is significant ($i, j = 1, \ldots, p$), the entries of (i, j) and (j, i) pairs in the estimated Θ can be assigned as 1 in the binary form. Otherwise both entries, i.e. (i, j) and (j, i), are set to 0. In biological speaking, it means that there is a relation between genes when the associated entry of Θ is 1 and there is no relation between genes when this entry equals to 0.

On the other side, we apply GGM and estimate its model parameters via the neighborhood selection [9] and glasso methods. In GGM with the neighborhood selection method, the inference is performed by fitting the lasso regression. Whereas in modelling via GGM with the glasso method, we implement the lasso regression under the penalized likelihood function. In the application of GGM, firstly, the true precision matrix Θ is estimated and then, the estimated Θ under the transformed data via the Bernstein operators' results are compared with the findings under the non-transformed datasets. In this comparison, as stated previously, we generate 500, 900 and 1000 dimensional scale-free networks. Lastly, we repeat this process via 1000 Monte Carlo runs.

In the evaluation of the outcomes based on the underlying dimensional systems, we calculate the F-score and the precision values for the measures of accuracy by using the following expressions.

$$\text{Precision} = \frac{\text{TP}}{\text{TP+FP}} \quad \text{and} \quad \text{F-measure} = 2\frac{\text{Precision} \times \text{Recall}}{\text{Precision} + \text{Recall}}, \quad (20)$$

where TP denotes the true positive measures correctly identified edges. FP shows the false positive value and computes misclassified edges that have zero entries in the estimate Θ. Moreover, FN presents the false negatives and measures the missclassified edges that have zero values in Θ and finally, Recall is calculated as Recall=TP/(TP+FN).

From the outcomes in Tables 1 and 2, it is observed that F-measure via MARS is not computable since the recalls are indefinite, resulting in indefinite F-measure.

Table 1 Comparison of the precision and F-measure value via MARS under 1000 Monte-Carlo runs based on systems with 500, 900 and 1000 dimensional networks

Accuracy measure	Number of nodes	Only MARS	MARS with bernstein	MARS with szasz
Precision	500	0.0017	0.0012	0.0013
	900	0.0000	0.0005	0.0007
	1000	0.0000	0.0004	0.0005
F-measure	500	0.0029	0.0025	0.0026
	900	Not computable	0.0013	0.0013
	1000	Not computable	0.0011	0.0012

Table 2 Comparison of the precision and F-measure values via GGM estimated by the neighborhood selection (NS) and glasso methods under 1000 Monte-Carlo runs based on 500, 900 and 1000 dimensional networks

Method	Accuracy measure	Number of nodes	Only MARS	MARS with bernstein	MARS with szasz
NS	Precision	500	Not computable	0.4768	0.4731
		900	Not computable	0.4679	0.4702
		1000	Not computable	0.4729	0.4700
NS	F-measure	500	0.0000	0.0179	0.0173
		900	0.0000	0.0092	0.0091
		1000	0.0000	0.0227	0.0289
glasso	Precision	500	Not computable	0.5002	0.4995
		900	Not computable	0.4968	0.4972
		1000	Not computable	0.4951	0.5342
glasso	F-measure	500	0.0000	0.1201	0.1531
		900	0.0000	0.1126	0.0830
		1000	0.0000	0.0775	0.1062

Whereas GGM with the neighborhood selection and the glasso methods can calculate F-value successfully. Moreover, it is seen that GGM overperforms MARS under the transformed datasets. If we compare the findings of both Bernstein operators, it is seen that the Szasz polynomials are more accurate for all cases. Furthermore, F-measure and precision values decrease when the dimension increases under all conditions. Additionally, we find that the accuracy of the estimates under MARS is higher when the data are not transformed via the Bernstein operators under relatively low dimensions. But when the dimension of the system raises, the transformed data have higher F-measure for both MARS and GGM models. On the contrary, when the dimension increases, the precision value decreases too.

5 Conclusion

In this study, two major lasso modelling approaches suggested for the biological networks, i.e. MARS and GGM models, are compared with the Bernstein operators under the Monte Carlo simulation. In GGM, we have performed the estimation by two main methods which are GGM with the neighborhood selection and GGM with the glasso techniques. In the comparison under the multivariate normally distributed data, the network structure and its precision matrix have evaluated based on the precision and F-measures in the Monte Carlo runs. From the analyses, we have detected that MARS gives more accurate results than GGM with/without Bernstein operators under realistically complex systems. On the other hand, the transformed data via the Bernstein operators, in particular, via the Szasz polynomials, have higher accuracy in the estimates. Therefore, we suggest that the Bernstein operators can be used to improve the accuracy under different types of lasso modelling.

As the extension of this study, we consider to perform other operator systems as the alternative of the Bernstein operators, which can specifically transform the data by taking into account their distributional features. Under such conditions, we believe that the operators based on binomial/multinomial, poisson or normal distributions can have better performance than their alternatives as they are more suitable for the description of the biochemical systems depending on the chemical master equations [27].

References

1. Sonka, M., Hlavac, V., Boyle, R.: Image Processing, Analysis and Machine Vision, 2nd edn. U.K. International Thomson, London (1999)
2. Dobra, A., Eicher, T., Lenkoski, A.: Modeling uncertainty in macroeconomic growth determinants using Gaussian graphical models. Stat. Method. **7**, 292–306 (2010)
3. Werhli, A., Grzegorczyk, M., Husmeier, D.: Comparative evaluation of reverse engineering gene regulatory networks with relevance networks, graphical Gaussian models and Bayesian networks. Bioinformatics **22**(20), 2523–2523 (2006)
4. Liu, Y., Kosut, O., Wilsky, A.: Sampling from gaussian graphical models using subgraph perturbations. In: Proceedings of the 2013 IEEE International Symposium on Information Theory (2013)
5. Li, H., Gui, J.: Gradient directed regularization for sparse Gaussian concentration graphs with applications to inference of genetic networks. Biostatistics **7**, 302–317 (2006)
6. Yuan, M., Lin, Y.: Model selection and estimation in the Gaussian graphical model. Biometrica **94**(10), 19–35 (2007)
7. Friedman, J., Hastie, R., Tibshirani, R.: S parse inverse covariance estimation with the graphical lasso. Biostatistics **9**, 432–441 (2007)
8. Witten, D.M., Friedman, J.H., Simon, N.: New insights and faster computations for the graphical lasso. J. Comput. Graph. Stat. **20**(4), 892–900 (2011)
9. Meinshaussen, N., Buhlmann, P.: High dimensional graphs and variable selection with the Lasso. Ann. Stat. **34**(3), 1436–1462 (2006)
10. Tibshirani, R.: Regression shrinkage and selection via the lasso. J. R. Stat. Soc. Ser. B **58**, 267–288 (1996)
11. Friedman, J.: Multivariate Adaptive regression splines. Ann. Stat. **19**(1), 1–67 (1991)
12. Deichmann, J., Esghi, A., Haughton, D., Sayek, S., Teebagy, N.: Application of multiple adaptive regression splines (MARS) in direct response modelling. J. Int. Mark. **16**, 15–27 (2002)
13. Andres, J.D., Sanchez, F., Lorca, P., Juez, F.A.: Hybrid device of self organizing maps and MARS for the forecasting of firms bankruptcy. J. Account. Manag. Inform. Syst. **10**(3), 351 (2011)
14. Tayyebia, B.A., Pijanowskib, B.C.: Modeling multiple land use changes using ANN, CART and MARS: Comparing tradeoffs in goodness of fit and explanatory power of data mining tools. Int. J. Appl. Earth Obs. Geoinf. **28** (2014)
15. Lewis, P., Stevens, J.: Nonlinear modelling of time series using MARS. J. Am. Stat. Assoc. **87**, 864–877 (1991)
16. Attoh-Okine, N.O., Cooger, K., Mensah, S.: Multivariate Adaptive Regression (MARS) and Hinged Hyperplanes (HHP) for Doweled Pavement Performance Modeling Construction and Building Materials. J. Constr. Build. Mater. **23**(9), 3020 (2009)
17. Babu, G.J., Canty, A.J., Chaubey, P.Y.: Application of Bernstein polynomials for smooth estimation of a distribution and density function. J. Stat. Plann. Infer. **105**, 377–392 (2001)

18. Phillips, G.M.: Bernstein polynomials based on the q-integers, the heritage of P. L. Chebyshev: a Festschrift in honor of the 70th birthday of T. J. Rivlin Ann. Numer. Math. **4**(1–4), 511–518 (1997)
19. Liao, C.W., Huang, J.S.: Stroke segmentation by bernstein-bezier curve fitting. Pattern Recogn. **23**(5), 475–484 (2001)
20. Belluci, M.: On the explicit representation of orthonormal Bernstein polynomials. arXiv:1404.2293v2 (2014)
21. Dempster, A.P.: Covariance selection. Biometrics **28**(1), 157–175 (1972)
22. Whittaker, J.: Graphical Models in Applied Multivariate Statistics. John Wiley and Sons, New York (1990)
23. Craven, P., Wahba, G.: Smoothing noisy data with spline functions. Numer. Math. **31**, 377–403 (1979)
24. Szasz, O.: Generalizations of S Bernstein polynomials to the infinite interval. J. Res. Nat. Bur. Stan. **45**, 239–245 (1950)
25. Mirakyan, G.M.: Approximation of continuous functions with the aid of polynomials of the form $e^{-nx} \sum_{k=0}^{M} c_{mn} C_{k,n} x^k$. Akad. Nauk SSSR **31**, 201–205 (1941)
26. Barabasi, A.L., Oltvai, Z.N.: Network biology: Understanding the cell's functional organization. Nat. Rev. Genet. **5**, 101–113 (2004)
27. Kampen, N.: Stochastic Processes in Physics and Chemistry, North Holland (1981)

Tracking the Interface of the Diffusion-Absorption Equation: Theoretical Analysis

Waleed S. Khedr

Abstract This work is devoted to the theoretical study of the Cauchy problem for the degenerate parabolic equation of the diffusion-absorption type $u_t = \Delta u^m - au^q$ with the exponents $m > 1$, $q > 0$, $m + q \geq 2$ and constant $a > 0$. We propose an algorithm for tracking the interface in the case of arbitrary $m > 1$ and $q > 0$. Based on the idea of Shmarev (Nonlinear Anal 53:791–828, 2003; Progr Nonlinear Diff Eqn Appl Birkhäuser, Basel 61:257–273, 2005), we transform the moving support of the solution into a time-independent domain by means of introduction of a local system of Lagrangian coordinates. In the new coordinate system the problem converts into a system of nonlinear differential equations, which describes the motion of a continuous medium. This system is solved by means of the modified Newton method, which allows one to reduce the nonlinear problem to a sequence of linear degenerate problems. We formulate the problem in the framework of Sobolev spaces and prove the convergence of the sequence of approximate solutions to the solution of the original problem.

1 Introduction

In this article, our main concern is the study of the free boundary problem

$$\begin{cases} U_t = \Delta U^m - aU^q & \text{in } (0, T] \times \mathbb{R}^n \\ U(X, 0) = U_0(X) \geq 0 & \text{in } \mathbb{R}^n \end{cases} \tag{1}$$

where $a, q > 0, m > 1$, are constants and $m + q \geq 2$. The solution $U(x, t)$ represents the density of some substance per unit volume and $T > 0$ is a fixed interval. It is known that the solutions of problem (1) possess the property of finite speed of propagation, which means that if the support of the initial function u_0 is compact,

W.S. Khedr (✉)
Erasmus-Mundus MEDASTAR Project, Universidad de Oviedo,
Oviedo, Asturias, Spain
e-mail: waleedshawki@yahoo.com

© Springer International Publishing Switzerland 2016
G.A. Anastassiou and O. Duman (eds.), *Intelligent Mathematics II:*
Applied Mathematics and Approximation Theory, Advances in Intelligent Systems
and Computing 441, DOI 10.1007/978-3-319-30322-2_25

so is the support of the solution at every instant t. The a priori unknown boundary of the support is the free boundary that has to be defined together with the solution $U(x, t)$. The question of regularity of interfaces has been studied by many authors. It is shown in [1] that in the case $n = 1$ the interfaces are real analytic plane curves. For the multi-dimensional case, the moving interface is shown to be a C^∞ hypersurface in [5, 15].

A special approach for the study of interfaces in the one-dimensional case was proposed in [11, 12]. In these papers the authors demonstrated the effect of different choices for the parameters m, q and a. In the case when $q \in (0, 1)$, they had shown that the right interface, which in the one dimensional case is a plane curve defined by the equality

$$\chi(t) = \sup\{x \in \mathbb{R} : u(x, t) > 0\},$$

which is governed by the first-order differential equation

$$\chi'(t) = -\frac{m}{m-1}(U^{m-1})_X(\chi(t), t) + \frac{a(1-q)}{(U^{1-q})_X(\chi(t), t)}, \tag{2}$$

which is a generalized version of Darcy law that relates the velocity of the particles with the system's pressure. The authors have shown that both terms on the right-hand side of equation (2) are well-defined as long as the solution is not identically zero but for $m + q > 2$ only one of them is different from zero. In the critical case $m + q = 2$ the motion of the interface is defined by an interaction of these two terms. This interaction can be observed in the behaviour of the solution as per the deduced explicit formula in that critical case, see [14].

In the multi-dimensional case of the diffusion-absorption equation, Shmarev used the transformation to the Lagrangian coordinates and proved that the solution exists in a special weighted Hölder space and that the solution itself and the corresponding interface are real analytic in time. He also provided explicit formulas that represent the interface $\gamma(t)$ as a bijection of the initial interface $\gamma(0)$ and demonstrated the improvement of the interface regularity with respect to the spatial variables compared to the initial instant [22]. However, these results were constrained to the case of $n = 1, 2, 3$. In [24], the author managed to eliminate that constraint by means of Helmholtz's orthogonal decomposition.

2 Objective

The above equation offers three main difficulties. First, it degenerates at some set of points (a hyper-surface) separating the zone where the solution is positive and the zone where the solution is identically zero. This degeneracy hides the information needed to accurately imitate the rate of vanishing of the solution, which causes any numerical simulation to fluctuate near the boundary. Multiple numerical schemes were suggested to simulate the solution of the above equation. The most remarkable results were presented at the 80s by Mimura et al. [19, 20, 25] and Dibenedetto and Hoff [8]. The proposed schemes were based on the utilization of Darcy law to approximate the location of the interface on a finite difference grid. All these schemes were designed for the one dimensional case, moreover, even in the cases when the convergence was proven, accuracy and stability remained unsolved issues.

The second obstacle is that the support of the solution is moving, which increases the difficulty of building a mesh which nodes are located exactly where the solution vanishes. To overcome such a problem, specific methods are employed, which allow one to control the vanishing rate of the solution in order to guarantee that the solution does not go negative. These methods succeed to preserve the physical nature of the solution, however, on the account of other mathematical or even physical measures. Solving the equation using Galerkin's methods produced unstable solutions and negativity started to appear due to the fact that we can't force the interface to where the nodes are located. Cockburn, and many others, had utilized the discontinuous Galerkin method with slope limiters to overcome this problem [3, 4, 26]. However, the difficulty to choose the proper fluxes, the complication of the implementation and the inaccurate gradient due to the application of the slope limiters were negative sides of such approach. The third obstacle is the presence of the low order term which implies the loss of the mass conservation property and consequently the loss of a very important mathematical tool.

Our goal is to handle these obstacles by constructing a transformation that allow for the exact location of the interface. To overcome this difficulty we reduce the initial domain into a ring near the boundary and pose a condition on the interior surface of that ring to guarantee that it moves in a way that compensates the loss of the mass. Hence, the mass within the initial ring is conserved for all times. Then we transfer the problem from Eulerian coordinates into Lagrangian coordinates so that the moving domain is converted into a fixed one. The most interesting advantage of the transformation to Lagrangian coordinates is that we can express the position of all particles in terms of the initial state's attributes. Hence, we know the position of the particles composing the interface at each time instant. Such information can be used to build a mesh with well known surfaces and then solve the original degenerate free-boundary problem as a sequence of elliptic problems endowed with the Dirichlet boundary condition. Since we know that the solution vanishes just by construction at each of those surfaces, we need no extra control on the rate of vanishing and we guarantee that the solution preserves its physical properties in the whole domain.

This transformation yields a system of nonlinear degenerate partial differential equations that describes different attributes of the fluid motion, and one of these equations describes the position of the fluid's particles. The existence of a solution for the degenerate system was proven in weighted Hölder spaces [22, 24]. However, these spaces are not suitable for numerical purposes. We propose a simple regularization scheme for the resultant system of equations, show that this system admits a weak solution, and prove the convergence of the regularized weak solution to the exact solution in the weak sense.

3 Background of the Problem

In this section we summarise what had been established in [22, 24]. According to
the above discussion, (1) was rewritten in the following form

$$\begin{cases} U_t = \Delta (U^m) - aU^q & \text{in } \Omega_t \times (0, T], \\ U = 0 & \text{on } \Gamma_T, \\ U(X, 0) = U_0(X) \geq 0 & \text{in } \overline{\Omega}_0, \end{cases} \tag{3}$$

where

$$\Omega_t = \Omega(t) = \{X \in \mathbb{R}^n, t \in [0, T] : U(X, t) > 0\} \equiv \text{Supp } U(X, t),$$

$$\Gamma_T = \{X \in \mathbb{R}^n, t \in [0, T] : U(X, t) = 0\} = \bigcup_{t \in [0, T]} \gamma_t,$$

and $\gamma_t = \gamma(t) \equiv \partial \Omega_t \; \forall t \in [0, T]$. The weak solution of problem (3) was defined in
the following sense.

Definition 1 A pair $(U(X, t), \Gamma_T)$ is said to be a weak solution of problem (3) if:

I. U is bounded, continuous, and nonnegative in $\overline{\Omega}_t \times (0, T]$ and $\nabla U^m \in L^2(\overline{\Omega}_t \times (0, T])$,
II. Γ_T is a surface of class C^1,
III. for every test function $\psi \in C^1(\overline{\Omega}_t \times (0, T])$, vanishing at $t = T$ and for all $(X, t) \in \Gamma_T$, the following identity holds:

$$\int_0^T \int_{\Omega_t} (U\psi_t - \nabla U^m . \nabla \psi - a\psi U^q) \, dX \, dt + \int_{\Omega_0} U_0 \psi(X, 0) \, dX = 0. \tag{4}$$

Thus a new variable was added to the problem that is Γ_T. The introduction of the
Lagrangian coordinates generates a time-independent domain with fixed boundaries
such that Γ_T can be expressed as a function in the new coordinates, which then can
be obtained by solving a system of differential equations on the plane of Lagrangian
coordinates. However, due to the presence of the low order term, the principle of the
mass conservation is no longer valid. The transformation to the Lagrangian coordinates entails a system with some fixed attribute. The author compensated the loss
of the mass by reducing the domain of the problem to some annular domain near
the interface. He introduced a simple-connected surface $S_T = \bigcup_{t \in [0, T]} s_t$ such that
$S_T \subset \Gamma_T$ and $S_T \cap \Gamma_T = \emptyset$. He defined $\omega_t \subset \Omega_t$ as the annular domain bounded
by the surfaces γ_t and s_t such that the problem's domain is $\mathcal{E} = \bigcup_{t \in [0, T]} \omega_t$ and
$\partial \mathcal{E} = S_T \cup \Gamma_T$. Denote by $\hat{U}(X, t)$ the solution inside the annular domain. The surface S_T is an artificial surface and it is chosen such that the following identity is

fulfilled:

$$\forall t \in (0, T] \quad \int_{\omega_t} \hat{U}(X, t) \, dX = \int_{\omega_0} U_0(X) \, dX = \text{const.} \tag{5}$$

Therefore a new auxiliary problem has been introduced as

$$\begin{cases} \hat{U}_t = \Delta \left(\hat{U}^m \right) - a \hat{U}^q & \text{in } \mathscr{E}, \\ \hat{U}|_{S_T} = f_{(s)}, \quad \hat{U}|_{\Gamma_T} = 0, \\ \hat{U}(X, 0) = U_0(X) \geq 0 & \text{in } \overline{\omega}_0, \end{cases} \tag{6}$$

where $f_{(s)}$ represents the dummy values of $\hat{U}$ on the inner surface which depend mainly on our choice of the boundary conditions on that surface. As per (5), our choice of the inner surface S_T guarantees that the mass is conserved in the new domain $\mathscr{E}$. Although this is a different problem, the author in [22, 24] proved that (6) and (3) coincide up to the motion of the boundary.

The general equation of the mass conservation is

$$\rho_t + div(\rho \, \mathscr{V}) = 0, \tag{7}$$

where $\mathscr{V}$ is the velocity of the fluid particles. Two terms define the velocity. The first is the gradient of the pressure denoted by $\hat{P}$, which is induced by the diffusive term, and the second is the gradient of some artificial pressure $\hat{\Pi}$ induced by the low order term, such that

$$\mathscr{V} = -\nabla \hat{P} + \nabla \hat{\Pi}, \tag{8}$$

where,

$$\hat{P} = \frac{m}{m-1} \hat{U}^{m-1}, \quad P_0 = \frac{m}{m-1} U_0^{m-1}, \tag{9}$$

and $\hat{\Pi}$ is the solution of the following degenerate elliptic problem:

$$\begin{cases} div(\hat{U} \nabla \hat{\Pi}) = a \hat{U}^q & \text{in } \mathscr{E}, \\ \hat{\Pi} = 0 & \text{on } \partial \mathscr{E}, \end{cases} \tag{10}$$

given that π_0 be the solution of the degenerate elliptic problem at the initial instant such that

$$\begin{cases} div(U_0 \nabla \pi_0) = a U_0^q & \text{in } \mathscr{E}, \\ \pi_0 = 0 & \text{on } \partial \mathscr{E}. \end{cases} \tag{11}$$

The mechanical problem of the fluid flow until now is described in terms of functions that depend on the time and on a Cartesian coordinate system not connected with the flow. As in [6, 7, 13, 22, 24] an alternative description was given by transforming the system into geometrical Lagrangian coordinates so that all the functions characterizing the motion became dependent on the initial positions of the particles and the time as a parameter. The cylinder Q with fixed lateral boundaries was defined as

$$Q = \omega_0 \times [0, T], \ \partial Q = \partial \omega_0 \times [0, T].$$

Let us denote the initial positions by η. In this way, the positions of the particles at any instant t are

$$X = x(\eta, t), \quad \eta \in \overline{\omega_0},$$

and consequently $\hat{U}[x(\eta, t), t] = \hat{u}(\eta, t)$ is the density corresponding to that particle, $\hat{P}[x(\eta, t), t] = \hat{p}(\eta, t)$ and $\hat{\Pi}[x(\eta, t), t] = \hat{\pi}(\eta, t)$ are the functions representing the pressure, and $\mathscr{V}[x(\eta, t), t] = v(\eta, t)$ is the velocity of that particle. We may then write the trajectory equation as

$$\begin{cases} x_t(\eta, t) = v(\eta, t) = -\nabla_x \hat{p}(\eta, t) + \nabla_x \hat{\pi}(\eta, t), \\ x(\eta, 0) = \eta, \quad \eta \in \overline{\omega_0}. \end{cases} \tag{12}$$

Let $J = [J_{ij}]$ be the Jacobi matrix of the mapping $\eta \to x(\eta, .)$ such that

$$\begin{cases} J_{ij} = \frac{\partial x_i}{\partial \eta_j}, \quad i, j = 1, 2, \\ |J| \equiv \det\left(\frac{\partial x}{\partial \eta}\right), \\ \frac{d|J|}{dt} = |J| div_x \mathscr{V}[x, t], \quad \text{(Cauchy identity)}. \end{cases}$$

The Lagrangian version of the mass conservation law was deduced to be:

$$\frac{d}{dt}\left(\hat{U}|J|\right) = 0,$$

or equivalently

$$\hat{U}[x, t]|J| = \hat{u}(\eta, t)|J| = \hat{u}(\eta, 0) = U_0 \ \text{ in } \overline{Q}. \tag{13}$$

The analysis in [24] leads to the following system on the plane of Lagrangian coordinates:

$$\begin{cases} div_\eta(J\nabla_\eta \hat{v}_t) + \Delta_\eta(\hat{p} - \hat{\pi}) = 0, \quad \text{in } Q, \\ \hat{p}|J|^{m-1} = P_0 \ \text{ in } \overline{Q}, \\ div_\eta\left(U_0(J^{-1})^2\nabla_\eta \hat{\pi}\right) = aU_0\hat{u}^{q-1} \ \text{ in } Q, \\ \hat{v}(\eta, 0) = 0, \ \hat{\pi}(\eta, 0) = \pi_0(\eta), \text{ and} \\ \hat{p}(\eta, 0) = P_0(\eta) \ \text{ in } \overline{\omega_0}, \\ \hat{p} = 0 \ \text{ on } \gamma_0 \times [0, T] \text{ and } \hat{v} = \hat{\pi} = 0 \ \text{ on } \partial Q. \end{cases} \tag{14}$$

Solving the above system to obtain the triad solution $(\hat{v}, \hat{p}, \hat{\pi})$ enables us to calculate the motion of the interface in a direction normal to the initial interface by the formula

$$x = \eta + \nabla_\eta \hat{v} \quad \text{in } \overline{Q}, \tag{15}$$

which is our target. The ability to solve the nonlinear system (14) and recovering the position of the particles x provides us with the necessary information to locate the interface Γ_T of (3). We need not worry about the solution or the motion of the particles in the rest of the domain $\mathcal{E}$. We only use the information we have regarding the position of the interface to solve (3) (which is equivalent to (1)) as a boundary-value problem with zero boundary condition on Γ_T, which is already defined.

The nonlinear system (14) is to be solved by means of an abstract version of the modified Newton method. Let us consider the three equations in (14) as the arguments of the functional equation $\mathcal{F}(\hat{z}) \equiv \{\mathcal{F}_1(\hat{z}), \mathcal{F}_2(\hat{z}), \mathcal{F}_3(\hat{z})\}$, where $\hat{z} = \{\hat{v}, \hat{p}, \hat{\pi}\}$. In the modified Newton method the differential of $\mathcal{F}$ is calculated once at the initial state. The solution is obtained then as the limit of the sequence

$$\hat{z}_{n+1} = \hat{z}_n - [\mathcal{G}(\hat{z}_0)]^{-1}\langle \mathcal{F}(\hat{z}_n)\rangle, \tag{16}$$

where in this case the operator $\mathcal{G}(\hat{z}_0)\langle z\rangle$ is the Frechét derivative of $\mathcal{F}$ at the initial instant $\hat{z}_0 = \{0, P_0, \pi_0\}$ and $z = \{v, p, \pi\}$ is the correction at each iteration. If the Frechét derivative is well defined at the initial instant and is Lipschitz-continuous, then it coincides with the Gateaux differential. More details regarding the Newton method and the modified Newton method can be found in [16].

Theorem 2 *Let $\mathcal{X}$ and $\mathcal{Y}$ be Banach spaces and assume that:*

I. *the operator $\mathcal{F} : \mathcal{X} \mapsto \mathcal{Y}$ admits a strong Frechét derivative $\mathcal{G}$ in a ball $\mathcal{B}_r(0) \subset \mathcal{X}$ of radius $r > 0$,*
II. *the differential $\mathcal{G}(\hat{z}) : \mathcal{X} \mapsto \mathcal{Y}$ is Lipschitz continuous in $\mathcal{B}_r(0)$*

$$\|\mathcal{G}(\hat{z}_2) - \mathcal{G}(\hat{z}_1)\| \le L\|\hat{z}_2 - \hat{z}_1\|, \quad L = const. \tag{17}$$

III. *there exists an inverse operator $[\mathcal{G}(\hat{z}_0)]^{-1}$ such that*

$$\left\|[\mathcal{G}(\hat{z}_0)]^{-1}\right\| = M \text{ and } \left\|[\mathcal{G}(\hat{z}_0)]^{-1}\langle \mathcal{F}(\hat{z}_0)\rangle\right\| = K. \tag{18}$$

Then, if $\lambda = MKL < \frac{1}{4}$, the equation $\mathcal{F}(\hat{z}) = 0$ has a unique solution $\hat{z}^*$ obtained as the limit of the sequence in (16), $\hat{z}^* \in \mathcal{B}_{Kh_0}(0)$, where h_0 is the least root of the equation $\lambda h^2 - h + 1 = 0$.

In our case, $\mathcal{X}$ and $\mathcal{Y}$ are Sobolev spaces such that

$$\mathcal{X} = \{\hat{v} \in L^2(0, T; H^4(\omega_0)), \hat{\pi} \in L^2(0, T; H^2(\omega_0)), \hat{p} \in L^2(0, T; H^2(\omega_0))\},$$

and

$$\mathcal{Y} = \{f \in L^\infty(0, T; H^2(\omega_0)), \Phi \in L^2(0, T; H^2(\omega_0)), H \in L^2(0, T; H^2(\omega_0))\}.$$

The ball $\mathscr{B}_r(0)$ is defined as $\mathscr{B}_r(0) := \{(\hat{v}, \hat{\pi}, \hat{p}) : \|\hat{v}\| + \|\hat{\pi} - \pi_0\| + \|\hat{p} - P_0\| < r\}$. In [22, 24] the modified Newton method was used in the scale of weighted Hölder spaces. For more details regarding the weighted Hölder spaces, see [15, 22–24]. If we follow the linearization techniques introduced in [22, 24] we get the following linearized system:

$$\begin{cases} v_t - (m-1)P_0\Delta v = f + \pi - \Phi & \text{in } Q, \\ div\left(U_0\nabla\pi - 2U_0 D^2(v).\nabla\pi_0\right) - a(1-q)U_0^q\Delta v = H & \text{in } Q, \\ v = \pi = 0 & \text{on } \partial Q. \end{cases} \quad (19)$$

We get the corrections z at each iteration by solving the system of linear equations in (19) given that all the assumptions posed in [21–24] are fulfilled. Yet, the weighted Hölder spaces are not suitable for finite element implementation. Hence, in the rest of this article we try to obtain the same results in some spaces more fit to the numerical implementation.

4 Regularization

We recall the definition of uniform elliptic and parabolic equations from [10, 17, 18]. The stated condition in these definitions is not only important to define a uniform equation, it is also a necessary condition for implementing a stable finite element scheme. In the linearized system (19), this condition is not satisfied neither for the elliptic equation nor for the parabolic one due to the degeneracy. Next, we explain how to satisfy the requirements for a stable finite element implementation.

4.1 Reduction to a Sequence of Linear Problems

Let us introduce the new functions w as a solution of the regularized parabolic equation and s as a solution for the regularized elliptic one, and introduce two small parameters $\delta > 0$ and $\mu > 0$ such that $P_{0\delta} = P_0 + \delta$ and $P_{0\mu} = P_0 + \mu$. Denote $\alpha = m + q - 2 \geq 0$ and $C_m = (m-2)/(m-1)$. We introduce also the following general assumptions

(A1) $\partial\omega_0$ is of class C^2.
(A2) $U_0 \in C^2(\overline{\omega}_0)$.
(A3) $P_0 \in C^1(\overline{\omega}_0)$, $\nabla P_0 \in C^0(\overline{\omega}_0)$, and $\Delta P_0 \in L^\infty(\overline{\omega}_0)$.
(A4) The annular domain is chosen initially such that $-C_0 \leq \Delta P_0 \leq -M_0$ a.e. in $\overline{\omega}_0$.
(A5) There exists some constant $0 < v < 1$ such that $v < |J| < v^{-1}$.

We introduce the regularized parabolic equation

$$\begin{cases} w_t - (m-1)P_{0\mu}\Delta w = f - \Phi + s & \text{in } Q, \\ w(\eta, 0) = 0, \\ w = 0 & \text{on } \partial Q. \end{cases} \tag{20}$$

We define p as

$$p = \Phi - (m-1)P_0\Delta w \quad \text{in } Q. \tag{21}$$

Finally, we introduce the regularized elliptic equation

$$\begin{cases} div\,(P_{0\delta}\nabla s) - C_m\nabla P_0\nabla s = div\,(2P_0D^2(w)\nabla s_0) - 2C_m\nabla P_0D^2(w)\nabla s_0 \\ \quad + a(1-q)U_0^\alpha\Delta w + U_0^{m-2}H \quad \text{in } Q, \\ s = 0 \quad \text{on } \partial Q, \end{cases} \tag{22}$$

where s_0 is the solution of

$$\begin{cases} div\,(P_{0\delta}\nabla s_0) - C_m\nabla P_0\nabla s_0 = aU_0^\alpha & \text{in } \omega_0, \\ s_0 = 0 & \text{on } \partial\omega_0. \end{cases} \tag{23}$$

Therefore, the solution of the nonlinear system is obtained as the limit of the sequence

$$\hat{w}_{n+1} = \hat{w}_n - w \quad \text{and} \quad \hat{s}_{n+1} = \hat{s}_n - s, \quad \text{for } n = 0, 1, 2, \ldots$$

Consequently, the position of the interface is obtained by

$$y = \eta + \nabla\hat{w}. \tag{24}$$

For every $\delta, \mu > 0$, the existence of a unique weak solution of the elliptic and the parabolic equations follows directly from the standard theories, review [10, 17, 18]. However, we need to investigate the effect of these small parameters on the solution and also we need to prove the convergence of this approach to the degenerate case when $\delta, \mu \to 0$. In order to perform this we need to pose an extra condition which is $\nabla P_0 \neq 0$ on ∂Q. This condition was necessary to prove the existence of a solution for the degenerate linear system in weighted Hölder spaces, see [22, 24].

Notation: Sometimes we will use the notation $L^p(Q)$ to shorten $L^p(0, T; L^p(\omega_0))$.

4.2 Linear Elliptic Equations

The solution of the elliptic equation (23) is understood in the following sense.

Definition 3 The function $s_0(\eta)$ is said to be a weak solution of the regularized problem (23) if it satisfies the following conditions:

I. $s_0 \in H^2(\omega_0) \cap H_0^1(\omega_0)$,
II. for any test function $\psi(\eta) \in H_0^1(\omega_0)$ the following identity holds:

$$\int_{\omega_0} \left(P_{0\delta} \nabla s_0 \nabla \psi + C_m \nabla P_0 \nabla s_0 \psi + U_0^\alpha \psi \right) d\eta = 0. \qquad (25)$$

By assumption **A3**, P_0 and ∇P_0 are continuous and bounded functions in $\overline{\omega}_0$. Moreover, $P_{0\delta} \geq \delta > 0$ and $U_0^\alpha \in L^2(\omega_0)$ since it is a continuous function as per assumption **A2**. Hence, problem (23) has a weak solution $s_0 \in H^2(\omega_0)$. For more details regarding solutions of linear uniformly elliptic equations, review [10, 17].

Lemma 4 *If assumptions (A2) and (A4) are fulfilled, then*

$$\|s_0\|_{L^2(\omega_0)}^2 + \delta \|\nabla s_0\|_{L^2(\omega_0)}^2 \leq C, \qquad (26)$$

where C is a constant that does not depend on δ.

Proof Multiply equation (23) by s_0 and integrate by parts over ω_0, apply Green's theorem, use assumption **A4** and Cauchy's inequality, and use the fact that $P_{0\delta} \geq \delta$ to get the required result.

Lemma 5 *Suppose that assumptions (A2) and (A3) are fulfilled. Then*

$$\|\Delta s_0\|_{L^2(\omega_0)} \leq C\delta^{-\frac{3}{2}}, \qquad (27)$$

where C is a constant that does not depend on δ.

Proof Let us rewrite the elliptic equation in the form

$$P_{0\delta} \Delta s_0 = aU_0^\alpha - \frac{1}{m-1} \nabla P_0 \nabla s_0.$$

Square and integrate both sides, use assumption **A3** and the fact that $P_{0\delta} \geq \delta$ then utilize Hölder's inequality and estimate (26) to obtain the above result.

Theorem 6 *Suppose that* s_0 *is the solution of the regularized problem (23) and* π_0 *is the solution of the degenerate problem (11). If assumptions (A1–A4) are fulfilled, then for some* $0 < \beta < \infty$, *we have the estimates*

$$\|\pi_0 - s_0\|_{L^\beta(\omega_0)}^\beta \leq C\delta^{\frac{1}{4}}, \qquad (28)$$

and

$$\|\nabla \pi_0 - \nabla s_0\|^2_{L^2(\omega_0)} \le C\delta^{\frac{1}{4}}, \tag{29}$$

where C is a constant that does not depend on δ.

Proof We use the technique of [2, 9]. First we need to highlight that the transformation of the initial elliptic equation (11) into the plane of Lagrangian coordinates will generate the same equation, since initially $J = I$, where I is the identity matrix. Let us multiply each of the Eqs. (11) and (23) by an arbitrary test function ψ that satisfies the conditions of Definition 3, integrate by parts, subtract and apply Green's theorem to get

$$\int_{\omega_0} (eP_0 \Delta \psi + (1 + C_m)e\nabla P_0 \nabla \psi + C_m e\Delta P_0 \psi)d\eta = \int_{\omega_0} \delta s_0 \Delta \psi \, d\eta, \tag{30}$$

where $e = \pi_0 - s_0$.

Let us introduce the following linear uniform elliptic equation

$$\begin{cases} (P_0 + \epsilon)\Delta y + (1 + C_m)\nabla P_0 \nabla y + C_m \Delta P_0 y = h & \text{in } \omega_0, \\ y = 0 & \text{on } \partial\omega_0, \end{cases} \tag{31}$$

where $\epsilon > 0$ is a constant, h is an arbitrary function, and both will be defined later. If $h \in L^2(\omega_0)$ and assumptions (A3) and (A4) are fulfilled, then there exists a unique solution $y \in H^2(\omega_0)$, review [10, 17]. We multiply (31) by y and integrate by parts, we use Green's theorem, assumption **A4**, Cauchy's inequality, and the assumption that $h \in L^2(\omega_0)$ to conclude

$$\epsilon \|\nabla y\|^2_{L^2(\omega_0)} + C\|y\|^2_{L^2(\omega_0)} \le C. \tag{32}$$

Now, taking for ψ the solution of (31) and plugging it into (30), we obtain

$$\int_{\omega_0} e h \, d\eta \le \int_{\omega_0} \delta |\nabla s_0 \cdot \nabla \psi| \, d\eta + \int_{\omega_0} \epsilon |\nabla e \cdot \nabla \psi| \, d\eta.$$

Applying Hölder's inequality to both terms on the right-hand side, using estimates (26) and (32), and choosing $\epsilon = \delta^{\frac{1}{2}}$ yields

$$\int_{\omega_0} e h \, d\eta \le C_1 \delta^{\frac{1}{4}} + C_2 \delta^{\frac{1}{4}} \|\nabla e\|_{L^2(\omega_0)}. \tag{33}$$

It remains to make a proper choice of h which guarantees that $h \in L^2(\omega_0)$. The degenerate problem (11) has a solution in the weighted Hölder spaces and this solution and its first two derivatives are bounded point-wise [22]. Moreover, the regularized problem (23) has a weak solution that belongs to $H^2(\omega_0)$, review [10, 17]. Therefore, we have $e, \Delta e \in L^2(\omega_0)$. Choose $h = \Delta e$, substitute it in (33) and consider the estimates of Lemmas 4 and 5 to obtain $\|\nabla e\|_{L^2(\omega_0)} \leq C\delta^{\frac{1}{8}}$. This is the second result of the theorem. The first result follows by setting $h = e^{\beta-1}$ with some power $0 < \beta < \infty$, thus $\|e\|_{L^\beta(\omega_0)}^\beta \leq C_1\delta^{\frac{1}{4}} + C_2\delta^{\frac{3}{8}} \leq C\delta^{\frac{1}{4}}$.

Corollary 7 *If assumptions (A1–A4) are fulfilled, then $s_0 \in H^2(\omega_0) \cap H_0^1(\omega_0)$. In particular, $s_0 \in C^{1,\gamma}(\overline{\omega}_0)$ for some $0 < \gamma < 1$, and we have the estimate*

$$\|\nabla s_0\|_{L^\infty(\omega_0)} \leq \|s_0\|_{H^2(\omega_0)} \leq C\delta^{-\frac{3}{2}}, \tag{34}$$

where C is a constant that depends on U_0, m, q, and the geometry but not on δ.

Proof The assertion follows from Sobolev Embedding Theorem, the definition of Sobolev norm, and the estimates (26) and (27). For more details, review Sobolev spaces in [10]. ∎

Now, let us consider the elliptic equation

$$\begin{cases} div\,(P_{0\delta}\nabla s) - C_m\nabla P_0\nabla s = g & \text{in } Q, \\ s = 0 & \text{on } \partial Q. \end{cases} \tag{35}$$

with an arbitrary function $g \in L^2(0, T; L^2(\omega_0))$. The solution $s(\eta, t)$ is a function of the variables η and depends on t as a parameter. The solution of the above elliptic equation is understood in the following sense.

Definition 8 The function $s(\eta, t)$ is said to be a weak solution of the regularized Eq. (35) if it satisfies the following conditions:

I. $s \in L^2(0, T; H^2(\omega_0)) \cap L^2(0, T; H_0^1(\omega_0))$,
II. for any test function $\psi \in L^2(0, T; H_0^1(\omega_0))$ the following identity holds:

$$\int_{t_1}^{t_2} \int_{\omega_0} P_{0\delta}\nabla s\nabla\psi + C_m\nabla P_0\nabla s\,\psi + g\psi \,d\eta dt = 0, \ \forall\, t_0 \leq t_1 < t_2 \leq T. \tag{36}$$

Since $P_{0\delta} \geq \delta > 0$, P_0 and ∇P_0 are bounded and continuous functions in $\overline{\omega}_0$ as per condition **A3**, then for every $g \in L^2(0, T; L^2(\omega_0))$ problem (35) has a weak solution $s \in L^2(0, T; H^2(\omega_0))$. For almost every $t \in (0, T)$ the solution belongs to $H^2(\omega_0)$, the inclusion $s \in L^2(0, T; H^2(\omega_0))$ follows by integrating the spatial estimate with respect to the time over $(0, T)$. For more details regarding solutions of linear uniformly elliptic equations, review [10, 17].

Corollary 9 *Suppose that assumption (A1–A4) are fulfilled, then*

$$\delta\|\nabla s\|_{L^2(Q)}^2 + C\|s\|_{L^2(Q)}^2 \leq C_1\|g\|_{L^2(Q)}^2, \tag{37}$$

and

$$\|\Delta s\|_{L^2(Q)} \leq C_2 \delta^{-\frac{3}{2}} \|g\|_{L^2(Q)}, \tag{38}$$

where C_1 and C_2 are constants that do not depend on δ.

Proof Using identical arguments to those used for the initial elliptic equation, we conclude that for each $t \in (0, T]$ we have the estimates

$$\delta \|\nabla s(\eta, t)\|^2_{L^2(\omega_0)} + C \|s(\eta, t)\|^2_{L^2(\omega_0)} \leq \|g(\eta, t)\|^2_{L^2(\omega_0)},$$

and

$$\delta^2 \|\Delta s(\eta, t)\|^2_{L^2(\omega_0)} \leq \delta^{-1} \|g(\eta, t)\|^2_{L^2(\omega_0)}.$$

Integrating these estimates with respect to t completes the proof.

Corollary 10 *Suppose that s is the solution of the regularized problem (35) and π is the solution of the degenerate problem (41). If assumptions (A1–A4) are fulfilled, then for some $0 < \beta < \infty$ we have the estimates*

$$\|\pi - s\|^\beta_{L^\beta(Q)} \leq C \delta^{\frac{1}{4}} \|g\|_{L^2(Q)}, \tag{39}$$

and

$$\|\nabla \pi - \nabla s\|^2_{L^2(Q)} \leq C \delta^{\frac{1}{4}} \|g\|_{L^2(Q)}, \tag{40}$$

where C is a constant that does not depend on δ.

Proof Consider the degenerate elliptic equation

$$\begin{cases} div\,(P_0 \nabla \pi) - C_m \nabla P_0 \nabla \pi = g & \text{in } Q, \\ \pi = 0 & \text{on } \partial Q, \end{cases} \tag{41}$$

and follow the proof of Theorem 26.

Corollary 11 *If assumptions (A1–A4) are fulfilled, then $s \in L^2(0, T; H^2(\omega_0)) \cap L^2(0, T; H_0^1(\omega_0))$. In particular, $s \in L^2(0, T; C^{1,\gamma}(\overline{\omega}_0))$ for some $0 < \gamma < 1$, and we have the estimate*

$$\|s\|_{L^2(0,T;H^2(\omega_0))} \leq C \delta^{-\frac{3}{2}} \|g\|_{L^2(0,T;L^2(\omega_0))}, \tag{42}$$

where C is a constant that does not depend on δ.

Proof Consider Sobolev Embedding Theorem, the definition of Sobolev norm, and the estimates of Corollary 9. For more details, review Sobolev spaces in [10].

4.3   Linear Parabolic Equation

Consider the regularized parabolic problem

$$\begin{cases} w_t - (m-1)P_{0\mu}\Delta w = \tilde{f} & \text{in } Q, \\ w(\eta,0) = 0, \\ w = 0 & \text{on } \partial Q, \end{cases} \tag{43}$$

where $\tilde{f}$ is some arbitrary function. We will assume for now that

$$\tilde{f} \in L^\infty(0,T;H^2(\omega_0)) \cap L^\infty(0,T;H_0^1(\omega_0)).$$

The solution of problem (43) is understood in the following sense.

Definition 12 The function $w(\eta,t)$ is said to be a weak solution of the regularized problem (43) if it satisfies the following conditions:

I. $w \in L^2(0,T;H^2(\omega_0)) \cap L^2(0,T;H_0^1(\omega_0))$,
II. for any test function $\psi(\eta,t)$, vanishing at $t=T$ and satisfying the conditions

$$\psi \in L^2(0,T;H_0^1(\omega_0)), \psi_t \in L^2(0,T;L^2(\omega_0)),$$

the following identity holds:

$$\int_{t_1}^{t_2}\int_{\omega_0}\Big((m-1)(P_{0\mu}\nabla w\nabla\psi + \nabla P_0\nabla w\,\psi) - (w\psi_t + \tilde{f}\psi)\Big)d\eta dt = 0, \tag{44}$$

$\forall\, t_0 \le t_1 < t_2 \le T$.

Since $P_{0\mu} \ge \mu > 0$ and it is a continuous and bounded function in $\bar{\omega}_0$ as per condition **A3**, then for every $\tilde{f} \in L^2(0,T;L^2(\omega_0))$ there exists a solution $w \in L^2(0,T;H^2(\omega_0))$. For more details regarding solutions of linear uniformly parabolic equations, review [10, 18].

Lemma 13 *Suppose (A1), (A3) and (A4) are fulfilled and $\tilde{f} \in L^\infty(0,T;H^2(\omega_0))$. Then*

$$\max_{0<t\le T}\|w(\eta,t)\|_{L^2(\omega_0)}^2 + \mu\|\nabla w\|_{L^2(Q)}^2 \le CT\|\tilde{f}\|_{L^\infty(0,T;L^2(\omega_0))}^2, \tag{45}$$

where C is a constant that does not depend on μ nor T.

Proof Let us multiply (43) by w and integrate by parts over ω_0, apply Green's theorem, use assumption **A4**, and apply Cauchy's inequality. Then, integrate the result over the interval $(0,t)$, use the facts that $P_{0\mu} \ge \mu$ and $w(\eta,0)=0$, and the assumption that $\tilde{f} \in L^\infty(0,T;H^2(\omega_0))$ to obtain the required result.

Lemma 14 *Suppose that assumptions (A1), (A3) and (A4) are fulfilled and $\tilde{f} \in L^\infty(0, T; H^2(\omega_0)) \cap L^\infty(0, T; H_0^1(\omega_0))$. Then,*

$$\max_{0<t\le T} \|\nabla w(\eta, t)\|^2_{L^2(\omega_0)} + \mu \|\Delta w\|^2_{L^2(Q)} \le C\mu^{-1} T \|\tilde{f}\|^2_{L^\infty(0,T;L^2(\omega_0))}, \tag{46}$$

where C is a constant that does not depend on μ nor T.

Proof Since $\tilde{f} \in L^\infty(0, T; H^2(\omega_0)) \cap L^\infty(0, T; H_0^1(\omega_0))$, then by virtue of [10, Sect. 7.1, Theorem 6], problem (43) has a weak solution $w \in L^2(0, T; H^4(\omega_0))$. We apply the Laplace operator to both sides of (43) and we use the initial and the boundary conditions of w to conclude the problem

$$\begin{cases} \Delta w_t - (m-1)\Delta(P_{0\mu}\Delta w) = \Delta\tilde{f} & \text{in } Q, \\ \Delta w(\eta, 0) = 0, \\ \Delta w = 0 & \text{on } \partial Q. \end{cases}$$

Note that the boundary condition ($\Delta w = 0$ on ∂Q) follows from the boundary conditions of problem (43) on both w and $\tilde{f}$. If we consider solving the above problem to obtain Δw, then by virtue of [10, Sect. 7.1, Theorem 5], the above problem has a weak solution $\Delta w \in L^2(0, T; H^2(\omega_0)) \cap L^2(0, T; H_0^1(\omega_0))$. Hence Δw satisfies the conditions of Definition 12 for test functions. We multiply (43) by Δw and integrate by parts over ω_0, we recall that $P_{0\mu} \ge \mu$ and $\nabla w(\eta, 0) = 0$, and we use Cauchy's inequality then we integrate over the interval $(0, t)$ to obtain the result.

Lemma 15 *Suppose that assumptions (A1), (A3) and (A4) are fulfilled and $\tilde{f} \in L^\infty(0, T; H^2(\omega_0))$. Then,*

$$\|w_t\|_{L^2(Q)} \le C\mu^{-1} T^{\frac{1}{2}} \|\tilde{f}\|_{L^\infty(0,T;L^2(\omega_0))}, \tag{47}$$

where C is a constant that does not depend on μ nor T.

Proof Write (43) in the form

$$w_t = \tilde{f} + (m-1)P_{0\mu}\Delta w.$$

Square and integrate both sides over Q, use the assumptions on $\tilde{f}$ and P_0, recall estimate (46), and utilize Hölder's inequality to get the required result.

Let us consider the degenerate parabolic equation in (19) combined with $\tilde{f}$ as the right hand side such that

$$\begin{cases} v_t - (m-1)P_0\Delta v = \tilde{f} & \text{in } Q, \\ v(\eta, 0) = 0, \\ v = 0 & \text{on } \partial Q. \end{cases} \tag{48}$$

Theorem 16 *Suppose that w is the solution of the regularized problem (43) and v is the solution of the degenerate problem (48). If the assumptions (A1–A4) are fulfilled and $\tilde{f} \in L^\infty(0, T; H^2(\omega_0))$, then*

$$\|v - w\|_{L^\beta(Q)}^\beta \leq C\mu^{\frac{1}{2}} T^{\frac{1}{4}} \|\tilde{f}\|_{L^\infty(0,T;L^2(\omega_0))}, \tag{49}$$

and

$$\|\nabla v - \nabla w\|_{L^2(Q)}^2 \leq C\mu^{\frac{1}{2}} T^{\frac{1}{4}} \|\tilde{f}\|_{L^\infty(0,T;L^2(\omega_0))}, \tag{50}$$

where C is a constant that does not depend on μ nor T.

Proof Multiply (43) and (48) by the same test function ψ that satisfies the conditions of Definition 12, subtract, and follow the technique used in the proof of Theorem 6.

Corollary 17 *If assumptions (A1–A4) are fulfilled and $\tilde{f} \in L^\infty(0, T; H^2(\omega_0))$, then $w \in L^2(0, T; H^2(\omega_0)) \cap L^2(0, T; H_0^1(\omega_0))$.*

Corollary 18 *If assumptions (A1–A4) are fulfilled and $\tilde{f} \in L^\infty(0, T; H^2(\omega_0))$, then $w \in L^2(0, T; H^4(\omega_0)) \cap L^2(0, T; H_0^1(\omega_0))$. In particular, $w \in C(0, T; C^{2,\gamma}(\overline{\omega}_0))$ and $\Delta w \in C(0, T; C^{0,\alpha}(\overline{\omega}_0))$ for some $0 < \gamma, \alpha < 1$, and we have the estimate*

$$\|w\|_{L^2(0,T;H^4(\omega_0))} \leq C\mu^{-\frac{3}{2}} T \|\Delta \tilde{f}\|_{L^\infty(0,T;L^2(\omega_0))}, \tag{51}$$

where C is a constant that does not depend on μ nor T.

Proof Let us apply the Laplace operator to (43) to get

$$\Delta w_t - (m-1)\Delta(P_{0\mu}\Delta w) = \Delta \tilde{f}.$$

Let $\tilde{w} = \Delta w$. Hence, we obtain

$$\begin{cases} \tilde{w}_t - (m-1)\Delta(P_{0\mu}\tilde{w}) = \Delta \tilde{f} & \text{in } Q, \\ \tilde{w}(\eta, 0) = \Delta w(\eta, 0) = 0, \\ \tilde{w} = \Delta w = 0 & \text{on } \partial Q. \end{cases} \tag{52}$$

By assumption $\tilde{f}, \Delta \tilde{f} \in L^\infty(0, T; L^2(\omega_0))$. Consequently, this equation has a weak solution $\tilde{w} \in L^2(0, T; H^2(\omega_0)) \cap L^2(0, T; H_0^1(\omega_0))$, review [10, 18]. But $\tilde{w} = \Delta w$ which implies the result. To obtain the estimate we follow the proofs of Lemmas 13, 14 and 15. The Hölder continuity of w and Δw with respect to the time and the spatial follows by virtue of [10, Sect. 5.6, Theorem 6] (Sobolev Embedding Theorem) and [10, Sect. 5.9, Theorem 4].

Corollary 19 *For a sufficiently small time interval T, the Jacobian $|J|$ is separated away from zero and infinity.*

Proof The Jacobian $|J|$ is defined as

$$|J| = |I + D^2(w)|.$$

Corollary 18 asserts the Hölder continuity of Δw, which implies the boundedness of the components of $|J|$ at *a.e.* $t \in (0, T]$. Moreover, initially $|J| = 1$ and since the estimate provided in the same corollary depends on T, then for sufficiently small time interval we guarantee that $|J|$ is separated away from zero.

4.4 Existence of a Solution

Up to this point we were investigating the set of equations that represents the inverse of the operator $\mathscr{G}(\hat{z}_0)$. According to our definition of the operator $\mathscr{G}$ (review [22, 24]) the functions f, Φ and H are calculated such that they represent the error in the nonlinear system (14) at each iteration, hence,

$$\begin{cases} \Delta f = div(J\nabla \hat{v}_t) + \Delta(\hat{p} - \hat{\pi}) & \text{in } Q, \\ f = 0 & \text{on } \partial Q, \\ \Phi = \hat{p}|J|^{m-1} - P_0 & \text{in } \overline{Q}, \\ H = div\left(U_0(J^{-1})^2\nabla\hat{\pi}\right) - aU_0\hat{u}^{q-1} & \text{in } \overline{Q}. \end{cases} \tag{53}$$

Notice that initially $f(\eta, 0) = P_0 - s_0$ and $\Phi = H = 0$. Consequently, by the assumptions on P_0 and due to the results obtained for s_0, the choice $\tilde{f} = f + s - \Phi$ fulfills the required assumptions on $\tilde{f}$ at the initial state. In turn, it validates all the results obtained for the parabolic equation. It remains to investigate the elliptic equation for right-hand side of a special form.

Lemma 20 *Let H be defined as per the last equation in (53) and suppose that assumptions (A1–A4) are fulfilled. Then the solution s of the regularized elliptic equation (22) satisfies the following estimate:*

$$\|s\|_{L^2(0,T;H^2(\omega_0))} \leq C_1\left(\delta^{-3}\|w\|_{L^2(0,T;H^4(\omega_0))} + \delta^{-\frac{3}{2}}\|H\|_{L^2(0,T;L^2(\omega_0))}\right). \tag{54}$$

Moreover, at the initial state we have the estimate

$$\|s\|_{L^2(0,T;H^2(\omega_0))} \leq C_2\delta^{-\frac{9}{2}}\mu^{-\frac{3}{2}}T, \tag{55}$$

where C_1 and C_2 are constants not depending on δ, μ nor T.

Proof The elliptic equation takes the form

$$div\,(P_{0\delta}\nabla s) - C_m\nabla P_0\nabla s = div\,\left(2P_0 D^2(w)\cdot\nabla s_0\right) - 2C_m\nabla P_0\cdot(D^2(w)\cdot\nabla s_0)$$
$$+a(1-q)U_0^\alpha \Delta w + U_0^{m-2}H.$$

Recall that the results of Corollary 9 were derived for an arbitrary function $g \in L^2(0, T; L^2(\omega_0))$. We can now take g as the right hand side of the above formula. If we expand the first term on the right hand side. Considering the continuity of P_0, $w \in L^2(0, T; H^4(\omega_0))$ and $s_0 \in H^2(\omega_0)$; we can then use standard estimation techniques to get

$$\int_Q g^2 d\eta dt \le C\delta^{-\frac{3}{2}}\|g\|_{L^2(Q)}\|w\|_{L^2(0,T;H^4(\omega_0))}.$$

By estimate (27) and the embedding theorem we get the first result assuming that H exists and that it is bounded in $L^2(Q)$. The results of Corollaries 11, 19 and 21 imply the boundedness of H in $L^2(Q)$. Initially $H = \Phi = 0$ and $\Delta f = \Delta(P_0 - s_0)$, thus, by virtue of the assumptions on P_0 and estimate (51) we have

$$\|g\|_{L^2(Q)} \le C\delta^{-\frac{3}{2}}\mu^{-\frac{3}{2}}T\|\Delta f + \Delta s - \Delta\Phi\|_{L^\infty(0,t;L^2(\omega_0))} \le C\delta^{-3}\mu^{-\frac{3}{2}}T. \quad (56)$$

Consequently, from the estimate of Corollary 11 we get the second estimate.

Corollary 21 *Let p be defined by (21), w be the solution of (20), and Φ be defined as per (53). If the conditions of Corollary 18 hold and assumption (A5) is fulfilled, then $p \in L^2(0, T; H^2(\omega_0))$, $p = 0$ on $\gamma_0 \times [0, T]$, and we have the estimate*

$$\|p\|_{L^2(0,T;H^2(\omega_0))} \le C_1\left(\|\Phi\|_{L^2(0,T;H^2(\omega_0))} + \|w\|_{L^2(0,T;H^4(\omega_0))}\right). \quad (57)$$

Moreover, at the initial state of the modified Newton method's iterations we have the estimate

$$\|p\|_{L^2(0,T;H^2(\omega_0))} \le C_2\delta^{-\frac{3}{2}}\mu^{-\frac{3}{2}}T. \quad (58)$$

where C_1 and C_2 are constants that do not depend on δ, μ nor T.

Proof Recalling that the linear system consisted originally of three linear equations, the reduction was made by eliminating the equation

$$p = \Phi - (m-1)P_0\Delta w.$$

It is straightforward to deduce the first result from the above formula assuming that Φ is bounded in $L^2(0, T; H^2(\omega_0))$. By the definition of Φ in (53), our assumptions on the initial data, and the result of Corollary 19; it is evident that $\Phi \in L^2(0, T; H^2(\omega_0))$

and $\Phi = 0$ on $\gamma_0 \times [0, T]$. Initially, $\Phi = 0$ and $\Delta f = \Delta(P_0 - s_0)$. Using estimates (27) and (51) completes the proof.

Corollary 22 *Let f and Φ be defined by (53) and s be the solution of the regularized elliptic equation (22). Suppose assumptions (A1–A4) are fulfilled. Then the solution w of the regularized parabolic equation (20) satisfies the following estimate:*

$$\|w\|_{L^2(0,T;H^4(\omega_0))} \leq C_1 \mu^{-\frac{3}{2}} T \left(\|s\|_{L^2(0,T;H^2(\omega_0))} + \|f\|_{L^2(0,T;H^2(\omega_0))} \right.$$
$$\left. + \|\Phi\|_{L^2(0,T;H^2(\omega_0))} \right).$$

Moreover, at the initial state, we have the estimate

$$\|w\|_{L^2(0,T;H^4(\omega_0))} \leq C_2 \delta^{-\frac{3}{2}} \mu^{-\frac{3}{2}} T, \tag{59}$$

where C_1 and C_2 are constants that do not depend on δ, μ nor T.

Corollary 23 $\|[\mathscr{G}(\hat{z}_0)]^{-1}\| = M \leq C(\delta^{-\frac{3}{2}} + \mu^{-\frac{3}{2}} T)$, *where C is the sum of the constants defined in Corollary 21, Lemma 20 and Corollary 22 respectively.*

Proof For details of the proof review [6, 7] and the references within.

Corollary 24 $\|[\mathscr{G}(\hat{z}_0)]^{-1} \langle \mathscr{F}(\hat{z}_0) \rangle\| = K \leq C \delta^{-\frac{9}{2}} \mu^{-\frac{3}{2}} T$.

Proof For details of the proof review [6, 7] and the references within.

We need now to check the Lipschitz continuity of the operator $\mathscr{G}$ at an arbitrary iteration. To do so we can follow the steps of the author in [22] to arrive at

$$\mathscr{G}(\hat{z}_n)\langle z \rangle = \begin{cases} div \left(D^2(w) \nabla \hat{w}_{n_t} + A \nabla w_t \right) + \Delta(p - s), \\ |A|^{m-1} \left(p + (m-1) \hat{p}_n \, trace \left(A^{-1} D^2(w) \right) \right), \\ div \left(U_0 (A^{-1})^2 \nabla s - U_0 \left(B(A^{-1})^2 + (A^{-1})^2 B \right) \nabla \hat{s}_n \right) \\ -(1-q) U_0^q |A|^{-q} \, trace(A^{-1} D^2(w)). \end{cases}$$

Since the above formula is linearized, we deduce that $\mathscr{G}(\hat{z}_{n+1})\langle z \rangle - \mathscr{G}(\hat{z}_n)\langle z \rangle$ is only given in terms of the corrections z and the quantities $\hat{w}_n$, $\hat{w}_{n+1}$, $\hat{s}_n$ and $\hat{s}_{n+1}$. The inverse matrices can be expressed in terms of nth-order polynomials of their algebraic adjoints and determinants which depend also on $\hat{w}_n$ and $\hat{w}_{n+1}$. Moreover, we have the estimates for the corrections w and s at each iteration. At the initial iteration $\hat{w} = w$ and $\hat{s} = s_0 - s$ which we have proven their boundedness and their dependence on T. By mathematical induction we can prove the validity of these estimates for each iteration and their continuous dependence on T. Therefore we have

$$\|\mathscr{G}(\hat{z}_{n+1})\langle z \rangle - \mathscr{G}(\hat{z}_n)\langle z \rangle\| \leq L\|\hat{z}_{n+1} - \hat{z}_n\| \leq L\|z\| \leq \hat{L}\|w\|_{L^2(0,T;H^4(\omega_0))}, \tag{60}$$

where $\hat{L} \equiv \hat{L}(m, q, a, \mu^\alpha, \delta^\beta, \omega_0, U_0, T^\sigma)$ for some real numbers $\{\alpha < 0, \beta < 0, \sigma > 0\}$. Choosing T sufficiently small is enough to fulfill the conditions of Theorem 2 and to prove the following theorem.

Theorem 25 *Suppose that assumptions (A1–A5) are fulfilled. Then there exists* $0 <$
$T^* \ll 1$ *such that* $\lambda = MK\hat{L} < \frac{1}{4}$, *where* M, K *and* $\hat{L}$ *are the constants defined
by Corollary 23, Corollary 24 and estimate (60), and the nonlinear regularization
of problem (14) has a unique solution* $\hat{z} \in \mathscr{B}_r(0)$ *such that* $\|\hat{z}\| < r$, *where* $r =$
$\frac{K}{2\lambda}\left(1 - \sqrt{1 - 4\lambda}\right) < 2K$. *Moreover, the dynamics of the interface* Γ_t *is defined by
formula (24).*

Proof For details of the proof review [6, 7] and the references within.

Theorem 26 *Let* y *be the trajectories calculated by (24) and* x *be the exact trajectories defined by (15). If assumptions (A1–A5) are fulfilled, then*

$$\lim_{\delta, \mu \to 0} \|x - y\|_{L^2(Q)} = 0. \tag{61}$$

Proof As per (24) and (15), the quantity $x - y$ is defined as

$$x - y = \nabla \hat{v} - \nabla \hat{w}.$$

The quantities $\nabla \hat{w}$ and $\nabla \hat{v}$ are calculated at each iteration as

$$\nabla \hat{w}_{n+1} = \nabla \hat{w}_n - \nabla w \quad \text{and} \quad \nabla \hat{v}_{n+1} = \nabla \hat{v}_n - \nabla v,$$

By recalling the estimates of Theorem 16 we conclude that

$$\lim_{\delta, \mu \to 0} \|\nabla v - \nabla w\|_{L^2(Q)} = 0,$$

and using induction we conclude the result of the theorem.

5   Conclusions and Suggestions

We adopted the framework of Sobolev spaces to approximate the solution of problem
(1), previously solved in weighted Hölder spaces [22, 24]. We proved the convergence of the approximated solution to the theoretical one in the weak sense. In that
way a suitable background was provided for the numerical implementation of the
interface tracking algorithm based on the introduction of Lagrangian coordinates,
which transforms the free boundary problem into a problem posed in a time independent domain.

It can be of a great benefit to consider the classical Newton method instead of
the modified one. Employing an iterative method with higher rate of convergence
can lead to a significant increase in the accuracy of the output. Yet, it will be on the
account of the complexity of the implementation.

References

1. Angenent, S.B.: Analyticity of the interface of the porous media equation after the waiting time. Proc. Amer. Math. Soc. **102**, 329–336 (1988)
2. Antontsev, S., Shmarev, S.: A model porous medium equation with variable exponent of non-linearity: existence, uniqueness and localization properties of solutions. Nonlinear Anal. **60**, 515–545 (2005)
3. Arnold, D., Brezzi, F., Cockburn, B., Marini, L.: Unified analysis of discontinuous Galerkin methods for elliptic problems. SIAM J. Numer. Anal. **39**(5), 1749–1779 (2002)
4. Cockburn, B., Shu, C.-W.: Runge-Kutta discontinuous Galerkin methods for convection-dominated problems. J. Sci. Comput. **16**(3) (2001)
5. Daskalopoulos, P., Hamilton, R.: Regularity of the free boundary for the porous medium equation. J. Amer. Math. Soc. **11**(4), 899–965 (1988)
6. Diaz, J., Shmarev, S.: Lagrangian approach to the study of level sets: application to a free boundary problem in climatolog. Arch. Ration. Mech. Anal. **194**, 75–103 (2009)
7. Diaz, J., Shmarev, S.: Lagrangian approach to the study of level sets. II. A quasilinear equation in climatology. J. Math. Anal. Appl. **352**, 475–495 (2009)
8. Dibenedetto, E., Hoff, D.: An interface tracking algorithm for the porous medium equation. Trans. Amer. Math. Soc. **284**(2) (1984)
9. Duque, J.: Método dos Elementos Finitos para Problemas com Fronteiras Livres. Ph.D. Thesis, Universidade da Beira Interior, Ciências (2013)
10. Evans, L.: Partial Differential Equations, vol. 19, 2nd edn. In: Graduate Studies in Mathematics. American Mathematical Society (2010)
11. Galaktionov, V., Shmarev, S., Vazquez, J.: Regularity of interfaces in diffusion processes under the influence of strong absorption. Arch. Ration. Mech. Anal. **149**, 183–212 (1999)
12. Galaktionov, V., Shmarev, S., Vazquez, J.: Regularity of solutions and interfaces to degenerate parabolic equations. The intersection comparison method. In: Free boundary problems: theory and applications, vol. 115–130. Chapman & Hall/CRC, Boca Raton (1999)
13. Galiano, G., Shmarev, S., Velasco, J.: Existence and nonuniqueness of segregated solutions to a class of cross-diffusion systems. arXiv:1311.3454v1 [math.AP] (2013)
14. Kersner, R.: The behaviour of temperature fronts in media with nonlinear heat conductivity under absorption. Moscow Univ. Math. Bull. **33**(5), 35–41 (1978). Translated from: Vestnik Moskov. Univ. Ser. I Mat. Mekh, **5**, 44–51 (1978)
15. Koch, H.: Non-Euclidean singular integrals and the porous medium equation, Habilitation Thesis, Univ. of Heidelberg (1999)
16. Kolmogorov, A.N., Fomin, S.V.: Elements of the theory of functions and functional analysis, measure, the Lebesgue integral, Hilbert, vol. 2. Translated from the first (1960) Russian ed. by Hyman Kamel and Horace Komm, Graylock Press, Albany, N.Y. (1961)
17. Ladyzhenskaya, O.A., Uralćtseva, N.A.: Linear and quasilinear elliptic equations. Translated from the Russian by Scripta Technica, Inc., Translation editor: Leon Ehrenpreis, Academic Press, New York-London (1968)
18. Ladyzhenskaya, O.A., Solonnikov, V.A., Uralćtseva, N.A.: Linear and quasilinear equations of parabolic type. Translated from the Russian by S. Smith. Translations of Mathematical Monographs, vol. 23. American Mathematical Society, Providence, R.I. (1968)
19. Mimura, M., Nakaki, T., Tomoeda, K.: A numerical approach to interface curves for some nonlinear diffusion equations. Japan J. Appl. Math. **1**, 93–139 (1984)
20. Nakaki, T.: Numerical interfaces in nonlinear diffusion equations with finite extinction phenomena. Hiroshima Math. J. **18**, 373–397 (1988)
21. Shmarev, S.: Interfaces in solutions of diffusion-absorption equations. RACSAM Rev. R. Acad. Cienc. Exactas Fís. Nat. Ser. A Mat. **96**, 129–134 (2002)
22. Shmarev, S.: Interfaces in multidimensional diffusion equations with absorption terms. Nonlinear Anal. **53**, 791–828 (2003)
23. Shmarev, S.: On a class of degenerate elliptic equations in weighted Hölder spaces. Diff. Integr. Eqn. **17**, 1123–1148 (2004)

24. Shmarev, S.: Interfaces in solutions of diffusion-absorption equations in arbitrary space dimension. In: Trends in partial differential equations of mathematical physics. Progr. Nonlinear Diff. Eqn. Appl. Birkhäuser, Basel **61**, 257–273 (2005)
25. Tomoeda, K.: Convergence of numerical interface curves for nonlinear diffusion equations, Advances in computational methods for boundary and interior layers. In: Lecture Notes of an International Short Course held in Association with the BAIL III Conference, Trinity College, Dublin, Ireland (1984)
26. Zhang, Q., Wu, Z.-L.: Numerical simulation for porous medium equation by local discontinuous Galerkin finite element method. J. Sci. Comput. **38**, 127–148 (2009)

Tracking the Interface of the Diffusion-Absorption Equation: Numerical Analysis

Waleed S. Khedr

Abstract Our concern in this article is the Cauchy problem for the degenerate parabolic equation of the diffusion-absorption type $u_t = \Delta u^m - au^q$ with the exponents $m > 1, q > 0, m + q \geq 2$ and constant $a > 0$. In a previous article (Khedr W.S. Tracking the Interface of the Diffusion-Absorption Equation: Theoretical Analysis) we investigated an algorithm for tracking the moving interface of the above model based on the idea of Shmarev (Nonlinear Anal 53:791–828, 2003; Interfaces in solutions of diffusion-absorption equations in arbitrary space dimension, 2005). By means of domain reduction, introduction of local system of Lagrangian coordinates, utilization of the modified Newton method, linearization and regularization we managed to transform the nonlinear problem into a system of linearized equations, and we proved the convergence of the approximated solution of the regularized problem to the solution of the original problem in the weak sense. The introduction of the regularization parameters provided the necessary requirement for a stable numerical implementation of the algorithm. In this article we investigate the numerical error at each Newtonian iteration in terms of the discretization and the regularization parameters. We also try to deduce the minimum order of the finite element's interpolating polynomials to be used in order to maintain the stability of the algorithm. Finally, we present a number of numerical experiments to validate the algorithm and to investigate its advantages and disadvantages. We will also illustrate the preference of adaptive implementation of the algorithm over direct implementation.

W.S. Khedr (✉)
Erasmus-Mundus MEDASTAR Project, Universidad de Oviedo,
Oviedo, Asturias, Spain
e-mail: waleedshawki@yahoo.com

© Springer International Publishing Switzerland 2016   379
G.A. Anastassiou and O. Duman (eds.), *Intelligent Mathematics II:*
Applied Mathematics and Approximation Theory, Advances in Intelligent Systems
and Computing 441, DOI 10.1007/978-3-319-30322-2_26

1 Introduction

In this article, our main concern is the study of the free boundary problem

$$
\begin{cases}
U_t = \Delta U^m - aU^q & \text{in } (0, T] \times \mathbb{R}^n \\
U(X, 0) = U_0(X) \geq 0 & \text{in } \mathbb{R}^n
\end{cases}
\tag{1}
$$

where $a, q > 0, m > 1$, are constants and $m + q \geq 2$. The solution $U(x, t)$ represents the density of some substance per unit volume and $T > 0$ is a fixed interval. It is known that the solutions of problem (1) possess the property of finite speed of propagation, which means that if the support of the initial function u_0 is compact, so is the support of the solution at every instant t. The a priori unknown boundary of the support is the free boundary that has to be defined together with the solution $U(x, t)$. In [10] Shmarev used the transformation to the Lagrangian coordinates and proved that the solution exists in a special weighted Hölder space and that the solution itself and the corresponding interface are real analytic in time. He also provided explicit formulas that represent the interface $\gamma(t)$ as a bijection of the initial interface $\gamma(0)$ and demonstrated the improvement of the interface regularity with respect to the spatial variables compared to the initial instant. However, these results were constrained to the case of $n = 1, 2, 3$. In [11] the author managed to eliminate that constraint by means of Helmholtz's orthogonal decomposition. More results regarding the above problem exist in the literature and they can be obtained by exploring [5] and the references within.

In [6] we followed the idea of Shmarev, but instead of dealing with the degenerate problem we used the regularization technique and we formulated the problem in the framework of Sobolev spaces. The convergence of the sequence of the approximate solutions of the regularized problem to the solution of the degenerate one was proven in the weak sense. The main advantage of the introduction of the regularization parameters is the replacement of the degenerate system of the linearized equations by a uniform one, which consequently fulfils the condition of a stable finite element's implementation of the linearized system of equations.

2 Objective

In [6] our concern was to investigate the analytical foundation of the algorithm and to prove the theoretical convergence of the approximated solution to the solution of the original problem. As mentioned earlier, the main goal was to facilitate a stable implementation of the algorithm which was not possible in the presence of the degeneracy.

Our objective in this article is to investigate the numerical implementation of the algorithm. We study the definition of the discrete solution of the linearized system of equations. We also try to find a rough estimate for the numerical error at each iteration of the modified Newton method in terms of the discretization and the regularization parameters. Additionally, we pose a condition on the minimum order of the interpolating polynomials to be used in order to maintain the stability of the algorithm. Finally, we present a number of numerical experiments to validate the algorithm and to illustrate different aspects of its functionality.

3 Background

In this section we introduce a quick review for the basic steps for establishing the algorithm, for more details see [6, 10, 11].

3.1 Statement of the Problem

Our main problem was formulated in the following form

$$
\begin{cases}
U_t = \Delta\left(U^m\right) - aU^q & \text{in } \Omega_t \times (0, T], \\
U = 0 & \text{on } \Gamma_T, \\
U(X, 0) = U_0(X) \geq 0 & \text{in } \overline{\Omega}_0,
\end{cases} \tag{2}
$$

where

$$
\Omega_t = \Omega(t) = \{X \in \mathbb{R}^n, t \in [0, T] : U(X, t) > 0\} \equiv \operatorname{Supp} U(X, t),
$$

$$
\Gamma_T = \{X \in \mathbb{R}^n, t \in [0, T] : U(X, t) = 0\} = \bigcup_{t \in [0, T]} \gamma_t,
$$

and $\gamma_t = \gamma(t) \equiv \partial \Omega_t \ \forall t \in [0, T]$. The weak solution of problem (2) was defined in the following sense.

Definition 1 A pair $(U(X, t), \Gamma_T)$ is said to be a weak solution of problem (2) if:

I. U is bounded, continuous, and nonnegative in $\overline{\Omega}_t \times (0, T]$ and $\nabla U^m \in L^2$
$(\overline{\Omega}_t \times (0, T])$,
II. Γ_T is a surface of class C^1,
III. for every test function $\psi \in C^1(\overline{\Omega}_t \times (0, T])$, vanishing at $t = T$ and for all $(X, t) \in \Gamma_T$, the following identity holds:

$$
\int_0^T \int_{\Omega_t} (U\psi_t - \nabla U^m . \nabla \psi - a\psi U^q) \, dX \, dt + \int_{\Omega_0} U_0 \psi(X, 0) \, dX = 0. \tag{3}
$$

The loss of the mass was compensated by reducing the domain of the problem to some annular domain near the interface. Consider a simple-connected surface $S_T = \bigcup_{t \in [0,T]} s_t$ such that $S_T \subset \Gamma_T$ and $S_T \cap \Gamma_T = \emptyset$. Let also $\omega_t \subset \Omega_t$ be the annular domain bounded by the surfaces γ_t and s_t such that the problem's domain is $\mathscr{E} = \bigcup_{t \in [0,T]} \omega_t$ and $\partial \mathscr{E} = S_T \cup \Gamma_T$. Denote by $\hat{U}(X,t)$ the solution inside the annular domain. The surface S_T is an artificial surface and it is chosen such that the following identity is fulfilled:

$$\forall t \in (0,T] \quad \int_{\omega_t} \hat{U}(X,t)\,dX = \int_{\omega_0} U_0(X)\,dX = \text{const.} \tag{4}$$

Therefore a new auxiliary problem has been introduced as

$$\begin{cases} \hat{U}_t = \Delta\left(\hat{U}^m\right) - a\hat{U}^q & \text{in } \mathscr{E}, \\ \hat{U}|_{S_T} = f_{(s)}, \quad \hat{U}|_{\Gamma_T} = 0, \\ \hat{U}(X,0) = U_0(X) \geq 0 & \text{in } \overline{\omega}_0, \end{cases} \tag{5}$$

where $f_{(s)}$ represents the dummy values of $\hat{U}$ on the inner surface which depend mainly on our choice of the boundary conditions on that surface. As per (4), our choice of the inner surface S_T guarantees that the mass is conserved in the new domain $\mathscr{E}$. Although this is a different problem, the author in [10, 11] proved that (5) and (2) coincide up to the motion of the boundary. The general equation of the mass conservation is

$$\rho_t + div(\rho \mathcal{V}) = 0, \tag{6}$$

where $\mathcal{V}$ is the velocity of the fluid particles. Two terms define the velocity. The first is the gradient of the pressure denoted by $\hat{P}$, which is induced by the diffusive term, and the second is the gradient of some artificial pressure $\hat{\Pi}$ induced by the low order term, such that

$$\mathcal{V} = -\nabla\hat{P} + \nabla\hat{\Pi}, \tag{7}$$

where,

$$\hat{P} = \frac{m}{m-1}\hat{U}^{m-1}, \quad P_0 = \frac{m}{m-1}U_0^{m-1}, \tag{8}$$

and $\hat{\Pi}$ is the solution of the following degenerate elliptic problem:

$$\begin{cases} div(\hat{U}\nabla\hat{\Pi}) = a\hat{U}^q & \text{in } \mathscr{E}, \\ \hat{\Pi} = 0 & \text{on } \partial\mathscr{E}, \end{cases} \tag{9}$$

given that π_0 be the solution of the degenerate elliptic problem at the initial instant such that

$$\begin{cases} div(U_0 \nabla \pi_0) = aU_0^q & \text{in } \mathscr{E}, \\ \pi_0 = 0 & \text{on } \partial \mathscr{E}. \end{cases} \qquad (10)$$

As in [1, 2, 4, 10, 11] an alternative description was given by transforming the system into geometrical Lagrangian coordinates so that all the functions character-izing the motion became dependent on the initial positions of the particles and the time as a parameter. The cylinder Q with fixed lateral boundaries was defined as

$$Q = \omega_0 \times [0, T], \ \partial Q = \partial \omega_0 \times [0, T].$$

Let us denote the initial positions by η. In this way, the positions of the particles at any instant t are

$$X = x(\eta, t), \quad \eta \in \overline{\omega_0},$$

and consequently $\hat{U}[x(\eta, t), t] = \hat{u}(\eta, t)$ is the density corresponding to that particle, $\hat{P}[x(\eta, t), t] = \hat{p}(\eta, t)$ and $\hat{\Pi}[x(\eta, t), t] = \hat{\pi}(\eta, t)$ are the functions representing the pressure, and $\mathscr{V}[x(\eta, t), t] = v(\eta, t)$ is the velocity of that particle. The Lagrangian version of the mass conservation law can be deduced as

$$\hat{U}[x, t]|J| = \hat{u}(\eta, t)|J| = \hat{u}(\eta, 0) = U_0 \quad \text{in } \overline{Q}. \qquad (11)$$

where $J = [J_{ij}]$ is the Jacobi matrix of the mapping $\eta \to x(\eta, .)$, for more details review [10, 11].

The analysis in [11] led us to the following system on the plane of Lagrangian coordinates:

$$\begin{cases} div_\eta(J\nabla_\eta \hat{v}_t) + \Delta_\eta(\hat{p} - \hat{\pi}) = 0, & \text{in } Q, \\ \hat{p}|J|^{m-1} = P_0 & \text{in } \overline{Q}, \\ div_\eta \left(U_0(J^{-1})^2 \nabla_\eta \hat{\pi} \right) = aU_0 \hat{u}^{q-1} & \text{in } Q, \\ \hat{v}(\eta, 0) = 0, \ \hat{\pi}(\eta, 0) = \pi_0(\eta), & \text{and} \\ \hat{p}(\eta, 0) = P_0(\eta) & \text{in } \overline{\omega_0}, \\ \hat{p} = 0 & \text{on } \gamma_0 \times [0, T] \text{ and } \hat{v} = \hat{\pi} = 0 \quad \text{on } \partial Q. \end{cases} \qquad (12)$$

Solving the above system to obtain the triad solution $(\hat{v}, \hat{p}, \hat{\pi})$ enables us to calculate the motion of the interface in a direction normal to the initial interface by the formula

$$x = \eta + \nabla_\eta \hat{v} \quad \text{in } \overline{Q}, \qquad (13)$$

which is our target. The ability to solve the nonlinear system (12) and recovering the position of the particles x provides us with the necessary information to locate the interface Γ_T of (2). We need not worry about the solution or the motion of the particles in the rest of the domain $\mathcal{E}$. We only use the information we have regarding the position of the interface to solve (2) (which is equivalent to (1)) as a boundary-value problem with zero boundary condition on Γ_T, which is already defined.

The nonlinear system (12) is to be solved by means of an abstract version of the modified Newton method. Let us consider the three equations in (12) as the arguments of the functional equation $\mathcal{F}(\hat{z}) \equiv \{\mathcal{F}_1(\hat{z}), \mathcal{F}_2(\hat{z}), \mathcal{F}_3(\hat{z})\}$, where $\hat{z} = \{\hat{v}, \hat{p}, \hat{\pi}\}$. In the modified Newton method the differential of $\mathcal{F}$ is calculated once at the initial state. The solution is obtained then as the limit of the sequence

$$\hat{z}_{n+1} = \hat{z}_n - [\mathcal{G}(\hat{z}_0)]^{-1}\langle\mathcal{F}(\hat{z}_n)\rangle, \tag{14}$$

where in this case the operator $\mathcal{G}(\hat{z}_0)\langle z\rangle$ is the Frechét derivative of $\mathcal{F}$ at the initial instant $\hat{z}_0 = \{0, P_0, \pi_0\}$ and $z = \{v, p, \pi\}$ is the correction at each iteration. More details regarding the Newton method and the modified Newton method can be found in [7].

Following the linearization techniques introduced in [10, 11] leads us to the following linearized system:

$$\begin{cases} v_t - (m-1)P_0\Delta v = f + \pi - \Phi & \text{in } Q, \\ div\left(U_0\nabla\pi - 2U_0D^2(v).\nabla\pi_0\right) - a(1-q)U_0^q\Delta v = H & \text{in } Q, \\ v = \pi = 0 & \text{on } \partial Q. \end{cases} \tag{15}$$

We get the corrections z at each iteration by solving the system of linear equations in (15) given that all the assumptions posed in [10–13] are fulfilled. We should note that f, Φ and H represent the error within the nonlinear system at each iteration such that

$$\begin{cases} \Delta f = div(J\nabla\hat{v}_t) + \Delta(\hat{p} - \hat{\pi}) & \text{in } Q, \\ f = 0 & \text{on } \partial Q, \\ \Phi = \hat{p}|J|^{m-1} - P_0 & \text{in } \overline{Q}, \\ H = div\left(U_0(J^{-1})^2\nabla\hat{\pi}\right) - aU_0\hat{u}^{q-1} & \text{in } \overline{Q}. \end{cases} \tag{16}$$

To overcome the degeneracy in the linearized system, we approximated it by a sequence of regularized linear problems [6].

3.2   Reduction to a Sequence of Linear Problems

We introduced the new functions w as a solution of the regularized parabolic equation and s as a solution for the regularized elliptic one, and introduced two small parameters $\delta > 0$ and $\mu > 0$ such that $P_{0\delta} = P_0 + \delta$ and $P_{0\mu} = P_0 + \mu$. Denote $\alpha = m + q - 2 \geq 0$ and $C_m = (m-2)/(m-1)$. We introduced also the following general assumptions

(A1) $\partial \omega_0$ is of class C^2.

(A2) $U_0 \in C^2(\overline{\omega}_0)$.

(A3) $P_0 \in C^1(\overline{\omega}_0)$, $\nabla P_0 \in C^0(\overline{\omega}_0)$, and $\Delta P_0 \in L^\infty(\overline{\omega}_0)$.

(A4) The annular domain is chosen initially such that $-C_0 \leq \Delta P_0 \leq -M_0$ a.e. in $\overline{\omega}_0$.

(A5) There exists some constant $0 < \nu < 1$ such that $\nu < |J| < \nu^{-1}$.

We introduced the regularized parabolic equation

$$\begin{cases} w_t - (m-1)P_{0\mu}\Delta w = f - \Phi + s & \text{in } Q, \\ w(\eta, 0) = 0, \\ w = 0 & \text{on } \partial Q. \end{cases} \tag{17}$$

We defined p as

$$p = \Phi - (m-1)P_0\Delta w \quad \text{in } Q. \tag{18}$$

Finally, we introduced the regularized elliptic equation

$$\begin{cases} div\,(P_{0\delta}\nabla s) - C_m\nabla P_0\nabla s = div\,(2P_0 D^2(w)\nabla s_0) - 2C_m\nabla P_0 D^2(w)\nabla s_0 \\ +a(1-q)U_0^\alpha\Delta w + U_0^{m-2}H \quad \text{in } Q, \\ s = 0 \quad \text{on } \partial Q, \end{cases} \tag{19}$$

where s_0 is the solution of

$$\begin{cases} div\,(P_{0\delta}\nabla s_0) - C_m\nabla P_0\nabla s_0 = aU_0^\alpha & \text{in } \omega_0, \\ s_0 = 0 & \text{on } \partial \omega_0. \end{cases} \tag{20}$$

Therefore, the solution of the nonlinear system is obtained as the limit of the sequence

$$\hat{w}_{n+1} = \hat{w}_n - w \quad \text{and} \quad \hat{s}_{n+1} = \hat{s}_n - s, \quad \text{for } n = 0, 1, 2, \ldots$$

Consequently, the position of the interface is obtained by

$$y = \eta + \nabla\hat{w}. \tag{21}$$

For every $\delta, \mu > 0$, the existence of a unique weak solution of the elliptic and the parabolic equations follows directly from the standard theories, review [3, 8, 9]. We investigated the effect of these small parameters on the solution and also we proved the convergence of this approach to the degenerate case when $\delta, \mu \to 0$. In the rest of this article we discuss numerical aspects related to the implementation of the regularized scheme.

4 Discretization of the Regularized Problem

First, let us decompose the cylinder $\overline{Q} = [0, T] \times \overline{\omega}_0$ using a set of time layers sharing a common base of closed triangulation $\mathscr{T}_h$ that consists of the triangles T_r with the diameters h_r. The intersection of any two distinct elements is empty, a common edge, or a vertex such that $\overline{\omega}_0 = \bigcup_{T_r \in \mathscr{T}_h} \overline{T}_r$. The maximal diameter of the mesh is

$$h = \max_{T_r \in \mathscr{T}_h} h_r.$$

We define the finite element space $\Psi_h = \{\psi_h \in H_0^1(\overline{\omega}_0) : \psi_h|_{T_r} \in \mathscr{P}_r(T_r) \, \forall \, T_r \in \mathscr{T}_h\}$ where $\mathscr{P}_r$ is the piecewise approximating polynomials. We make the assumption that U_0, U_0^α, $P_0 \in \Psi_h$. We use the index $j \in \mathbb{N}$ with a fixed time step $k > 0$ such that each instant of time is denoted by $t_j = jk$ and $0 \le t_0 < t_1 < t_2 < \ldots t_N$, where $N = 1 + T/k$ is the total number of instants. Sometimes in the numerical analysis we will need what is so called *the inverse estimate*. For some numerical function f and some constant C, it states that $\|\nabla f\| \le C h^{-1} \|f\|$.

4.1 The Discrete Initial Elliptic Equation

The discrete solution s_{0h} of the regularized initial elliptic equation (20) is understood in the following sense.

Definition 2 A function $s_{0h} \in \Psi_h$ is said to be a discrete solution of (20) and (10) if $s_{0h} = 0$ on $\partial\omega_0$ and it satisfies the integral identity:

$$\int_{\omega_0} \left(P_{0\delta} \nabla s_{0h} \nabla \psi + C_m \nabla P_0 \nabla s_{0h} \psi + U_0^\alpha \psi\right) d\eta = 0. \tag{22}$$

Theorem 3 *Let s_0 be the solution of* (20), π_0 *be the solution of* (10) *and s_{0h} be their discrete solution. If the conditions of* [6, Theorem 25.6] *hold, then for an arbitrary* $0 < \beta < \infty$

$$\|s_0 - s_{0h}\|_{L^\beta(\omega_0)}^\beta \le C\delta^{-2} h^r, \tag{23}$$

and

$$\|\pi_0 - s_{0h}\|_{L^\beta(\omega_0)}^\beta \le C(\delta^{-2} h^r + \delta^{\frac{1}{4}}),$$

where C is a constant that does not depend on δ nor h.

Proof Choosing $\psi = \psi_h \in \Psi_h$ as a test function and subtracting the discrete identity (22) from the original identity [6, Eq. 25.25] yields

$$\int_{\omega_0} (P_{0\delta}(\nabla s_0 - \nabla s_{0h})\nabla \psi_h + C_m \nabla P_0(\nabla s_0 - \nabla s_{0h})\psi_h) \, d\eta = 0.$$

We define a representative $\tilde{s}_0$ such that

$$\|s_0 - \tilde{s}_0\| = \inf_{\psi_h \in \Psi_h} \|s_0 - \psi_h\|.$$

We denote $\rho = s_0 - \tilde{s}_0$. By virtue of the previous estimates and standard finite element arguments [14], we have

$$\|\rho\|_{L^2(\omega_0)} \leq Ch^{r+1}\|s_0\|_{H^{r+1}(\omega_0)} \leq C\delta^{-\frac{3}{2}}h^{r+1}. \tag{24}$$

We denote $e = s_0 - s_{0h} = \rho + \theta$, where $\theta = \tilde{s}_0 - s_{0h}$ and for which we need to derive some estimate. Returning to the last subtraction, substituting and rearranging we obtain

$$\int_{\omega_0} (P_{0\delta}\nabla\theta\nabla\psi_h + C_m \nabla P_{0h}\nabla\theta\psi_h)d\eta = \int_{\omega_0} -(P_{0\delta}\nabla\rho\nabla\psi_h + C_m\nabla P_0\nabla\rho\psi_h)d\eta$$

$$= -(I_1 + I_2).$$

We apply Green's theorem to both terms of the left-hand side to obtain

$$\int_{\omega_0} \theta \, (P_{0\delta}\Delta\psi_h + (1 + C_m)\nabla P_0\nabla\psi_h + C_m\Delta P_0\psi_h) \, d\eta = I_1 + I_2. \tag{25}$$

Let us consider the elliptic problem

$$\begin{cases} P_{0\delta}\Delta y + (1 + C_m)\nabla P_0\nabla y + C_m\Delta P_0 y = \phi, & \text{in } \omega_0 \\ y = 0 & \text{on } \partial\omega_0. \end{cases} \tag{26}$$

We have a nondegenerate linear elliptic equation, for which if we have $\phi \in L^2(\omega_0)$, then $y \in H^2(\omega_0)$, review [3, 8]. By standard estimation techniques involving the use of Green's theorem, Cauchy's inequality, assumption **A4** stating that $\Delta P_0 \leq -M_0$, and assuming $\phi \in L^2(\omega_0)$ we can conclude that

$$\delta\|\nabla y\|^2_{L^2(\omega_0)} + C\|y\|^2_{L^2(\omega_0)} \leq C. \tag{27}$$

If ψ_h is the discrete solution of (26), then

$$\int_{\omega_0} \theta\phi d\eta \leq I_1 + I_2.$$

By assumption **A3** on P_0 and Hölder's inequality, we get for the right-hand side

$$|I_1| \leq \int_{\omega_0} |P_{0\delta}\nabla\rho\nabla\psi_h| d\eta$$

$$\leq C\|P_0\|_{L^\infty(\omega_0)} \int_{\omega_0} |\nabla\rho\nabla\psi_h| d\eta \leq C\|\nabla\rho\|_{L^2(\omega_0)} \|\nabla\psi_h\|_{L^2(\omega_0)}.$$

Using estimates (24), (27) and the inverse estimate, we conclude

$$|I_1| \leq C\delta^{-2}h^r, \quad \text{and} \quad |I_2| \leq C\delta^{-\frac{3}{2}}h^r.$$

Setting $\phi = \theta^{\beta-1}$ yields

$$\int_{\omega_0} |\theta|^\beta d\eta \leq C\delta^{-2}h^r.$$

Combining this estimate with the estimate (24) provides the first result. Using the triangle inequality with the estimate of [6, Theorem 25.6], we get the second result.

4.2   The Discrete Elliptic Equation

We consider the discretization of the regularized elliptic equation (19), for which we know that

$$g = div(2P_0 D^2(w) \cdot \nabla s_0) - 2C_m \nabla P_0 \cdot (D^2(w) \cdot \nabla s_0)$$
$$+ a(1-q)U_0^\alpha \Delta w + U_0^{m-2} H,$$

and H is the error function defined by (16). We assume that $H \in \Psi_h$. We introduce s_h as the discrete solution of the elliptic problem which is understood in the following sense.

Definition 4 A function $s_h \in \Psi_h$ is said to be a discrete solution of the regularized elliptic equation (19) and the degenerate elliptic equation in (15) if $s_h = 0$ on $\partial\omega_0$ and it satisfies the integral identity:

$$\int\limits_{t_1}^{t_2} \int\limits_{\omega_0} (P_{0\delta}\nabla s_h \nabla \psi + C_m \nabla P_0 \nabla s_h \, \psi + g_h \psi)\, d\eta dt = 0, \tag{28}$$

$\forall \, t_0 \le t \le T.$

Theorem 5 *Let s be the solution of (19), π be the solution of the elliptic equation in (15) and s_h be their discrete solution. If the conditions of [6, Corollary 25.10] hold, then for an arbitrary $0 < \beta < \infty$*

$$\|s - s_h\|_{L^\beta(Q)}^\beta \le C_1 \left(\delta^{-2} \sum_{i,j=1}^{2} \|D_{ij}^2(w - w_h)\|_{L^2(Q)} + (\delta\mu)^{-\frac{3}{2}} T h^{r-1} \right), \tag{29}$$

and

$$\|\pi - s_h\|_{L^\beta(Q)}^\beta \le C_2 \left(\|s - s_h\|_{L^\beta(Q)}^\beta + \delta^{\frac{1}{4}} \|g\|_{L^2(Q)} \right), \tag{30}$$

where C_1 and C_2 are constants that do not depend on δ, μ, h nor T.

Proof To prove this result we repeat the same steps as in the proof of the previous theorem; however, we add an extra integral I_3 in (25) which is expressed as

$$|I_3| = \left| \int\limits_Q (g - g_h)\psi_h d\eta dt \right| \le \int\limits_Q |(g - g_h)\psi_h| d\eta dt,$$

Expanding g and g_h, simplifying the produced terms, performing standard estimation techniques, assuming that ψ_h is the discrete solution of (26) at each instant t_j, applying the inverse estimate to (23), and recalling the estimates of [6, Corollary 25.7, Corollary 25.18] provides us with multiple terms. We consider the term with the highest order to get

$$|I_3| \le C \left(\delta^{-2} \sum_{i,j=1}^{2} \|D_{ij}^2(w - w_h)\|_{L^2(Q)} + \delta^{-\frac{3}{2}} \mu^{-\frac{3}{2}} T h^{r-1} \right).$$

Using the triangle inequality with the estimate of [6, Corollary 25.10] we conclude the second result.

4.3 The Discrete Parabolic Equation

We consider the function w_h as the discrete solution of the regularized parabolic equation (17) which is understood in the following sense.

Definition 6 A function $w_h \in \Psi_h$ is said to be a discrete solution of the regularized equation (17) and the degenerate parabolic equation in the system (15) if $w_h = 0$ on $\partial \omega_0$ and it satisfies the integral identity:

$$\int\limits_{t_1}^{t_2} \int\limits_{\omega_0} \left(-w_h \psi_t + (m-1)(P_{0\mu} \nabla w_h \nabla \psi + \nabla P_0 \nabla w_h \, \psi) - \tilde{f}_h \psi \right) d\eta dt = 0, \quad (31)$$

$\forall \, t_0 \le t_1 < t_2 \le T$.

We will consider the general case when $\tilde{f} = f + s - \Phi$, where s is the solution of the regularized elliptic equation (19) and f and Φ represent the error in the nonlinear system as per (16). We will assume that $\Phi(\eta, t_j) \in \Psi_h$ for every $t_0 \le t_j \le T$. Moreover, since in (16) we actually calculate Δf, then the projection $f_h(\eta, t_j)$ is calculated as $\Delta_h^{-1}(\Delta f(\eta, t_j))$. Hence, it is standard to consider the error $\| f - f_h \|_{L^2(\omega_0)}$ of order $\mathcal{O}(h^{r+1})$ for every $t_0 \le t_j \le T$, see [14].

Theorem 7 *Let w be the solution of the regularized parabolic equation (17), v be the solution of the degenerate parabolic equation in (15) and w_h be their discrete solution. If the conditions of [6, Theorem 25.16] hold, then for an arbitrary $0 < \beta < \infty$*

$$\| w - w_h \|^{\beta}_{L^{\beta}(Q)} \le C_1 \left(\mu^{-\frac{1}{2}} \| s - s_h \|_{L^2(Q)} + \mu^{-2} T (k^2 + h^r) \right),$$

and

$$\| v - w_h \|^{\beta}_{L^{\beta}(Q)} \le C_2 \left(\| w - w_h \|^{\beta}_{L^{\beta}(Q)} + \mu^{\frac{1}{2}} T^{\frac{1}{4}} \| f + s - \Phi \|_{L^2(Q)} \right),$$

where C_1 and C_2 are constants that do not depend on δ, μ, T, k nor h.

Proof We choose the test function $\psi = \psi_h \in \Psi_h$. We integrate by parts the first term of both identities [6, Eq. 25.44] and (31), then we subtract to get

$$\int\limits_{t_1}^{t_2} \int\limits_{\omega_0} \frac{\partial(w - w_h)}{\partial t} \psi_h d\eta dt + (m-1) \int\limits_{t_1}^{t_2} \int\limits_{\omega_0} P_{0\mu} \nabla (w - w_h) \nabla \psi_h d\eta dt$$

$$+ (m-1) \int\limits_{t_1}^{t_2} \int\limits_{\omega_0} \nabla P_0 \nabla (w - w_h) \psi_h d\eta dt$$

$$= \int\limits_{t_1}^{t_2} \int\limits_{\omega_0} (\tilde{f} - \tilde{f}_h) \psi_h d\eta dt.$$

We denote the approximation error by $e = w - w_h = \rho + \theta$, where $\rho = w - \tilde{w}$, $\theta = \tilde{w} - w_h$ and $\tilde{w}$ is a representative for w chosen as the elliptic projection of w which is defined as

$$b(\tilde{w}, \psi_h) = b(w, \psi_h),$$

where the bilinear form $b(x, y)$ is defined in the following way:

$$b(x, y) = (m - 1) \int_{\omega_0} (-P_{0\mu} \nabla x \nabla y - \nabla P_0 \nabla x \, y) d\eta.$$

We use the well known Cranck-Nicolson time discretization scheme. Using the previous estimates and assuming that w is three times differentiable with respect to t, we can verify by standard finite element argument that [14]

$$\max_{0 < j \le N} \|\rho\|_{L^2(\omega_0)} \le C \mu^{-\frac{3}{2}} T (k^2 + h^{r+1}). \tag{32}$$

Since we assume the continuity of the discrete solution with respect to t, we extend the time integral over Q. We rewrite the difference equation, make all substitutions and rearrange to obtain

$$\int_Q (e_t \psi_h + (m - 1)(P_{0\mu} \nabla e \nabla \psi_h + \nabla P_0 \nabla e \, \psi_h)) d\eta dt = \int_Q (f - f_h + s - s_h) \psi_h d\eta dt,$$

and by decomposing e, we arrive at

$$\int_Q (\theta_t \psi_h + (m - 1)(P_{0\mu} \nabla \theta \nabla \psi_h + \nabla P_0 \nabla \theta \, \psi_h)) d\eta dt$$

$$= \int_Q (f - f_h) \psi_h d\eta dt + \int_Q (s - s_h) \psi_h d\eta dt$$

$$- (m - 1) \left(\int_Q P_{0\mu} \nabla \rho \nabla \psi_h d\eta dt + \int_Q \nabla P_0 \nabla \rho \, \psi_h d\eta dt \right)$$

$$= I_1 + I_2 - I_3 - I_4.$$

We highlight that ρ_t were dropped because ρ was defined as an elliptic projection. We proceed trying to find an estimate for θ. Assuming that $\psi_h(\eta, T) = 0$, we integrate the first term on the left-hand side by parts to get

$$\int_Q (\theta \psi_{th} - (m - 1)(P_{0\mu} \nabla \theta \nabla \psi_h + \nabla P_0 \nabla \theta \, \psi_h)) d\eta dt = -I_1 - I_2 + I_3 + I_4.$$

By applying Green's theorem to the second and the third terms on the left-hand side we can deduce that

$$\int_Q \theta(\psi_{th} + (m-1)(P_{0\mu}\Delta\psi_h + 2\nabla P_0\nabla\psi_h + \Delta P_0\psi_h))d\eta dt \le \sum_i |I_i| \quad (33)$$

We introduce the uniformly parabolic equation

$$\begin{cases} y_t - (m-1)(P_{0\mu}\Delta y + 2\nabla P_0\nabla y + \Delta P_0 y) = \phi & \text{in } Q, \\ y(\eta, 0) = 0, & \text{in } Q \\ y = 0 & \text{on } \partial Q, \end{cases}$$

for which, if $\phi \in L^2(0, T, L^2(\omega_0))$, then $y \in L^2(0, T, H^2(\omega_0))$, review [3, 9]. Assuming that ϕ is bounded in $L^2(Q)$, we can use standard estimation techniques to conclude that

$$\|y\|_{L^2(Q)} \le \|\nabla y\|_{L^2(Q)} \le C\mu^{-\frac{1}{2}}. \quad (34)$$

If ψ_h is the discrete solution of this problem in the sense that $\psi_h(\eta, t_j) = y(\eta, T - t_j)$ for every $0 < t_j \le T$, then we can plug it in (33) to obtain

$$\int_Q \theta\phi d\eta \le \sum_i |I_i|.$$

Using the previous estimates and assumptions, applying Hölder's inequality, applying the inverse estimate to (32), and choosing $\phi = \theta^{\beta-1}$ we get the first result,

$$\int_Q |\theta|^\beta d\eta dt \le C\left(\mu^{-\frac{1}{2}}\|s - s_h\|_{L^2(Q)} + \mu^{-2}T(k^2 + h^r)\right),$$

and the second result follows by virtue of [6, Theorem 25.16] and the triangle inequality.

Theorem 8 *Suppose that the conditions and assumptions of Theorems 5 and 7 are fulfilled. Then, the minimum order of interpolating polynomials to be used in order to maintain the stability of the finite element scheme is $r \ge 3$.*

Proof We consider the mutual dependence of the errors in the elliptic and the parabolic equations. The error in the elliptic equation depends on the error in the projection of the second derivative of w. But the error in projecting w is of order $\mathcal{O}(h^r)$. The result is an immediate consequence of applying the inverse estimate twice to the estimate of Theorem 7.

5   Experiments

In all the experiments we employ an unstructured mesh with extra refinement near the boundaries and we construct the output of the algorithm using standard finite element implementation. We use $r = 3$ as a minimum order of the Lagrange polynomials. The initial instant is always considered as $t_0 = 1$. The solutions generated by the tracking algorithm will be denoted by u_c. Whenever needed to measure the error in the convergence of the iteration scheme we shall use u_n and u_{n-1} to denote the error in the last iteration and for that purpose only we will use $x_n - x_{n-1}$ to denote the convergence error in the position of the interface.

5.1   Experiment 1

We start with a benchmark experiment considering the standard Barenblatt solution. In this case we set $a = 0$ and no absorption should be present. The parameters of the experiment are provided by Table 1 and the exact formula that governs the solution's evolution is

$$u^{m-1} = \max\left(0, \frac{L}{t + t_0}\left(R^2(t + t_0)^\alpha - x^2 - y^2\right)\right),$$

where $t \in [0, T]$ and $t_0 = 1$. The initial data are calculated by the same formula at $t = 0$ considering $R = 4$. The constructed trajectories of the interface's particles are illustrated in Fig. 1. The relative error between the exact solution and the constructed solution is

$$\frac{\|u - u_c\|_{L^\infty(Q)}}{\|u\|_{L^\infty(Q)}} = 0.012.$$

In Fig. 2 we provide a combination of the exact profile at the last instant and the trajectories of the particles to illustrate the high accuracy of approximating the interface's position. The gradient of the potential function ∇v, which represents the normal motion of the interface as per (13), is shown in Fig. 3. We recall the error functional defined by (16); the norm of the error at the 25th Newton's iteration is provided for different quantities in Table 2. Note that $H = 0$ since we have no low order term and consequently $\pi = 0$.

Table 1 Parameters of Experiment 1

m	a	c_0	α	L	T
2	0	–	0.5	0.0625	1

(a)   **(b)**

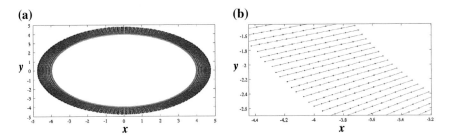

Fig. 1 Illustration of the trajectories of the interface's particles. **a** The motion of the interface. **b** Zoom on the trajectories

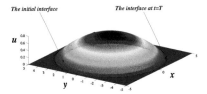

Fig. 2 The match between the constructed interface and the exact one

(a)   **(b)**

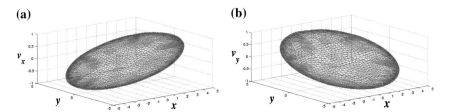

Fig. 3 Illustration of both components of the potential's gradient ∇v. **a** The x-component of ∇v. **b** The y-component of ∇v

Table 2 The convergence error of Newton's iterations for Experiment 1

$\|u_n - u_{n-1}\|_{L^\infty(Q)}$	$\|x_n - x_{n-1}\|_{L^\infty(Q)}$	$\|\Delta f\|_{L^\infty(Q)}$	$\|\Phi\|_{L^\infty(Q)}$	$\|H\|_{L^\infty(Q)}$
0.0013719	0.0052311	0.019367	0.0016108	0

5.2 Experiment 2

In the previous experiment we established a strong verification on the accuracy of the algorithm when we considered a suitable choice of the initial data. In this experiment we choose the initial function in an ellipsoidal domain according to the following formula:

$$u_0 = \max\left(0,\ L\left(R^2 - (x-y)^2 - 16(x+y)^2\right)\right)^{\frac{1}{m-1}},$$

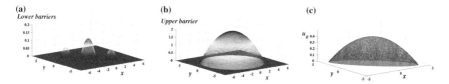

Fig. 4 The initial function u_0 and its lower and upper barriers. **a** The lower barriers at $t = 0$. **b** The upper barrier at $t = 0$. **c** The initial data u_0

Table 3 Parameters of Experiment 2

m	a	c_0	α	L	T
2	0	–	0.5	0.0625	1

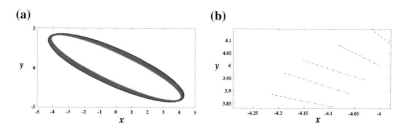

Fig. 5 Illustration of the trajectories of the interface's particles. **a** The motion of the interface. **b** Zoom on the trajectories

where $R = 8$. We can not use Barenblatt solution in a direct way as in the previous experiment. However, we can use the profiles admitted by his explicit formula as lower and upper barriers. Then we consider the maximum principle to establish another way of verification. The pictures in Fig. 4 illustrate the shape of the initial data, the inner and the outer barriers. We design the experiment with the parameters shown in Table 3.

The output of the interface tracking algorithm is illustrated in Fig. 5. We establish the verification by comparing the position of the interface at any instant with the position of the barriers' interfaces at the same instant. We choose the last instant and the verification is established as shown in Figs. 6 and 7. Since we have no exact solution to compare with we will only present the convergence error in the iteration scheme by Table 4.

5.3 Experiment 3

In this experiment we introduce a more complicated type of initial data. We will consider a square domain with edge length $2R$, over which the initial function u_0 is defined as

Fig. 6 Lower bound for the
interface location imposed
by the lower barriers

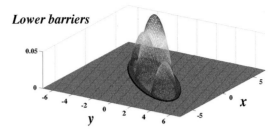

Fig. 7 Upper bound for the
interface location imposed
by the explicit solution

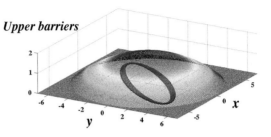

Table 4 The convergence error of Newton's iterations for Experiment 2

$\|u_n - u_{n-1}\|_{L^\infty(Q)}$	$\|x_n - x_{n-1}\|_{L^\infty(Q)}$	$\|\Delta f\|_{L^\infty(Q)}$	$\|\Phi\|_{L^\infty(Q)}$	$\|H\|_{L^\infty(Q)}$
6.2708×10^{-5}	0.0002442	0.0007	5.1818×10^{-5}	0

$$u_0 = \max(0, \ (R^2 - x^2)(R^2 - y^2)),$$

where $R = 4$. The difficulty with such type of initial data is the ability to correctly
simulate the evolution of the solution or the interface at the corners. The reason is
that at the corner points not only the solution vanishes, but also the gradient of the
solution does. Since the gradient is zero initially at this point we expect no motion
for a certain amount of time called the waiting time. During that time the profile of
the solution keeps adjusting itself until the moment when the gradient is non zero
and the corner point starts moving.

We will keep the configuration of the last experiment which means that no absorption will be present. We will employ the technique of the barriers as in the previous
experiment, and we will use the ellipsoidal initial function which was introduced and
verified in the previous experiment as an inner barrier to verify the evolution of the
corners. The upper barrier will be taken as the profile admitted by the explicit formula
and will be designed to touch the corners from outside. Moreover, we will consider
small time steps so that we can catch the moment when the corners start moving such
that $dt = 0.001$ and $T = 0.01$. Since the data is completely irregular we will only
consider $m = 2$, the rest of the parameters are meaningless in this context except
maybe for choosing a proper upper barrier. The initial data, the lower and the upper
barriers are shown in Fig. 8.

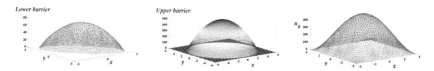

Fig. 8 The initial function u_0 and its lower and upper barriers. **a** The lower barriers at $t = 0$. **b** The upper barrier at $t = 0$. **c** The initial data u_0

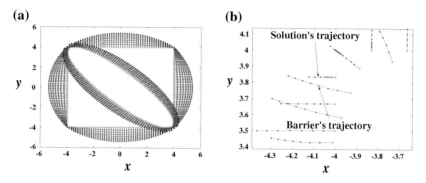

Fig. 9 Non-agreement with the maximum principle. **a** The trajectories at $t = 0.007$. **b** Inaccurate interface positioning

We start by showing the output of the direct implementation of the algorithm accompanied with the evolution of the interface of the inner ellipsoidal barrier in Fig. 9. We note that the algorithm fails to complete after $t = 0.007$. We recall that the mass conservation law in the Lagrangian coordinates takes the form

$$u = \frac{u_0}{|J|}.$$

As we explained earlier, the profile of the solution near the corners keeps raising which means that $|J|$ around the corners keeps decreasing. Actually, the moment when the corner starts moving is when $|J|$ approaches zero, which consequently generates explosion in the output of the algorithm, see Fig. 10. Another drawback of

Fig. 10 The profile of $|J|$ approaching zero at the *corners*

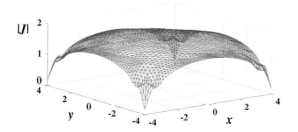

the direct implementation of the algorithm is that it keeps tracking the motion of the interface in the normal direction to the interface of the initial data. The changes in the normal vector itself is not considered which leads to inaccurate tracking of the interface's position. As shown in Fig. 9b, the trajectories of the lower barriers exceeds those of the solution, which means that there exists a region where the profile of the lower barrier is actually higher than the solution. This indicates that the maximum principle is not fulfilled in this case.

To overcome both problems we implement the algorithm in an adaptive way. Instead of constructing the solution in the whole cylinder we only construct the solution for three time steps at once. In other words, we perform the iteration process considering three time steps, then we set the solution at the third time step as the initial data of the next three time steps. In that way we keep tracking the changes in the interface's normal direction instead of being stuck to the normal direction of the initial interface. Moreover, when we reset the algorithm every three time steps, we actually reset J to the identity matrix and thus we keep $|J|$ separated away from the zero level.

The reason for choosing three time steps is that we approximate the time derivative of the potential v_t using three points scheme, so this is the minimum number of instants required to initiate the algorithm. It can be reduced to two instants in the case of approximating the time derivatives using two points scheme but not less than that. The only drawback of such scheme is the accumulation of the convergence error of the iteration process in each realization specially at the corners in our case, see the convergence error in the last realization in Table 5. We illustrate the accurate motion of the interface together with a focus on the corner motion in Fig. 11. The interface's trajectories generated by the algorithm are bounded from below by the interface of

Table 5 The convergence error of Newton's iterations for Experiment 3

$\|u_n - u_{n-1}\|_{L^\infty(Q)}$	$\|x_n - x_{n-1}\|_{L^\infty(Q)}$	$\|\Delta f\|_{L^\infty(Q)}$	$\|\Phi\|_{L^\infty(Q)}$	$\|H\|_{L^\infty(Q)}$
0.31521	0.0011882	9.3055	0.079621	0

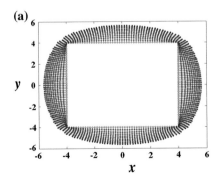
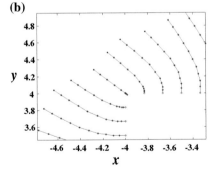

Fig. 11 The interface's particles trajectories with a zoom on the corners. **a** The interface's trajectories. **b** Zoom on the corner's motion

Fig. 12 Lower bound for the interface location imposed by the lower barrier

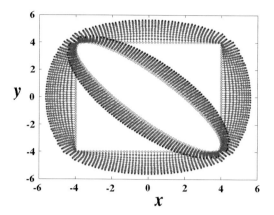

Fig. 13 Upper bound for the interface location imposed by the explicit solution

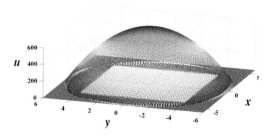

the lower barrier and from above by the interface of the upper barrier. We represent this result at the last instant as illustrated in Figs. 12 and 13.

5.4 Experiment 4

In the previous experiments we were considering the Porous Medium Equation with different choices of the initial data and the low order term was not present. In this experiment we consider the diffusion-absorption equation (5) with the presence of a linear sink, in other words we set $q = 1$. To provide a way of verification we perform a couple of transformations, first we hide the linear term away then we scale the time and thus we can easily retrieve the Porous Medium Equation again. To clarify, let us consider the following diffusion-absorption equation with a linear sink

$$u_t + au = \Delta u^m.$$

Multiply both sides by e^{at}, let $v = e^{at}u$, and consider a new time scale τ such that $e^{at(m-1)}\partial_t = \partial_\tau$. By integrating both sides we conclude that

$$\tau = \tau_0 + \frac{1 - e^{-at(m-1)}}{a(m-1)}, \tag{35}$$

where we choose $\tau_0 = 1$. Thus we can obtain

$$v_\tau = \Delta v^m, \tag{36}$$

which is the Porous Medium Equation for the new variable v on the time scale denoted by τ. The purpose of this experiment is to run the algorithm in the presence of the linear low order term over the interval $t \in [0, T]$. We choose the initial data as

$$u_0 = \max\left(0, L\left(R^2 - x^2 - y^2\right)\right)^{\frac{1}{m-1}},$$

where $R = 4$. Since the initial data are chosen from the set of Barenblatt profiles, then we can verify the output of the algorithm at $t = T$ by comparing it with the exact Barenblatt profile

$$v^{m-1} = \max\left(0, \frac{L}{\tau}\left(R^2\tau^\alpha - x^2 - y^2\right)\right),$$

calculated at $\tau = \mathscr{T}$, where

$$\mathscr{T} = \tau_0 + \frac{1 - e^{-aT(m-1)}}{a(m-1)}.$$

Note that $v_0 = u_0$. We will consider three cases in this experiment. The parameters of the first case are shown in Table 6.

We recall from [6] that in the presence of the low order term we need to reduce the domain in order to compensate the loss of the mass due to the presence of the sink. Moreover, we will consider zero Dirichlet boundary conditions for π on both the exterior and the interior boundaries. The profile of u_0 and the corresponding pressure π_0 in the reduced domain are shown in Fig. 14. The calculated motion of both the exterior and the interior interfaces is illustrated in Fig. 15.

In Fig. 16 we represent a comparison between the location of the interface produced by the algorithm and the location calculated as per Barenblatt profile at $\tau = \mathscr{T}$. We note in the zoom provided in Fig. 16b that the calculated location of the interface is not accurately matching the interface of the exact profile–at

Table 6 Parameters of Experiment 4a

m	a	c_0	α	L	T	$\mathscr{T}$
2	0.1	–	0.5	0.0625	0.2	1.198

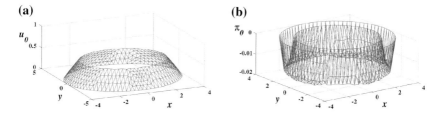

Fig. 14 The initial functions u_0 and π_0 in the reduced domain. The initial data u_0. **b** The initial pressure π_0

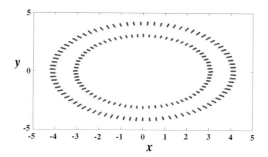

Fig. 15 The interfaces motion in the case when $a = 0.1$ and $\pi = 0$ on both interfaces

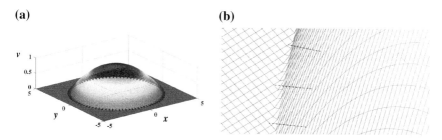

Fig. 16 Comparing the exact location of the exterior interface and the calculated one. **a** Verification of the match. **b** Zoom on the exterior interfaces

least not at all the points. To understand this behaviour we recall from [6] that the problem in the reduced domain is a new problem and completely different than the original one, and that the solutions of both problems coincides only on the exterior interface. We need also to recall the formula of the velocity (7). The velocity of the particles depends on the gradient of the pressure induced by the diffusion (∇p) and the gradient of the pressure induced by the low order term $(\nabla \pi)$. As shown in Fig. 17, the slope of the profile of π supports a motion in the outward direction on the side of the exterior interface. On the other hand, the motion of the interior boundary should maintain the condition of the mass conservation regardless the direction or the rate of this motion. The choice of zero Dirichlet boundary conditions for π on both interfaces restricts the profile of π from being corrected quickly. However, as

(a)   **(b)**

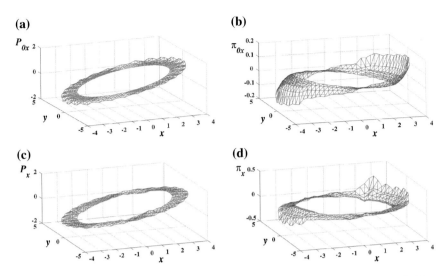

Fig. 17 Illustration of both components of the velocity in x direction. **a** The x-component of $-\nabla p_0$. **b** The x-component of $\nabla \pi_0$. **c** The x-component of $-\nabla p(x, T)$. **d** The x-component of $\nabla \pi x$, T

per [6, Theorem 25.25], the convergence of the algorithm in the reduced domain entails a small time interval, which in turn does not allow sufficient corrections of π.

To overcome this problem we keep the zero Dirichlet boundary condition on the exterior interface only and we leave the interior interface free of any conditions. However, such approach was found to diverge when we choose large values of a. The reason can be seen by observing the sharp slope of π on the exterior boundary. We need to keep in mind that even if the profile of π implies a motion in the outward direction with respect to the exterior interface, it is not necessarily implying the same motion with the same rate in the interior of the domain. Larger values of a can easily lead to the deformation of the mesh, which in this case entails the re-initiation of the algorithm. Accordingly, we choose a smaller value $a = 0.01$ and we rerun the algorithm. Liberating π from any conditions on the interior interface improves the accuracy of the algorithm and a better match can be seen in Fig. 18. The profile of π with the free interior surface is shown in Fig. 19. Yet, analytical justification of this approach is still not revealed.

In the third case we reduce the value of a, consider the parameters in Table 7. In Fig. 20 we can note that this motion is slightly slower compared to the above case with the higher value of a. That should not be confusing; in the reduced domain there is no absorption and the mass is conserved. The motion of the interface is controlled according to a new system of equations and the function π is meant only to translate the dynamics of the absorption from the original domain into the reduced one without losing the mass conservation. By performing several experiments one can deduce that, in the case of a linear sink, the diffusion process is dominating and the motion of the interface is always in the outward direction only with different rates. The strength

(a)   **(b)**

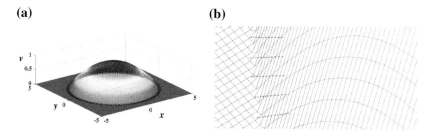

Fig. 18 Comparing the exact location of the exterior interface and the calculated one. **a** Verification of the match. **b** Zoom on the exterior interfaces

(a)   **(b)**

(c)   **(d)**

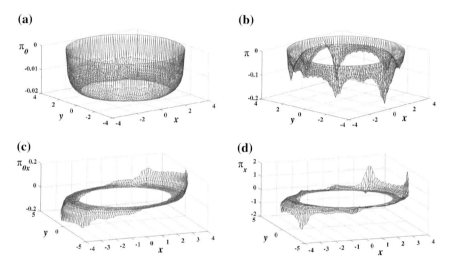

Fig. 19 The profile and the gradient of π in the case of no condition on the interior surface. **a** The profile of π_0. **b** The profile of $\pi(x, T)$. **c** The x-component of $\nabla \pi_0$. **d** The x-component of $\nabla \pi(x, T)$

Table 7 Parameters of Experiment 4b

m	a	c_0	α	L	T	$\mathcal{T}$
2	0.001	–	0.5	0.0625	0.2	1.2

of the low order term affects the motion and such effect is implicitly translated by π which is defined in terms of u in the first place.

To clarify our findings more, let us observe the profile of π in Figs. 17d and 19d. Both profiles share a sharp rate at the exterior interface which implies an outward motion in both cases. On the other hand, the flatness of the second profile (Fig. 19d) in the interior of the domain implies a slower motion inside the ring. In spite of having no clear motion in the inward direction, the dynamics of the absorption can be seen in the reduction of the rate with witch the inner particles move even if such motion is in the outward direction.

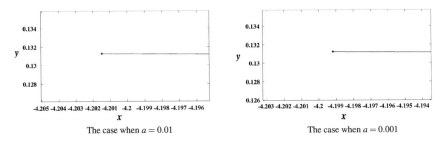

Fig. 20 Comparison of the displacements of the exterior interface for different values of a. **a** The case when $a = 0.01$. **b** The case when $a = 0.001$

Table 8 The convergence error of Newton's iterations for Experiment 5 ($a = 0.01$)

$\|u_n - u_{n-1}\|_{L^\infty(Q)}$	$\|x_n - x_{n-1}\|_{L^\infty(Q)}$	$\|\Delta f\|_{L^\infty(Q)}$	$\|\Phi\|_{L^\infty(Q)}$	$\|H\|_{L^\infty(Q)}$
0.1152	0.1332	0.5234	0.0518	0.0377

Table 9 The convergence error of Newton's iterations for Experiment 5 ($a = 0.001$)

$\|u_n - u_{n-1}\|_{L^\infty(Q)}$	$\|x_n - x_{n-1}\|_{L^\infty(Q)}$	$\|\Delta f\|_{L^\infty(Q)}$	$\|\Phi\|_{L^\infty(Q)}$	$\|H\|_{L^\infty(Q)}$
0.0117	0.0105	0.1085	8.7339e-04	0.0021

Finally we need to highlight the necessity to implement the algorithm over a small time intervals, refer to [6, Theorem 25.25]. In the case of large time intervals direct implementation of the algorithm fails and Newton's iterations diverge. This can be fixed by sequential implementation over successive small intervals as explained in *Experiment 3*. However, with each iteration process a certain amount of error remains due to the incomplete convergence, and these amounts of error accumulates each time the algorithm is re-initiated. See Tables 8 and 9 for details on the level of the convergence errors due to a single iteration process for each case of the last two cases.

6 Failures

In spite of the satisfactory results of the tracking algorithm in detecting the motion of the interface they were only tangible in the case of the forward motion. The algorithm fails to detect the backward motion of the interface. In other words it can not represent the process of the absorption. The reason is not related to the mathematical model upon which the algorithm works. It is more likely the numerical environment itself. In the case when the diffusion process takes place, the particles move outside and the worst scenario is the deformation of the mesh which can be overcome by reconstructing the mesh and re-initiating the algorithm as we explained in the last experiment.

On the other hand, when the absorption process occurs the particles of the interface move inside with a rate much faster than the rate of the motion of the particles inside the domain, which eventually causes the intersection of the particles' trajectories. These collisions are not allowed in the Lagrangian coordinates as the inverse transformation becomes impossible. Imposing restrictions on the rate of the motion of the interface is not a solution since it contradicts the nature of the problem and the target of the study which aims at maintaining both the physical and the mathematical measures of the solution. Nonetheless, the particles inside the domain move with a rate much slower than the rate of the motion of the interface's particles and not necessarily in the inward direction; especially at the initial instant.

7 Conclusions

In this article we managed to deduce a rough estimate for the numerical error at each Newtonian iteration in terms of the discretization and the regularization parameters. We also concluded the minimum order of the finite element's interpolating polynomials to be used in order to preserve the stability of the algorithm. We validated the output of the interface tracking algorithm in the case when the diffusion is dominating over the absorption. We illustrated the drawbacks of the direct implementation of the algorithm and proposed an adaptive scheme which enhances the accuracy of the interface tracking.

It was discovered that the numerical implementation failed to work when the absorption prevails over the diffusion. The reason is that the algorithm, based on the mechanical idea of the possibility of two counterpart representation of the motion of a continuum, can not proceed when the particles' trajectories cross and the particles collide. The nature of the problem imposes the possibility of such collisions and the numerical environment can not prevent it.

We shall design some kind of monitor-control-correction scheme to prevent the particles from colliding in the presence of absorption forces. The proper choice of the boundary conditions on the artificial free boundary also can greatly enhance the output of the algorithm. Another promising possibility is to use a generalization of the interface equation [6, Eq. 25.2] to the case of several space dimensions in order to correct the interface location at every time step.

References

1. Diaz, J., Shmarev, S.: Lagrangian approach to the study of level sets: application to a free boundary problem in climatolog. Arch. Ration. Mech. Anal. **194**, 75–103 (2009)
2. Diaz, J., Shmarev, S.: Lagrangian approach to the study of level sets. II. A quasilinear equation in climatology. J. Math. Anal. Appl. **352**, 475–495 (2009)
3. Evans, L.: Partial Differential Equations, vol. 19, 2nd edn. In: Graduate Studies in Mathematics. American Mathematical Society (2010)

4. Galiano, G., Shmarev, S., Velasco, J.: Existence and nonuniqueness of segregated solutions to a class of cross-diffusion systems. arXiv:1311.3454v1 [math.AP] (2013)

5. Kalashinkov, A.: Some problems of the qualitative theory of non-linear degenerate second-order parabolic equations. Russian Math. Surveys **42**(2), 169–222 (1987)

6. Khedr, W.S.: Tracking the Interface of the Diffusion-Absorption Equation: Theoretical Analysis, to appear as Chapter 25 in this proceedings

7. Kolmogorov, A.N., Fomin, S.V.: Elements of the theory of functions and functional analysis, measure, the Lebesgue integral, Hilbert, vol. 2.Translated from the first (1960) Russian ed. by Hyman Kamel and Horace Komm. Graylock Press, Albany (1961)

8. Ladyzhenskaya, O.A., Uralćtseva, N.A.: Linear and quasilinear elliptic equations. Translated from the Russian by Scripta Technica, Inc., Translation editor: Leon Ehrenpreis. Academic Press, New York-London (1968)

9. Ladyzhenskaya, O.A., Solonnikovm, V.A., Uralćtseva, N.A.: Linear and quasilinear equations of parabolic type. Translated from the Russian by S. Smith. Translations of Mathematical Monographs, vol. 23. American Mathematical Society, Providence, R.I. (1968)

10. Shmarev, S.: Interfaces in multidimensional diffusion equations with absorption terms. Nonlinear Anal. **53**, 791–828 (2003)

11. Shmarev, S.: Interfaces in solutions of diffusion-absorption equations in arbitrary space dimension. In: Trends in partial differential equations of mathematical physics. Progr. Nonlinear Differential Equations Appl. Birkhäuser, Basel **61**, 257–273 (2005)

12. Shmarev, S.: Interfaces in solutions of diffusion-absorption equations. RACSAM Rev. R. Acad. Cienc. Exactas Fís. Nat. Ser. A Mat. **96**, 129–134 (2002)

13. Shmarev, S.: On a class of degenerate elliptic equations in weighted Hölder spaces. Diff. Integr. Eqn. **17**, 1123–1148 (2004)

14. Thomée, V.: Galerkin finite element methods for parabolic problems. Springer, Secaucus (2006)

Analyzing Multivariate Cross-Sectional Poisson Count Using a Quasi-Likelihood Approach: The Case of Trivariate Poisson

Naushad Mamode Khan, Yuvraj Sunecher and Vandna Jowaheer

Abstract In the last few years, the modelling of multivariate count data has been a topic of concern for many researchers in the field of epidemiology, agriculture, economics and finance. The most recent findings in the analysis of such data illustrate that it is easier to specify the likelihood function of these multivariate count responses through the use of copula constructors such as Clayton and Frank copulas. However, in a regression set-up, the resulting maximum likelihood estimation equation involves huge computationally intensive expressions that make the methodology almost unfeasible. This raises the need to explore some other parsimonious estimation methodologies. In this context, this paper proposes an alternative estimation scheme based on the quasi-likelihood (QL) approach. The performance of the QL method is assessed through a simulation study that is based on an observation-driven generating process for trivariate and multivariate Poisson counts.

1 Background and Motivation

In the last five years, the modelling of multivariate count data has preoccupied the interest of many researchers as such data appear in various disciplines related to sports, epidemiology, criminology among many others [5, 14]. Under the simple case of bivariate Poisson, the analysis of counts do not pose any modelling and

The original version of this chapter was revised: The erratum to this chapter is available at DOI 10.1007/978-3-319-30322-2_30.

N. Mamode Khan · V. Jowaheer
University of Mauritius, Le Reduit, Moka, Mauritius
e-mail: n.mamodekhan@uom.ac.mu

V. Jowaheer
e-mail: vandnaj@uom.ac.mu

Y. Sunecher (✉)
University of Technology, La Tour Koenig, Port Louis, Mauritius
e-mail: yuvisun@yahoo.com

© Springer International Publishing Switzerland 2016
G.A. Anastassiou and O. Duman (eds.), *Intelligent Mathematics II:
Applied Mathematics and Approximation Theory*, Advances in Intelligent Systems
and Computing 441, DOI 10.1007/978-3-319-30322-2_27

estimation challenges since the likelihood function can easily be expressed [6]. However, in trivariate or multivariate set-up, the expression for the likelihood function becomes tedious even though copula is used [3, 10, 12] and thus necessitates numerical approximations for the application of the maximum likelihood approach (MLE) [13]. In the same context, Nikoloulopoulos and Karlis [11] have developed the multivariate Poisson model but in terms of applications, these authors have restricted their analysis to a simple bivariate data set-up, thus not portraying the real difficulties. In this paper, we review their derivation of the trivariate Poisson model and particularly highlight the difficulties in obtaining the derivatives of the log-likelihood function under Frank Copula assumption. Our concern here is to propose an alternative approach that would reduce the computational difficulties in obtaining the regression and correlation estimates whilst at the same time maintain the level of efficiency of the estimates. Until now, the only alternative methodology that omits the specification of the likelihood function is the quasi-likelihood based approach (QL). This is an approach that has been expounded by Wedderburn [15] and McCullagh and Nelder [8] and lately in longitudinal system by Jowaheer and Sutradhar [4] as a perfect substitute to MLE. Overall, QL equation has a parsimonious structure constituting of the derivative and covariance components only.

As for the covariance components, the only work carried out so far in this direction is the formulation of the dependence structure by Nikoloulopoulos and Karlis [11] which is based on the use of copulas. Their covariance formulation has yielded consistent estimates but is very complicated. This complexity arises because of the structure of the copula and as a result this influences the time and computational involvements. To overcome this challenge, this paper proposes to model the dependence among the variates through an observation-driven modelling strategy [1, 9] and the moment estimating equations. Thus, to summarize, our contribution in this paper is to first develop an observation-driven model that connects the multivariate counts and thereafter obtain the covariance expressions. The second contribution of this paper is to assess the QL method in the estimation of parameters mainly the regression and dependence parameters under trivariate and multivariate cross-sectional set-ups which has yet not been covered in statistics literature. The outline of this paper is as follows: In the next section, the MLE approach under trivariate Poisson regressions are discussed followed by the development of observation-driven process and QL methodology under trivariate and multivariate set-ups in Sects. 4 and 5, respectively. Section 6 focuses on simulation studies to assess the performance of QL under trivariate set-ups based under different parameter assumptions. The conclusion and the scope for future work are presented in the last section.

2   The Cross-Sectional Set-Up

This section provides an overview on the notations used in this paper along with the basic assumptions. Under a multivariate cross-sectional set-up, where we assume $k = 1, 2, 3, \ldots, p$ variates, the vector of responses is represented by

$$Y = \begin{pmatrix} Y_1^{[1]} & Y_1^{[2]} & \cdots & Y_1^{[k]} & \cdots & Y_1^{[p]} \\ Y_2^{[1]} & Y_2^{[2]} & \cdots & Y_2^{[k]} & \cdots & Y_2^{[p]} \\ \vdots & \vdots & \ddots & \vdots & \vdots & \vdots \\ \cdots & \cdots & Y_i^{[k]} & \cdots & Y_i^{[k']} & \cdots \\ Y_I^{[1]} & Y_I^{[2]} & \cdots & Y_I^{[k]} & \cdots & Y_I^{[p]} \end{pmatrix}$$

where $i = 1, 2, \ldots, I$ independent individuals. Under this representation, there is a clear dependence between $y_i^{[k]}$ and $y_i^{[k']}$ but obviously there exists no association between measurements $y_i^{[k]}$ and $y_j^{[k]}$ and $y_i^{[k]}$ and $y_j^{[k']}$, that is,

$$Cov(Y_i^{[k]}, Y_j^{[k]}) = 0, \, Cov(Y_i^{[k]}, Y_j^{[k']}) = 0,$$

Under the first dependence, Nikoloulopoulos and Karlis [11] formulated the dependence coefficient between any variate Y_1 and Y_2 as

$$\tau(Y_1, Y_2) = \sum_{y_1=0}^{\infty} \sum_{y_2=0}^{\infty} h(y_1, y_2)$$
$$*4C(F_1(y_1 - 1), F_2(y_2 - 1)) - h(y_1, y_2)$$
$$+ \sum_{y_1=0}^{\infty} (f_1^2(y_1) + f_2^2(y_1)) - 1$$

Suppose these responses are influenced by a common set of J explanatory variables represented by the structure

$$X = \begin{pmatrix} X_{11} & X_{12} & \cdots & X_{1J} \\ X_{21} & X_{22} & \cdots & X_{2J} \\ \vdots & \vdots & & \vdots \\ X_{i1} & X_{i2} & \cdots & X_{iJ} \\ \vdots & \vdots & & \vdots \\ X_{I1} & X_{I2} & \cdots & X_{IJ} \end{pmatrix}$$

and the vector of the kth variate regression parameter

$$\beta^{[k]} = [\beta_1^{[k]}, \beta_2^{[k]}, \ldots, \beta_i^{[k]}, \ldots, \beta_J^{[k]}]$$

3 A Note on the MLE Approach

To proceed to the multivariate set-up, we first consider the situation under the trivariate mode. Thus, in this section, the trivariate Poisson likelihood functions are developed under the Frank copula assumption, that is,

$$C(u, v) = \frac{1}{\theta} ln[1 + \frac{(exp(-\theta u) - 1)(exp(-\theta v) - 1)}{exp(-\theta) - 1}]$$

For this, we assume the marginal Poisson density function for the kth variate where $k = 1, 2, 3$ is given by

$$f_k(y_i) = exp(-\mu_i^{[k]}) \frac{[\mu_i^{[k]}]^{y_i}}{y_i!}$$

where $i = 1, 2, 3, \ldots, I$ subjects and where, following Sect. 2,

$$\mu_i^{[k]} = exp(x_i \beta^{[k]})$$

and the marginal distribution function is given by

$$F_k(y_i) = \sum_{s=0}^{y_i} exp(-\mu_i^{[k]}) \frac{[\mu_i^{[k]}]^s}{s!}$$

The trivariate Poisson probability function based on copula representation is then expressed as

$$
\begin{aligned}
-h(y_1, y_2, y_3) = {} & C(F_1(y_1), F_2(y_2), F_3(y_3)) \\
& + C(F_1(y_1 - 1), F_2(y_2) F_3(y_3)) \\
& + C(F_1(y_1), F_2(y_2 - 1) F_3(y_3)) \\
& + C(F_1(y_1), F_2(y_2) F_3(y_3 - 1)) \\
& - C(F_1(y_1 - 1), F_2(y_2 - 1) F_3(y_3)) \\
& - C(F_1(y_1 - 1), F_2(y_2) F_3(y_3 - 1)) \\
& - C(F_1(y_1), F_2(y_2 - 1) F_3(y_3 - 1)) \\
& + C(F_1(y_1 - 1), F_2(y_2 - 1) F_3(y_3 - 1))
\end{aligned}
$$

From the above, it is clear that taking the derivative of the log-likelihood function with respect to the vector of regression parameters will not be a straightforward task. Obviously, further computational problems will arise when assuming a high-dimensional space of variates.

4 Developing an Observation-Driven Model Under Trivariate Assumption and Its Extension to a Multivariate Set-Up

Following the drawbacks of the copula approach, we aim to establish a model whereby under the trivariate assumption, the variates $[y_i^{[1]}, y_i^{[2]}, y_i^{[3]}]$ are related by a certain dependence structure of the form To proceed with the derivation of the observation-driven approach, assume

$$y_i^{[1]} \sim P(\mu_i^{[1]}) \tag{1}$$

$$y_i^{[2]} = \alpha_1 * y_i^{[1]} + d_i^{[1]} \tag{2}$$

$$y_i^{[3]} = \alpha_1 * y_i^{[1]} + \alpha_2 * y_i^{[2]} + d_i^{[2]} \tag{3}$$

where $y_i^{[2]} \sim P(\mu_i^{[2]})$, $y_i^{[3]} \sim P(\mu_i^{[3]})$, $d_i^{[1]} \sim P(\mu_i^{[2]} - \alpha_1 \mu_i^{[1]})$ and $d_i^{[2]} \sim P(\mu_i^{[3]} - \alpha_1 \mu_i^{[1]} - \alpha_2 \mu_i^{[2]})$. Note here, the dependence between $y_i^{[1]}, y_i^{[2]} y_i^{[3]}$ are induced by α_1, α_2 and the error terms $d_i^{[k]}$. Under these assumptions,

$$
\begin{aligned}
E(Y_i^{[2]}) &= E_{Y_i^{[1]}} E[\alpha_1 * y_i^{[1]} + d_i^{[1]} | Y_i^{[1]}] \\
&= E_{Y_i^{[1]}}[\alpha_1 y_i^{[1]} + \mu_i^{[2]} - \alpha_1 \mu_i^{[1]}] \\
&= \alpha_1 \mu_i^{[1]} + \mu_i^{[2]} - \alpha_1 \mu_i^{[1]} \\
&= \mu_i^{[2]}
\end{aligned}
$$

Under the same principle

$$E(Y_i^{[2]}) = \mu_i^{[2]} \text{ and } E(Y_i^{[3]}) = \mu_i^{[3]}$$

$$
\begin{aligned}
Var(Y_t^{[2]}) &= Var[\alpha_1 * y_i^{[1]}] + Var(d_i^{[1]}) \\
&= E_{Y_i^{[1]}}[Var(\alpha_1 * y_i^{[1]} | y_i^{[1]})] + Var_{Y_i^{[1]}}[E(\alpha_1 * y_i^{[1]} | y_i^{[1]})] \\
&\quad + Var[d_i^{[1]}] \\
&= E_{Y_i^{[1]}}[\alpha_1(1 - \alpha_1) y_i^{[1]} | y_i^{[1]}] + Var_{Y_i^{[1]}}[\alpha_1 y_i^{[1]} | y_i^{[1]}] \\
&\quad + Var[d_i^{[1]}] \\
&= \alpha_1(1 - \alpha_1) \mu_i^{[1]} + \alpha_1^2 \mu_i^{[1]} + \mu_i^{[2]} - \alpha_1 \mu_i^{[1]} \\
&= \mu_i^{[2]}
\end{aligned}
$$

Hence,

$$Var(Y_i^{[1]}) = \mu_i^{[1]}, \ Var(Y_i^{[2]}) = \mu_i^{[2]}, \ Var(Y_i^{[3]}) = \mu_i^{[3]}$$

We formulate expressions for $E(Y_i^{[1]}Y_i^{[2]})$

$$
\begin{aligned}
E(Y_i^{[1]}Y_i^{[2]}) &= E[y_i^{[1]}(\alpha_1 * y_i^{[1]} + d_i^{[1]})] \\
&= E[Y_i^{[1]}(\alpha_1 * Y_i^{[1]}) + Y_i^{[1]})d_i^{[1]}]
\end{aligned}
$$

From above,

$$
\begin{aligned}
E[Y_i^{[1]}(\alpha_1 * Y_i^{[1]})] &= E_{Y^{[1]}}E[Y_i^{[1]}(\alpha_1 * Y_i^{[1]})|Y_i^{[1]}] \\
&= E[Y_i^{[1]}(\alpha_1 Y_i^{[1]})] \\
&= \alpha_1 E[Y_i^{[1]^2}] \\
&= \alpha_1[\mu_i^{[1]} + \mu_i^{[1]^2}]
\end{aligned}
$$

Hence,

$$
\begin{aligned}
E(Y_i^{[1]}Y_i^{[2]}) &= \alpha_1[\mu_i^{[1]} + \mu_i^{[1]^2}] + \mu_i^{[1]}(\mu_i^{[2]} - \alpha_1\mu_i^{[1]}) \\
&= \alpha_1\mu_i^{[1]} + \mu_i^{[1]}\mu_i^{[2]}
\end{aligned}
$$

Since $E(y_i^{[1]}) = \mu_i^{[1]}$ and $E(y_i^{[2]}) = \mu_i^{[2]}$,

$$Cov(Y_i^{[1]}, Y_i^{[2]}) = E(Y_i^{[1]}Y_i^{[2]}) - E(Y_i^{[1]})E(Y_i^{[2]}) = \alpha_1\mu_i^{[1]} \qquad (4)$$

Similarly,

$$
\begin{aligned}
Cov(Y_i^{[1]}, Y_i^{[3]}) &= \alpha_2 Cov[Y_i^{[1]}Y_i^{[2]}] + \alpha_1\mu_i^{[1]} \\
&= \alpha_1\alpha_2\mu_i^{[1]} + \alpha_1\mu_i^{[1]} \qquad (5)
\end{aligned}
$$

$$
\begin{aligned}
Cov(Y_i^{[2]}, Y_i^{[3]}) &= \alpha_1\alpha_2 Cov[Y_i^{[1]}Y_i^{[2]}] + \alpha_1^2\mu_i^{[1]} + Cov(d_i^{[1]}d_i^{[2]}) \\
&= \alpha_1^2\alpha_2\mu_i^{[1]} + \alpha_1^2\mu_i^{[1]} + Cov(d_i^{[1]}, d_i^{[2]}) \qquad (6)
\end{aligned}
$$

At this stage, the dependence parameters $\alpha_1, \alpha_2, \phi_{12} = Cov(d_i^{[1]}, d_i^{[2]})$ are estimated using the method of moments and this yields: Given

$$\hat{\rho}_{jk} = \frac{\sum_{i=1}^{I}\sum_{i=1}^{I}(y_i^{[j]} - \mu_i^{[j]})(y_i^{[k]} - \mu_i^{[k]})}{\sqrt{\sum_{i=1}^{I}(y_i^{[j]} - \mu_i^{[j]})^2}\sqrt{\sum_{i=1}^{I}(y_i^{[k]} - \mu_i^{[k]})^2}}$$

Then, by equating (4), (5) and (6) with the sample moment estimators,

$$\hat{\alpha}_1 = \frac{\hat{\rho}_{12}\sqrt{\mu_i^{[2]}}}{\sqrt{\mu_i^{[1]}}},$$

$$\hat{\alpha}_2 = \frac{\hat{\rho}_{13}\sqrt{\mu_i^{[3]}}}{\hat{\rho}_{12}\sqrt{\mu_i^{[2]}}} - 1,$$

$$\hat{\phi} = (\hat{\rho}_{23} - \hat{\rho}_{13}\hat{\rho}_{12})\sqrt{\mu_i^{[2]}}\sqrt{\mu_i^{[3]}}$$

Now, for a general p-variate set-up,

$$y_i^{[1]} \sim P(\mu_i^{[1]})$$
$$y_i^{[2]} = \alpha_1 * y_i^{[1]} + d_i^{[1]}$$
$$y_i^{[3]} = \alpha_1 * y_i^{[1]} + \alpha_2 * y_i^{[2]} + d_i^{[2]}$$
$$\cdots$$
$$y_i^{[p]} = \alpha_1 * y_i^{[1]} + \alpha_2 * y_i^{[2]} + \cdots + \alpha_{p-1} * y_i^{[p-1]} + d_i^{[p-1]}$$

where

$$d_i^{[l]} \sim P(\mu_i^{[l+1]} - \sum_{j=1}^{l} \alpha_j \mu_i^{[l]})$$

Following the moment estimation procedures developed above, here, as well, for any p variate set-up, the sample correlations between the different variates will be used for estimating the dependence parameters.

5 The QL Approach for the Multivariate Cross-Sectional Set-Up

The QL approach [8, 15] is an estimation methodology that does not require the specification of the likelihood function but only depends on the moments of the distribution. It is a very appealing technique particularly where the density function is cumbersome to express. In literature, the QL approach has been intensively used in longitudinal studies as the likelihood function for longitudinal models cannot be expressed explicitly because of multiple integrals and necessitate complex approaches like the MCMC approach and EM algorithms [2]. While, on the other hand, QL reduces this burden of estimating the regression and correlation parameters through simply the notion of the moments and the appropriate modelling of the covariances [4, 7]. In similar lines, Pedeli and Karlis [13] claimed that MLE yields huge complicated expressions and necessitate numerical approximations. However, these authors have noted these computational problems at a bivariate stage (Refer to Nikouloulopoulos and Karlis [12]) and have not investigated the status of MLE at trivariate or multivariate stages. Following the derivation of the trivariate models in the previous two sections, it is clear that MLE is almost unfeasible. To overcome these challenges, this paper motivates the use of QL as an alternate option to MLE. Given that the QL approach has never been explored under multivariate domains, this section provides the details of how to construct the different components in the QL equation given by

$$\sum_{i=1}^{I} D_i^T \Sigma_i^{-1}(y_i - \mu_i) = 0 \tag{7}$$

where the components of GQL under a general multivariate set-up comprising of J explanatory variables and p variates are: The derivative matrix has a block structure of the form

$$D_i^{[k]} = \begin{pmatrix} \frac{\partial \mu_i^{[1]}}{\partial \hat{\beta}_1^{[k]}} & \frac{\partial \mu_i^{[2]}}{\partial \hat{\beta}_1^{[k]}} & \cdots & \frac{\partial \mu_i^{[P]}}{\partial \hat{\beta}_1^{[k]}} \\ \frac{\partial \mu_i^{[1]}}{\partial \hat{\beta}_2^{[k]}} & \frac{\partial \mu_i^{[2]}}{\partial \hat{\beta}_2^{[k]}} & \cdots & \frac{\partial \mu_i^{[P]}}{\partial \hat{\beta}_2^{[k]}} \\ \vdots & \vdots & & \vdots \\ \frac{\partial \mu_i^{[1]}}{\partial \hat{\beta}_J^{[k]}} & \frac{\partial \mu_i^{[2]}}{\partial \hat{\beta}_J^{[k]}} & \cdots & \frac{\partial \mu_i^{[P]}}{\partial \hat{\beta}_J^{[k]}} \end{pmatrix}_{J \times P}$$

$$D_i^T = [[D^T]_i^{[1]}, [D^T]_i^{[2]}, \ldots, [D^T]_i^{[k]}, \ldots, [D^T]_i^{[P]}]_{JP \times P}^T$$

where the general term

$$\frac{\partial \mu_i^{[k]}}{\partial \beta_j^{[k]}} = \mu_i^{[k]} \times x_{ij}$$

and

$$\Sigma_i = \begin{pmatrix} V(Y_i^{[1]}) & Cov(Y_i^{[1]}, Y_i^{[2]}) & \cdots & & Cov(Y_i^{[1]}, Y_i^{[p]}) \\ \vdots & V(Y_i^{[2]}) & \cdots & & Cov(Y_i^{[2]}, Y_i^{[p]}) \\ \vdots & \vdots & \ddots & & \vdots \\ \vdots & \vdots & & V(Y_i^{[k]}) & \cdots & Cov(Y_i^{[k]}, Y_i^{[p]}) \\ \vdots & \vdots & \cdots & & \ddots & \vdots \\ \vdots & \vdots & & \cdots & & V(Y_i^{[p]}) \end{pmatrix}$$

and

$$y_i = \begin{pmatrix} y_i^{[1]} \\ y_i^{[2]} \\ y_i^{[3]} \\ \vdots \\ y_i^{[P]} \end{pmatrix}, \mu_i = \begin{pmatrix} \mu_i^{[1]} \\ \mu_i^{[2]} \\ \mu_i^{[3]} \\ \vdots \\ \mu_i^{[P]} \end{pmatrix}$$

By letting

$$\hat{\beta}_r^{[k]} = [\hat{\beta}_{1,r}^{[k]}, \hat{\beta}_{2,r}^{[k]}, \ldots, \hat{\beta}_{J,r}^{[k]}]^T$$

The GQL equation is solved using the Newton-Raphson iterative scheme that yields

$$\begin{pmatrix} \hat{\beta}_{r+1}^{[1]} \\ \vdots \\ \hat{\beta}_{r+1}^{[k]} \\ \vdots \\ \hat{\beta}_{r+1}^{[P]} \end{pmatrix} = \begin{pmatrix} \hat{\beta}_r^{[1]} \\ \vdots \\ \hat{\beta}_r^{[k]} \\ \vdots \\ \hat{\beta}_r^{[P]} \end{pmatrix} + [\sum_{i=1}^{I} D_i^T \Sigma_i^{-1} D_i]_r^{-1} [\sum_{i=1}^{I} D_i^T \Sigma_i^{-1} (y_i - \mu_i)]_r$$

where $[.]_r$ represents the expressions at the rth iteration. The algorithm works as follows: For an initial set of vector of regression parameters of the p variates, we calculate the means, variances and the dependence parameters and replace these estimates in the Hessian and score parts to obtain an update of the set of regression vectors under the different variates. This iterative process is stopped when the absolute difference between the updated and previous regression estimates becomes less than a tolerance 10^{-10}. The estimators are consistent and under mild regularity conditions, for $I \to \infty$, it may be shown that $I^{\frac{1}{2}}((\hat{\beta}^1, \ldots, \hat{\beta}^P) - (\beta^1, \ldots, \beta^P))^T$ has an asymptotic normal distribution with mean 0 and covariance matrix

$$[D^T \Sigma^{-1} D]^{-1} [D^T \Sigma^{-1} (\tilde{y} - \tilde{\mu})(\tilde{y} - \tilde{\mu})^T \Sigma^{-1} D][D^T \Sigma^{-1} D]^{-1}$$

and the

$$V(\hat{\beta}) = [\sum_{i=1}^{I} D_i^T \Sigma_i^{-1} D_i]^{-1}$$

Table 1 Estimates of the parameters and standard errors under Trivariate Poisson model

α_1	α_2	ϕ	T	$\hat{\beta}_1^{[1]}$	$\hat{\beta}_2^{[1]}$	$\hat{\beta}_1^{[2]}$	$\hat{\beta}_2^{[2]}$	$\hat{\beta}_1^{[3]}$	$\hat{\beta}_2^{[3]}$	$\hat{\alpha}_1$	$\hat{\alpha}_2$	$\hat{\phi}$
0.9	0.9	0.9	30	0.9151	0.9272	0.9582	0.9645	0.9726	0.9288	0.8915	0.8924	0.8336
				(0.1117)	(0.2411)	(0.2089)	(0.1919)	(0.1881)	(0.1745)			
			60	0.9442	0.9741	0.9769	0.9305	0.9288	0.9247	0.8905	0.8323	0.8224
				(0.1340)	(0.2468)	(0.2325)	(0.2442)	(0.1793)	(0.1714)			
			100	0.9164	0.9231	0.9427	0.9857	0.9560	0.9611	0.8291	0.8140	0.8831
				(0.1812)	(0.2137)	(0.2667)	(0.1773)	(0.1251)	(0.1365)			
			500	0.9411	0.9693	0.9301	0.9710	0.9755	0.9445	0.8622	0.8679	0.8741
				(0.1346)	(0.1626)	(0.2523)	(0.1131)	(0.01312)	(0.1184)			
0.3	0.3	0.3	30	0.9367	0.9134	0.9868	0.9898	0.9404	0.9316	0.2666	0.2202	0.2537
				(0.1941)	(0.2143)	(0.2167)	(0.1519)	(0.1135)	(0.1065)			
			60	0.9055	0.9013	0.9165	0.9609	0.9292	0.9766	0.2168	0.2123	0.2677
				(0.1568)	(0.2163)	(0.2103)	(0.1444)	(0.1462)	(0.1414)			
			100	0.9545	0.9850	0.9913	0.9222	0.9132	0.9627	0.2106	0.2197	0.2779
				(0.1319)	(0.2131)	(0.2419)	(0.1288)	(0.1241)	(0.1238)			
			500	0.9551	0.9811	0.9861	0.9438	0.9582	0.9342	0.2448	0.2862	0.2872
				(0.1881)	(0.2088)	(0.2298)	(0.1658)	(0.1773)	(0.1441)			
0.5	0.5	0.5	30	0.9131	0.9419	0.9288	0.9142	0.9238	0.9551	0.4811	0.4861	0.4834
				(0.1582)	(0.2342)	(0.2484)	(0.1861)	(0.1278)	(0.1188)			
			60	0.9678	0.9269	0.9581	0.9773	0.9144	0.9252	0.4411	0.4845	0.4275
				(0.1375)	(0.2013)	(0.2028)	(0.1201)	(0.1346)	(0.1029)			
			100	0.9535	0.9852	0.9767	0.9777	0.9752	0.9873	0.4418	0.4299	0.4603
				(0.1116)	(0.2130)	(0.2036)	(0.1366)	(0.1686)	(0.1652)			
			500	0.9578	0.9605	0.9566	0.9504	0.9533	0.9149	0.4877	0.4337	0.4142
				(0.1259)	(0.2196)	(0.2538)	(0.1769)	(0.1114)	(0.1203)			

(continued)

Table 1 (continued)

α_1	α_2	ϕ	T	$\hat{\beta}_1^{[1]}$	$\hat{\beta}_2^{[1]}$	$\hat{\beta}_1^{[2]}$	$\hat{\beta}_2^{[2]}$	$\hat{\beta}_1^{[3]}$	$\hat{\beta}_2^{[3]}$	$\hat{\alpha}_1$	$\hat{\alpha}_2$	$\hat{\phi}$
0.5	0.9	0.9	30	0.9331	0.9651	0.9559	0.9342	0.9162	0.9486	0.4551	0.8566	0.8567
				(0.1519)	(0.2627)	(0.2117)	(0.1674)	(0.1787)	(0.1164)			
			60	0.9699	0.9249	0.9451	0.9310	0.9241	0.9148	0.4156	0.8511	0.8255
				(0.1698)	(0.2298)	(0.1778)	(0.2147)	(0.1254)	(0.1617)			
			100	0.9663	0.9228	0.9203	0.9411	0.9579	0.9538	0.4196	0.8259	0.8632
				(0.1034)	(0.2149)	(0.2057)	(0.1799)	(0.1178)	(0.1336)			
			500	0.9386	0.9889	0.9025	0.9208	0.9555	0.9701	0.4050	0.8804	0.8882
				(0.1389)	(0.1506)	(0.2152)	(0.1134)	(0.1069)	(0.1253)			
0.9	0.5	0.9	30	0.9265	0.9272	0.9407	0.9543	0.9830	0.9167	0.8237	0.4134	0.8198
				(0.1942)	(0.2464)	(0.1233)	(0.1081)	(0.1104)	(0.1183)			
			60	0.9811	0.9698	0.9821	0.9128	0.9365	0.9147	0.8164	0.497	0.8912
				(0.1323)	(0.2362)	(0.2137)	(0.1339)	(0.1245)	(0.1590)			
			100	0.9311	0.9302	0.9986	0.9556	0.9367	0.9749	0.8214	0.4571	0.8961
				(0.1252)	(0.2131)	(0.2035)	(0.1151)	(0.1752)	(0.1458)			
			500	0.9421	0.9292	0.9404	0.9487	0.9157	0.9658	0.8119	0.4292	0.8274
				(0.1098)	(0.2159)	(0.2088)	(0.1135)	(0.1500)	(0.1789)			
0.5	0.5	0.9	30	0.9202	0.9175	0.9117	0.9918	0.9103	0.9821	0.4654	0.4222	0.8153
				(0.1165)	(0.2189)	(0.2058)	(0.1759)	(0.1258)	(0.1155)			
			60	0.9776	0.9090	0.9307	0.9316	0.9563	0.9482	0.4451	0.4621	0.8874
				(0.1273)	(0.2245)	(0.2159)	(0.1896)	(0.1907)	(0.1801)			
			100	0.9115	0.9678	0.9212	0.9124	0.9102	0.9153	0.4244	0.4235	0.8667
				(0.1356)	(0.2142)	(0.2067)	(0.1210)	(0.1487)	(0.1462)			
			500	0.9521	0.9688	0.9771	0.9509	0.9120	0.9216	0.4518	0.4678	0.8587
				(0.1251)	(0.2172)	(0.2167)	(0.1188)	(0.1193)	(0.1666)			

Table 2 Estimates of the parameters and standard errors under Trivariate Poisson model

α_1	α_2	ϕ	T	$\hat{\beta}_1^{[1]}$	$\hat{\beta}_2^{[1]}$	$\hat{\beta}_1^{[2]}$	$\hat{\beta}_2^{[2]}$	$\hat{\beta}_1^{[3]}$	$\hat{\beta}_2^{[3]}$	$\hat{\alpha}_1$	$\hat{\alpha}_2$	$\hat{\phi}$
0.3	0.3	0.9	30	0.9256	0.9674	0.9981	0.94002	0.9982	0.9666	0.2557	0.2412	0.8111
				(0.1246)	(0.2282)	(0.2661)	(0.1956)	(0.1851)	(0.1205)			
			60	0.9601	0.9215	0.9444	0.9799	0.9149	0.9651	0.2299	0.2687	0.8656
				(0.1111)	(0.2116)	(0.2195)	(0.1289)	(0.1365)	(0.1009)			
			100	0.9109	0.9249	0.9239	0.9581	0.9629	0.9249	0.2101	0.2502	0.8288
				(0.1177)	(0.2158)	(0.2135)	(0.1089)	(0.1405)	(0.1333)			
			500	0.9181	0.9575	0.9622	0.9277	0.9155	0.9897	0.2243	0.2384	0.8108
				(0.1901)	(0.2116)	(0.2154)	(0.1991)	(0.1367)	(0.1525)			
0.9	0.3	0.9	30	0.9331	0.9683	0.9854	0.9240	0.9386	0.9495	0.8922	0.2701	0.8943
				(0.1093)	(0.2116)	(0.2411)	(0.1742)	(0.1692)	(0.1812)			
			60	0.9333	0.9513	0.9091	0.9612	0.9681	0.9178	0.8342	0.2945	0.8479
				(0.1526)	(0.2287)	(0.2155)	(0.1479)	(0.1337)	(0.1289)			
			100	0.9708	0.9283	0.9668	0.9178	0.9765	0.9046	0.8296	0.2740	0.8233
				(0.1247)	(0.2161)	(0.2134)	(0.1122)	(0.1138)	(0.1245)			
			500	0.9622	0.9951	0.9930	0.9478	0.9911	0.9932	0.8358	0.2632	0.8818
				(0.1025)	(0.2286)	(0.2239)	(0.1917)	(0.1345)	(0.1668)			
0.3	0.3	0.9	30	0.9545	0.9035	0.9366	0.9951	0.9162	0.9724	0.2362	0.2964	0.8295
				(0.1896)	(0.2232)	(0.2174)	(0.1276)	(0.1481)	(0.1175)			
			60	0.9557	0.9095	0.9861	0.9684	0.9340	0.9307	0.2925	0.2118	0.8828
				(0.1768)	(0.2152)	(0.2132)	(0.1909)	(0.1942)	(0.1549)			
			100	0.9651	0.9146	0.9552	0.9266	0.9134	0.9178	0.2721	0.2841	0.8232
				(0.1249)	(0.2188)	(0.2342)	(0.1156)	(0.1871)	(0.1090)			
			500	0.9654	0.9789	0.9632	0.9185	0.9557	0.9742	0.2132	0.2704	0.8211
				(0.1552)	(0.2140)	(0.2551)	(0.1182)	(0.1507)	(0.1352)			

(continued)

Table 2 (continued)

α_1	α_2	ϕ	T	$\hat{\beta}_1^{[1]}$	$\hat{\beta}_2^{[1]}$	$\hat{\beta}_1^{[2]}$	$\hat{\beta}_2^{[2]}$	$\hat{\beta}_1^{[3]}$	$\hat{\beta}_2^{[3]}$	$\hat{\alpha}_1$	$\hat{\alpha}_2$	$\hat{\phi}$
0.3	0.5	0.5	30	0.9781	0.9785	0.9780	0.9219	0.9728	0.9159	0.2372	0.4425	0.4151
				(0.1108)	(0.2457)	(0.2222)	(0.1756)	(0.1423)	(0.1526)			
			60	0.9820	0.9862	0.9851	0.9877	0.9787	0.9872	0.2840	0.4961	0.4825
				(0.1289)	(0.2154)	(0.2186)	(0.1598)	(0.1812)	(0.1952)			
			100	0.9511	0.9268	0.9801	0.9500	0.9459	0.9259	0.2212	0.4511	0.4424
				(0.1277)	(0.2188)	(0.2131)	(0.5081)	(0.1108)	(0.1114)			
			500	0.9109	0.9788	0.9255	0.9288	0.9522	0.9675	0.2710	0.4918	0.4937
				(0.1652)	(0.2177)	(0.2428)	(0.1279)	(0.1189)	(0.1247)			
0.5	0.3	0.5	30	0.9159	0.9247	0.9598	0.9781	0.9111	0.9237	0.4881	0.2155	0.4981
				(0.1265)	(0.2010)	(0.2124)	(0.1247)	(0.1465)	(0.1155)			
			60	0.9182	0.9378	0.9686	0.9398	0.9818	0.9507	0.4851	0.2111	0.4850
				(0.1411)	(0.2105)	(0.2128)	(0.1399)	(0.1178)	(0.1010)			
			100	0.9511	0.9812	0.9618	0.9532	0.9273	0.9478	0.4362	0.2545	0.4821
				(0.1563)	(0.2136)	(0.2141)	(0.1199)	(0.1208)	(0.1173)			
			500	0.9852	0.9759	0.9584	0.9982	0.9411	0.9654	0.4854	0.2651	0.4626
				(0.1008)	(0.2179)	(0.2198)	(0.1111)	(0.1549)	(0.1225)			
0.5	0.5	0.3	30	0.9110	0.9225	0.9221	0.9811	0.9241	0.9701	0.4189	0.4449	0.2784
				(0.1358)	(0.2189)	(0.2138)	(0.1285)	(0.1466)	(0.1473)			
			60	0.9753	0.9145	0.9128	0.9123	0.9457	0.9753	0.4258	0.4211	0.2281
				(0.1578)	(0.21912)	(0.2125)	(0.1115)	(0.1025)	(0.1721)			
			100	0.9218	0.9178	0.9727	0.9711	0.9821	0.9618	0.4271	0.4108	0.2987
				(0.1085)	(0.2091)	(0.2155)	(0.1585)	(0.1726)	(0.1216)			
			500	0.9612	0.9321	0.9459	0.9899	0.9148	0.9186	0.4871	0.4511	0.2927
				(0.1171)	(0.2141)	(0.2089)	(0.1910)	(0.1121)	(0.1211)			

6 Simulations and Results

In this section, a simulation design is developed under the assumption $\hat{\beta}_1^{[k]} = \hat{\beta}_2^{[k]} = 1$ for $k = 1, 2, 3$ and for values of $\alpha_1 = \alpha_2 = \phi = 0.3, 0.5, 0.9$ under $I = 30, 60, 100$ and 500. Under each combination of parameters, 5000 simulations were run. Note that the mean parameters $\mu_i^{[k]}$ were obtained assuming two sets of covariates $[x_{i1}, x_{i2}]$ as follows:

$$x_{i1} = \begin{cases} -1 & (i = 1, \ldots, I/3), \\ 0 & (i = (I/3) + 1, \ldots, 2I), \\ 1 & (i = 2I + 1, \ldots, 3I), \end{cases}$$

and $x_{i2} = rnorm(1, 0, 1)$. Using these information, the values of $y_i^{[k]}$ were generating using Eqs. (1), (2) and (3) and thereafter, the QL Eq. (7) was solved to obtain the following results.

The results in the Tables 1 and 2 were obtained assuming initial values of the regression parameters under the different variates. The main remarks are the simulation results indicate that the generating process with the various assumed parameters $\hat{\rho}_1, \hat{\rho}_2, \hat{\alpha}_1$ and $\hat{\phi}$ were reliable for all sample sizes. In fact, as the number of time points increase, the estimates come closer to the population values. On the other hand, for small sample size, some non-convergent simulations were noted. For instance, for $T = 30$, when $\hat{\rho}_1 = \hat{\rho}_2 = 0.9$ and $\hat{\alpha}_1 = 0.3$, the simulation failed in 1500 out of 5,000 simulations. Moreover, as the sample size increases to $T = 100$ with same parameters, the number of non-convergent simulations increases to 2100. These failure rates were noted across almost all clusters. The non-convergence was mainly due to the ill-conditioned nature of the covariance matrix which is inevitable here. However, overall, at a cross-sectional level, apart from this computational issue, the estimation procedure works satisfactory well and yields consistent parameter estimates (Tables 1 and 2).

7 Conclusion

This paper introduces an observation-driven approach to generate multivariate cross-sectional Poisson counts where the count observations were influenced by a number of explanatory variables. As illustrated in previous studies, the MLE approach is quite unfeasible under the multivariate cross-sectional set-up. This paper overcomes this shortcoming through the application of the QL approach which requires only the moments of the responses under the different variates. The QL estimates of the regression parameters were obtained using the Newton-Raphson scheme while the dependence parameters were estimated using the method of moment estimation technique. Through a simulation study, it is noted that this estimation procedure yielded consistent parameter estimates but with some computational issues in inverting the covariance matrix. Overall, QL works satisfactorily and can be an emerging solution in more complex multivariate solutions.

References

1. Al-Osh, M.A., Alzaid, A.A.: First-order integer-valued autoregressive (INAR(1)) process. J. Time Ser. Anal. **8**, 261–275 (1987)
2. Fahrmeir, L., Tutz, G.: Multivariate Statistical Modelling Based on Generalized Linear Models. Springer (2001)
3. Genest, C., Neslehova, J.: A Primer on copulas for count data. Astin Bull. **37**, 475–515 (2007)
4. Jowaheer, V., Sutradhar, B.C.: Analysing longitudinal count data with overdipsersion. Biometrika **89**, 389–399 (2002)
5. Karlis, D., Ntzoufras, I.: Analysis of sports data by using bivariate Poisson models. J. R. Stat. Soc. **52**, 381–393 (2003)
6. Kocherlakota, S., Kocherlakota, K.: Regression in the bivariate Poisson distribution. Commun. Stat. Theory Methods **30**, 815–827 (2001)
7. Mamode Khan, N., Jowaheer, V.: Comparing joint GQL estimation and GMM adaptive estimation in COM-Poisson longitudinal regression model. Commun. Stat. Simul. Comput. **42**, 755–770 (2013)
8. McCullagh, P. Nelder, J.A.: Generalized Linear Models. Chapman and Hall (1999)
9. McKenzie, E.: Autoregressive moving-average processes with negative binomial and geometric marginal distrbutions. Adv. Appl. Probab. **18**, 679–705 (1986)
10. Nelsen, R.B.: An Introduction to Copulas. Springer (2006)
11. Nikoloulopoulos, A.K., Karlis, D.: Modeling multivariate count data using copulas. Commun. Stat. Simul. Comput. **39**(1), 172–187 (2009)
12. Nikoloulopoulos, A.K., Karlis, D.: Regression in a copula model for bivariate count data. J. Appl. Stat. **139**, 1555–1568 (2010)
13. Pedeli, X., Karlis, D.: Some properties of multivariate INAR(1) processes. Comput. Stat. Data Anal. **67**, 213–225 (2013)
14. Rodrigues-Motta, m., Pinheiro, H.P., Martins, E.G., Araujo, M.S., Reis, S.F.D., Harstad, K. Bellan, J.: Multivariate models for correlated count data. J. Appl. Stat. (2013)
15. Wedderburn, R.: Quasi-likelihood functions, generalized linear models, and the Gauss-Newton method. Biometrika **61**, 439–447 (1974)

Generalized Caputo Type Fractional Inequalities

George A. Anastassiou

Abstract We establish here generalized Caputo type fractional inequalities of the following kinds: Opial, Ostrowski, Comparison of means, Grüss and Csiszar's f-divergence. Prior to these we derive an interesting generalized fractional representation formula, using it we prove our inequalities.

1 Background

Here $AC([a, b])$ stands for the space of absolutely continuous functions from $[a, b]$ into $\mathbb{R}$. Also $f \in AC^n([a, b])$, $n \in \mathbb{N}$, means that $f^{(n-1)} \in AC([a, b])$.

We need

Definition 1 (*see also* [10, p. 99]) The left and right fractional integrals, respectively, of a function f with respect to given function g are defined as follows:

Let $a, b \in \mathbb{R}$, $a < b$, $\alpha > 0$. Here $g \in AC([a, b])$ and is strictly increasing, $f \in L_\infty([a, b])$. We set

$$\left(I_{a+;g}^\alpha f\right)(x) = \frac{1}{\Gamma(\alpha)} \int_a^x (g(x) - g(t))^{\alpha-1} g'(t) f(t) \, dt, \, x \geq a, \qquad (1)$$

where Γ is the gamma function, clearly $\left(I_{a+;g}^\alpha f\right)(a) = 0$, $I_{a+;g}^0 f := f$ and

$$\left(I_{b-;g}^\alpha f\right)(x) = \frac{1}{\Gamma(\alpha)} \int_x^b (g(t) - g(x))^{\alpha-1} g'(t) f(t) \, dt, \, x \leq b, \qquad (2)$$

G.A. Anastassiou (✉)
Department of Mathematical Sciences, University of Memphis, Memphis, TN 38152, USA
e-mail: ganastss@memphis.edu

© Springer International Publishing Switzerland 2016
G.A. Anastassiou and O. Duman (eds.), *Intelligent Mathematics II:*
Applied Mathematics and Approximation Theory, Advances in Intelligent Systems
and Computing 441, DOI 10.1007/978-3-319-30322-2_28

clearly $\left(I_{b-;g}^{\alpha}f\right)(b)=0,\ I_{b-;g}^{0}f:=f.$

When g is the identity function id, we get that $I_{a+;id}^{\alpha}=I_{a+}^{\alpha}$, and $I_{b-;id}^{\alpha}=I_{b-}^{\alpha}$, the ordinary left and right Riemann-Liouville fractional integrals, where

$$\left(I_{a+}^{\alpha}f\right)(x)=\frac{1}{\Gamma(\alpha)}\int_{a}^{x}(x-t)^{\alpha-1}f(t)\,dt,\ x\geq a, \tag{3}$$

$\left(I_{a+}^{\alpha}f\right)(a)=0$ and

$$\left(I_{b-}^{\alpha}f\right)(x)=\frac{1}{\Gamma(\alpha)}\int_{x}^{b}(t-x)^{\alpha-1}f(t)\,dt,\ x\leq b, \tag{4}$$

$\left(I_{b-}^{\alpha}f\right)(b)=0.$

We also need

Definition 2 (*see* [1]) Let $\alpha>0$, $\lceil\alpha\rceil=n$, $\lceil\cdot\rceil$ the ceiling of the number. Again here $g\in AC([a,b])$ and strictly increasing. We assume that $\left(f\circ g^{-1}\right)^{(n)}\circ g\in L_{\infty}([a,b])$. We define the left generalized g-fractional derivative of f of order α as follows:

$$\left(D_{a+;g}^{\alpha}f\right)(x):=\frac{1}{\Gamma(n-\alpha)}\int_{a}^{x}(g(x)-g(t))^{n-\alpha-1}g'(t)\left(f\circ g^{-1}\right)^{(n)}(g(t))\,dt, \tag{5}$$

$x\geq a.$

If $\alpha\notin\mathbb{N}$, by [1], we have that $D_{a+;g}^{\alpha}f\in C([a,b])$.

We see that

$$\left(I_{a+;g}^{n-\alpha}\left(\left(f\circ g^{-1}\right)^{(n)}\circ g\right)\right)(x)=\left(D_{a+;g}^{\alpha}f\right)(x),\ x\geq a. \tag{6}$$

We set

$$D_{a+;g}^{n}f(x):=\left(\left(f\circ g^{-1}\right)^{(n)}\circ g\right)(x), \tag{7}$$

$$D_{a+;g}^{0}f(x)=f(x),\ \forall x\in[a,b].$$

When $g=id$, then

$$D_{a+;g}^{\alpha}f=D_{a+;id}^{\alpha}f=D_{*a}^{\alpha}f, \tag{8}$$

the usual left Caputo fractional derivative.

We will use the following g-left fractional generalized Taylor's formula:

Theorem 3 (see [1]) *Let g be strictly increasing function and $g \in AC([a, b])$. We assume that $\left(f \circ g^{-1}\right) \in AC^n([g(a), g(b)])$, where $\mathbb{N} \ni n = \lceil \alpha \rceil$, $\alpha > 0$. Also we assume that $\left(f \circ g^{-1}\right)^{(n)} \circ g \in L_\infty([a, b])$. Then*

$$f(x) = f(a) + \sum_{k=1}^{n-1} \frac{\left(f \circ g^{-1}\right)^{(k)}(g(a))}{k!}(g(x) - g(a))^k$$

$$+ \frac{1}{\Gamma(\alpha)} \int_a^x (g(x) - g(t))^{\alpha-1} g'(t) \left(D_{a+;g}^\alpha f\right)(t) \, dt, \tag{9}$$

$\forall x \in [a, b]$. *Calling $R_n(a, x)$ the remainder of (9), we get that*

$$R_n(a, x) = \frac{1}{\Gamma(\alpha)} \int_{g(a)}^{g(x)} (g(x) - z)^{\alpha-1} \left(\left(D_{a+;g}^\alpha f\right) \circ g^{-1}\right)(z) \, dz, \tag{10}$$

$\forall x \in [a, b]$.

Remark 4 By [1], $R_n(a, x)$ is a continuous function in $x \in [a, b]$. Also, by [9], change of variable in Lebesgue integrals, (10) is valid.

We need

Definition 5 (*see* [1]) Here we assume that $\left(f \circ g^{-1}\right)^{(n)} \circ g \in L_\infty([a, b])$, where $\mathbb{N} \ni n = \lceil \alpha \rceil$, $\alpha > 0$. We define the right generalized g-fractional derivative of f of order α as follows:

$$\left(D_{b-;g}^\alpha f\right)(x) := \frac{(-1)^n}{\Gamma(n-\alpha)} \int_x^b (g(t) - g(x))^{n-\alpha-1} g'(t) \left(f \circ g^{-1}\right)^{(n)}(g(t)) \, dt, \tag{11}$$

all $x \in [a, b]$.

If $\alpha \notin \mathbb{N}$, by [1], we get that $\left(D_{b-;g}^\alpha f\right) \in C([a, b])$.

We see that

$$I_{b-;g}^{n-\alpha}\left((-1)^n \left(f \circ g^{-1}\right)^{(n)} \circ g\right)(x) = \left(D_{b-;g}^\alpha f\right)(x), \ a \le x \le b. \tag{12}$$

We set

$$D_{b-;g}^n f(x) = (-1)^n \left(\left(f \circ g^{-1}\right)^{(n)} \circ g\right)(x), \tag{13}$$

$$D_{b-;g}^0 f(x) = f(x), \ \forall x \in [a, b].$$

When $g = id$, then

$$D_{b-;g}^{\alpha} f (x) = D_{b-;id}^{\alpha} f (x) = D_{b-}^{\alpha} f, \tag{14}$$

the usual right Caputo fractional derivative.

We will use the g-right generalized fractional Taylor's formula:

Theorem 6 (see [1]) *Let g be strictly increasing function and $g \in AC ([a, b])$. We assume that $(f \circ g^{-1}) \in AC^n ([g (a), g (b)])$, where $\mathbb{N} \ni n = \lceil \alpha \rceil$, $\alpha > 0$. Also we assume that $(f \circ g^{-1})^{(n)} \circ g \in L_{\infty} ([a, b])$. Then*

$$f (x) = f (b) + \sum_{k=1}^{n-1} \frac{(f \circ g^{-1})^{(k)} (g (b))}{k!} (g (x) - g (b))^k$$

$$+ \frac{1}{\Gamma (\alpha)} \int_x^b (g (t) - g (x))^{\alpha-1} g' (t) (D_{b-;g}^{\alpha} f) (t) \, dt, \tag{15}$$

all $a \leq x \leq b$. Calling $R_n (b, x)$ the remainder in (15), we get that

$$R_n (b, x) = \frac{1}{\Gamma (\alpha)} \int_{g(x)}^{g(b)} (z - g (x))^{\alpha-1} ((D_{b-;g}^{\alpha} f) \circ g^{-1}) (z) \, dz, \tag{16}$$

$\forall x \in [a, b]$.

Remark 7 By [1], $R_n (b, x)$ is a continuous function in $x \in [a, b]$. Also, by [9], change of variable in Lebesgue integrals, (16) is valid.

2   Main Results

We give the following representation formula:

Theorem 8 *Let $f \in C ([a, b])$. Let g be strictly increasing function and $g \in AC ([a, b])$. We assume that $(f \circ g^{-1}) \in AC^n ([g (a), g (b)])$, where $\mathbb{N} \ni n = \lceil \alpha \rceil$, $\alpha > 0$. We also assume that $(f \circ g^{-1})^{(n)} \circ g \in L_{\infty} ([a, b])$. Then*

$$f(y) = \frac{1}{b-a} \int_a^b f(x)\, dx \tag{17}$$

$$- \sum_{k=1}^{n-1} \frac{\left(f \circ g^{-1}\right)^{(k)} (g(y))}{k!\,(b-a)} \int_a^b (g(x) - g(y))^k\, dx + R_1(y),$$

for any $y \in [a, b]$, where

$$R_1(y) = \frac{1}{\Gamma(\alpha)(b-a)} \left[\int_a^b \left(\int_x^y |g(x) - g(t)|^{\alpha-1}\, g'(t) \right. \right. \tag{18}$$

$$\cdot \left. \left. \left[\chi_{[y,b]}(x) \left(D_{y+;g}^\alpha f\right)(t) - \chi_{[a,y)}(x) \left(D_{y-;g}^\alpha f\right)(t) \right] dt \right) dx \right],$$

here χ_A stands for the characteristic function set A, where A is an arbitrary set.
   One may write also that

$$R_1(y) = -\frac{1}{\Gamma(\alpha)(b-a)} \left[\int_a^y \left(\int_x^y (g(t) - g(x))^{\alpha-1}\, g'(t) \left(D_{y-;g}^\alpha f\right)(t)\, dt \right) dx \right.$$

$$+ \left. \int_y^b \left(\int_y^x (g(x) - g(t))^{\alpha-1}\, g'(t) \left(D_{y+;g}^\alpha f\right)(t)\, dt \right) dx \right], \tag{19}$$

for any $y \in [a, b]$.
   Putting things together, one has

$$f(y) = \frac{1}{b-a} \int_a^b f(x)\, dx - \sum_{k=1}^{n-1} \frac{\left(f \circ g^{-1}\right)^{(k)} (g(y))}{k!\,(b-a)} \int_a^b (g(x) - g(y))^k\, dx$$

$$+ \frac{1}{\Gamma(\alpha)(b-a)} \left[\int_a^b \left(\int_x^y |g(x) - g(t)|^{\alpha-1}\, g'(t) \cdot \right. \right. \tag{20}$$

$$\left. \left. \left[\chi_{[y,b]}(x) \left(D_{y+;g}^\alpha f\right)(t) - \chi_{[a,y)}(x) \left(D_{y-;g}^\alpha f\right)(t) \right] dt \right) dx \right].$$

In particular, one has

$$f(y) - \frac{1}{b-a} \int_a^b f(x)\, dx + \sum_{k=1}^{n-1} \frac{\left(f \circ g^{-1}\right)^{(k)} (g(y))}{k!\,(b-a)} \int_a^b (g(x) - g(y))^k\, dx$$

$$= R_1(y), \tag{21}$$

for any $y \in [a, b]$.

Proof Here $x, y \in [a, b]$. We keep y as fixed. By Theorem 3 we get:

$$f(x) = f(y) + \sum_{k=1}^{n-1} \frac{\left(f \circ g^{-1}\right)^{(k)}(g(y))}{k!} (g(x) - g(y))^k \tag{22}$$

$$+ \frac{1}{\Gamma(\alpha)} \int_y^x (g(x) - g(t))^{\alpha-1} g'(t) \left(D_{y+;g}^\alpha f\right)(t) \, dt,$$

for any $x \geq y$. By Theorem 6 we get:

$$f(x) = f(y) + \sum_{k=1}^{n-1} \frac{\left(f \circ g^{-1}\right)^{(k)}(g(y))}{k!} (g(x) - g(y))^k \tag{23}$$

$$+ \frac{1}{\Gamma(\alpha)} \int_x^y (g(t) - g(x))^{\alpha-1} g'(t) \left(D_{y-;g}^\alpha f\right)(t) \, dt,$$

for any $x \leq y$. By (22), (23) we notice that

$$\int_a^b f(x) \, dx = \int_a^y f(x) \, dx + \int_y^b f(x) \, dx \tag{24}$$

$$= \int_a^y f(y) \, dx + \sum_{k=1}^{n-1} \frac{\left(f \circ g^{-1}\right)^{(k)}(g(y))}{k!} \int_a^y (g(x) - g(y))^k \, dx$$

$$+ \frac{1}{\Gamma(\alpha)} \int_a^y \left(\int_x^y (g(t) - g(x))^{\alpha-1} g'(t) \left(D_{y-;g}^\alpha f\right)(t) \, dt \right) dx$$

$$+ \int_y^b f(y) \, dx + \sum_{k=1}^{n-1} \frac{\left(f \circ g^{-1}\right)^{(k)}(g(y))}{k!} \int_y^b (g(x) - g(y))^k \, dx$$

$$+ \frac{1}{\Gamma(\alpha)} \int_y^b \left(\int_y^x (g(x) - g(t))^{\alpha-1} g'(t) \left(D_{y+;g}^\alpha f\right)(t) \, dt \right) dx.$$

Hence it holds

$$\frac{1}{b-a} \int_a^b f(x)\, dx$$

$$= f(y) + \sum_{k=1}^{n-1} \frac{\left(f \circ g^{-1}\right)^{(k)}(g(y))}{k!(b-a)} \int_a^b (g(x) - g(y))^k\, dx \qquad (25)$$

$$+ \frac{1}{\Gamma(\alpha)(b-a)} \left[\int_a^y \left(\int_x^y |g(x) - g(t)|^{\alpha-1} g'(t) \left(D_{y-;g}^\alpha f\right)(t)\, dt \right) dx \right.$$

$$\left. + \int_y^b \left(\int_y^x |g(x) - g(t)|^{\alpha-1} g'(t) \left(D_{y+;g}^\alpha f\right)(t)\, dt \right) dx \right].$$

Therefore we obtain

$$f(y) = \frac{1}{b-a} \int_a^b f(x)\, dx \qquad (26)$$

$$- \sum_{k=1}^{n-1} \frac{\left(f \circ g^{-1}\right)^{(k)}(g(y))}{k!(b-a)} \int_a^b (g(x) - g(y))^k\, dx$$

$$- \frac{1}{\Gamma(\alpha)(b-a)} \left[\int_a^y \left(\int_x^y |g(x) - g(t)|^{\alpha-1} g'(t) \left(D_{y-;g}^\alpha f\right)(t)\, dt \right) dx \right.$$

$$\left. + \int_y^b \left(\int_y^x |g(x) - g(t)|^{\alpha-1} g'(t) \left(D_{y+;g}^\alpha f\right)(t)\, dt \right) dx \right].$$

Hence the remainder

$$R_1(y)$$

$$:= -\frac{1}{\Gamma(\alpha)(b-a)}\left[\int_a^y\left(\int_x^y |g(x)-g(t)|^{\alpha-1} g'(t)\left(D_{y-;g}^\alpha f\right)(t)\,dt\right)dx\right.$$

$$\left.+\int_y^b\left(\int_y^x |g(x)-g(t)|^{\alpha-1} g'(t)\left(D_{y+;g}^\alpha f\right)(t)\,dt\right)dx\right] \tag{27}$$

$$= -\frac{1}{\Gamma(\alpha)(b-a)}\left[\int_a^b \chi_{[a,y)}(x)\left(\int_x^y |g(x)-g(t)|^{\alpha-1} g'(t)\left(D_{y-;g}^\alpha f\right)(t)\,dt\right)dx\right.$$

$$\left.+\int_a^b \chi_{[y,b]}(x)\left(\int_y^x |g(x)-g(t)|^{\alpha-1} g'(t)\left(D_{y+;g}^\alpha f\right)(t)\,dt\right)dx\right]$$

$$= \frac{1}{\Gamma(\alpha)(b-a)}\left[-\int_a^b\left(\int_x^y \chi_{[a,y)}(x) |g(x)-g(t)|^{\alpha-1} g'(t)\left(D_{y-;g}^\alpha f\right)(t)\,dt\right)dx\right.$$

$$\left.+\int_a^b\left(\int_x^y \chi_{[y,b]}(x) |g(x)-g(t)|^{\alpha-1} g'(t)\left(D_{y+;g}^\alpha f\right)(t)\,dt\right)dx\right]. \tag{28}$$

The theorem is proved.

Next we present a left fractional Opial type inequality:

Theorem 9 *All as in Theorem 8. Additionally assume that $\alpha \geq 1$, $g \in C^1([a,b])$, and $(f\circ g^{-1})^{(k)}(g(a)) = 0$, for $k = 0, 1, \ldots, n-1$. Let $p,q > 1 : \frac{1}{p}+\frac{1}{q} = 1$. Then*

$$\int_a^x |f(w)|\left|\left(D_{a+;g}^\alpha f\right)(w)\right| g'(w)\,dw$$

$$\leq \frac{1}{\Gamma(\alpha)2^{\frac{1}{q}}}\left(\int_a^x\left(\int_a^w (g(w)-g(t))^{p(\alpha-1)}\,dt\right)dw\right)^{\frac{1}{p}} \tag{29}$$

$$\cdot\left(\int_a^x (g'(w))^q\left|\left(D_{a+;g}^\alpha f\right)(w)\right|^q dw\right)^{\frac{2}{q}},$$

$\forall x \in [a,b]$.

Proof By Theorem 3, we have that

$$f(x) = \frac{1}{\Gamma(\alpha)} \int_a^x (g(x) - g(t))^{\alpha-1} g'(t) \left(D_{a+;g}^\alpha f\right)(t) \, dt, \ \forall \, x \in [a, b]. \quad (30)$$

Then, by Hölder's inequality we obtain,

$$|f(x)| \leq \frac{1}{\Gamma(\alpha)} \left(\int_a^x (g(x) - g(t))^{p(\alpha-1)} \, dt \right)^{\frac{1}{p}} \left(\int_a^x (g'(t))^q \left| \left(D_{a+;g}^\alpha f\right)(t) \right|^q \, dt \right)^{\frac{1}{q}}. \quad (31)$$

Call

$$z(x) := \int_a^x (g'(t))^q \left| \left(D_{a+;g}^\alpha f\right)(t) \right|^q \, dt, \quad (32)$$

$z(a) = 0.$

Thus

$$z'(x) = (g'(x))^q \left| \left(D_{a+;g}^\alpha f\right)(x) \right|^q \geq 0, \quad (33)$$

and

$$(z'(x))^{\frac{1}{q}} = g'(x) \left| \left(D_{a+;g}^\alpha f\right)(x) \right| \geq 0, \ \forall \, x \in [a, b]. \quad (34)$$

Consequently, we get

$$|f(w)| \, g'(w) \left| \left(D_{a+;g}^\alpha f\right)(w) \right| \quad (35)$$

$$\leq \frac{1}{\Gamma(\alpha)} \left(\int_a^w (g(w) - g(t))^{p(\alpha-1)} \, dt \right)^{\frac{1}{p}} (z(w) z'(w))^{\frac{1}{q}},$$

$\forall w \in [a, b]$. Then

$$\int_a^x |f(w)| \left| \left(D_{a+;g}^\alpha f\right)(w) \right| g'(w) \, dw$$

$$\leq \frac{1}{\Gamma(\alpha)} \int_a^x \left(\int_a^w (g(w) - g(t))^{p(\alpha-1)} \, dt \right)^{\frac{1}{p}} (z(w) z'(w))^{\frac{1}{q}} \, dw \quad (36)$$

$$\leq \frac{1}{\Gamma(\alpha)} \left(\int_a^x \left(\int_a^w (g(w) - g(t))^{p(\alpha-1)} \, dt \right) dw \right)^{\frac{1}{p}} \left(\int_a^x z(w) z'(w) \, dw \right)^{\frac{1}{q}}$$

$$= \frac{1}{\Gamma(\alpha)} \left(\int_a^x \left(\int_a^w (g(w) - g(t))^{p(\alpha-1)} dt \right) dw \right)^{\frac{1}{p}} \left(\frac{z^2(x)}{2} \right)^{\frac{1}{q}} \quad (37)$$

$$= \frac{1}{\Gamma(\alpha)} \left(\int_a^x \left(\int_a^w (g(w) - g(t))^{p(\alpha-1)} dt \right) dw \right)^{\frac{1}{p}} \quad (38)$$

$$\cdot \left(\int_a^x (g'(t))^q \left| (D_{a+;g}^\alpha f)(t) \right|^q dt \right)^{\frac{2}{q}} \cdot 2^{-\frac{1}{q}}.$$

The theorem is proved.

We also give a right fractional Opial type inequality:

Theorem 10 *All as in Theorem 8. Additionally assume that $\alpha \geq 1$, $g \in C^1([a,b])$, and $(f \circ g^{-1})^{(k)}(g(b)) = 0$, $k = 0, 1, \ldots, n-1$. Let $p, q > 1 : \frac{1}{p} + \frac{1}{q} = 1$. Then*

$$\int_x^b |f(w)| \left| (D_{b-;g}^\alpha f)(w) \right| g'(w) dw$$

$$\leq \frac{1}{2^{\frac{1}{q}} \Gamma(\alpha)} \left(\int_x^b \left(\int_w^b (g(t) - g(w))^{p(\alpha-1)} dt \right) dw \right)^{\frac{1}{p}} \quad (39)$$

$$\cdot \left(\int_x^b (g'(w))^q \left| (D_{b-;g}^\alpha f)(w) \right|^q dw \right)^{\frac{2}{q}},$$

all $a \leq x \leq b$.

Proof By Theorem 6, we have that

$$f(x) = \frac{1}{\Gamma(\alpha)} \int_x^b (g(t) - g(x))^{\alpha-1} g'(t) (D_{b-;g}^\alpha f)(t) dt, \text{ all } a \leq x \leq b. \quad (40)$$

Then, by Hölder's inequality we obtain,

$$|f(x)| \leq \frac{1}{\Gamma(\alpha)} \left(\int_x^b (g(t) - g(x))^{p(\alpha-1)} dt \right)^{\frac{1}{p}} \left(\int_x^b (g'(t))^q \left| (D_{b-;g}^\alpha f)(t) \right|^q dt \right)^{\frac{1}{q}}.$$

$$(41)$$

Call

$$z(x) := \int_x^b \left(g'(t)\right)^q \left|\left(D_{b-;g}^\alpha f\right)(t)\right|^q dt, \qquad (42)$$

$$z(b) = 0.$$

Hence

$$z'(x) = -\left(g'(x)\right)^q \left|\left(D_{b-;g}^\alpha f\right)(x)\right|^q \le 0, \qquad (43)$$

and

$$-z'(x) = \left(g'(x)\right)^q \left|\left(D_{b-;g}^\alpha f\right)(x)\right|^q \ge 0, \qquad (44)$$

and

$$\left(-z'(x)\right)^{\frac{1}{q}} = g'(x)\left|\left(D_{b-;g}^\alpha f\right)(x)\right| \ge 0, \ \forall x \in [a,b]. \qquad (45)$$

Consequently, we get

$$|f(w)| g'(w)\left|\left(D_{b-;g}^\alpha f\right)(w)\right| \qquad (46)$$

$$\le \frac{1}{\Gamma(\alpha)}\left(\int_w^b (g(t)-g(w))^{p(\alpha-1)} dt\right)^{\frac{1}{p}} \left(z(w)\left(-z'(w)\right)\right)^{\frac{1}{q}},$$

$\forall w \in [a,b]$. Then

$$\int_x^b |f(w)|\left|\left(D_{b-;g}^\alpha f\right)(w)\right| g'(w)\, dw \qquad (47)$$

$$\le \frac{1}{\Gamma(\alpha)} \int_x^b \left(\int_w^b (g(t)-g(w))^{p(\alpha-1)} dt\right)^{\frac{1}{p}} \left(-z(w) z'(w)\right)^{\frac{1}{q}} dw$$

$$\le \frac{1}{\Gamma(\alpha)}\left(\int_x^b \left(\int_w^b (g(t)-g(w))^{p(\alpha-1)} dt\right) dw\right)^{\frac{1}{p}} \left(-\int_x^b z(w) z'(w)\, dw\right)^{\frac{1}{q}}$$

$$
= \frac{1}{\Gamma(\alpha)} \left(\int_x^b \left(\int_w^b (g(t) - g(w))^{p(\alpha-1)} \, dt \right) dw \right)^{\frac{1}{p}} \left(\frac{z^2(x)}{2} \right)^{\frac{1}{q}} \tag{48}
$$

$$
= \frac{1}{2^{\frac{1}{q}} \Gamma(\alpha)} \left(\int_x^b \left(\int_w^b (g(t) - g(w))^{p(\alpha-1)} \, dt \right) dw \right)^{\frac{1}{p}} \tag{49}
$$

$$
\cdot \left(\int_x^b (g'(t))^q \left| \left(D_{b-;g}^{\alpha} f \right)(t) \right|^q \, dt \right)^{\frac{2}{q}}.
$$

The theorem is proved.

A left fractional reverse Opial type inequality follows:

Theorem 11 *All as in Theorem 8. Additionally assume that* $\alpha \geq 1$, $g \in C^1([a, b])$, *and* $\left(f \circ g^{-1} \right)^{(k)} (g(a)) = 0$, *for* $k = 0, 1, \dots, n - 1$. *Let* $0 < p < 1$ *and* $q < 0$: $\frac{1}{p} + \frac{1}{q} = 1$. *We assume that* $\left(D_{a+;g}^{\alpha} f \right)$ *is of fixed sign and nowhere zero. Then*

$$
\int_a^x |f(w)| \left| \left(D_{a+;g}^{\alpha} f \right)(w) \right| g'(w) \, dw
$$

$$
\geq \frac{1}{2^{\frac{1}{q}} \Gamma(\alpha)} \left(\int_a^x \left(\int_a^w (g(w) - g(t))^{p(\alpha-1)} \, dt \right) dw \right)^{\frac{1}{p}} \tag{50}
$$

$$
\cdot \left(\int_a^x (g'(w))^q \left| \left(D_{a+;g}^{\alpha} f \right)(w) \right|^q \, dw \right)^{\frac{2}{q}},
$$

for all $a < x \leq b$.

Proof For every $x \in [a, b]$, we get again

$$
f(x) = \frac{1}{\Gamma(\alpha)} \int_a^x (g(x) - g(t))^{\alpha-1} g'(t) \left(D_{a+;g}^{\alpha} f \right)(t) \, dt. \tag{51}
$$

We have by assumption that

$$
|f(w)| = \frac{1}{\Gamma(\alpha)} \int_a^w (g(w) - g(t))^{\alpha-1} g'(t) \left| \left(D_{a+;g}^{\alpha} f \right)(t) \right| \, dt, \tag{52}
$$

for $a < w \leq x$. Then, by reverse Hölder's inequality we obtain

$$|f(w)| \geq \frac{1}{\Gamma(\alpha)} \left(\int_a^w (g(w) - g(t))^{(\alpha-1)p} \, dt \right)^{\frac{1}{p}}$$

$$\cdot \left(\int_a^w \left(g'(t) \right)^q \left| \left(D_{a+;g}^{\alpha} f \right)(t) \right|^q dt \right)^{\frac{1}{q}}, \quad (53)$$

for $a < w \leq x$. Call

$$z(w) := \int_a^w \left(g'(t) \right)^q \left| \left(D_{a+;g}^{\alpha} f \right)(t) \right|^q dt, \quad (54)$$

$z(a) = 0$, all $a \leq w \leq x$; $z(w) > 0$ on $(a, x]$. Thus

$$z'(w) = \left(g'(w) \right)^q \left| \left(D_{a+;g}^{\alpha} f \right)(w) \right|^q > 0, \quad (55)$$

all $a < w \leq x$. Hence

$$\left(z'(w) \right)^{\frac{1}{q}} = g'(w) \left| \left(D_{a+;g}^{\alpha} f \right)(w) \right| > 0, \quad (56)$$

all $a < w \leq x$. Consequently we get

$$|f(w)| \, g'(w) \left| \left(D_{a+;g}^{\alpha} f \right)(w) \right| \quad (57)$$

$$\geq \frac{1}{\Gamma(\alpha)} \left(\int_a^w (g(w) - g(t))^{p(\alpha-1)} \, dt \right)^{\frac{1}{p}} \left(z(w) \, z'(w) \right)^{\frac{1}{q}},$$

all $a < w \leq x$. Let $a < \theta \leq w \leq x$ and $\theta \downarrow a$. Then it holds

$$\int_a^x |f(w)| \, g'(w) \left| \left(D_{a+;g}^{\alpha} f \right)(w) \right| dw$$

$$= \lim_{\theta \downarrow a} \int_\theta^x |f(w)| \, g'(w) \left| \left(D_{a+;g}^{\alpha} f \right)(w) \right| dw \quad (58)$$

$$\geq \frac{1}{\Gamma(\alpha)} \lim_{\theta \downarrow a} \int_\theta^x \left(\int_a^w (g(w) - g(t))^{p(\alpha-1)} \, dt \right)^{\frac{1}{p}} \left(z(w) \, z'(w) \right)^{\frac{1}{q}} dw$$

$$\geq \frac{1}{\Gamma(\alpha)} \lim_{\theta \downarrow a} \left(\int_\theta^x \left(\int_a^w (g(w) - g(t))^{p(\alpha-1)} \, dt \right) dw \right)^{\frac{1}{p}} \lim_{\theta \downarrow a} \left(\int_\theta^x z(w) \, z'(w) \, dw \right)^{\frac{1}{q}}$$

$$= 2^{-\frac{1}{q}} \frac{1}{\Gamma(\alpha)} \left(\int_a^x \left(\int_a^w (g(w) - g(t))^{p(\alpha-1)} dt \right) dw \right)^{\frac{1}{p}} \lim_{\theta \downarrow a} \left(z^2(x) - z^2(\theta) \right)^{\frac{1}{q}}$$

$$= 2^{-\frac{1}{q}} \frac{1}{\Gamma(\alpha)} \left(\int_a^x \left(\int_a^w (g(w) - g(t))^{p(\alpha-1)} dt \right) dw \right)^{\frac{1}{p}} (z(x))^{\frac{2}{q}}. \tag{59}$$

The theorem is proved.

A right fractional reverse Opial type inequality follows:

Theorem 12 *All as in Theorem 8. Additionally assume that $\alpha \geq 1$, $g \in C^1([a,b])$, and $(f \circ g^{-1})^{(k)} (g(b)) = 0$, for $k = 0, 1, \ldots, n-1$. Let $0 < p < 1$ and $q < 0$: $\frac{1}{p} + \frac{1}{q} = 1$. We assume that $\left(D^\alpha_{b-;g} f \right)$ is of fixed sign and nowhere zero. Then*

$$\int_x^b |f(w)| \left| (D^\alpha_{b-;g} f)(w) \right| g'(w) \, dw$$

$$\geq \frac{1}{2^{\frac{1}{q}} \Gamma(\alpha)} \left(\int_x^b \left(\int_w^b (g(t) - g(w))^{p(\alpha-1)} dt \right) dw \right)^{\frac{1}{p}} \tag{60}$$

$$\left(\int_x^b (g'(w))^q \left| (D^\alpha_{b-;g} f)(w) \right|^q dw \right)^{\frac{2}{q}},$$

for all $a \leq x < b$.

Proof As before it holds

$$f(x) = \frac{1}{\Gamma(\alpha)} \int_x^b (g(t) - g(x))^{\alpha-1} g'(t) (D^\alpha_{b-;g} f)(t) \, dt, \tag{61}$$

$a \leq x \leq b$. We also have that

$$|f(w)| = \frac{1}{\Gamma(\alpha)} \int_x^b (g(t) - g(w))^{\alpha-1} g'(t) \left| (D^\alpha_{b-;g} f)(t) \right| dt, \tag{62}$$

$x \leq w < b$. Then it holds

$$|f(w)| \geq \frac{1}{\Gamma(\alpha)} \left(\int_w^b (g(t) - g(w))^{p(\alpha-1)} \, dt \right)^{\frac{1}{p}}$$

$$\cdot \left(\int_w^b (g'(t))^q \left| (D_{b-;g}^\alpha f)(t) \right|^q dt \right)^{\frac{1}{q}}, \qquad (63)$$

$x \leq w < b$. Call

$$z(w) := \int_w^b (g'(t))^q \left| (D_{b-;g}^\alpha f)(t) \right|^q dt = - \int_b^w (g'(t))^q \left| (D_{b-;g}^\alpha f)(t) \right|^q dt, \quad (64)$$

$z(b) = 0$, all $x \leq w \leq b$; $z(w) > 0$ on $[x, b)$. Thus

$$z'(w) = - (g'(w))^q \left| (D_{b-;g}^\alpha f)(w) \right|^q, \qquad (65)$$

and

$$- z'(w) = (g'(w))^q \left| (D_{b-;g}^\alpha f)(w) \right|^q > 0, \qquad (66)$$

all $x \leq w < b$, and

$$\left(-z'(w) \right)^{\frac{1}{q}} = g'(w) \left| (D_{b-;g}^\alpha f)(w) \right| > 0, \qquad (67)$$

all $x \leq w < b$. Consequently we get

$$|f(w)| \, g'(w) \left| (D_{b-;g}^\alpha f)(w) \right| \qquad (68)$$

$$\geq \frac{1}{\Gamma(\alpha)} \left(\int_w^b (g(t) - g(w))^{p(\alpha-1)} \, dt \right)^{\frac{1}{p}} \left(-z(w) z'(w) \right)^{\frac{1}{q}},$$

all $x \leq w < b$. Let $x \leq w \leq \theta < b$ and $\theta \uparrow b$. Then we obtain

$$\int_x^b |f(w)| \, g'(w) \left| (D_{b-;g}^\alpha f)(w) \right| dw$$

$$= \lim_{\theta \uparrow b} \int_x^\theta |f(w)| \, g'(w) \left| (D_{b-;g}^\alpha f)(w) \right| dw \qquad (69)$$

$$\geq \frac{1}{\Gamma(\alpha)} \lim_{\theta \uparrow b} \int_x^\theta \left(\int_w^b (g(t) - g(w))^{p(\alpha-1)} dt \right)^{\frac{1}{p}} (-z(w) z'(w))^{\frac{1}{q}} dw$$

$$\geq \frac{1}{\Gamma(\alpha)} \lim_{\theta \uparrow b} \left(\int_x^\theta \left(\int_w^b (g(t) - g(w))^{p(\alpha-1)} dt \right)^{\frac{1}{p}} dw \right)$$

$$\cdot \lim_{\theta \uparrow b} \left(\int_x^\theta (-z(w) z'(w)) dw \right)^{\frac{1}{q}}$$

$$= 2^{-\frac{1}{q}} \frac{1}{\Gamma(\alpha)} \left(\int_x^b \left(\int_w^b (g(t) - g(w))^{p(\alpha-1)} dt \right) dw \right)^{\frac{1}{p}} \tag{70}$$

$$\cdot \lim_{\theta \uparrow b} \left(z^2(x) - z^2(\theta) \right)^{\frac{1}{q}}$$

$$= 2^{-\frac{1}{q}} \frac{1}{\Gamma(\alpha)} \left(\int_x^b \left(\int_w^b (g(t) - g(w))^{p(\alpha-1)} dt \right) dw \right)^{\frac{1}{p}} \tag{71}$$

$$\cdot \left(z^2(x) - z^2(b) \right)^{\frac{1}{q}}$$

$$= \frac{1}{\Gamma(\alpha) 2^{\frac{1}{q}}} \left(\int_x^b \left(\int_w^b (g(t) - g(w))^{p(\alpha-1)} dt \right) dw \right)^{\frac{1}{p}} (z(x))^{\frac{2}{q}}. \tag{72}$$

The theorem is proved.

Two extreme fractional Opial type inequalities follow (case $p = 1, q = \infty$).

Theorem 13 *All as in Theorem 8. Assume that $\left(f \circ g^{-1} \right)^{(k)} (g(a)) = 0, k = 0, 1, \ldots, n - 1$. Then*

$$\int_a^x |f(w)| \left| D_{a+;g}^\alpha f(w) \right| dw \leq \frac{\left\| D_{a+;g}^\alpha f \right\|_\infty^2}{\Gamma(\alpha+1)} \left(\int_a^x (g(w) - g(a))^\alpha dw \right), \tag{73}$$

all $a \leq x \leq b$.

Proof For any $w \in [a, b]$, we have that

$$f(x) = \frac{1}{\Gamma(\alpha)} \int_a^w (g(w) - g(t))^{\alpha-1} g'(t) \left(D_{a+;g}^\alpha f \right)(t) dt, \tag{74}$$

and

$$|f(x)| \leq \frac{1}{\Gamma(\alpha)} \left(\int_a^w (g(w) - g(t))^{\alpha-1} g'(t) \, dt \right) \left\| D^{\alpha}_{a+;g} f \right\|_{\infty}$$

$$= \frac{\left\| D^{\alpha}_{a+;g} f \right\|_{\infty}}{\Gamma(\alpha+1)} (g(w) - g(a))^{\alpha}. \tag{75}$$

Hence we obtain

$$|f(w)| \left| D^{\alpha}_{a+;g} f(w) \right| \leq \frac{\left\| D^{\alpha}_{a+;g} f \right\|^2_{\infty}}{\Gamma(\alpha+1)} (g(w) - g(a))^{\alpha}. \tag{76}$$

Integrating (76) over $[a, x]$ we derive (73).

Theorem 14 *All as in Theorem 8. Assume that $\left(f \circ g^{-1}\right)^{(k)} (g(b)) = 0$, $k = 0, 1$, $\ldots, n - 1$. Then*

$$\int_x^b |f(w)| \left| D^{\alpha}_{b-;g} f(w) \right| \, dw \leq \frac{\left\| D^{\alpha}_{b-;g} f \right\|^2_{\infty}}{\Gamma(\alpha+1)} \left(\int_x^b (g(b) - g(w))^{\alpha} \, dw \right), \tag{77}$$

all $a \leq x \leq b$.

Proof For any $w \in [a, b]$, we have

$$f(x) = \frac{1}{\Gamma(\alpha)} \int_w^b (g(t) - g(w))^{\alpha-1} g'(t) \left(D^{\alpha}_{b-;g} f \right)(t) \, dt, \tag{78}$$

and

$$|f(x)| \leq \frac{1}{\Gamma(\alpha)} \left(\int_w^b (g(t) - g(w))^{\alpha-1} g'(t) \, dt \right) \left\| D^{\alpha}_{b-;g} f \right\|_{\infty}$$

$$= \frac{\left\| D^{\alpha}_{b-;g} f \right\|_{\infty}}{\Gamma(\alpha+1)} (g(b) - g(w))^{\alpha}. \tag{79}$$

Hence we obtain

$$|f(w)| \left| D_{b-;g}^{\alpha} f(w) \right| \leq \frac{\left\| D_{b-;g}^{\alpha} f \right\|_{\infty}^{2}}{\Gamma(\alpha+1)} (g(b) - g(w))^{\alpha}. \tag{80}$$

Integrating (80) over $[x, b]$ we derive (77).

Next we present three fractional Ostrowski type inequalities:

Theorem 15 *All as in Theorem* 8. *Then*

$$\left| f(y) - \frac{1}{b-a} \int_a^b f(x)\,dx + \sum_{k=1}^{n-1} \frac{\left(f \circ g^{-1} \right)^{(k)} (g(y))}{k!\,(b-a)} \int_a^b (g(x) - g(y))^k\,dx \right|$$

$$\leq \frac{1}{\Gamma(\alpha+1)(b-a)} \tag{81}$$

$$\cdot \left[(g(y) - g(a))^{\alpha} (y-a) \left\| D_{y-;g}^{\alpha} f \right\|_{\infty} + (g(b) - g(y))^{\alpha} (b-y) \left\| D_{y+;g}^{\alpha} f \right\|_{\infty} \right],$$

any $y \in [a, b]$.

Proof Define

$$\left(D_{y+;g}^{\alpha} f \right)(t) = 0, \text{ for } t < y, \tag{82}$$

$$\text{and}$$

$$\left(D_{y-;g}^{\alpha} f \right)(t) = 0, \text{ for } t > y.$$

Notice for $0 < \alpha \notin \mathbb{N}$ that

$$\left| \left(D_{a+;g}^{\alpha} f \right)(x) \right|$$

$$\leq \frac{1}{\Gamma(n-\alpha)} \left(\int_a^x (g(x) - g(t))^{n-\alpha-1} g'(t)\,dt \right) \left\| \left(f \circ g^{-1} \right)^{(n)} \circ g \right\|_{\infty}$$

$$= \frac{1}{\Gamma(n-\alpha)} \frac{(g(x) - g(a))^{n-\alpha}}{(n-\alpha)} \left\| \left(f \circ g^{-1} \right)^{(n)} \circ g \right\|_{\infty} \tag{83}$$

$$= \frac{1}{\Gamma(n-\alpha+1)} (g(x) - g(a))^{n-\alpha} \left\| \left(f \circ g^{-1} \right)^{(n)} \circ g \right\|_{\infty}.$$

Hence

$$\left(D_{a+;g}^{\alpha} f \right)(a) = 0. \tag{84}$$

Similarly it holds

$$\left|\left(D^\alpha_{b-;g}f\right)(x)\right|$$

$$\leq \frac{1}{\Gamma(n-\alpha)}\left(\int_x^b (g(t)-g(x))^{n-\alpha-1}g'(t)\,dt\right)\left\|\left(f\circ g^{-1}\right)^{(n)}\circ g\right\|_\infty$$

$$= \frac{1}{\Gamma(n-\alpha)}\frac{(g(b)-g(x))^{n-\alpha}}{(n-\alpha)}\left\|\left(f\circ g^{-1}\right)^{(n)}\circ g\right\|_\infty \qquad (85)$$

$$= \frac{1}{\Gamma(n-\alpha+1)}(g(b)-g(x))^{n-\alpha}\left\|\left(f\circ g^{-1}\right)^{(n)}\circ g\right\|_\infty.$$

Hence

$$\left(D^\alpha_{b-;g}f\right)(b)=0. \qquad (86)$$

I.e.

$$\left(D^\alpha_{y+;g}f\right)(y)=0, \quad \left(D^\alpha_{y-;g}f\right)(y)=0, \qquad (87)$$

$0<\alpha\notin\mathbb{N}$, any $y\in[a,b]$. We observe that

$$|R_1(y)| \qquad (88)$$

$$\overset{(19)}{\leq} \frac{1}{\Gamma(\alpha)(b-a)}\left[\left(\int_a^y\left(\int_x^y(g(t)-g(x))^{\alpha-1}g'(t)\,dt\right)dx\right)\left\|D^\alpha_{y-;g}f\right\|_\infty\right.$$

$$\left.+\left(\int_y^b\left(\int_y^x(g(x)-g(t))^{\alpha-1}g'(t)\,dt\right)dx\right)\left\|D^\alpha_{y+;g}f\right\|_\infty\right]$$

$$= \frac{1}{\Gamma(\alpha)(b-a)}\left[\left(\int_a^y\frac{(g(y)-g(x))^\alpha}{\alpha}\,dx\right)\left\|D^\alpha_{y-;g}f\right\|_\infty\right.$$

$$\left.+\left(\int_y^b\frac{(g(x)-g(y))^\alpha}{\alpha}\,dx\right)\left\|D^\alpha_{y+;g}f\right\|_\infty\right],$$

$$\leq \frac{1}{\Gamma(\alpha+1)(b-a)}\left[(g(y)-g(a))^\alpha(y-a)\left\|D^\alpha_{y-;g}f\right\|_\infty\right. \qquad (89)$$

$$\left.+(g(b)-g(y))^\alpha(b-y)\left\|D^\alpha_{y+;g}f\right\|_\infty\right].$$

We have proved that

$$|R_1(y)| \le \frac{1}{\Gamma(\alpha+1)(b-a)}\left[(g(y)-g(a))^\alpha (y-a)\left\|D^\alpha_{y-;g}f\right\|_\infty\right.$$
$$\left. + (g(b)-g(y))^\alpha (b-y)\left\|D^\alpha_{y+;g}f\right\|_\infty\right], \tag{90}$$

any $y \in [a,b]$. We have established the theorem.

Theorem 16 *All as in Theorem 8. Here we take $\alpha \ge 1$. Then*

$$\left| f(y) - \frac{1}{b-a}\int_a^b f(x)\,dx + \sum_{k=1}^{n-1}\frac{\left(f\circ g^{-1}\right)^{(k)}(g(y))}{k!(b-a)}\int_a^b (g(x)-g(y))^k\,dx\right|$$

$$\le \frac{1}{\Gamma(\alpha)(b-a)}\left[\left\|\left(D^\alpha_{y-;g}f\right)\circ g^{-1}\right\|_{1,[g(a),g(y)]}(y-a)(g(y)-g(a))^{\alpha-1}\right.$$
$$\left. + \left\|\left(D^\alpha_{y+;g}f\right)\circ g^{-1}\right\|_{1,[g(y),g(b)]}(b-y)(g(b)-g(y))^{\alpha-1}\right]. \tag{91}$$

Proof We can rewrite

$$R_1(y) \tag{92}$$

$$= -\frac{1}{\Gamma(\alpha)(b-a)}\left[\int_a^y\left(\int_{g(x)}^{g(y)}(z-g(x))^{\alpha-1}\left(\left(D^\alpha_{y-;g}f\right)\circ g^{-1}\right)(z)\,dz\right)dx\right.$$

$$\left. + \int_y^b\left(\int_{g(y)}^{g(x)}(g(x)-z)^{\alpha-1}\left(\left(D^\alpha_{y+;g}f\right)\circ g^{-1}\right)(z)\,dz\right)dx\right].$$

We assumed $\alpha \ge 1$, then

$$|R_1(y)| \tag{93}$$

$$\le \frac{1}{\Gamma(\alpha)(b-a)}\left[\int_a^y\left(\int_{g(x)}^{g(y)}(z-g(x))^{\alpha-1}\left|\left(\left(D^\alpha_{y-;g}f\right)\circ g^{-1}\right)(z)\right|dz\right)dx\right.$$

$$\left. + \int_y^b\left(\int_{g(y)}^{g(x)}(g(x)-z)^{\alpha-1}\left|\left(\left(D^\alpha_{y+;g}f\right)\circ g^{-1}\right)(z)\right|dz\right)dx\right]$$

$$\le \frac{1}{\Gamma(\alpha)(b-a)}\left[\left(\int_a^y\left(\int_{g(x)}^{g(y)}\left|\left(\left(D^\alpha_{y-;g}f\right)\circ g^{-1}\right)(z)\right|dz\right)dx\right)(g(y)-g(a))^{\alpha-1}\right.$$

$$\left. + \left(\int_y^b\left(\int_{g(y)}^{g(x)}\left|\left(\left(D^\alpha_{y+;g}f\right)\circ g^{-1}\right)(z)\right|dz\right)dx\right)(g(b)-g(y))^{\alpha-1}\right]$$

$$
\leq \frac{1}{\Gamma(\alpha)(b-a)} \left[\left\| \left(D_{y-;g}^{\alpha} f \right) \circ g^{-1} \right\|_{1,[g(a),g(y)]} (y-a)(g(y)-g(a))^{\alpha-1} \right.
$$
$$
\left. + \left\| \left(D_{y+;g}^{\alpha} f \right) \circ g^{-1} \right\|_{1,[g(y),g(b)]} (b-y)(g(b)-g(y))^{\alpha-1} \right]. \tag{94}
$$

So when $\alpha \geq 1$, we obtained

$$
|R_1(y)| \tag{95}
$$
$$
\leq \frac{1}{\Gamma(\alpha)(b-a)} \left[\left\| \left(D_{y-;g}^{\alpha} f \right) \circ g^{-1} \right\|_{1,[g(a),g(y)]} (y-a)(g(y)-g(a))^{\alpha-1} \right.
$$
$$
\left. + \left\| \left(D_{y+;g}^{\alpha} f \right) \circ g^{-1} \right\|_{1,[g(y),g(b)]} (b-y)(g(b)-g(y))^{\alpha-1} \right].
$$

Clearly here g^{-1} is continuous, thus $\left(D_{y-;g}^{\alpha} f \right) \circ g^{-1} \in C([g(a), g(y)])$, and $\left(D_{y+;g}^{\alpha} f \right) \circ g^{-1} \in C([g(y), g(b)])$. Therefore

$$
\left\| \left(D_{y-;g}^{\alpha} f \right) \circ g^{-1} \right\|_{1,[g(a),g(y)]}, \ \left\| \left(D_{y+;g}^{\alpha} f \right) \circ g^{-1} \right\|_{1,[g(y),g(b)]} < \infty. \tag{96}
$$

The proof of the theorem now is complete.

Theorem 17 *All as in Theorem 8. Let $p, q > 1 : \frac{1}{p} + \frac{1}{q} = 1$, $\alpha > \frac{1}{q}$. Then*

$$
\left| f(y) - \frac{1}{b-a} \int_a^b f(x) \, dx + \sum_{k=1}^{n-1} \frac{\left(f \circ g^{-1} \right)^{(k)} (g(y))}{k!(b-a)} \int_a^b (g(x) - g(y))^k \, dx \right|
$$
$$
\leq \frac{1}{\Gamma(\alpha)(b-a)(p(\alpha-1)+1)^{\frac{1}{p}}} \tag{97}
$$
$$
\cdot \left[(g(y) - g(a))^{\alpha-1+\frac{1}{p}} (y-a) \left\| \left(D_{y-;g}^{\alpha} f \right) \circ g^{-1} \right\|_{q,[g(a),g(y)]} \right.
$$
$$
\left. + (g(b) - g(y))^{\alpha-1+\frac{1}{p}} (b-y) \left\| \left(D_{y+;g}^{\alpha} f \right) \circ g^{-1} \right\|_{q,[g(y),g(b)]} \right].
$$

Proof Here we use (92). We get that

$$
|R_1(y)|
$$
$$
\leq \frac{1}{\Gamma(\alpha)(b-a)} \left[\int_a^y \left(\int_{g(x)}^{g(y)} (z - g(x))^{p(\alpha-1)} \, dz \right)^{\frac{1}{p}} \right.
$$
$$
\cdot \left(\int_{g(x)}^{g(y)} \left| \left(\left(D_{y-;g}^{\alpha} f \right) \circ g^{-1} \right)(z) \right|^q \, dz \right)^{\frac{1}{q}} dx + \int_y^b \left(\int_{g(y)}^{g(x)} (g(x) - z)^{p(\alpha-1)} \, dz \right)^{\frac{1}{p}}
$$
$$
\left. \cdot \left(\int_{g(y)}^{g(x)} \left| \left(\left(D_{y+;g}^{\alpha} f \right) \cup g^{-1} \right)(z) \right|^q \, dz \right)^{\frac{1}{q}} dx \right]
$$

$$\leq \frac{1}{\Gamma(\alpha)(b-a)} \left[\left(\int_a^y \frac{(g(y)-g(x))^{(\alpha-1)+\frac{1}{p}}}{(p(\alpha-1)+1)^{\frac{1}{p}}} dx \right) \left\| \left(D_{y-;g}^\alpha f\right) \circ g^{-1} \right\|_{q,[g(a),g(y)]} \right.$$

$$\left. + \left(\int_a^y \frac{(g(x)-g(y))^{(\alpha-1)+\frac{1}{p}}}{(p(\alpha-1)+1)^{\frac{1}{p}}} dx \right) \left\| \left(D_{y+;g}^\alpha f\right) \circ g^{-1} \right\|_{q,[g(y),g(b)]} \right] \tag{98}$$

(here it is $\alpha - 1 + \frac{1}{p} > 0$). Hence it holds

$$|R_1(y)|$$

$$\leq \frac{1}{\Gamma(\alpha)(b-a)(p(\alpha-1)+1)^{\frac{1}{p}}} \tag{99}$$

$$\cdot \left[(g(y)-g(a))^{\alpha-1+\frac{1}{p}}(y-a) \left\| \left(D_{y-;g}^\alpha f\right) \circ g^{-1} \right\|_{q,[g(a),g(y)]} \right.$$

$$\left. + (g(b)-g(y))^{\alpha-1+\frac{1}{p}}(b-y) \left\| \left(D_{y+;g}^\alpha f\right) \circ g^{-1} \right\|_{q,[g(y),g(b)]} \right].$$

Clearly here

$$\left\| \left(D_{y-;g}^\alpha f\right) \circ g^{-1} \right\|_{q,[g(a),g(y)]}, \quad \left\| \left(D_{y+;g}^\alpha f\right) \circ g^{-1} \right\|_{q,[g(y),g(b)]} < \infty.$$

We have proved the theorem.

We make

Remark 18 Let μ be a finite positive measure of mass $m > 0$ on $([c,d], \mathscr{P}([c,d]))$, with $[c,d] \subseteq [a,b]$, where $\mathscr{P}$ stands for the power set.

We found that (by Theorem 8)

$$f(y) - \frac{1}{b-a} \int_a^b f(x) dx + \sum_{k=1}^{n-1} \frac{\left(f \circ g^{-1}\right)^{(k)}(g(y))}{k!(b-a)} \int_a^b (g(x)-g(y))^k dx$$

$$= -\frac{1}{\Gamma(\alpha)(b-a)} \left[\int_a^y \left(\int_{g(x)}^{g(y)} (z-g(x))^{\alpha-1} \left((D_{y-;g}^\alpha f) \circ g^{-1}\right)(z) dz \right) dx \right.$$

$$\left. + \int_y^b \left(\int_{g(y)}^{g(x)} (g(x)-z)^{\alpha-1} \left((D_{y+;g}^\alpha f) \circ g^{-1}\right)(z) dz \right) dx \right]. \tag{100}$$

Then we have

$$M_n(f,g)$$

$$:= \frac{1}{m} \int_{[c,d]} f(y)\, d\mu(y) - \frac{1}{b-a} \int_a^b f(x)\, dx$$

$$+ \sum_{k=1}^{n-1} \frac{1}{k!\,(b-a)} \frac{1}{m} \int_{[c,d]} \left(\int_a^b \left(f \circ g^{-1} \right)^{(k)} (g(y))\, (g(x) - g(y))^k\, dx \right) d\mu(y)$$

$$= -\frac{1}{\Gamma(\alpha)(b-a)\,m} \left[\int_{[c,d]} \left\{ \int_a^y \left(\int_{g(x)}^{g(y)} (z - g(x))^{\alpha-1} \left(\left(D_{y-;g}^{\alpha} f \right) \circ g^{-1} \right)(z)\, dz \right) dx \right. \right.$$

$$\left. \left. + \int_y^b \left(\int_{g(y)}^{g(x)} (g(x) - z)^{\alpha-1} \left(\left(D_{y+;g}^{\alpha} f \right) \circ g^{-1} \right)(z)\, dz \right) dx \right\} d\mu(y) \right] \qquad (101)$$

$$= \frac{1}{m} \int_{[c,d]} R_1(y)\, d\mu(y). \qquad (102)$$

We present the following fractional comparison of means inequalities:

Theorem 19 *All as in Theorem* 8. *Then*

1.

$$|M_n(f,g)|$$

$$\leq \frac{1}{\Gamma(\alpha+1)(b-a)\,m} \left[\int_{[c,d]} \left[(g(y) - g(a))^{\alpha} (y-a) \left\| D_{y-;g}^{\alpha} f \right\|_{\infty} \right. \right.$$

$$\left. \left. + (g(b) - g(y))^{\alpha} (b-y) \left\| D_{y+;g}^{\alpha} f \right\|_{\infty} \right] d\mu(y) \right], \qquad (103)$$

2. if $\alpha \geq 1$, we get:

$$|M_n(f,g)|$$

$$\leq \frac{1}{\Gamma(\alpha)(b-a)\,m} \left[\int_{[c,d]} \left[\left\| \left(D_{y-;g}^{\alpha} f \right) \circ g^{-1} \right\|_{1,[g(a),g(y)]} (y-a)(g(y) - g(a))^{\alpha-1} \right. \right.$$

$$\left. \left. + \left\| \left(D_{y+;g}^{\alpha} f \right) \circ g^{-1} \right\|_{1,[g(y),g(b)]} (b-y)(g(b) - g(y))^{\alpha-1} \right] d\mu(y) \right], \qquad (104)$$

3. if $p, q > 1 : \frac{1}{p} + \frac{1}{q} = 1$, $\alpha > \frac{1}{q}$, we get:

$$|M_n(f,g)|$$

$$\leq \frac{1}{\Gamma(\alpha)(b-a)m}\left[\int_{[c,d]}\left[\int_a^y\left(\int_{g(x)}^{g(y)}(z-g(x))^{p(\alpha-1)}\,dz\right)^{\frac{1}{p}}\right.\right.$$

$$\cdot\left(\int_{g(x)}^{g(y)}\left|\left(\left(D_{y-;g}^{\alpha}f\right)\circ g^{-1}\right)(z)\right|^q\,dz\right)^{\frac{1}{q}}dx+\int_y^b\left(\int_{g(y)}^{g(x)}(g(x)-z)^{p(\alpha-1)}\,dz\right)^{\frac{1}{p}}$$

$$\cdot\left.\left.\left(\int_{g(y)}^{g(x)}\left|\left(\left(D_{y+;g}^{\alpha}f\right)\circ g^{-1}\right)(z)\right|^q\,dz\right)^{\frac{1}{q}}dx\right]d\mu(y)\right]. \tag{105}$$

Proof By Theorems 15, 16, 17, for (103), (104), (105), respectively.

Next we give some fractional Grüss type inequalities:

Theorem 20 *Let f,h as in Theorem 8. Here $R_1(y)$ will be renamed as $R_1(f,y)$, so we can consider $R_1(h,y)$. Then*

1.

$$\Delta_n(f,h):=\frac{1}{b-a}\int_a^b f(x)h(x)\,dx-\frac{\left(\int_a^b f(x)\,dx\right)\left(\int_a^b h(x)\,dx\right)}{(b-a)^2}$$

$$+\frac{1}{2(b-a)^2}\sum_{k=1}^{n-1}\frac{1}{k!}\left[\int_a^b\left(\int_a^b\left(h(y)\left(f\circ g^{-1}\right)^{(k)}(g(y))\right.\right.\right.$$

$$+\left.\left.\left. f(y)\left(h\circ g^{-1}\right)^{(k)}(g(y))\right)(g(x)-g(y))^k\,dx\right)dy\right]$$

$$=\frac{1}{2(b-a)}\left[\int_a^b(h(y)R_1(f,y)+f(y)R_1(h,y))\,dy\right] \tag{106}$$

$$=:K_n(f,h),$$

2. it holds

$$|\Delta_n(f,h)| \tag{107}$$

$$\leq\frac{(g(b)-g(a))^{\alpha}}{2\Gamma(\alpha+1)}\left[\|h\|_{\infty}\left(\sup_{y\in[a,b]}\left(\left\|D_{y-;g}^{\alpha}f\right\|_{\infty}+\left\|D_{y+;g}^{\alpha}f\right\|_{\infty}\right)\right)\right.$$

$$+\left.\|f\|_{\infty}\left(\sup_{y\in[a,b]}\left(\left\|D_{y-;g}^{\alpha}h\right\|_{\infty}+\left\|D_{y+;g}^{\alpha}h\right\|_{\infty}\right)\right)\right],$$

3. *if $\alpha \geq 1$, we get:*

$$|\Delta_n (f, h)| \tag{108}$$

$$\leq \frac{1}{2\Gamma(\alpha)(b-a)} (g(b) - g(a))^{\alpha-1}$$

$$\cdot \left\{ \|h\|_1 \left(\sup_{y\in[a,b]} \left(\left\| \left(D^{\alpha}_{y-;g} f\right) \circ g^{-1} \right\|_{1,[g(a),g(b)]} + \left\| \left(D^{\alpha}_{y+;g} f\right) \circ g^{-1} \right\|_{1,[g(a),g(b)]} \right) \right) \right.$$

$$\left. + \|f\|_1 \left(\sup_{y\in[a,b]} \left(\left\| \left(D^{\alpha}_{y-;g} h\right) \circ g^{-1} \right\|_{1,[g(a),g(b)]} + \left\| \left(D^{\alpha}_{y+;g} h\right) \circ g^{-1} \right\|_{1,[g(a),g(b)]} \right) \right) \right\},$$

4. *if $p, q > 1 : \frac{1}{p} + \frac{1}{q} = 1$, $\alpha > \frac{1}{q}$, we get:*

$$|\Delta_n (f, h)| \tag{109}$$

$$\leq \frac{(g(b) - g(a))^{\alpha-1+\frac{1}{p}}}{2\Gamma(\alpha)(p(\alpha-1)+1)^{\frac{1}{p}}} \cdot$$

$$\left\{ \|h\|_\infty \left(\sup_{y\in[a,b]} \left(\left\| \left(D^{\alpha}_{y-;g} f\right) \circ g^{-1} \right\|_{q,[g(a),g(b)]} + \left\| \left(D^{\alpha}_{y+;g} f\right) \circ g^{-1} \right\|_{q,[g(a),g(b)]} \right) \right) \right.$$

$$\left. + \|f\|_\infty \left(\sup_{y\in[a,b]} \left(\left\| \left(D^{\alpha}_{y-;g} h\right) \circ g^{-1} \right\|_{q,[g(a),g(b)]} + \left\| \left(D^{\alpha}_{y+;g} h\right) \circ g^{-1} \right\|_{q,[g(a),g(b)]} \right) \right) \right\}.$$

All right hand sides of (107)–(109) *are finite.*

Proof By Theorem 8 we have

$$h(y) f(y)$$

$$= \frac{h(y)}{b-a} \int_a^b f(x)\,dx \tag{110}$$

$$- \sum_{k=1}^{n-1} \frac{h(y) \left(f \circ g^{-1}\right)^{(k)} (g(y))}{k!(b-a)} \int_a^b (g(x) - g(y))^k\,dx + h(y)\, R_1(f, y),$$

and

$$f(y) h(y)$$

$$= \frac{f(y)}{b-a} \int_a^b h(x)\,dx \tag{111}$$

$$- \sum_{k=1}^{n-1} \frac{f(y) \left(h \circ g^{-1}\right)^{(k)} (g(y))}{k!(b-a)} \int_a^b (g(x) - g(y))^k\,dx + f(y)\, R_1(h, y),$$

$\forall\, y \in [a, b]$. Then integrating (110) we find

$$
\int_a^b h(y) f(y)\, dy
$$

$$
= \frac{\left(\int_a^b h(y)\, dy\right)}{b-a} \left(\int_a^b f(x)\, dx\right) \tag{112}
$$

$$
- \sum_{k=1}^{n-1} \frac{1}{k!(b-a)} \int_a^b \int_a^b h(y) \left(f \circ g^{-1}\right)^{(k)} (g(y)) (g(x) - g(y))^k\, dx\, dy
$$

$$
+ \int_a^b h(y) R_1(f, y)\, dy,
$$

and integrating (111) we obtain

$$
\int_a^b f(y) h(y)\, dy
$$

$$
= \frac{\left(\int_a^b f(y)\, dy\right) \left(\int_a^b h(x)\, dx\right)}{b-a} \tag{113}
$$

$$
- \sum_{k=1}^{n-1} \frac{1}{k!(b-a)} \int_a^b \int_a^b f(y) \left(h \circ g^{-1}\right)^{(k)} (g(y)) (g(x) - g(y))^k\, dx\, dy
$$

$$
+ \int_a^b f(y) R_1(h, y)\, dy.
$$

Adding the last two equalities (112) and (113), we get:

$$
2 \int_a^b f(x) h(x)\, dx
$$

$$
= \frac{2 \left(\int_a^b f(x)\, dx\right) \left(\int_a^b h(x)\, dx\right)}{b-a} \tag{114}
$$

$$
- \sum_{k=1}^{n-1} \frac{1}{k!(b-a)} \left[\int_a^b \int_a^b [h(y) \left(f \circ g^{-1}\right)^{(k)} (g(y)) + f(y) \left(h \circ g^{-1}\right)^{(k)} (g(y))] \right.
$$

$$
\left. \cdot (g(x) - g(y))^k\, dx\, dy \right] + \int_a^b (h(y) R_1(f, y) + f(y) R_1(h, y))\, dy.
$$

Divide the last (114) by $2(b-a)$ to obtain (106).

Then, we upper bound $K_n(f, h)$ using Theorems 15, 16, 17, to obtain (107), (108) and (109), respectively.

We use also that a norm is a continuous function. The theorem is proved.

Background Next we follow [10]. This is related to Information theory. Let f be a convex function from $(0, +\infty)$ into $\mathbb{R}$, which is strictly convex at 1 with $f(1) = 0$. Let $(X, \mathscr{A}, \lambda)$ be a measure space, where λ is a finite or a σ-finite measure on $(X, \mathscr{A})$. And let μ_1, μ_2 be two probability measures on $(X, \mathscr{A})$ such that $\mu_1 \ll \lambda$, $\mu_2 \ll \lambda$ (absolutely continuous), e.g. $\lambda = \mu_1 + \mu_2$. Denote by $p = \frac{d\mu_1}{d\lambda}, q = \frac{d\mu_2}{d\lambda}$ the Radon-Nikodym derivatives of μ_1, μ_2 with respect to λ (densities). Here, we assume that

$$0 < a \leq \frac{p}{q} \leq b, \text{ a.e. on } X \text{ and } a \leq 1 \leq b.$$

The quantity

$$\Gamma_f(\mu_1, \mu_2) = \int_X q(x) f\left(\frac{p(x)}{q(x)}\right) d\lambda(x) \tag{115}$$

was introduced by Csiszar in 1967 (see [7]), and is called the f-divergence of the probability measures μ_1 and μ_2. By Lemma 1.1 of [7], the integral (115) is well defined, and $\Gamma_f(\mu_1, \mu_2) \geq 0$, with equality only when $\mu_1 = \mu_2$. Furthermore $\Gamma_f(\mu_1, \mu_2)$ does not depend on the choice of λ. Here, by assuming $f(1) = 0$, we can consider $\Gamma_f(\mu_1, \mu_2)$ the f-divergence, as a measure of the difference between the probability measures μ_1, μ_2.

Here we give a representation and estimates for $\Gamma_f(\mu_1, \mu_2)$.

Furthermore, we assume f as in Theorem 8, where $[a, b] \subset (0, +\infty)$ as above. We make

Remark 21 By Theorem 8 we have found that

$$f(y)$$

$$= \frac{1}{b-a} \int_a^b f(x) \, dx + \sum_{k=1}^{n-1} \frac{\left(f \circ g^{-1}\right)^{(k)}(g(y))}{k!(b-a)} \int_a^b (g(x) - g(y))^k \, dx \tag{116}$$

$$- \frac{1}{\Gamma(\alpha)(b-a)} \left[\int_a^y \left(\int_{g(x)}^{g(y)} (\rho - g(x))^{\alpha-1} \left(\left(D_{y-;g}^\alpha f\right) \circ g^{-1}\right)(\rho) \, d\rho \right) dx \right.$$

$$\left. + \int_y^b \left(\int_{g(y)}^{g(x)} (g(x) - \rho)^{\alpha-1} \left(\left(D_{y+;g}^\alpha f\right) \circ g^{-1}\right)(\rho) \, d\rho \right) dx \right].$$

We get the equality:

$$q(z) f\left(\frac{p(z)}{q(z)}\right)$$

$$= \frac{q(z)}{b-a} \int_a^b f(x)\,dx \tag{117}$$

$$+ \sum_{k=1}^{n-1} \frac{\left(f \circ g^{-1}\right)^{(k)}\left(g\left(\frac{p(z)}{q(z)}\right)\right)}{k!\,(b-a)} q(z) \int_a^b \left(g(x) - g\left(\frac{p(z)}{q(z)}\right)\right)^k dx$$

$$- \frac{q(z)}{\Gamma(\alpha)(b-a)}\left[\int_a^{\frac{p(z)}{q(z)}}\left(\int_{g(x)}^{g\left(\frac{p(z)}{q(z)}\right)} (\rho - g(x))^{\alpha-1}\left(\left(D^\alpha_{\frac{p(z)}{q(z)}-;g}f\right)\circ g^{-1}\right)(\rho)\,d\rho\right)dx\right.$$

$$\left.+ \int_{\frac{p(z)}{q(z)}}^b\left(\int_{g\left(\frac{p(z)}{q(z)}\right)}^{g(x)} (g(x) - \rho)^{\alpha-1}\left(\left(D^\alpha_{\frac{p(z)}{q(z)}+;g}f\right)\circ g^{-1}\right)(\rho)\,d\rho\right)dx\right],$$

a.e. on X. Hence, it holds the representation

$$\Gamma_f(\mu_1, \mu_2)$$

$$= \int_X q(z) f\left(\frac{p(z)}{q(z)}\right)d\lambda(z)$$

$$= \frac{1}{b-a}\int_a^b f(x)\,dx + \sum_{k=1}^{n-1} \frac{1}{k!\,(b-a)}\int_X \left[q(z)\left(f \circ g^{-1}\right)^{(k)}\left(g\left(\frac{p(z)}{q(z)}\right)\right)\right.$$

$$\left.\cdot\left(\int_a^b \left(g(x) - g\left(\frac{p(z)}{q(z)}\right)\right)^k dx\right)\right]d\lambda(z) - \frac{1}{\Gamma(\alpha)(b-a)} \tag{118}$$

$$\cdot\left\{\int_X q(z)\left[\int_a^{\frac{p(z)}{q(z)}}\left(\int_{g(x)}^{g\left(\frac{p(z)}{q(z)}\right)} (\rho - g(x))^{\alpha-1}\left(\left(D^\alpha_{\frac{p(z)}{q(z)}-;g}f\right)\circ g^{-1}\right)(\rho)\,d\rho\right)dx\right.\right.$$

$$\left.\left.+ \int_{\frac{p(z)}{q(z)}}^b\left(\int_{g\left(\frac{p(z)}{q(z)}\right)}^{g(x)} (g(x) - \rho)^{\alpha-1}\left(\left(D^\alpha_{\frac{p(z)}{q(z)}+;g}f\right)\circ g^{-1}\right)(\rho)\,d\rho\right)dx\right]d\lambda(z)\right\}.$$

Clearly here $R_1(y)$ is continuous function in $y \in [a,b]$.
Let here

$$\left(D^\alpha_{x_0+;g} f\right)(x) = \frac{1}{\Gamma(n-\alpha)} \int_{x_0}^{x} (g(x) - g(t))^{n-\alpha-1} g'(t) \left(f \circ g^{-1}\right)^{(n)} (g(t)) \, dt,$$

(119)

$x \geq x_0$;

$$\left(D^\alpha_{y_0+;g} f\right)(x) = \frac{1}{\Gamma(n-\alpha)} \int_{y_0}^{x} (g(x) - g(t))^{n-\alpha-1} g'(t) \left(f \circ g^{-1}\right)^{(n)} (g(t)) \, dt,$$

(120)

$x \geq y_0$; where fixed $x : x \geq y_0 \geq x_0$; $x, x_0, y_0 \in [a,b]$. Then we obtain

$$\left| \left(D^\alpha_{x_0+;g} f\right)(x) - \left(D^\alpha_{y_0+;g} f\right)(x) \right|$$

$$= \frac{1}{\Gamma(n-\alpha)} \left| \int_{x_0}^{y_0} (g(x) - g(t))^{n-\alpha-1} g'(t) \left(f \circ g^{-1}\right)^{(n)} (g(t)) \, dt \right| \quad (121)$$

$$\leq \frac{\left\| \left(f \circ g^{-1}\right)^{(n)} \circ g \right\|_{\infty,[a,b]}}{\Gamma(n-\alpha)} \left(\int_{x_0}^{y_0} (g(x) - g(t))^{n-\alpha-1} g'(t) \, dt \right)$$

(by [11, p. 107] and [9])

$$= \frac{\left\| \left(f \circ g^{-1}\right)^{(n)} \circ g \right\|_{\infty,[a,b]}}{\Gamma(n-\alpha)} \left(\int_{g(x_0)}^{g(y_0)} (g(x) - \rho)^{n-\alpha-1} \, d\rho \right) \quad (122)$$

$$= \frac{\left\| \left(f \circ g^{-1}\right)^{(n)} \circ g \right\|_{\infty,[a,b]}}{\Gamma(n-\alpha+1)} \left((g(x) - g(y_0))^{n-\alpha} - (g(x) - g(x_0))^{n-\alpha} \right)$$

$$\to 0,$$

as $y_0 \to x_0$, proving continuity of $\left(D^\alpha_{x_0+;g} f\right)(x)$ in $x_0 \in [a,b]$. Similarly $\left(D^\alpha_{x_0-;g} f\right)(x)$ is continuous in $x_0 \in [a,b]$. We want to estimate

$$I := \int_X q(z) R_1 \left(\frac{p(z)}{q(z)} \right) d\lambda(z).$$

That is

$$|I| = \left| \int_X q(z) R_1 \left(\frac{p(z)}{q(z)} \right) d\lambda(z) \right| \leq \int_X q(z) \left| R_1 \left(\frac{p(z)}{q(z)} \right) \right| d\lambda(z). \qquad (123)$$

Notice, by (118), that

$$I = \Gamma_f(\mu_1, \mu_2) - \frac{1}{b-a} \int_a^b f(x)\, dx \qquad (124)$$

$$- \sum_{k=1}^{n-1} \frac{1}{k!\,(b-a)} \int_X \left[g(z) \left(f \circ g^{-1} \right)^{(k)} \left(g \left(\frac{p(z)}{q(z)} \right) \right) \right.$$

$$\left. \cdot \left(\int_a^b \left(g(x) - g \left(\frac{p(z)}{q(z)} \right) \right)^k dx \right) \right] d\lambda(z).$$

We give

Theorem 22 *All as above. Then*

$$|I| \qquad\qquad\qquad\qquad\qquad\qquad\qquad\qquad (125)$$

$$\leq \frac{1}{\Gamma(\alpha+1)(b-a)} \left[\int_X q(z) \left[\left(g \left(\frac{p(z)}{q(z)} \right) - g(a) \right)^\alpha \left(\frac{p(z)}{q(z)} - a \right) \right. \right.$$

$$\left. \left. \cdot \left\| D^\alpha_{\frac{p(z)}{q(z)}-;g} f \right\|_\infty + \left(g(b) - g \left(\frac{p(z)}{q(z)} \right) \right)^\alpha \left(b - \frac{p(z)}{q(z)} \right) \left\| D^\alpha_{\frac{p(z)}{q(z)}+;g} f \right\|_\infty \right] d\lambda(z) \right].$$

Proof By Theorem 14.

Theorem 23 *All as above and $\alpha \geq 1$. Then*

$$|I|$$

$$\leq \frac{1}{\Gamma(\alpha)(b-a)} \left[\int_X q(z) \left[\left\| \left(D^\alpha_{\frac{p(z)}{q(z)}-;g} f \right) \circ g^{-1} \right\|_{1,[g(a),g(b)]} \right. \right. \qquad (126)$$

$$\cdot \left(\frac{p(z)}{q(z)} - a \right) \left(g \left(\frac{p(z)}{q(z)} \right) - g(a) \right)^{\alpha-1} + \left\| \left(D^\alpha_{\frac{p(z)}{q(z)}+;g} f \right) \circ g^{-1} \right\|_{1,[g(a),g(b)]}$$

$$\left. \left. \cdot \left(b - \frac{p(z)}{q(z)} \right) \left(g(b) - g \left(\frac{p(z)}{q(z)} \right) \right)^{\alpha-1} \right] d\lambda(z) \right].$$

Proof By Theorem 16, and $\left(D^{\alpha}_{y\pm;g}f\right) \circ g^{-1}$ are continuous functions in $y \in [a, b]$.

Theorem 24 *All as above. Let* $p^{*}, q^{*} > 1 : \frac{1}{p^{*}} + \frac{1}{q^{*}} = 1, \alpha > \frac{1}{q^{*}}.$ *Then*

$$|I|$$

$$\leq \frac{1}{\Gamma(\alpha)(b-a)(p^{*}(\alpha-1)+1)^{\frac{1}{p^{*}}}} \tag{127}$$

$$\cdot \left[\int_X q(z) \left[\left(g\left(\frac{p(z)}{q(z)}\right) - g(a) \right)^{\alpha-1+\frac{1}{p^{*}}} \left(\frac{p(z)}{q(z)} - a \right) \right. \right.$$

$$\left\| \left(D^{\alpha}_{\frac{p(z)}{q(z)}-;g}f \right) \circ g^{-1} \right\|_{q^{*},[g(a),g(b)]} + \left(g(b) - g\left(\frac{p(z)}{q(z)}\right) \right)^{\alpha-1+\frac{1}{p^{*}}}$$

$$\left. \cdot \left(b - \frac{p(z)}{q(z)} \right) \left\| \left(D^{\alpha}_{\frac{p(z)}{q(z)}+;g}f \right) \circ g^{-1} \right\|_{q^{*},[g(a),g(b)]} \right] d\lambda(z) \right].$$

Proof By Theorem 17.

Remark 25 Some examples for g follow:

$$g(x) = e^x, x \in [a, b] \subset \mathbb{R},$$
$$g(x) = \sin x,$$
$$g(x) = \tan x,$$
$$\text{where } x \in \left[-\frac{\pi}{2} + \varepsilon, \frac{\pi}{2} - \varepsilon \right], \text{ where } \varepsilon > 0 \text{ small.}$$

Indeed, the above examples of g are strictly increasing and absolutely continuous functions.

One can apply all of our results here for the above specific choices of g. We choose to omit this lengthy job.

References

1. Anastassiou, G.A.: Advanced fractional Taylor's formulae. J. Comput. Anal. Appl. (2015)
2. Anastassiou, G.: On right fractional calculus. Chaos, Solitons Fractals **42**, 365–376 (2009)
3. Anastassiou, G.: Fractional Differentiation Inequalities. Springer, New York (2009)
4. Anastassiou, G.: Basic inequalities, revisited. Math. Balkanica, New Ser. **24**(1–2), 59–84 (2010)
5. Anastassiou, G.: Fractional representation formulae and right fractional inequalities. Math. Comput. Model. **54**(11–12), 3098–3115 (2011)
6. Anastassiou, G.: The reduction method in fractional calculus and fractional Ostrowski type Inequalities. Indian J. Math. **56**(3), 333–357 (2014)
7. Csiszar, I.: Information-type measures of difference of probability distributions. Stud. Scientiarum Mathematicarum Hung. **2**, 299–318 (1967)

8. Diethelm, K.: The Analysis of Fractional Differential Equations, 1st edition, vol. 2004. Lecture Notes in Mathematics, New York, Heidelberg (2010)
9. Jia, R.-Q.: Chapter 3. Absolutely Continuous Functions. https://www.ualberta.ca/~jia/Math418/Notes/Chap.3.pdf
10. Kilbas, A.A., Srivastava, H.M., Tujillo, J.J.: Theory and Applications of Fractional Differential Equations. North-Holland mathematics studies, vol. 204. Elsevier, New York, USA (2006)
11. Royden, H.L.: Real Analysis, 2nd edn. Macmillan Publishing Co., Inc, New York (1968)

Basic Iterated Fractional Inequalities

George A. Anastassiou

Abstract Using fundamental formulae of iterated fractional Caputo type calculus, we establish several important fractional representation formulae, included iterated ones. Based on these we derive: a whole family of fractional Opial type inequalities, Hilbert-Pachpatte type fractional inequalities, Ostrowski type fractional inequalities, Poincaré and Sobolev type fractional inequalities, finally we give Grüss type fractional inequalities.

1 Background

Let $0 < \alpha < 1$, $f : [a, b] \to \mathbb{R}$ such that $f' \in L_\infty ([a, b])$, the left Caputo fractional derivative of order α is defined as follows:

$$\left(D_a^\alpha f \right)(x) = \frac{1}{\Gamma (1 - \alpha)} \int_a^x (x - t)^{-\alpha} f'(t)\, dt, \tag{1}$$

where Γ is the gamma function, $\forall\, x \in [a, b]$.

We observe that

$$\left| \left(D_a^\alpha f \right)(x) \right| \leq \frac{1}{\Gamma (1 - \alpha)} \int_a^x (x - t)^{-\alpha} \left| f'(t) \right| dt$$

$$\leq \frac{\| f' \|_\infty}{\Gamma (1 - \alpha)} \left(\int_a^x (x - t)^{-\alpha} dt \right)$$

$$= \frac{\| f' \|_\infty}{\Gamma (1 - \alpha)} \frac{(x - a)^{1-\alpha}}{(1 - \alpha)} = \frac{\| f' \|_\infty (x - a)^{1-\alpha}}{\Gamma (2 - \alpha)}.$$

G.A. Anastassiou (✉)
Department of Mathematical Sciences, University of Memphis, Memphis 38152, USA
e-mail: ganastss@memphis.edu

© Springer International Publishing Switzerland 2016
G.A. Anastassiou and O. Duman (eds.), *Intelligent Mathematics II:*
Applied Mathematics and Approximation Theory, Advances in Intelligent Systems
and Computing 441, DOI 10.1007/978-3-319-30322-2_29

That is

$$\left|\left(D_a^\alpha f\right)(x)\right| \leq \frac{\|f'\|_\infty}{\Gamma(2-\alpha)}(x-a)^{1-\alpha}, \quad \forall x \in [a,b]. \tag{2}$$

We obtain

$$\left|\left(D_a^\alpha f\right)(a)\right| = 0, \tag{3}$$

and

$$\|D_a^\alpha f\|_\infty \leq \frac{\|f'\|_\infty}{\Gamma(2-\alpha)}(b-a)^{1-\alpha}. \tag{4}$$

Denote for $n \in \mathbb{N}$:

$$D_a^{n\alpha} = D_a^\alpha D_a^\alpha \ldots D_a^\alpha \ (n\text{-times}). \tag{5}$$

Assume that $D_a^{k\alpha} f \in C([a,b])$, $k = 0, 1, \ldots, n+1$; $n \in \mathbb{N}$, $0 < \alpha < 1$.
   Then, by [7, 9], we get the left fractional Taylor formula:

$$f(x) = f(a) + \sum_{i=2}^{n} \frac{(x-a)^{i\alpha}}{\Gamma(i\alpha+1)}\left(D_a^{i\alpha} f\right)(a)$$

$$+ \frac{1}{\Gamma((n+1)\alpha)} \int_a^x (x-t)^{(n+1)\alpha-1}\left(D_a^{(n+1)\alpha} f\right)(t)\,dt, \tag{6}$$

$\forall x \in (a,b]$.
   Let fixed $y \in [a,b]$, we define

$$\left(D_y^\alpha f\right)(x) = \frac{1}{\Gamma(1-\alpha)} \int_y^x (x-t)^{-\alpha} f'(t)\,dt, \tag{7}$$

for any $x \geq y$; $x \in [a,b]$, and

$$\left(D_y^\alpha f\right)(x) = 0, \quad \text{for } x < y. \tag{8}$$

Of course $\left(D_y^\alpha f\right)(y) = 0$.
   By [3], p. 388, we get that $\left(D_y^\alpha f\right) \in C([a,b])$.
   We assume that $f' \in L_\infty([a,b])$, and $D_y^{k\alpha} f \in C([a,b])$, $k = 0, 1, 2, 3, \ldots,$
$n+1$, $n \in \mathbb{N}$; $0 < \alpha < 1$.

Then, by (6), we obtain

$$f(x) = f(y) + \sum_{i=2}^{n} \frac{(x-y)^{i\alpha}}{\Gamma(i\alpha+1)} \left(D_y^{i\alpha} f\right)(y) \tag{9}$$

$$+ \frac{1}{\Gamma((n+1)\alpha)} \int_y^x (x-t)^{(n+1)\alpha-1} \left(D_y^{(n+1)\alpha} f\right)(t)\, dt,$$

$\forall\, x \geq y;\ x, y \in [a, b],\ 0 < \alpha < 1$.

Let again $0 < \alpha < 1$, $f : [a, b] \to \mathbb{R}$ such that $f' \in L_\infty([a, b])$, the right Caputo fractional derivative of order α is defined as follows:

$$\left(D_{b-}^{\alpha} f\right)(x) = \frac{-1}{\Gamma(1-\alpha)} \int_x^b (z-x)^{-\alpha} f'(z)\, dz, \ \forall\, x \in [a, b]. \tag{10}$$

We notice that

$$\left|\left(D_{b-}^{\alpha} f\right)(x)\right| \tag{11}$$

$$\leq \frac{1}{\Gamma(1-\alpha)} \int_x^b (z-x)^{-\alpha} \left|f'(z)\right| dz \leq \frac{\|f'\|_\infty}{\Gamma(1-\alpha)} \left(\int_x^b (z-x)^{-\alpha}\, dz\right)$$

$$= \frac{\|f'\|_\infty}{\Gamma(1-\alpha)} \frac{(b-x)^{1-\alpha}}{(1-\alpha)} = \frac{\|f'\|_\infty}{\Gamma(2-\alpha)} (b-x)^{1-\alpha}.$$

That is

$$\left|\left(D_{b-}^{\alpha} f\right)(x)\right| \leq \frac{\|f'\|_\infty}{\Gamma(2-\alpha)} (b-x)^{1-\alpha}, \ \forall\, x \in [a, b]. \tag{12}$$

We derive

$$\left(D_{b-}^{\alpha} f\right)(b) = 0, \tag{13}$$

and

$$\|D_{b-}^{\alpha} f\|_\infty \leq \frac{\|f'\|_\infty}{\Gamma(2-\alpha)} (b-a)^{1-\alpha}. \tag{14}$$

Denote for $n \in \mathbb{N}$:

$$D_{b-}^{n\alpha} = D_{b-}^{\alpha} D_{b-}^{\alpha} \ldots D_{b-}^{\alpha} \ (n\text{-times}). \tag{15}$$

Assume that $D_{b-}^{k\alpha} f \in C([a, b])$, $k = 0, 1, \ldots, n+1$; $n \in \mathbb{N}$, $0 < \alpha < 1$.

Then, by [6], we get the right fractional Taylor formula:

$$f(x) = f(b) + \sum_{i=2}^{n} \frac{(b-x)^{i\alpha}}{\Gamma(i\alpha+1)} \left(D_{b-}^{i\alpha} f\right)(b) \tag{16}$$

$$+ \frac{1}{\Gamma((n+1)\alpha)} \int_x^b (z-x)^{(n+1)\alpha-1} \left(D_{b-}^{(n+1)\alpha} f\right)(z)\,dz,$$

$\forall\, x \in [a, b]$.

Let fixed $y \in [a, b]$, we define

$$\left(D_{y-}^{\alpha} f\right)(x) = -\frac{1}{\Gamma(1-\alpha)} \int_x^y (z-x)^{-\alpha} f'(z)\,dz, \tag{17}$$

$\forall\, x \le y;\ x \in [a, b]$, and

$$\left(D_{y-}^{\alpha} f\right)(x) = 0,\ \text{for } x > y. \tag{18}$$

Of course $\left(D_{y-}^{\alpha} f\right)(y) = 0$.

By [4], we get that $\left(D_{y-}^{\alpha} f\right) \in C([a, b])$.

We assume that $f' \in L_\infty([a, b])$, and $D_{y-}^{k\alpha} f \in C([a, b]), k = 0, 1, 2, 3, \ldots, n+1, n \in \mathbb{N}; 0 < \alpha < 1$.

Then, by (16), we obtain

$$f(x) = f(y) + \sum_{i=2}^{n} \frac{(y-x)^{i\alpha}}{\Gamma(i\alpha+1)} \left(D_{y-}^{i\alpha} f\right)(y) \tag{19}$$

$$+ \frac{1}{\Gamma((n+1)\alpha)} \int_x^y (z-x)^{(n+1)\alpha-1} \left(D_{y-}^{(n+1)\alpha} f\right)(z)\,dz,$$

$\forall\, x \le y;\ x, y \in [a, b], 0 < \alpha < 1$.

2 Main Results

We give the following representation formula:

Theorem 1 Let $0 < \alpha < 1$, $f : [a, b] \to \mathbb{R}$ such that $f' \in L_\infty([a, b])$, $y \in [a, b]$ is fixed. We assume that $D_{y}^{k\alpha} f, D_{y-}^{k\alpha} f \in C([a, b]), k = 0, 1, 2, 3, \ldots, n+1; n \in \mathbb{N}$. Then

$$f(y) = \frac{1}{b-a} \int_a^b f(x)\,dx \tag{20}$$

$$-\sum_{i=2}^n \frac{1}{(b-a)\,\Gamma(i\alpha+2)} \left[\left(D_{y-}^{i\alpha} f\right)(y)(y-a)^{i\alpha+1} + \left(D_y^{i\alpha} f\right)(y)(b-y)^{i\alpha+1} \right]$$

$$+ R_1(y),$$

where

$$R_1(y)$$

$$= -\frac{1}{\Gamma((n+1)\alpha)(b-a)} \left[\int_a^y \left(\int_x^y (t-x)^{(n+1)\alpha-1} \left(D_{y-}^{(n+1)\alpha} f\right)(t)\,dt \right) dx \right.$$

$$\left. + \int_y^b \left(\int_y^x (x-t)^{(n+1)\alpha-1} \left(D_y^{(n+1)\alpha} f\right)(t)\,dt \right) dx \right]. \tag{21}$$

Proof Here $x, y \in [a, b]$. We keep y as fixed. By (9) and (19), we obtain

$$\frac{1}{b-a} \int_a^b f(x)\,dx = \frac{1}{b-a} \left(\int_a^y f(x)\,dx + \int_y^b f(x)\,dx \right) \tag{22}$$

$$= \frac{1}{b-a} \left\{ \int_a^y \left[f(y) + \sum_{i=2}^n \frac{(y-x)^{i\alpha}}{\Gamma(i\alpha+1)} \left(D_{y-}^{i\alpha} f\right)(y) \right. \right.$$

$$\left. + \frac{1}{\Gamma((n+1)\alpha)} \int_x^y (z-x)^{(n+1)\alpha-1} \left(D_{y-}^{(n+1)\alpha} f\right)(z)\,dz \right] dx$$

$$+ \int_y^b \left[f(y) + \sum_{i=2}^n \frac{(x-y)^{i\alpha}}{\Gamma(i\alpha+1)} \left(D_y^{i\alpha} f\right)(y) \right.$$

$$\left. \left. + \frac{1}{\Gamma((n+1)\alpha)} \int_y^x (x-t)^{(n+1)\alpha-1} \left(D_y^{(n+1)\alpha} f\right)(t)\,dt \right] dx \right\}$$

$$= f(y) + \sum_{i=2}^n \frac{1}{(b-a)\,\Gamma(i\alpha+1)} \left[\left(D_{y-}^{i\alpha} f\right)(y)(y-a)^{i\alpha+1} + \left(D_y^{i\alpha} f\right)(y)(b-y)^{i\alpha+1} \right]$$

$$+ \frac{1}{\Gamma((n+1)\alpha)(b-a)} \left[\int_a^y \left(\int_x^y (z-x)^{(n+1)\alpha-1} \left(D_{y-}^{(n+1)\alpha} f\right)(z)\,dz \right) dx \right.$$

$$\left. + \int_y^b \left(\int_y^x (x-t)^{(n+1)\alpha-1} \left(D_y^{(n+1)\alpha} f\right)(t)\,dt \right) dx \right]. \tag{23}$$

Solving (22), (23) for $f(y)$, we have proved theorem.

Next we present a left fractional Opial type inequality:

Theorem 2 *Let* $0 < \alpha < 1$, $f : [a, b] \to \mathbb{R}$ *such that* $f' \in L_\infty([a, b])$. *Assume that* $D_a^{k\alpha} f \in C([a, b])$, $k = 0, 1, \ldots, n + 1$; $n \in \mathbb{N}$. *Suppose that* $\left(D_a^{i\alpha} f\right)(a) = 0$, *for* $i = 0, 2, 3, \ldots, n$. *Let* $p, q > 1 : \frac{1}{p} + \frac{1}{q} = 1$, *such that* $\alpha > \frac{1}{(n+1)q}$. *Then*

$$\int_a^x |f(w)| \left| D_a^{(n+1)\alpha} f(w) \right| dw \tag{24}$$

$$\leq \frac{(x-a)^{\left((n+1)\alpha + \frac{1}{p} - \frac{1}{q}\right)}}{2^{\frac{1}{q}} \Gamma((n+1)\alpha) [(p((n+1)\alpha - 1) + 1)(p((n+1)\alpha - 1) + 2)]^{\frac{1}{p}}}$$

$$\cdot \left(\int_a^x \left| D_a^{(n+1)\alpha} f(w) \right|^q dw \right)^{\frac{2}{q}},$$

$\forall\, x \in [a, b]$.

Proof By [7, 9], our assumption, and (6), we obtain

$$f(x) = \frac{1}{\Gamma((n+1)\alpha)} \int_a^x (x-t)^{(n+1)\alpha - 1} \left(D_a^{(n+1)\alpha} f\right)(t)\, dt, \tag{25}$$

$\forall\, x \in [a, b]$. Then, by Hölder's inequality we obtain,

$$|f(x)| \leq \frac{1}{\Gamma((n+1)\alpha)} \int_a^x (x-t)^{(n+1)\alpha - 1} \left| D_a^{(n+1)\alpha} f(t) \right| dt$$

$$\leq \frac{1}{\Gamma((n+1)\alpha)} \left(\int_a^x (x-t)^{p((n+1)\alpha - 1)} dt \right)^{\frac{1}{p}} \left(\int_a^x \left| D_a^{(n+1)\alpha} f(t) \right|^q dt \right)^{\frac{1}{q}}$$

$$= \frac{1}{\Gamma((n+1)\alpha)} \frac{(x-a)^{((n+1)\alpha - 1) + \frac{1}{p}}}{(p((n+1)\alpha - 1) + 1)^{\frac{1}{p}}} \left(\int_a^x \left| D_a^{(n+1)\alpha} f(t) \right|^q dt \right)^{\frac{1}{q}}$$

$$= \frac{(x-a)^{(n+1)\alpha - \frac{1}{q}}}{\Gamma((n+1)\alpha)(p((n+1)\alpha - 1) + 1)^{\frac{1}{p}}} \left(\int_a^x \left| D_a^{(n+1)\alpha} f(t) \right|^q dt \right)^{\frac{1}{q}}.$$

$$\tag{26}$$

That is

$$|f(x)| \leq \frac{(x-a)^{(n+1)\alpha-\frac{1}{q}}}{\Gamma((n+1)\alpha)(p((n+1)\alpha-1)+1)^{\frac{1}{p}}} \left(\int_a^x \left| D_a^{(n+1)\alpha} f(t) \right|^q dt \right)^{\frac{1}{q}},$$

(27)

$\forall\, x \in [a,b]$. Call

$$z(x) := \int_a^x \left| D_a^{(n+1)\alpha} f(t) \right|^q dt,$$

(28)

$z(a) = 0$. Thus

$$z'(x) := \left| D_a^{(n+1)\alpha} f(x) \right|^q \geq 0,$$

(29)

and

$$\left(z'(x) \right)^{\frac{1}{q}} := \left| D_a^{(n+1)\alpha} f(x) \right| \geq 0, \ \forall x \in [a,b].$$

Consequently, we get

$$|f(w)| \left| D_a^{(n+1)\alpha} f(w) \right| \leq \frac{(w-a)^{(n+1)\alpha-\frac{1}{q}}}{\Gamma((n+1)\alpha)(p((n+1)\alpha-1)+1)^{\frac{1}{p}}} \left(z(w) z'(w) \right)^{\frac{1}{q}},$$

(30)

$\forall\, w \in [a,b]$. Then it holds

$$\int_a^x |f(w)| \left| D_a^{(n+1)\alpha} f(w) \right| dw$$

$$\leq \frac{1}{\Gamma((n+1)\alpha)(p((n+1)\alpha-1)+1)^{\frac{1}{p}}} \int_a^x (w-a)^{\frac{(p((n+1)\alpha-1)+1)}{p}} \left(z(w) z'(w) \right)^{\frac{1}{q}} dw$$

$$\leq \frac{1}{\Gamma((n+1)\alpha)(p((n+1)\alpha-1)+1)^{\frac{1}{p}}}$$

(31)

$$\cdot \left(\int_a^x (w-a)^{(p((n+1)\alpha-1)+1)} dw \right)^{\frac{1}{p}} \left(\int_a^x z(w) z'(w) dw \right)^{\frac{1}{q}}$$

$$= \frac{(x-a)^{((n+1)\alpha-1)+\frac{2}{p}}}{\Gamma((n+1)\alpha)\left[(p((n+1)\alpha-1)+1)(p((n+1)\alpha-1)+2)\right]^{\frac{1}{p}}}\left(\frac{z^2(x)}{2}\right)^{\frac{1}{q}}$$

$$= \frac{(x-a)^{(n+1)\alpha+\frac{1}{p}-\frac{1}{q}}}{2^{\frac{1}{q}}\Gamma((n+1)\alpha)\left[(p((n+1)\alpha-1)+1)(p((n+1)\alpha-1)+2)\right]^{\frac{1}{p}}}$$

$$\cdot\left(\int_a^x \left|D_a^{(n+1)\alpha}f(t)\right|^q dt\right)^{\frac{2}{q}}. \tag{32}$$

The theorem is proved.

It follows a right fractional Opial type inequality:

Theorem 3 *Let* $0<\alpha<1$, $f:[a,b]\to\mathbb{R}$ *such that* $f'\in L_\infty([a,b])$. *Assume that* $D_{b-}^{k\alpha}f\in C([a,b])$, *for* $k=0,1,\ldots,n+1$; $n\in\mathbb{N}$. *Suppose that* $\left(D_{b-}^{i\alpha}f\right)(b)=0$, *for* $i=0,2,3,\ldots,n$. *Let* $p,q>1:\frac{1}{p}+\frac{1}{q}=1$, *such that* $\alpha>\frac{1}{(n+1)q}$. *Then*

$$\int_x^b |f(w)|\left|\left(D_{b-}^{(n+1)\alpha}f\right)(w)\right|dw$$

$$\leq \frac{(b-x)^{\left((n+1)\alpha+\frac{1}{p}-\frac{1}{q}\right)}}{2^{\frac{1}{q}}\Gamma((n+1)\alpha)\left[(p((n+1)\alpha-1)+1)(p((n+1)\alpha-1)+2)\right]^{\frac{1}{p}}}$$

$$\cdot\left(\int_x^b \left|\left(D_{b-}^{(n+1)\alpha}f\right)(w)\right|^q dw\right)^{\frac{2}{q}}, \tag{33}$$

$\forall\, x\in[a,b]$.

Proof By [6], our assumption, and (16), we obtain

$$f(x)=\frac{1}{\Gamma((n+1)\alpha)}\int_x^b (z-x)^{(n+1)\alpha-1}\left(D_{b-}^{(n+1)\alpha}f\right)(z)\,dz, \tag{34}$$

$\forall\, x\in[a,b]$. Then, by Hölder's inequality we find,

$$|f(x)|$$
$$\leq \frac{1}{\Gamma((n+1)\alpha)}\int_x^b (z-x)^{(n+1)\alpha-1}\left|\left(D_{b-}^{(n+1)\alpha}f\right)(z)\right|dz \tag{35}$$

$$\leq \frac{1}{\Gamma\left((n+1)\alpha\right)} \left(\int_x^b (z-x)^{p((n+1)\alpha-1)}\, dz\right)^{\frac{1}{p}} \left(\int_x^b \left|\left(D_{b-}^{(n+1)\alpha} f\right)(z)\right|^q dz\right)^{\frac{1}{q}}$$

$$= \frac{1}{\Gamma\left((n+1)\alpha\right)} \frac{(b-x)^{\frac{(p((n+1)\alpha-1)+1)}{p}}}{(p((n+1)\alpha-1)+1)^{\frac{1}{p}}} \left(\int_x^b \left|\left(D_{b-}^{(n+1)\alpha} f\right)(t)\right|^q dt\right)^{\frac{1}{q}}.$$

That is, we derive

$$|f(x)| \leq \frac{(b-x)^{\frac{(p((n+1)\alpha-1)+1)}{p}}}{(p((n+1)\alpha-1)+1)^{\frac{1}{p}}\, \Gamma((n+1)\alpha)} \left(\int_x^b \left|\left(D_{b-}^{(n+1)\alpha} f\right)(t)\right|^q dt\right)^{\frac{1}{q}}, \qquad (36)$$

$\forall\, x \in [a, b]$. Call

$$z(x) := \int_x^b \left|\left(D_{b-}^{(n+1)\alpha} f\right)(t)\right|^q dt, \qquad (37)$$

$z(b) = 0$. Hence

$$z'(x) = -\left|\left(D_{b-}^{(n+1)\alpha} f\right)(x)\right|^q, \qquad (38)$$

and

$$-z'(x) = \left|\left(D_{b-}^{(n+1)\alpha} f\right)(x)\right|^q \geq 0, \qquad (39)$$

and

$$\left(-z'(x)\right)^{\frac{1}{q}} = \left|\left(D_{b-}^{(n+1)\alpha} f\right)(x)\right| \geq 0, \ \forall x \in [a, b]. \qquad (40)$$

Consequently, we get

$$|f(w)| \left|\left(D_{b-}^{(n+1)\alpha} f\right)(w)\right| \qquad (41)$$

$$\leq \frac{(b-w)^{\frac{((p(n+1)\alpha-1)+1)}{p}}}{\Gamma\left((n+1)\alpha\right) \left(p\left((n+1)\alpha-1\right)+1\right)^{\frac{1}{p}}} \left(z(w)\left(-z'(w)\right)\right)^{\frac{1}{q}},$$

$\forall\, w \in [a, b]$. Then, it holds

$$\int_x^b |f(w)| \left| \left(D_{b-}^{(n+1)\alpha} f \right)(w) \right| dw$$

$$\leq \frac{1}{\Gamma((n+1)\alpha)(p((n+1)\alpha-1)+1)^{\frac{1}{p}}} \tag{42}$$

$$\cdot \int_x^b (b-w)^{\frac{(p((n+1)\alpha-1)+1)}{p}} \left(z(w) \left(-z'(w) \right) \right)^{\frac{1}{q}} dw$$

$$\leq \frac{1}{\Gamma((n+1)\alpha)(p((n+1)\alpha-1)+1)^{\frac{1}{p}}}$$

$$\cdot \left(\int_x^b (b-w)^{(p((n+1)\alpha-1)+1)} dw \right)^{\frac{1}{p}} \left(\int_x^b z(w) \left(-z'(w) \right) dw \right)^{\frac{1}{q}}$$

$$= \frac{(b-x)^{\frac{(p((n+1)\alpha-1)+2)}{p}}}{\Gamma((n+1)\alpha)[(p((n+1)\alpha-1)+1)(p((n+1)\alpha-1)+2)]^{\frac{1}{p}}} \left(\frac{z^2(x)}{2} \right)^{\frac{1}{q}} \tag{43}$$

$$= \frac{(b-x)^{\left((n+1)\alpha+\frac{1}{p}-\frac{1}{q}\right)}}{2^{\frac{1}{q}} \Gamma((n+1)\alpha)[(p((n+1)\alpha-1)+1)(p((n+1)\alpha-1)+2)]^{\frac{1}{p}}}$$

$$\cdot \left(\int_x^b \left| \left(D_{b-}^{(n+1)\alpha} f \right)(w) \right|^q dw \right)^{\frac{2}{q}}, \tag{44}$$

proving the claim.

We need
Background Let $v > 0$; the operator J_a^v, defined on $L_1([a, b])$ is given by

$$J_a^v f(x) = \frac{1}{\Gamma(v)} \int_a^x (x-t)^{v-1} f(t) \, dt, \tag{45}$$

for $a \leq x \leq b$, is called the left Riemann-Liouville fractional integral operator of order v. We set $J_a^0 = I$, the identity operator; see [3], p. 392, also [8].

From [1], p. 543, when $f \in C([a, b])$, $\mu, v > 0$, we get that

$$J_a^\mu \left(J_a^v f \right) = J_a^{\mu+v} f = J_a^v \left(J_a^\mu f \right), \tag{46}$$

which is the semigroup property.

Let now $\gamma > 0, m := \lceil \gamma \rceil$ ($\lceil \cdot \rceil$ ceiling of the number), $f \in AC^m([a, b])$ (it means $f^{(m-1)} \in AC([a, b])$ (absolutely continuous functions)). The left Caputo fractional derivative of order γ is given by

$$D_a^\gamma f(x) = \frac{1}{\Gamma(m - \gamma)} \int_a^x (x - t)^{m-\gamma-1} f^{(m)}(t)\, dt = \left(J_a^{m-\gamma} f^{(m)}\right)(x), \quad (47)$$

and it exists almost everywhere for $x \in [a, b]$. See Corollary 16.8, p. 394, of [3], and [8], pp. 49–50.

We set $D_a^m f = f^{(m)}, m \in \mathbb{N}$.

We need

Remark 4 Let $0 < \alpha < 1$, $f : [a, b] \to \mathbb{R}$ such that $f' \in L_\infty([a, b])$. Assume that $D_a^{k\alpha} f \in C([a, b])$, $k = 0, 1, \ldots, n + 1$; $n \in \mathbb{N}$. Suppose that $\left(D_a^{i\alpha} f\right)(a) = 0$, for $i = 0, 2, 3, \ldots, n$. By [7, 9] and (6), and our assumption here, we obtain

$$f(x) = \frac{1}{\Gamma((n + 1)\alpha)} \int_a^x (x - t)^{(n+1)\alpha-1} \left(D_a^{(n+1)\alpha} f\right)(t)\, dt, \quad (48)$$

$\forall x \in [a, b]$.

By [3], Theorem 7.7, p. 117, when $(n + 1)\alpha - 1 > 0$, equivalently when $\alpha > \frac{1}{n+1}$, we get that there exists

$$f'(x) = \frac{((n + 1)\alpha - 1)}{\Gamma((n + 1)\alpha)} \int_a^x (x - t)^{(n+1)\alpha-2} \left(D_a^{(n+1)\alpha} f\right)(t)\, dt, \quad (49)$$

$\forall x \in [a, b]$.

If $(n + 1)\alpha - 2 > 0$, equivalently, if $\alpha > \frac{2}{n+1}$, we get that there exists

$$f''(x) = \frac{((n+1)\alpha-1)((n+1)\alpha-2)}{\Gamma((n+1)\alpha)} \int_a^x (x - t)^{(n+1)\alpha-3} \left(D_a^{(n+1)\alpha} f\right)(t)\, dt, \quad (50)$$

$\forall x \in [a, b]$.

In general, if $(n + 1)\alpha - m > 0$, equivalently, if $\alpha > \frac{m}{n+1}$, we get that there exists

$$f^{(m)}(x) = \frac{\prod_{j=1}^m ((n + 1)\alpha - j)}{\Gamma((n + 1)\alpha)} \int_a^x (x - t)^{(n+1)\alpha-m-1} \left(D_a^{(n+1)\alpha} f\right)(t)\, dt, \quad (51)$$

$\forall x \in [a, b]$.

By [3], p. 388, we get that $f^{(m)} \in C([a, b])$.

By (45), we derive

$$f^{(m)}(x) = \frac{\prod_{j=1}^{m}((n+1)\alpha - j)\Gamma((n+1)\alpha - m)}{\Gamma((n+1)\alpha)} \left(J_a^{((n+1)\alpha - m)}\left(D_a^{(n+1)\alpha}f\right)\right)(x)$$
$$= \left(J_a^{((n+1)\alpha - m)}\left(D_a^{(n+1)\alpha}f\right)\right)(x). \tag{52}$$

We have proved that

$$f^{(m)}(x) = \left(J_a^{((n+1)\alpha - m)}\left(D_a^{(n+1)\alpha}f\right)\right)(x), \tag{53}$$

$\forall\, x \in [a, b]$.

We have that (case of $\gamma < m$)

$$\left(D_a^\gamma f\right)(x) = \left(J_a^{m-\gamma}f^{(m)}\right)(x) \tag{54}$$
$$= \left(J_a^{m-\gamma}J_a^{((n+1)\alpha - m)}\left(D_a^{(n+1)\alpha}f\right)\right)(x)$$
$$= \left(J_a^{(n+1)\alpha - \gamma}\left(D_a^{(n+1)\alpha}f\right)\right)(x).$$

That is

$$\left(D_a^\gamma f\right)(x) = \left(J_a^{(n+1)\alpha - \gamma}\left(D_a^{(n+1)\alpha}f\right)\right)(x), \tag{55}$$

$\forall\, x \in [a, b]$.

I.e. we have found the representation formula:

$$\left(D_a^\gamma f\right)(x) = \frac{1}{\Gamma((n+1)\alpha - \gamma)} \int_a^x (x-t)^{(n+1)\alpha - \gamma - 1}\left(D_a^{(n+1)\alpha}f\right)(t)\,dt, \tag{56}$$

$\forall x \in [a, b]$.

The last formula (56) is true under the assumption $(n+1)\alpha > m$, and since $m \geq \gamma$, it implies $(n+1)\alpha > \gamma$ and $(n+1)\alpha - \gamma > 0$. Furthermore, by [3], p. 388, we get that $\left(D_a^\gamma f\right) \in C([a, b])$.

We have proved the following left fractional representation theorem:

Theorem 5 *Let $0 < \alpha < 1$, $f : [a, b] \to \mathbb{R}$ such that $f' \in L_\infty([a, b])$. Assume that $D_a^{k\alpha}f \in C([a, b])$, $k = 0, 1, \ldots, n+1$; $n \in \mathbb{N}$. Suppose that $\left(D_a^{i\alpha}f\right)(a) = 0$, for $i = 0, 2, 3, \ldots, n$. Let $\gamma > 0$ with $\lceil \gamma \rceil = m < n+1$, such that $m < (n+1)\alpha$, equivalently, $\alpha > \frac{m}{n+1}$. Then*

$$\left(D_a^\gamma f\right)(x) = \frac{1}{\Gamma((n+1)\alpha - \gamma)} \int_a^x (x-t)^{(n+1)\alpha - \gamma - 1}\left(D_a^{(n+1)\alpha}f\right)(t)\,dt, \tag{57}$$

$\forall\, x \in [a, b]$. *Furthermore it holds $\left(D_a^\gamma f\right) \in C([a, b])$.*

We make

Remark 6 Call $\lambda := (n+1)\alpha - \gamma - 1$, i.e. $\lambda + 1 = (n+1)\alpha - \gamma$, and call $\delta :=$ $(n+1)\alpha$. Then we can write

$$\left(D_a^\gamma f\right)(x) = \frac{1}{\Gamma(\lambda+1)} \int_a^x (x-t)^\lambda \left(D_a^\delta f\right)(t)\, dt, \; \forall\, x \in [a, b]. \qquad (58)$$

If $\lambda > 0$, then

$$\left(D_a^\gamma f\right)'(x) = \frac{\lambda}{\Gamma(\lambda+1)} \int_a^x (x-t)^{\lambda-1} \left(D_a^\delta f\right)(t)\, dt, \; \forall x \in [a, b]. \qquad (59)$$

If $\lambda - 1 > 0$, then

$$\left(D_a^\gamma f\right)''(x) = \frac{\lambda(\lambda-1)}{\Gamma(\lambda+1)} \int_a^x (x-t)^{\lambda-2} \left(D_a^\delta f\right)(t)\, dt, \; \forall x \in [a, b]. \qquad (60)$$

If $\lambda - 2 > 0$, then

$$\left(D_a^\gamma f\right)^{(3)}(x) = \frac{\lambda(\lambda-1)(\lambda-2)}{\Gamma(\lambda+1)} \int_a^x (x-t)^{\lambda-3} \left(D_a^\delta f\right)(t)\, dt, \; \forall x \in [a, b], \qquad (61)$$

etc.

In general, if $\lambda - m + 1 > 0$, then

$$\left(D_a^\gamma f\right)^{(m)}(x) \qquad (62)$$

$$= \frac{\lambda(\lambda-1)(\lambda-2)\ldots(\lambda-m+1)}{\Gamma(\lambda+1)} \int_a^x (x-t)^{(\lambda-m+1)-1} \left(D_a^\delta f\right)(t)\, dt$$

$$= \frac{\lambda(\lambda-1)(\lambda-2)\ldots(\lambda-m+1)\Gamma(\lambda-m+1)\left(J_a^{(\lambda-m+1)}\left(D_a^\delta f\right)\right)(x)}{\Gamma(\lambda+1)}$$

$$= \left(J_a^{(\lambda-m+1)}\left(D_a^\delta f\right)\right)(x), \qquad (63)$$

$\forall x \in [a, b]$.

That is, if $\lambda - m + 1 > 0$, then

$$\left(D_a^\gamma f\right)^{(m)}(x) = \left(J_a^{(\lambda-m+1)}\left(D_a^\delta f\right)\right)(x), \qquad (64)$$

$\forall x \in [a, b]$.

We notice that

$$\left(D_a^{2\gamma} f\right)(x) = \left(D_a^{\gamma} \left(D_a^{\gamma} f\right)\right)(x) = \left(J_a^{m-\gamma} \left(D_a^{\gamma} f\right)^{(m)}\right)(x) \tag{65}$$

$$= \left(J_a^{m-\gamma} J_a^{\lambda-m+1} \left(D_a^{\delta} f\right)\right)(x) = \left(J_a^{\lambda-\gamma+1} \left(D_a^{\delta} f\right)\right)(x)$$

$$= \left(J_a^{(n+1)\alpha-\gamma-1-\gamma+1} \left(D_a^{\delta} f\right)\right)(x) = \left(J_a^{(n+1)\alpha-2\gamma} \left(D_a^{\delta} f\right)\right)(x). \tag{66}$$

That is

$$\left(D_a^{2\gamma} f\right)(x) = \left(J_a^{(n+1)\alpha-2\gamma} \left(D_a^{(n+1)\alpha} f\right)\right)(x), \tag{67}$$

$\forall x \in [a, b]$, under the condition $\frac{\gamma+m}{n+1} < \alpha < 1$.

We give

Theorem 7 *Under the assumptions of Theorem 5, and when $\frac{\gamma+m}{n+1} < \alpha < 1$, we get that*

$$\left(D_a^{2\gamma} f\right)(x) = \frac{1}{\Gamma\left((n+1)\alpha - 2\gamma\right)} \int_a^x (x-t)^{(n+1)\alpha-2\gamma-1} \left(D_a^{(n+1)\alpha} f\right)(t)\, dt, \tag{68}$$

$\forall x \in [a, b]$, *and* $\left(D_a^{2\gamma} f\right) \in C\left([a, b]\right)$.

We make

Remark 8 Call $\rho := (n+1)\alpha - 2\gamma - 1$, i.e. $\rho + 1 = (n+1)\alpha - 2\gamma$, and call again $\delta := (n+1)\alpha$. Then we can write

$$\left(D_a^{2\gamma} f\right)(x) = \frac{1}{\Gamma(\rho+1)} \int_a^x (x-t)^{\rho} \left(D_a^{\delta} f\right)(t)\, dt, \tag{69}$$

$\forall x \in [a, b]$.

If $\rho > 0$, then

$$\left(D_a^{2\gamma} f\right)'(x) = \frac{\rho}{\Gamma(\rho+1)} \int_a^x (x-t)^{\rho-1} \left(D_a^{\delta} f\right)(t)\, dt, \tag{70}$$

$\forall x \in [a, b]$.

If $\rho - 1 > 0$, then

$$\left(D_a^{2\gamma} f\right)'' (x) = \frac{\rho (\rho - 1)}{\Gamma (\rho + 1)} \int_a^x (x - t)^{\rho - 2} \left(D_a^{\delta} f\right) (t) \, dt, \tag{71}$$

$\forall x \in [a, b]$.

If $\rho - 2 > 0$, then

$$\left(D_a^{2\gamma} f\right)^{(3)} (x) = \frac{\rho (\rho - 1)(\rho - 2)}{\Gamma (\rho + 1)} \int_a^x (x - t)^{\rho - 3} \left(D_a^{\delta} f\right) (t) \, dt, \tag{72}$$

$\forall x \in [a, b]$, etc.

In general, if $\rho - m + 1 > 0$, then

$$\left(D_a^{2\gamma} f\right)^{(m)} (x) \tag{73}$$

$$= \frac{\rho (\rho - 1)(\rho - 2) \dots (\rho - m + 1)}{\Gamma (\rho + 1)} \int_a^x (x - t)^{(\rho - m + 1) - 1} \left(D_a^{\delta} f\right) (t) \, dt$$

$$= \frac{\rho (\rho - 1)(\rho - 2) \dots (\rho - m + 1) \Gamma (\rho - m + 1) \left(J_a^{(\rho - m + 1)} \left(D_a^{\delta} f\right)\right) (x)}{\Gamma (\rho + 1)}$$

$$= \left(J_a^{(\rho - m + 1)} \left(D_a^{\delta} f\right)\right) (x), \tag{74}$$

$\forall x \in [a, b]$.

That is, if $\rho - m + 1 > 0$, then

$$\left(D_a^{2\gamma} f\right)^{(m)} (x) = \left(J_a^{(\rho - m + 1)} \left(D_a^{\delta} f\right)\right) (x), \tag{75}$$

$\forall x \in [a, b]$.

We notice that

$$\left(D_a^{3\gamma} f\right) (x) = \left(D_a^{\gamma} \left(D_a^{2\gamma} f\right)\right) (x) = \left(J_a^{m-\gamma} \left(D_a^{2\gamma} f\right)^{(m)}\right) (x) \tag{76}$$

$$= \left(J_a^{m-\gamma} J_a^{\rho-m+1} \left(D_a^{\delta} f\right)\right) (x) = \left(J_a^{\rho-\gamma+1} \left(D_a^{\delta} f\right)\right) (x)$$

$$= \left(J_a^{(n+1)\alpha - 2\gamma - 1 - \gamma + 1} \left(D_a^{\delta} f\right)\right) (x) = \left(J_a^{(n+1)\alpha - 3\gamma} \left(D_a^{(n+1)\alpha} f\right)\right) (x). \tag{77}$$

That is, if $\frac{m + 2\gamma}{n + 1} < \alpha < 1$, we get

$$\left(D_a^{3\gamma} f\right) (x) = \left(J_a^{(n+1)\alpha - 3\gamma} \left(D_u^{(n+1)\alpha} f\right)\right) (x), \tag{78}$$

$\forall x \in [a, b]$.

We have proved

Theorem 9 *Under the assumptions of Theorem 5, and when* $\frac{m+2\gamma}{n+1} < \alpha < 1$, *we obtain that*

$$\left(D_a^{3\gamma} f\right)(x) = \frac{1}{\Gamma\left((n+1)\alpha - 3\gamma\right)} \int_a^x (x-t)^{(n+1)\alpha - 3\gamma - 1} \left(D_a^{(n+1)\alpha} f\right)(t)\, dt, \quad (79)$$

$\forall x \in [a, b]$, *and* $\left(D_a^{3\gamma} f\right) \in C\left([a, b]\right)$.

In general, we derive:

Theorem 10 *Under the assumptions of Theorem 5, and when* $\frac{m+(k-1)\gamma}{n+1} < \alpha < 1$, $k \in \mathbb{N}$, *we obtain that*

$$\left(D_a^{k\gamma} f\right)(x) = \frac{1}{\Gamma\left((n+1)\alpha - k\gamma\right)} \int_a^x (x-t)^{(n+1)\alpha - k\gamma - 1} \left(D_a^{(n+1)\alpha} f\right)(t)\, dt, \quad (80)$$

$\forall x \in [a, b]$, *and* $\left(D_a^{k\gamma} f\right) \in C\left([a, b]\right)$.

We need
Background Let $f \in L_1\left([a, b]\right)$, $v > 0$. The right Riemann-Liouville fractional integral operator of order v is defined by

$$\left(I_{b-}^v f\right)(x) = \frac{1}{\Gamma(v)} \int_x^b (z-x)^{v-1} f(z)\, dz, \quad (81)$$

$\forall x \in [a, b]$. We set $I_{b-}^0 = I$, the identity operator.
Let now $f \in AC^m\left([a, b]\right)$, $m \in \mathbb{N}$, with $m = \lceil \gamma \rceil$. We define the right Caputo fractional derivative of order $\gamma > 0$, by

$$\left(D_{b-}^\gamma f\right)(x) = \frac{(-1)^m}{\Gamma(m-\gamma)} \int_x^b (z-x)^{m-\gamma-1} f^{(m)}(z)\, dz, \quad (82)$$

i.e.

$$\left(D_{b-}^\gamma f\right)(x) = (-1)^m I_{b-}^{m-\gamma} f^{(m)}(x), \quad (83)$$

$\forall x \in [a, b]$. We set $D_{b-}^0 f = f$, and $\left(D_{b-}^m f\right)(x) = (-1)^m f^{(m)}(x)$, for $m \in \mathbb{N}, \forall x \in [a, b]$.

By [2], when $f \in C([a, b])$ and $\mu, \nu > 0$, we get that

$$I_{b-}^{\mu} I_{b-}^{\nu} f = I_{b-}^{\mu+\nu} f = I_{b-}^{\nu} I_{b-}^{\mu} f, \tag{84}$$

which is the semigroup property.

We need

Remark 11 Let $0 < \alpha < 1$, $f : [a, b] \to \mathbb{R}$ such that $f' \in L_{\infty}([a, b])$. Assume that $D_{b-}^{k\alpha} f \in C([a, b])$, for $k = 0, 1, \ldots, n + 1$; $n \in \mathbb{N}$. Suppose that $\left(D_{b-}^{i\alpha} f \right)(b) = 0$, $i = 0, 2, 3, \ldots, n$.

By (16) we obtain

$$f(x) = \frac{1}{\Gamma((n+1)\alpha)} \int_{x}^{b} (z - x)^{(n+1)\alpha - 1} \left(D_{b-}^{(n+1)\alpha} f \right)(z) \, dz, \tag{85}$$

$\forall x \in [a, b]$.

Call $\delta := (n + 1)\alpha$, then we have

$$f(x) = \frac{1}{\Gamma(\delta)} \int_{x}^{b} (z - x)^{\delta - 1} \left(D_{b-}^{\delta} f \right)(z) \, dz, \tag{86}$$

$\forall x \in [a, b]$.

By [5], when $\delta - 1 > 0$, we get that there exists

$$f'(x) = \frac{(-1)(\delta - 1)}{\Gamma(\delta)} \int_{x}^{b} (z - x)^{\delta - 2} \left(D_{b-}^{\delta} f \right)(z) \, dz, \tag{87}$$

$\forall x \in [a, b]$.

If $\delta - 2 > 0$, then

$$f''(x) = \frac{(-1)^{2}(\delta - 1)(\delta - 2)}{\Gamma(\delta)} \int_{x}^{b} (z - x)^{\delta - 3} \left(D_{b-}^{\delta} f \right)(z) \, dz, \tag{88}$$

$\forall x \in [a, b]$.

In general, if $\delta - m > 0$, equivalently, if $\alpha > \frac{m}{n+1}$, we get that there exists

$$f^{(m)}(x) = \frac{(-1)^{m} \prod_{j=1}^{m} (\delta - j)}{\Gamma(\delta)} \int_{x}^{b} (z - x)^{\delta - m - 1} \left(D_{b-}^{\delta} f \right)(z) \, dz, \tag{89}$$

$\forall x \in [a, b]$.

By [4], we get $f^{(m)} \in C\left([a, b]\right)$.

By (81), we derive

$$f^{(m)}(x) = \frac{(-1)^m \prod_{j=1}^m (\delta - j)\, \Gamma\left(\delta - m\right)\left(I_{b-}^{\delta-m}\left(D_{b-}^{\delta} f\right)\right)(x)}{\Gamma\left(\delta\right)}$$

$$= (-1)^m \left(I_{b-}^{\delta-m}\left(D_{b-}^{\delta} f\right)\right)(x), \tag{90}$$

$\forall x \in [a, b]$.

We have proved that

$$f^{(m)}(x) = (-1)^m \left(I_{b-}^{\delta-m}\left(D_{b-}^{\delta} f\right)\right)(x), \tag{91}$$

$\forall x \in [a, b]$.

We have that (case of $\gamma < m$)

$$\left(D_{b-}^{\gamma} f\right)(x) = (-1)^m \left(I_{b-}^{m-\gamma} f^{(m)}\right)(x)$$

$$= (-1)^{2m}\left(I_{b-}^{m-\gamma}\left(I_{b-}^{\delta-m}\left(D_{b-}^{\delta} f\right)\right)\right)(x)$$

$$= \left(I_{b-}^{\delta-\gamma}\left(D_{b-}^{\delta} f\right)\right)(x), \tag{92}$$

$\forall x \in [a, b]$.

That is

$$\left(D_{b-}^{\gamma} f\right)(x) = \left(I_{b-}^{(n+1)\alpha-\gamma}\left(D_{b-}^{(n+1)\alpha} f\right)\right)(x), \tag{93}$$

$\forall x \in [a, b]$.

I.e. we have found the representation formula:

$$\left(D_{b-}^{\gamma} f\right)(x) = \frac{1}{\Gamma\left((n+1)\alpha - \gamma\right)} \int_x^b (z-x)^{(n+1)\alpha-\gamma-1}\left(D_{b-}^{(n+1)\alpha} f\right)(z)\, dz, \tag{94}$$

$\forall x \in [a, b]$.

The last formula (94) is true under the assumption $(n+1)\alpha > m$, and since $m \geq \gamma$, it implies $(n+1)\alpha > \gamma$ and $(n+1)\alpha - \gamma > 0$. Furthermore, by [4] , we get that $\left(D_{b-}^{\gamma} f\right) \in C\left([a, b]\right)$.

We have proved the following right fractional representation theorem:

Theorem 12 *Let $0 < \alpha < 1$, $f : [a, b] \to \mathbb{R}$ such that $f' \in L_\infty\left([a, b]\right)$. Assume that $D_{b-}^{k\alpha} f \in C\left([a, b]\right)$, for $0, 1, \ldots, n+1$; $n \in \mathbb{N}$. Suppose that $\left(D_{b-}^{i\alpha} f\right)(b) = 0$, $i = 0, 2, 3, \ldots, n$. Let $\gamma > 0$ with $\lceil \gamma \rceil = m < n+1$, such that $m < (n+1)\alpha$, equivalently, $\alpha > \frac{m}{n+1}$. Then*

$$\left(D_{b-}^{\gamma} f\right)(x) = \frac{1}{\Gamma\left((n+1)\alpha - \gamma\right)} \int_{x}^{b} (z-x)^{(n+1)\alpha - \gamma - 1} \left(D_{b-}^{(n+1)\alpha} f\right)(z)\, dz, \quad (95)$$

$\forall x \in [a, b]$. *Furthermore it holds* $\left(D_{b-}^{\gamma} f\right) \in C\left([a, b]\right)$.

We make

Remark 13 Call $\lambda := (n+1)\alpha - \gamma - 1$, i.e. $\lambda + 1 = (n+1)\alpha - \gamma$, and call $\delta := (n+1)\alpha$. Then we can write

$$\left(D_{b-}^{\gamma} f\right)(x) = \frac{1}{\Gamma\left(\lambda + 1\right)} \int_{x}^{b} (z-x)^{\lambda} \left(D_{b-}^{\delta} f\right)(z)\, dz, \quad (96)$$

$\forall\, x \in [a, b]$.
If $\lambda > 0$, then

$$\left(D_{b-}^{\gamma} f\right)'(x) = \frac{(-1)\lambda}{\Gamma\left(\lambda + 1\right)} \int_{x}^{b} (z-x)^{\lambda - 1} \left(D_{b-}^{\delta} f\right)(z)\, dz, \quad (97)$$

$\forall\, x \in [a, b]$.
If $\lambda - 1 > 0$, then

$$\left(D_{b-}^{\gamma} f\right)''(x) = \frac{(-1)^{2}\lambda\,(\lambda - 1)}{\Gamma\left(\lambda + 1\right)} \int_{x}^{b} (z-x)^{\lambda - 2} \left(D_{b-}^{\delta} f\right)(z)\, dz, \quad (98)$$

$\forall\, x \in [a, b]$.
If $\lambda - 2 > 0$, then

$$\left(D_{b-}^{\gamma} f\right)^{(3)}(x) = \frac{(-1)^{3}\lambda\,(\lambda - 1)\,(\lambda - 2)}{\Gamma\left(\lambda + 1\right)} \int_{x}^{b} (z-x)^{\lambda - 3} \left(D_{b-}^{\delta} f\right)(z)\, dz, \quad (99)$$

$\forall\, x \in [a, b]$, etc.
In general, if $\lambda - m + 1 > 0$, then

$$\left(D_{b-}^{\gamma} f\right)^{(m)}(x)$$

$$= \frac{(-1)^{m}\lambda\,(\lambda - 1)\,(\lambda - 2)\dots(\lambda - m + 1)}{\Gamma\left(\lambda + 1\right)} \int_{x}^{b} (z-x)^{(\lambda - m + 1) - 1} \left(D_{b-}^{\delta} f\right)(z)\, dz$$

$$= \frac{(-1)^m \lambda (\lambda-1)(\lambda-2)\ldots(\lambda-m+1)\Gamma(\lambda-m+1)\left(I_{b-}^{(\lambda-m+1)}\left(D_{b-}^{\delta}f\right)\right)(x)}{\Gamma(\lambda+1)}$$

$$= (-1)^m \left(I_{b-}^{(\lambda-m+1)}\left(D_{b-}^{\delta}f\right)(x)\right), \tag{100}$$

$\forall\, x \in [a,b]$.

That is, if $\lambda - m + 1 > 0$, then

$$\left(D_{b-}^{\gamma}f\right)^{(m)}(x) = (-1)^m \left(I_{b-}^{(\lambda-m+1)}\left(D_{b-}^{\delta}f\right)\right)(x), \tag{101}$$

$\forall\, x \in [a,b]$.

We notice that

$$\left(D_{b-}^{2\gamma}f\right)(x) = \left(D_{b-}^{\gamma}\left(D_{b-}^{\gamma}f\right)\right)(x) = (-1)^m \left(I_{b-}^{m-\gamma}\left(D_{b-}^{\gamma}f\right)^{(m)}\right)(x)$$

$$= (-1)^{2m}\left(I_{b-}^{m-\gamma}I_{b-}^{\lambda-m+1}\left(D_{b-}^{\delta}f\right)\right)(x) = \left(I_{b-}^{\lambda-\gamma+1}\left(D_{b-}^{\delta}f\right)\right)(x) \tag{102}$$

$$= \left(I_{b-}^{(n+1)\alpha-\gamma-1-\gamma+1}\left(D_{b-}^{\delta}f\right)\right)(x) = \left(I_{b-}^{(n+1)\alpha-2\gamma}\left(D_{b-}^{\delta}f\right)\right)(x),$$

$\forall\, x \in [a,b]$.

That is

$$\left(D_{b-}^{2\gamma}f\right)(x) = \left(I_{b-}^{(n+1)\alpha-2\gamma}\left(D_{b-}^{(n+1)\alpha}f\right)\right)(x), \tag{103}$$

$\forall\, x \in [a,b]$, under the condition $\frac{\gamma+m}{n+1} < \alpha < 1$.

We have proved

Theorem 14 *Under the assumptions of Theorem 12, and when $\frac{\gamma+m}{n+1} < \alpha < 1$, we get that*

$$\left(D_{b-}^{2\gamma}f\right)(x) = \frac{1}{\Gamma((n+1)\alpha-2\gamma)}\int_x^b (z-x)^{(n+1)\alpha-2\gamma-1}\left(D_{b-}^{(n+1)\alpha}f\right)(z)\,dz, \tag{104}$$

$\forall\, x \in [a,b]$. *Furthermore it holds* $\left(D_{b-}^{2\gamma}f\right) \in C([a,b])$.

We make

Remark 15 Call $\rho := (n+1)\alpha - 2\gamma - 1$, i.e. $\rho + 1 = (n+1)\alpha - 2\gamma$, and call again $\delta := (n+1)\alpha$. Then we can write

$$\left(D_{b-}^{2\gamma}f\right)(x) = \frac{1}{\Gamma(\rho+1)}\int_x^b (z-x)^{\rho}\left(D_{b-}^{\delta}f\right)(z)\,dz, \tag{105}$$

$\forall\, x \in [a,b]$.

If $\rho > 0$, then

$$\left(D_{b-}^{2\gamma} f\right)'(x) = \frac{(-1)\,\rho}{\Gamma\,(\rho+1)} \int_x^b (z-x)^{\rho-1}\left(D_{b-}^{\delta} f\right)(z)\, dz, \qquad (106)$$

$\forall\, x \in [a,b]$.

If $\rho - 1 > 0$, then

$$\left(D_{b-}^{2\gamma} f\right)''(x) = \frac{(-1)^2\,\rho\,(\rho-1)}{\Gamma\,(\rho+1)} \int_x^b (z-x)^{\rho-2}\left(D_{b-}^{\delta} f\right)(z)\, dz, \qquad (107)$$

$\forall\, x \in [a,b]$.

If $\rho - 2 > 0$, then

$$\left(D_{b-}^{2\gamma} f\right)^{(3)}(x) = \frac{(-1)^3\rho(\rho-1)(\rho-2)}{\Gamma(\rho+1)} \int_x^b (z-x)^{\rho-3}\left(D_{b-}^{\delta} f\right)(z)\, dz, \qquad (108)$$

$\forall\, x \in [a,b]$, etc.

In general, if $\rho - m + 1 > 0$, then

$$\left(D_{b-}^{2\gamma} f\right)^{(m)}(x) \qquad (109)$$

$$= \frac{(-1)^m\,\rho\,(\rho-1)\,(\rho-2)\ldots(\rho-m+1)}{\Gamma\,(\rho+1)} \int_x^b (z-x)^{(\rho-m+1)-1}\left(D_{b-}^{\delta} f\right)(z)\, dz$$

$$= \frac{(-1)^m\,\rho\,(\rho-1)\,(\rho-2)\ldots(\rho-m+1)\,\Gamma\,(\rho-m+1)\left(I_{b-}^{(\rho-m+1)}\left(D_{b-}^{\delta} f\right)\right)(x)}{\Gamma\,(\rho+1)}$$

$$= (-1)^m \left(I_{b-}^{(\rho-m+1)}\left(D_{b-}^{\delta} f\right)(x)\right), \qquad (110)$$

$\forall\, x \in [a,b]$.

That is, if $\rho - m + 1 > 0$, then

$$\left(D_{b-}^{2\gamma} f\right)^{(m)}(x) = (-1)^m \left(I_{b-}^{(\rho-m+1)}\left(D_{b-}^{\delta} f\right)\right)(x), \qquad (111)$$

$\forall\, x \in [a,b]$.

We notice that

$$\left(D_{b-}^{3\gamma} f\right)(x) = \left(D_{b-}^{\gamma} \left(D_{b-}^{2\gamma} f\right)\right)(x) = (-1)^m \left(I_{b-}^{m-\gamma} \left(D_{b-}^{2\gamma} f\right)^{(m)}\right)(x)$$

$$= (-1)^{2m} \left(I_{b-}^{m-\gamma} I_{b-}^{\rho-m+1} \left(D_{b-}^{\delta} f\right)\right)(x) = \left(I_{b-}^{\rho-\gamma+1} \left(D_{b-}^{\delta} f\right)\right)(x) \quad (112)$$

$$= \left(I_{b-}^{(n+1)\alpha-2\gamma-1-\gamma+1} \left(D_{b-}^{(n+1)\alpha} f\right)\right)(x) = \left(I_{b-}^{(n+1)\alpha-3\gamma} \left(D_{b-}^{(n+1)\alpha} f\right)\right)(x),$$

$\forall\, x \in [a,b]$.
   That is, if $\frac{m+2\gamma}{n+1} < \alpha < 1$, we get

$$\left(D_{b-}^{3\gamma} f\right)(x) = \left(I_{b-}^{(n+1)\alpha-3\gamma} \left(D_{b-}^{(n+1)\alpha} f\right)\right)(x), \quad (113)$$

$\forall\, x \in [a,b]$.

   We have proved

Theorem 16 *Under the assumptions of Theorem 12, and when $\frac{m+2\gamma}{n+1} < \alpha < 1$, we get that:*

$$\left(D_{b-}^{3\gamma} f\right)(x) = \frac{1}{\Gamma((n+1)\alpha-3\gamma)} \int_x^b (z-x)^{(n+1)\alpha-3\gamma-1} \left(D_{b-}^{(n+1)\alpha} f\right)(z)\, dz, \quad (114)$$

$\forall\, x \in [a,b]$, and $\left(D_{b-}^{3\gamma} f\right) \in C([a,b])$.

   In general, we derive:

Theorem 17 *Under the assumptions of Theorem 12, and when $\frac{m+(k-1)\gamma}{n+1} < \alpha < 1$, $k \in \mathbb{N}$, we get that:*

$$\left(D_{b-}^{k\gamma} f\right)(x) = \frac{1}{\Gamma((n+1)\alpha-k\gamma)} \int_x^b (z-x)^{(n+1)\alpha-k\gamma-1} \left(D_{b-}^{(n+1)\alpha} f\right)(z)\, dz, \quad (115)$$

$\forall\, x \in [a,b]$, and $\left(D_{b-}^{k\gamma} f\right) \in C([a,b])$.

   Next we present a very general left fractional Opial type inequality:

Theorem 18 *Let $0 < \alpha < 1$, $f : [a,b] \to \mathbb{R}$ such that $f' \in L_\infty([a,b])$. Assume that $D_a^{\bar{k}\alpha} f \in C([a,b])$, $\bar{k} = 0,1,\ldots,n+1$; $n \in \mathbb{N}$. Suppose that $\left(D_a^{i\alpha} f\right)(a) = 0$, for $i = 0,2,3,\ldots,n$. Let $\gamma > 0$ with $\lceil \gamma \rceil = m$, and $p,q > 1 : \frac{1}{p} + \frac{1}{q} = 1$. We further assume that $(k \in \mathbb{N})$*

$$1 > \alpha > \max\left(\frac{m + (k-1)\gamma}{n+1}, \frac{k\gamma q + 1}{(n+1)q}\right). \tag{116}$$

Then

$$\int_a^x \left|\left(D_a^{k\gamma} f\right)(w)\right| \left|D_a^{(n+1)\alpha} f(w)\right| dw$$

$$\leq \frac{(x-a)^{((n+1)\alpha - k\gamma - 1) + \frac{2}{p}}}{2^{\frac{1}{q}} \Gamma((n+1)\alpha - k\gamma)\left[(p((n+1)\alpha - k\gamma - 1) + 1)(p((n+1)\alpha - k\gamma - 1) + 2)\right]^{\frac{1}{p}}}$$

$$\cdot \left(\int_a^x \left|\left(D_a^{(n+1)\alpha} f\right)(w)\right|^q dw\right)^{\frac{2}{q}}, \tag{117}$$

$\forall\, x \in [a, b]$.

Proof Assumption (116) implies

$$p((n+1)\alpha - k\gamma - 1) + 1 > 0. \tag{118}$$

By (80) and Hölder's inequality we have

$$\left|\left(D_a^{k\gamma} f\right)(x)\right|$$

$$\leq \frac{1}{\Gamma((n+1)\alpha - k\gamma)} \int_a^x (x-t)^{(n+1)\alpha - k\gamma - 1} \left|\left(D_a^{(n+1)\alpha} f\right)(t)\right| dt$$

$$\leq \frac{1}{\Gamma((n+1)\alpha - k\gamma)} \left(\int_a^x (x-t)^{p((n+1)\alpha - k\gamma - 1)} dt\right)^{\frac{1}{p}} \left(\int_a^x \left|\left(D_a^{(n+1)\alpha} f\right)(t)\right|^q dt\right)^{\frac{1}{q}}$$

$$= \frac{1}{\Gamma((n+1)\alpha - k\gamma)} \frac{(x-a)^{\frac{p((n+1)\alpha - k\gamma - 1) + 1}{p}}}{(p((n+1)\alpha - k\gamma - 1) + 1)^{\frac{1}{p}}} \left(\int_a^x \left|\left(D_a^{(n+1)\alpha} f\right)(t)\right|^q dt\right)^{\frac{1}{q}}, \tag{119}$$

$\forall\, x \in [a, b]$. We have proved that

$$\left|\left(D_a^{k\gamma} f\right)(x)\right| \tag{120}$$

$$\leq \frac{(x-a)^{\frac{p((n+1)\alpha - k\gamma - 1) + 1}{p}}}{\Gamma((n+1)\alpha - k\gamma)(p((n+1)\alpha - k\gamma - 1) + 1)^{\frac{1}{p}}} \left(\int_a^x \left|\left(D_a^{(n+1)\alpha} f\right)(t)\right|^q dt\right)^{\frac{1}{q}},$$

$\forall\, x \in [a, b]$. Call

$$z(x) := \int_a^x \left| \left(D_a^{(n+1)\alpha} f \right)(t) \right|^q dt, \tag{121}$$

$z(a) = 0$. Thus

$$z'(x) = \left| \left(D_a^{(n+1)\alpha} f \right)(x) \right|^q \geq 0, \tag{122}$$

and

$$\left(z'(x) \right)^{\frac{1}{q}} = \left| \left(D_a^{(n+1)\alpha} f \right)(x) \right| \geq 0, \ \forall x \in [a, b].$$

Consequently, we get

$$\left| \left(D_a^{k\gamma} f \right)(w) \right| \left| D_a^{(n+1)\alpha} f(w) \right| \tag{123}$$

$$\leq \frac{(w-a)^{\frac{p((n+1)\alpha - k\gamma - 1) + 1}{p}}}{\Gamma((n+1)\alpha - k\gamma)(p((n+1)\alpha - k\gamma - 1) + 1)^{\frac{1}{p}}} \left(z(w) z'(w) \right)^{\frac{1}{q}},$$

$\forall w \in [a, b]$. Then it holds

$$\int_a^x \left| \left(D_a^{k\gamma} f \right)(w) \right| \left| D_a^{(n+1)\alpha} f(w) \right| dw$$

$$\leq \frac{1}{\Gamma((n+1)\alpha - k\gamma)(p((n+1)\alpha - k\gamma - 1) + 1)^{\frac{1}{p}}}$$

$$\cdot \int_a^x (w-a)^{\frac{p((n+1)\alpha - k\gamma - 1) + 1}{p}} \left(z(w) z'(w) \right)^{\frac{1}{q}} dw$$

$$\leq \frac{1}{\Gamma((n+1)\alpha - k\gamma)(p((n+1)\alpha - k\gamma - 1) + 1)^{\frac{1}{p}}} \tag{124}$$

$$\cdot \left(\int_a^x (w-a)^{(p((n+1)\alpha - k\gamma - 1) + 1)} dw \right)^{\frac{1}{p}} \left(\int_a^x z(w) z'(w) dw \right)^{\frac{1}{q}}$$

$$= \frac{(x-a)^{\frac{p((n+1)\alpha - k\gamma - 1) + 2}{p}}}{\Gamma((n+1)\alpha - k\gamma) \left[(p((n+1)\alpha - k\gamma - 1) + 1)(p((n+1)\alpha - k\gamma - 1) + 2) \right]^{\frac{1}{p}}}$$

$$\cdot \left(\frac{z^2(x)}{2} \right)^{\frac{1}{q}}$$

$$= \frac{(x-a)^{((n+1)\alpha-k\gamma-1)+\frac{2}{p}}}{2^{\frac{1}{q}}\Gamma((n+1)\alpha-k\gamma)\left[(p((n+1)\alpha-k\gamma-1)+1)(p((n+1)\alpha-k\gamma-1)+2)\right]^{\frac{1}{p}}}$$

$$\cdot \left(\int_a^x \left| \left(D_a^{(n+1)\alpha} f \right)(w) \right|^q dw \right)^{\frac{2}{q}}, \tag{125}$$

$\forall\, x \in [a,b]$, proving the claim.

Next we present a very general right fractional Opial type inequality:

Theorem 19 *Let* $0 < \alpha < 1$, $f : [a,b] \to \mathbb{R}$ *such that* $f' \in L_\infty([a,b])$. *Assume that* $D_{b-}^{\bar{k}\alpha} f \in C([a,b])$, $\bar{k} = 0, 1, \ldots, n+1$; $n \in \mathbb{N}$. *Suppose that* $\left(D_{b-}^{i\alpha} f \right)(b) = 0$, *for* $i = 0, 2, 3, \ldots, n$. *Let* $\gamma > 0$ *with* $\lceil \gamma \rceil = m$, *and* $p, q > 1 : \frac{1}{p} + \frac{1}{q} = 1$. *We further assume that* $(k \in \mathbb{N})$

$$1 > \alpha > \max \left(\frac{m+(k-1)\gamma}{n+1}, \frac{k\gamma q+1}{(n+1)q} \right). \tag{126}$$

Then

$$\int_x^b \left| \left(D_{b-}^{k\gamma} f \right)(w) \right| \left| D_{b-}^{(n+1)\alpha} f(w) \right| dw$$

$$\leq \frac{(b-x)^{((n+1)\alpha-k\gamma-1)+\frac{2}{p}}}{2^{\frac{1}{q}}\Gamma((n+1)\alpha-k\gamma)\left[(p((n+1)\alpha-k\gamma-1)+1)(p((n+1)\alpha-k\gamma-1)+2)\right]^{\frac{1}{p}}}$$

$$\cdot \left(\int_x^b \left| \left(D_{b-}^{(n+1)\alpha} f \right)(w) \right|^q dw \right)^{\frac{2}{q}}, \tag{127}$$

$\forall\, x \in [a,b]$.

Proof Assumption (126) implies

$$p((n+1)\alpha - k\gamma - 1) + 1 > 0. \tag{128}$$

By (115) and Hölder's inequality we have

$$\left| \left(D_{b-}^{k\gamma} f \right)(x) \right|$$

$$\leq \frac{1}{\Gamma((n+1)\alpha - k\gamma)} \int_x^b (t-x)^{(n+1)\alpha-k\gamma-1} \left| \left(D_{b-}^{(n+1)\alpha} f \right)(t) \right| dt$$

$$\leq \frac{1}{\Gamma((n+1)\alpha-k\gamma)} \left(\int_x^b (t-x)^{p((n+1)\alpha-k\gamma-1)}\, dt \right)^{\frac{1}{p}} \left(\int_x^b \left| \left(D_{b-}^{(n+1)\alpha} f \right)(t) \right|^q dt \right)^{\frac{1}{q}}$$

$$= \frac{1}{\Gamma((n+1)\alpha-k\gamma)} \frac{(b-x)^{\frac{p((n+1)\alpha-k\gamma-1)+1}{p}}}{(p((n+1)\alpha-k\gamma-1)+1)^{\frac{1}{p}}} \left(\int_x^b \left| \left(D_{b-}^{(n+1)\alpha} f \right)(t) \right|^q dt \right)^{\frac{1}{q}}, \quad (129)$$

$\forall\, x \in [a, b]$. We have proved that

$$\left| \left(D_{b-}^{k\gamma} f \right)(x) \right| \quad (130)$$

$$\leq \frac{(b-x)^{\frac{p((n+1)\alpha-k\gamma-1)+1}{p}}}{\Gamma\left((n+1)\alpha-k\gamma\right)\left(p\left((n+1)\alpha-k\gamma-1\right)+1\right)^{\frac{1}{p}}} \left(\int_x^b \left| \left(D_{b-}^{(n+1)\alpha} f \right)(t) \right|^q dt \right)^{\frac{1}{q}},$$

$\forall\, x \in [a, b]$. Call

$$z(x) := \int_x^b \left| \left(D_{b-}^{(n+1)\alpha} f \right)(t) \right|^q dt, \quad (131)$$

$z(b) = 0$. Thus

$$z'(x) = - \left| \left(D_{b-}^{(n+1)\alpha} f \right)(x) \right|^q \leq 0, \quad (132)$$

and

$$\left(-z'(x)\right)^{\frac{1}{q}} = \left| \left(D_{b-}^{(n+1)\alpha} f \right)(x) \right| \geq 0, \ \forall x \in [a, b].$$

Consequently, we get

$$\left| \left(D_{b-}^{k\gamma} f \right)(w) \right| \left| D_{b-}^{(n+1)\alpha} f(w) \right| \quad (133)$$

$$\leq \frac{(b-w)^{\frac{p((n+1)\alpha-k\gamma-1)+1}{p}}}{\Gamma\left((n+1)\alpha-k\gamma\right)\left(p\left((n+1)\alpha-k\gamma-1\right)+1\right)^{\frac{1}{p}}} \left(z(w)\left(-z'(w)\right)\right)^{\frac{1}{q}},$$

$\forall\, w \in [a, b]$. Then it holds

$$\int_x^b \left|\left(D_{b-}^{k\gamma}f\right)(w)\right| \left|D_{b-}^{(n+1)\alpha}f(w)\right| dw$$

$$\leq \frac{1}{\Gamma\left((n+1)\alpha - k\gamma\right)\left(p\left((n+1)\alpha - k\gamma - 1\right) + 1\right)^{\frac{1}{p}}}$$

$$\cdot \int_x^b (b-w)^{\frac{p((n+1)\alpha - k\gamma - 1)+1}{p}} \left(z(w)\left(-z'(w)\right)\right)^{\frac{1}{q}} dw$$

$$\leq \frac{1}{\Gamma\left((n+1)\alpha - k\gamma\right)\left(p\left((n+1)\alpha - k\gamma - 1\right) + 1\right)^{\frac{1}{p}}} \tag{134}$$

$$\cdot \left(\int_x^b (b-w)^{(p((n+1)\alpha - k\gamma - 1)+1)} dw\right)^{\frac{1}{p}} \left(-\int_x^b z(w) z'(w) dw\right)^{\frac{1}{q}}$$

$$= \frac{(b-x)^{\frac{p((n+1)\alpha - k\gamma - 1)+2}{p}}}{\Gamma((n+1)\alpha - k\gamma)\left[(p((n+1)\alpha - k\gamma - 1)+1)(p((n+1)\alpha - k\gamma - 1)+2)\right]^{\frac{1}{p}}}$$

$$\cdot \left(\frac{z^2(x)}{2}\right)^{\frac{1}{q}}$$

$$= \frac{(b-x)^{((n+1)\alpha - k\gamma - 1)+\frac{2}{p}}}{2^{\frac{1}{q}} \Gamma((n+1)\alpha - k\gamma)\left[(p((n+1)\alpha - k\gamma - 1)+1)(p((n+1)\alpha - k\gamma - 1)+2)\right]^{\frac{1}{p}}} \tag{135}$$

$$\cdot \left(\int_x^b \left|\left(D_{b-}^{(n+1)\alpha}f\right)(w)\right|^q dw\right)^{\frac{2}{q}},$$

$\forall\, x \in [a, b]$, proving the claim.

It follows a reverse general left fractional Opial type inequality:

Theorem 20 *Let $0 < \alpha < 1$, $f : [a, b] \to \mathbb{R}$ such that $f' \in L_\infty ([a, b])$. Assume that $D_a^{\bar{k}\alpha} f \in C([a, b])$, $\bar{k} = 0, 1, \ldots, n + 1$; $n \in \mathbb{N}$. Suppose that $\left(D_a^{i\alpha} f\right)(a) = 0$, for $i = 0, 2, 3, \ldots, n$. Let $\gamma > 0$ with $\lceil \gamma \rceil = m$, and $0 < p < 1, q < 0 : \frac{1}{p} + \frac{1}{q} = 1$. We assume that $\left(D_a^{(n+1)\alpha} f\right)$ is of fixed sign and nowhere zero. We finally assume that $(k \in \mathbb{N})$*

$$1 > \alpha > \frac{m + (k - 1)\gamma}{n + 1}. \tag{136}$$

Then

$$\int_a^x \left|\left(D_a^{k\gamma} f\right)(w)\right| \left|D_a^{(n+1)\alpha} f(w)\right| dw$$

$$\geq \frac{(x-a)^{((n+1)\alpha-k\gamma-1)+\frac{2}{p}}}{2^{\frac{1}{q}} \Gamma((n+1)\alpha-k\gamma) \left[(p((n+1)\alpha-k\gamma-1)+1)(p((n+1)\alpha-k\gamma-1)+2)\right]^{\frac{1}{p}}}$$

$$\cdot \left(\int_a^x \left|\left(D_a^{(n+1)\alpha} f\right)(w)\right|^q dw\right)^{\frac{2}{q}}, \tag{137}$$

$\forall\, x \in [a, b]$.

Proof Clearly we have

$$p((n+1)\alpha - k\gamma - 1) + 1 > 0.$$

By (80), we get that

$$\left|\left(D_a^{k\gamma} f\right)(x)\right| = \frac{1}{\Gamma((n+1)\alpha-k\gamma)} \int_a^x (x-t)^{(n+1)\alpha-k\gamma-1} \left|\left(D_a^{(n+1)\alpha} f\right)(t)\right| dt, \tag{138}$$

$\forall\, x \in [a, b]$. Then, by reverse Hölder's inequality we obtain

$$\left|\left(D_a^{k\gamma} f\right)(x)\right|$$

$$\geq \frac{1}{\Gamma((n+1)\alpha-k\gamma)} \left(\int_a^x (x-t)^{p((n+1)\alpha-k\gamma-1)} dt\right)^{\frac{1}{p}} \left(\int_a^x \left|\left(D_a^{(n+1)\alpha} f\right)(t)\right|^q dt\right)^{\frac{1}{q}}$$

$$= \frac{1}{\Gamma((n+1)\alpha-k\gamma)} \frac{(x-a)^{\frac{p((n+1)\alpha-k\gamma-1)+1}{p}}}{(p((n+1)\alpha-k\gamma-1)+1)^{\frac{1}{p}}} \left(\int_a^x \left|\left(D_a^{(n+1)\alpha} f\right)(t)\right|^q dt\right)^{\frac{1}{q}}, \tag{139}$$

$\forall\, x \in [a, b]$. Call

$$z(w) := \int_a^w \left|\left(D_a^{(n+1)\alpha} f\right)(t)\right|^q dt, \tag{140}$$

all $a \leq w \leq x$; $z(a) = 0$, $z(w) > 0$ on $(a, x]$. Thus

$$z'(w) = \left|\left(D_a^{(n+1)\alpha} f\right)(w)\right|^q > 0, \tag{141}$$

all $a < w \leq x$. Hence

$$\left(z'(w)\right)^{\frac{1}{q}} = \left|\left(D_a^{(n+1)\alpha} f\right)(w)\right| > 0, \tag{142}$$

all $a < w \leq x$. Consequently we get

$$\left|\left(D_a^{k\gamma} f\right)(w)\right| \left|D_a^{(n+1)\alpha} f(w)\right| \tag{143}$$

$$\geq \frac{1}{\Gamma\left((n+1)\alpha - k\gamma\right)} \frac{(w-a)^{\frac{p((n+1)\alpha - k\gamma - 1)+1}{p}}}{\left(p\left((n+1)\alpha - k\gamma - 1\right)+1\right)^{\frac{1}{p}}} \left(z(w) z'(w)\right)^{\frac{1}{q}},$$

all $a < w \leq x$. Let $a < \theta \leq w \leq x$ and $\theta \downarrow a$. Then, it holds

$$\int_a^x \left|\left(D_a^{k\gamma} f\right)(w)\right| \left|D_a^{(n+1)\alpha} f(w)\right| dw$$

$$= \lim_{\theta \downarrow a} \int_\theta^x \left|\left(D_a^{k\gamma} f\right)(w)\right| \left|D_a^{(n+1)\alpha} f(w)\right| dw$$

$$\geq \frac{1}{\Gamma\left((n+1)\alpha - k\gamma\right)\left(p\left((n+1)\alpha - k\gamma - 1\right)+1\right)^{\frac{1}{p}}}$$

$$\cdot \lim_{\theta \downarrow a} \int_\theta^x (w-a)^{\frac{p((n+1)\alpha - k\gamma - 1)+1}{p}} \left(z(w) z'(w)\right)^{\frac{1}{q}} dw$$

$$\geq \frac{1}{\Gamma\left((n+1)\alpha - k\gamma\right)\left(p\left((n+1)\alpha - k\gamma - 1\right)+1\right)^{\frac{1}{p}}} \tag{144}$$

$$\cdot \lim_{\theta \downarrow a} \left(\int_\theta^x (w-a)^{(p((n+1)\alpha - k\gamma - 1)+1)} dw\right)^{\frac{1}{p}} \lim_{\theta \downarrow a} \left(\int_\theta^x z(w) z'(w) dw\right)^{\frac{1}{q}}$$

$$= \frac{1}{\Gamma\left((n+1)\alpha - k\gamma\right)\left(p\left((n+1)\alpha - k\gamma - 1\right)+1\right)^{\frac{1}{p}} 2^{\frac{1}{q}}}$$

$$\cdot \left(\int_a^x (w-a)^{(p((n+1)\alpha - k\gamma - 1)+1)} dw\right)^{\frac{1}{p}} \lim_{\theta \downarrow a} \left(z^2(x) - z^2(\theta)\right)^{\frac{1}{q}}$$

$$= \frac{1}{2^{\frac{1}{q}} \Gamma\left((n+1)\alpha - k\gamma\right)\left(p\left((n+1)\alpha - k\gamma - 1\right)+1\right)^{\frac{1}{p}}}$$

$$\cdot \frac{(x-a)^{\frac{p((n+1)\alpha - k\gamma - 1)+2}{p}}}{\left(p\left((n+1)\alpha - k\gamma - 1\right)+2\right)^{\frac{1}{p}}} \left(z(x)\right)^{\frac{2}{q}}. \tag{145}$$

The theorem is proved.

It follows a reverse general right fractional Opial type inequality:

Theorem 21 *Let* $0 < \alpha < 1$, $f : [a, b] \rightarrow \mathbb{R}$ *such that* $f' \in L_\infty ([a, b])$. *Assume that* $D_{b-}^{\bar{k}\alpha} f \in C ([a, b])$, $\bar{k} = 0, 1, \ldots, n + 1$; $n \in \mathbb{N}$. *Suppose that* $\left(D_{b-}^{i\alpha} f \right) (b) = 0$, *for* $i = 0, 2, 3, \ldots, n$. *Let* $\gamma > 0$ *with* $\lceil \gamma \rceil = m$, *and* $0 < p < 1, q < 0 : \frac{1}{p} + \frac{1}{q} = 1$. *We assume that* $\left(D_{b-}^{(n+1)\alpha} f \right)$ *is of fixed sign and nowhere zero. We finally assume that* $(k \in \mathbb{N})$

$$1 > \alpha > \frac{m + (k - 1) \gamma}{n + 1}. \tag{146}$$

Then

$$\int_x^b \left| \left(D_{b-}^{k\gamma} f \right) (w) \right| \left| D_{b-}^{(n+1)\alpha} f (w) \right| dw$$

$$\geq \frac{(b-x)^{((n+1)\alpha - k\gamma - 1) + \frac{2}{p}}}{2^{\frac{1}{q}} \Gamma((n+1)\alpha - k\gamma) \left[(p((n+1)\alpha - k\gamma - 1) + 1)(p((n+1)\alpha - k\gamma - 1) + 2) \right]^{\frac{1}{p}}} \tag{147}$$

$$\cdot \left(\int_x^b \left| \left(D_{b-}^{(n+1)\alpha} f \right) (w) \right|^q dw \right)^{\frac{2}{q}},$$

$\forall x \in [a, b]$.

Proof Similar to proof of Theorem 20, using (115). As such it is omitted.

Two extreme fractional Opial type inequalities follow (case $p = 1, q = \infty$).

Theorem 22 *Let* $0 < \alpha < 1$, $f : [a, b] \rightarrow \mathbb{R}$ *such that* $f' \in L_\infty ([a, b])$. *Assume that* $D_a^{\bar{k}\alpha} f \in C ([a, b])$, $\bar{k} = 0, 1, \ldots, n + 1$; $n \in \mathbb{N}$. *Suppose that* $\left(D_a^{i\alpha} f \right) (a) = 0$, *for* $i = 0, 2, 3, \ldots, n$. *Let* $\gamma > 0$ *with* $\lceil \gamma \rceil = m$. *We further assume that* $(k \in \mathbb{N})$

$$1 > \alpha > \frac{m + (k - 1) \gamma}{n + 1}. \tag{148}$$

Then

$$\int_a^x \left| \left(D_a^{k\gamma} f \right) (w) \right| \left| \left(D_a^{(n+1)\alpha} f \right) (w) \right| dw \leq \frac{(x-a)^{(n+1)\alpha - k\gamma + 1}}{\Gamma((n+1)\alpha - k\gamma + 2)} \left\| D_a^{(n+1)\alpha} f \right\|_\infty^2, \tag{149}$$

$\forall x \in [a, b]$.

Proof By (80), we get

$$\left| \left(D_a^{k\gamma} f \right)(x) \right|$$

$$\leq \frac{1}{\Gamma((n+1)\alpha - k\gamma)} \left(\int_a^x (x-t)^{(n+1)\alpha - k\gamma - 1} \, dt \right) \left\| D_a^{(n+1)\alpha} f \right\|_\infty$$

$$= \frac{1}{\Gamma((n+1)\alpha - k\gamma)} \frac{(x-a)^{(n+1)\alpha - k\gamma}}{((n+1)\alpha - k\gamma)} \left\| D_a^{(n+1)\alpha} f \right\|_\infty \qquad (150)$$

$$= \frac{(x-a)^{(n+1)\alpha - k\gamma}}{\Gamma((n+1)\alpha - k\gamma + 1)} \left\| D_a^{(n+1)\alpha} f \right\|_\infty.$$

Hence we obtain

$$\left| \left(D_a^{k\gamma} f \right)(w) \right| \left| \left(D_a^{(n+1)\alpha} f \right)(w) \right| \leq \frac{(w-a)^{(n+1)\alpha - k\gamma}}{\Gamma((n+1)\alpha - k\gamma + 1)} \left\| D_a^{(n+1)\alpha} f \right\|_\infty^2, \qquad (151)$$

$\forall\, w \in [a, x]$. Integrating (151) over $[a, x]$, we derive (149).

Theorem 23 *Let* $0 < \alpha < 1$, $f : [a, b] \to \mathbb{R}$ *such that* $f' \in L_\infty([a, b])$. *Assume that* $D_{b-}^{\bar{k}\alpha} f \in C([a, b])$, $\bar{k} = 0, 1, \ldots, n+1$; $n \in \mathbb{N}$. *Suppose that* $\left(D_{b-}^{i\alpha} f \right)(b) = 0$, *for* $i = 0, 2, 3, \ldots, n$. *Let* $\gamma > 0$ *with* $\lceil \gamma \rceil = m$. *We further assume that* $(k \in \mathbb{N})$

$$1 > \alpha > \frac{m + (k-1)\gamma}{n+1}. \qquad (152)$$

Then

$$\int_x^b \left| \left(D_{b-}^{k\gamma} f \right)(w) \right| \left| \left(D_{b-}^{(n+1)\alpha} f \right)(w) \right| \, dw \leq \frac{(b-x)^{(n+1)\alpha - k\gamma + 1}}{\Gamma((n+1)\alpha - k\gamma + 2)} \left\| D_{b-}^{(n+1)\alpha} f \right\|_\infty^2, \qquad (153)$$

$\forall x \in [a, b]$.

Proof Similar to the proof of Theorem 22, by using (115). As such it is omitted.

Next we present a left fractional Hilbert-Pachpatte type inequality:

Theorem 24 *Here* $i = 1, 2$. *Let* $0 < \alpha_i < 1$, $f_i : [a_i, b_i] \to \mathbb{R}$ *such that* $f_i' \in L_\infty$ $([a_i, b_i])$. *Assume that* $D_{a_i}^{\bar{k}_i \alpha_i} f_i \in C([a_i, b_i])$, $\bar{k}_i = 0, 1, \ldots, n_i + 1$; $n_i \in \mathbb{N}$. *Suppose that* $\left(D_{a_i}^{j_i \alpha_i} f_i \right)(a_i) = 0$, *for* $j_i = 0, 2, 3, \ldots, n_i$. *Let* $\gamma_i > 0$ *with* $\lceil \gamma_i \rceil = m_i$, *and* $p, q > 1 : \frac{1}{p} + \frac{1}{q} = 1$. *We further assume that* $(k_i \in \mathbb{N})$

$$1 > \alpha_1 > \max\left(\frac{m_1 + (k_1 - 1)\,\gamma_1}{n_1 + 1}, \frac{k_1\gamma_1 q + 1}{(n_1 + 1)\,q}\right). \tag{154}$$

and

$$1 > \alpha_2 > \max\left(\frac{m_2 + (k_2 - 1)\,\gamma_2}{n_2 + 1}, \frac{k_2\gamma_2 p + 1}{(n_2 + 1)\,p}\right). \tag{155}$$

Then

$$\int_{a_1}^{b_1} \int_{a_2}^{b_2} \frac{\left|\left(D_{a_1}^{k_1\gamma_1} f_1\right)(x_1)\right| \left|\left(D_{a_2}^{k_2\gamma_2} f_2\right)(x_2)\right| dx_1 dx_2}{\left[\frac{(x_1-a_1)^{p((n_1+1)\alpha_1-k_1\gamma_1-1)+1}}{p(p((n_1+1)\alpha_1-k_1\gamma_1-1)+1)} + \frac{(x_2-a_2)^{q((n_2+1)\alpha_2-k_2\gamma_2-1)+1}}{q(q((n_2+1)\alpha_2-k_2\gamma_2-1)+1)}\right]}$$

$$\leq \frac{(b_1 - a_1)(b_2 - a_2)}{\Gamma\left((n_1 + 1)\,\alpha_1 - k_1\gamma_1\right)\Gamma\left((n_2 + 1)\,\alpha_2 - k_2\gamma_2\right)} \tag{156}$$

$$\cdot \left(\int_{a_1}^{b_1} \left|D_{a_1}^{(n_1+1)\alpha_1} f_1(t_1)\right|^q dt_1\right)^{\frac{1}{q}} \left(\int_{a_2}^{b_2} \left|D_{a_2}^{(n_2+1)\alpha_2} f_2(t_2)\right|^p dt_2\right)^{\frac{1}{p}}.$$

Proof We have from (80) that

$$\left(D_{a_i}^{k_i\gamma_i} f_i\right)(x_i) \tag{157}$$

$$= \frac{1}{\Gamma\left((n_i + 1)\,\alpha_i - k_i\gamma_i\right)} \int_{a_i}^{x_i} (x_i - t_i)^{(n_i+1)\alpha_i - k_i\gamma_i - 1} \left(D_{a_i}^{(n_i+1)\alpha_i} f_i\right)(t_i)\, dt_i,$$

where $i = 1, 2$; $x_i \in [a_i, b_i]$, with $D_{a_i}^{k_i\gamma_i} f_i \in C\left([a_i, b_i]\right)$. Hence by Hölder's inequality we get:

$$\left|\left(D_{a_1}^{k_1\gamma_1} f_1\right)(x_1)\right| \tag{158}$$

$$\leq \frac{1}{\Gamma\left((n_1 + 1)\,\alpha_1 - k_1\gamma_1\right)} \left(\int_{a_1}^{x_1} (x_1 - t_1)^{p((n_1+1)\alpha_1 - k_1\gamma_1 - 1)}\, dt_1\right)^{\frac{1}{p}}$$

$$\cdot \left(\int_{a_1}^{x_1} \left|D_{a_1}^{(n_1+1)\alpha_1} f_1(t_1)\right|^q dt_1\right)^{\frac{1}{q}},$$

and

$$\left|\left(D_{a_2}^{k_2\gamma_2} f_2\right)(x_2)\right| \tag{159}$$

$$\leq \frac{1}{\Gamma\left((n_2+1)\alpha_2 - k_2\gamma_2\right)} \left(\int_{a_2}^{x_2} (x_2-t_2)^{q((n_2+1)\alpha_2 - k_2\gamma_2 - 1)}\, dt_2\right)^{\frac{1}{q}}$$

$$\cdot \left(\int_{a_2}^{x_2} \left|D_{a_2}^{(n_2+1)\alpha_2} f_2(t_2)\right|^p dt_2\right)^{\frac{1}{p}}.$$

So we have

$$\left|\left(D_{a_1}^{k_1\gamma_1} f_1\right)(x_1)\right| \tag{160}$$

$$\leq \frac{1}{\Gamma\left((n_1+1)\alpha_1 - k_1\gamma_1\right)} \left(\frac{(x_1-a_1)^{p((n_1+1)\alpha_1-k_1\gamma_1-1)+1}}{(p\left((n_1+1)\alpha_1 - k_1\gamma_1 - 1\right)+1)}\right)^{\frac{1}{p}}$$

$$\cdot \left(\int_{a_1}^{x_1} \left|D_{a_1}^{(n_1+1)\alpha_1} f_1(t_1)\right|^q dt_1\right)^{\frac{1}{q}},$$

and

$$\left|\left(D_{a_2}^{k_2\gamma_2} f_2\right)(x_2)\right| \tag{161}$$

$$\leq \frac{1}{\Gamma\left((n_2+1)\alpha_2 - k_2\gamma_2\right)} \left(\frac{(x_2-a_2)^{q((n_2+1)\alpha_2-k_2\gamma_2-1)+1}}{(q\left((n_2+1)\alpha_2 - k_2\gamma_2 - 1\right)+1)}\right)^{\frac{1}{q}}$$

$$\cdot \left(\int_{a_2}^{x_2} \left|D_{a_2}^{(n_2+1)\alpha_2} f_2(t_2)\right|^p dt_2\right)^{\frac{1}{p}}.$$

Hence

$$\left|\left(D_{a_1}^{k_1\gamma_1} f_1\right)(x_1)\right| \left|\left(D_{a_2}^{k_2\gamma_2} f_2\right)(x_2)\right|$$

$$\leq \frac{1}{\Gamma\left((n_1+1)\alpha_1 - k_1\gamma_1\right)\Gamma\left((n_2+1)\alpha_2 - k_2\gamma_2\right)} \tag{162}$$

$$\cdot \left(\frac{(x_1-a_1)^{p((n_1+1)\alpha_1-k_1\gamma_1-1)+1}}{(p\left((n_1+1)\alpha_1 - k_1\gamma_1 - 1\right)+1)}\right)^{\frac{1}{p}} \left(\frac{(x_2-a_2)^{q((n_2+1)\alpha_2-k_2\gamma_2-1)+1}}{(q\left((n_2+1)\alpha_2 - k_2\gamma_2 - 1\right)+1)}\right)^{\frac{1}{q}}$$

$$\cdot \left(\int_{a_1}^{x_1} \left|D_{a_1}^{(n_1+1)\alpha_1} f_1(t_1)\right|^q dt_1\right)^{\frac{1}{q}} \left(\int_{a_2}^{x_2} \left|D_{a_2}^{(n_2+1)\alpha_2} f_2(t_2)\right|^p dt_2\right)^{\frac{1}{p}}$$

(using Young's inequality for $a, b \geq 0$, $a^{\frac{1}{p}} b^{\frac{1}{q}} \leq \frac{a}{p} + \frac{b}{q}$)

$$\leq \frac{1}{\Gamma\left((n_1+1)\alpha_1 - k_1\gamma_1\right)\Gamma\left((n_2+1)\alpha_2 - k_2\gamma_2\right)} \quad (163)$$

$$\cdot \left[\left(\frac{(x_1-a_1)^{p((n_1+1)\alpha_1 - k_1\gamma_1 - 1)+1}}{p\left(p\left((n_1+1)\alpha_1 - k_1\gamma_1 - 1\right) + 1\right)}\right) + \left(\frac{(x_2-a_2)^{q((n_2+1)\alpha_2 - k_2\gamma_2 - 1)+1}}{q\left(q\left((n_2+1)\alpha_2 - k_2\gamma_2 - 1\right) + 1\right)}\right)\right]$$

$$\cdot \left(\int_{a_1}^{x_1}\left|D_{a_1}^{(n_1+1)\alpha_1} f_1(t_1)\right|^q dt_1\right)^{\frac{1}{q}} \left(\int_{a_2}^{x_2}\left|D_{a_2}^{(n_2+1)\alpha_2} f_2(t_2)\right|^p dt_2\right)^{\frac{1}{p}}.$$

So far we have

$$\frac{\left|\left(D_{a_1}^{k_1\gamma_1} f_1\right)(x_1)\right|\left|\left(D_{a_2}^{k_2\gamma_2} f_2\right)(x_2)\right|}{\left[\left(\frac{(x_1-a_1)^{p((n_1+1)\alpha_1 - k_1\gamma_1 - 1)+1}}{p(p((n_1+1)\alpha_1 - k_1\gamma_1 - 1)+1)}\right) + \left(\frac{(x_2-a_2)^{q((n_2+1)\alpha_2 - k_2\gamma_2 - 1)+1}}{q(q((n_2+1)\alpha_2 - k_2\gamma_2 - 1)+1)}\right)\right]}$$

$$\leq \frac{1}{\Gamma\left((n_1+1)\alpha_1 - k_1\gamma_1\right)\Gamma\left((n_2+1)\alpha_2 - k_2\gamma_2\right)} \quad (164)$$

$$\cdot \left(\int_{a_1}^{x_1}\left|D_{a_1}^{(n_1+1)\alpha_1} f_1(t_1)\right|^q dt_1\right)^{\frac{1}{q}} \left(\int_{a_2}^{x_2}\left|D_{a_2}^{(n_2+1)\alpha_2} f_2(t_2)\right|^p dt_2\right)^{\frac{1}{p}}.$$

The denominator in (164) can be zero only when $x_1 = a_1$ and $x_2 = a_2$. Therefore we obtain

$$\int_{a_1}^{b_1}\int_{a_2}^{b_2} \frac{\left|\left(D_{a_1}^{k_1\gamma_1} f_1\right)(x_1)\right|\left|\left(D_{a_2}^{k_2\gamma_2} f_2\right)(x_2)\right| dx_1 dx_2}{\left[\left(\frac{(x_1-a_1)^{p((n_1+1)\alpha_1 - k_1\gamma_1 - 1)+1}}{p(p((n_1+1)\alpha_1 - k_1\gamma_1 - 1)+1)}\right) + \left(\frac{(x_2-a_2)^{q((n_2+1)\alpha_2 - k_2\gamma_2 - 1)+1}}{q(q((n_2+1)\alpha_2 - k_2\gamma_2 - 1)+1)}\right)\right]}$$

$$\leq \frac{1}{\Gamma\left((n_1+1)\alpha_1 - k_1\gamma_1\right)\Gamma\left((n_2+1)\alpha_2 - k_2\gamma_2\right)} \quad (165)$$

$$\cdot \left(\int_{a_1}^{b_1}\left(\int_{a_1}^{x_1}\left|D_{a_1}^{(n_1+1)\alpha_1} f_1(t_1)\right|^q dt_1\right)^{\frac{1}{q}} dx_1\right)$$

$$\cdot \left(\int_{a_2}^{b_2}\left(\int_{a_2}^{x_2}\left|D_{a_2}^{(n_2+1)\alpha_2} f_2(t_2)\right|^p dt_2\right)^{\frac{1}{p}} dx_2\right)$$

$$\leq \frac{1}{\Gamma\left((n_1+1)\alpha_1 - k_1\gamma_1\right)\Gamma\left((n_2+1)\alpha_2 - k_2\gamma_2\right)} \quad (166)$$

$$\cdot \left(\int_{a_1}^{b_1}\left(\int_{a_1}^{b_1}\left|D_{a_1}^{(n_1+1)\alpha_1} f_1(t_1)\right|^q dt_1\right)^{\frac{1}{q}} dx_1\right)$$

$$\cdot \left(\int_{a_2}^{b_2} \left(\int_{a_2}^{b_2} \left| D_{a_2}^{(n_2+1)\alpha_2} f_2(t_2) \right|^p dt_2 \right) dx_2 \right)^{\frac{1}{p}}$$

$$= \frac{(b_1 - a_1)(b_2 - a_2)}{\Gamma((n_1+1)\alpha_1 - k_1\gamma_1)\Gamma((n_2+1)\alpha_2 - k_2\gamma_2)} \tag{167}$$

$$\cdot \left(\int_{a_1}^{b_1} \left| D_{a_1}^{(n_1+1)\alpha_1} f_1(t_1) \right|^q dt_1 \right)^{\frac{1}{q}} \left(\int_{a_2}^{b_2} \left| D_{a_2}^{(n_2+1)\alpha_2} f_2(t_2) \right|^p dt_2 \right)^{\frac{1}{p}}.$$

The theorem is proved.

Next we present a right fractional Hilbert-Pachpatte type inequality:

Theorem 25 *Here $i = 1, 2$. Let $0 < \alpha_i < 1$, $f_i : [a_i, b_i] \to \mathbb{R}$ such that $f_i' \in L_\infty$ ($[a_i, b_i]$). Assume that $D_{b_i-}^{\overline{k_i}\alpha_i} f_i \in C([a_i, b_i])$, $\overline{k_i} = 0, 1, \ldots, n_i + 1$; $n_i \in \mathbb{N}$. Suppose that $\left(D_{b_i-}^{j_i\alpha_i} f_i \right)(b_i) = 0$, for $j_i = 0, 2, 3, \ldots, n_i$. Let $\gamma_i > 0$ with $\lceil \gamma_i \rceil = m_i$, and $p, q > 1 : \frac{1}{p} + \frac{1}{q} = 1$. We further assume that $(k_i \in \mathbb{N})$*

$$1 > \alpha_1 > \max \left(\frac{m_1 + (k_1 - 1)\gamma_1}{n_1 + 1}, \frac{k_1\gamma_1 q + 1}{(n_1 + 1)q} \right). \tag{168}$$

and

$$1 > \alpha_2 > \max \left(\frac{m_2 + (k_2 - 1)\gamma_2}{n_2 + 1}, \frac{k_2\gamma_2 p + 1}{(n_2 + 1)p} \right). \tag{169}$$

Then

$$\int_{a_1}^{b_1} \int_{a_2}^{b_2} \frac{\left| \left(D_{b_1-}^{k_1\gamma_1} f_1 \right)(x_1) \right| \left| \left(D_{b_2-}^{k_2\gamma_2} f_2 \right)(x_2) \right| dx_1 dx_2}{\left[\frac{(b_1-x_1)^{p((n_1+1)\alpha_1 - k_1\gamma_1 - 1) + 1}}{p(p((n_1+1)\alpha_1 - k_1\gamma_1 - 1) + 1)} + \frac{(b_2-x_2)^{q((n_2+1)\alpha_2 - k_2\gamma_2 - 1) + 1}}{q(q((n_2+1)\alpha_2 - k_2\gamma_2 - 1) + 1)} \right]}$$

$$\leq \frac{(b_1 - a_1)(b_2 - a_2)}{\Gamma((n_1+1)\alpha_1 - k_1\gamma_1)\Gamma((n_2+1)\alpha_2 - k_2\gamma_2)} \tag{170}$$

$$\cdot \left(\int_{a_1}^{b_1} \left| D_{b_1-}^{(n_1+1)\alpha_1} f_1(t_1) \right|^q dt_1 \right)^{\frac{1}{q}} \left(\int_{a_2}^{b_2} \left| D_{b_2-}^{(n_2+1)\alpha_2} f_2(t_2) \right|^p dt_2 \right)^{\frac{1}{p}}.$$

Proof Similar to the proof of Theorem 24, using (115). As such it is omitted.

We give the following fractional Ostrowski type inequalities:

Theorem 26 *Let* $0 < \alpha < 1$, $f : [a, b] \to \mathbb{R}$ *such that* $f' \in L_\infty ([a, b])$, $y \in [a, b]$ *is fixed. We assume that* $D_y^{k\alpha} f$, $D_{y-}^{k\alpha} f \in C ([a, b])$, $k = 0, 1, 2, \ldots, n + 1$; $n \in \mathbb{N}$. *Let* $p, q > 1 : \frac{1}{p} + \frac{1}{q} = 1$. *Set*

$$
\Delta (f, y) := f (y) - \frac{1}{b - a} \int_a^b f (x) \, dx + \sum_{i=2}^n \frac{1}{(b - a) \, \Gamma (i\alpha + 2)}
$$
$$
\cdot \left[\left(D_{y-}^{i\alpha} f \right) (y) (y - a)^{i\alpha+1} + \left(D_y^{i\alpha} f \right) (y) (b - y)^{i\alpha+1} \right]. \quad (171)
$$

Then

(i)

$$
|\Delta (f, y)|
$$
$$
\leq \frac{1}{\Gamma ((n + 1) \alpha + 2) (b - a)} \quad (172)
$$
$$
\cdot \left[(b - y)^{(n+1)\alpha+1} \left\| D_y^{(n+1)\alpha} f \right\|_\infty + (y - a)^{(n+1)\alpha+1} \left\| D_{y-}^{(n+1)\alpha} f \right\|_\infty \right],
$$

(ii) *if* $\frac{1}{n+1} \leq \alpha < 1$, *we derive:*

$$
|\Delta (f, y)|
$$
$$
\leq \frac{1}{\Gamma ((n + 1) \alpha) (b - a)} \quad (173)
$$
$$
\cdot \left[(b - y)^{(n+1)\alpha} \left\| D_y^{(n+1)\alpha} f \right\|_{1,[y,b]} + (y - a)^{(n+1)\alpha} \left\| D_{y-}^{(n+1)\alpha} f \right\|_{1,[a,y]} \right],
$$

(iii) *if* $\frac{1}{(n+1)q} < \alpha < 1$, *we obtain:*

$$
|\Delta (f, y)| \quad (174)
$$
$$
\leq \frac{1}{\Gamma ((n + 1) \alpha) (b - a) \left((n + 1) \alpha + \frac{1}{p} \right) (p ((n + 1) \alpha - 1) + 1)^{\frac{1}{p}}}
$$
$$
\cdot \left[(b - y)^{(n+1)\alpha+\frac{1}{p}} \left\| D_y^{(n+1)\alpha} f \right\|_{q,[y,b]} + (y - a)^{(n+1)\alpha+\frac{1}{p}} \left\| D_{y-}^{(n+1)\alpha} f \right\|_{q,[a,y]} \right].
$$

Proof *(i)* By (20) and (21), we notice that

$$
\Delta (f, y) = R_1 (y)
$$

and

$$|R_1(y)|$$

$$\leq \frac{1}{\Gamma((n+1)\alpha)(b-a)} \left[\left(\int_a^y \left(\int_x^y (t-x)^{(n+1)\alpha-1} dt \right) dx \right) \left\| D_{y-}^{(n+1)\alpha} f \right\|_\infty \right.$$

$$+ \left. \left(\int_y^b \left(\int_y^x (x-t)^{(n+1)\alpha-1} dt \right) dx \right) \left\| D_y^{(n+1)\alpha} f \right\|_\infty \right]$$

$$= \frac{1}{\Gamma((n+1)\alpha)(b-a)} \left[\frac{(y-a)^{(n+1)\alpha+1}}{((n+1)\alpha)((n+1)\alpha+1)} \left\| D_{y-}^{(n+1)\alpha} f \right\|_\infty \right.$$

$$+ \left. \frac{(b-y)^{(n+1)\alpha+1}}{((n+1)\alpha)((n+1)\alpha+1)} \left\| D_y^{(n+1)\alpha} f \right\|_\infty \right]$$

$$= \frac{1}{\Gamma((n+1)\alpha+2)(b-a)} \tag{175}$$

$$\cdot \left[(b-y)^{(n+1)\alpha+1} \left\| D_y^{(n+1)\alpha} f \right\|_\infty + (y-a)^{(n+1)\alpha+1} \left\| D_{y-}^{(n+1)\alpha} f \right\|_\infty \right],$$

proving (172).

(ii) We use here that $\frac{1}{n+1} \leq \alpha < 1$. We have

$$|R_1(y)| \tag{176}$$

$$\leq \frac{1}{\Gamma((n+1)\alpha)(b-a)}$$

$$\cdot \left[\int_a^y \left(\int_x^y (t-x)^{(n+1)\alpha-1} \left| \left(D_{y-}^{(n+1)\alpha} f \right)(t) \right| dt \right) dx \right.$$

$$+ \left. \int_y^b \left(\int_y^x (x-t)^{(n+1)\alpha-1} \left| \left(D_y^{(n+1)\alpha} f \right)(t) \right| dt \right) dx \right]$$

$$< \frac{1}{\Gamma((n+1)\alpha)(b-a)}$$

$$\cdot \left[\left(\int_a^y \left(\int_x^y \left| \left(D_{y-}^{(n+1)\alpha} f \right)(t) \right| dt \right) dx \right) (y-a)^{(n+1)\alpha-1} \right.$$

$$+ \left. \left(\int_y^b \left(\int_y^x \left| \left(D_y^{(n+1)\alpha} f \right)(t) \right| dt \right) dx \right) (b-y)^{(n+1)\alpha-1} \right]$$

$$\leq \frac{1}{\Gamma((n+1)\alpha)(b-a)} \tag{177}$$

$$\cdot \left[\left\| D_{y-}^{(n+1)\alpha} f \right\|_{1,[a,y]} (y-a)^{(n+1)\alpha} + \left\| D_y^{(n+1)\alpha} f \right\|_{1,[y,b]} (b-y)^{(n+1)\alpha} \right],$$

proving (173).

(iii) We use here that $\frac{1}{(n+1)q} < \alpha < 1$. We observe that

$$|R_1(y)|$$

$$\leq \frac{1}{\Gamma((n+1)\alpha)(b-a)} \tag{178}$$

$$\cdot \left[\int_a^y \left(\int_x^y (t-x)^{p((n+1)\alpha-1)} dt \right)^{\frac{1}{p}} \left(\int_x^y \left| \left(D_{y-}^{(n+1)\alpha} f \right)(t) \right|^q dt \right)^{\frac{1}{q}} dx \right.$$

$$\left. + \int_y^b \left(\int_y^x (x-t)^{p((n+1)\alpha-1)} dt \right)^{\frac{1}{p}} \left(\int_y^x \left| (D_y^{(n+1)\alpha} f)(t) \right|^q dt \right)^{\frac{1}{q}} dx \right]$$

$$\leq \frac{1}{\Gamma((n+1)\alpha)(b-a)}$$

$$\cdot \left[\left(\int_a^y \frac{(y-x)^{(n+1)\alpha-1+\frac{1}{p}}}{(p((n+1)\alpha-1)+1)^{\frac{1}{p}}} dx \right) \left\| D_{y-}^{(n+1)\alpha} f \right\|_{q,[a,y]} \right.$$

$$\left. + \left(\int_y^b \frac{(x-y)^{(n+1)\alpha-1+\frac{1}{p}}}{(p((n+1)\alpha-1)+1)^{\frac{1}{p}}} dx \right) \left\| D_y^{(n+1)\alpha} f \right\|_{q,[y,b]} \right] \tag{179}$$

$$= \frac{1}{\Gamma((n+1)\alpha)(b-a)(p((n+1)\alpha-1)+1)^{\frac{1}{p}}} \tag{180}$$

$$\cdot \left[\frac{(y-a)^{(n+1)\alpha+\frac{1}{p}}}{\left((n+1)\alpha+\frac{1}{p}\right)} \left\| D_{y-}^{(n+1)\alpha} f \right\|_{q,[a,y]} + \frac{(b-y)^{(n+1)\alpha+\frac{1}{p}}}{\left((n+1)\alpha+\frac{1}{p}\right)} \left\| D_y^{(n+1)\alpha} f \right\|_{q,[y,b]} \right]$$

$$= \frac{1}{\Gamma((n+1)\alpha)(b-a)\left((n+1)\alpha+\frac{1}{p}\right)(p((n+1)\alpha-1)+1)^{\frac{1}{p}}} \tag{181}$$

$$\cdot \left[(b-y)^{(n+1)\alpha+\frac{1}{p}} \left\| D_y^{(n+1)\alpha} f \right\|_{q,[y,b]} + (y-a)^{(n+1)\alpha+\frac{1}{p}} \left\| D_{y-}^{(n+1)\alpha} f \right\|_{q,[a,y]} \right]$$

proving the claim (174).

The theorem is proved.

We present the following very general left fractional Poincaré type inequality:

Theorem 27 *Let $0 < \alpha < 1$, $f : [a, b] \to \mathbb{R}$ such that $f' \in L_\infty([a, b])$. Assume that $D_a^{\bar{k}\alpha} f \in C([a, b])$, $\bar{k} = 0, 1, \ldots, n + 1$; $n \in \mathbb{N}$. Suppose that $\left(D_a^{i\alpha} f\right)(a) = 0$, for $i = 0, 2, 3, \ldots, n$. Let $\gamma > 0$ with $\lceil \gamma \rceil = m$, and $p, q > 1 : \frac{1}{p} + \frac{1}{q} = 1$. We further assume that $(k \in \mathbb{N})$*

$$1 > \alpha > \max\left(\frac{m + (k - 1)\gamma}{n + 1}, \frac{k\gamma q + 1}{(n + 1)q}\right). \tag{182}$$

Then

$$\left\| D_a^{k\gamma} f \right\|_q$$
$$\leq \frac{1}{\Gamma((n + 1)\alpha - k\gamma)} \tag{183}$$
$$\cdot \frac{(b - a)^{(n+1)\alpha - k\gamma}}{(p((n + 1)\alpha - k\gamma - 1) + 1)^{\frac{1}{p}} q^{\frac{1}{q}} ((n + 1)\alpha - k\gamma)^{\frac{1}{q}}} \left\| D_a^{(n+1)\alpha} f \right\|_q.$$

Proof We use (80). We observe that

$$\left|\left(D_a^{k\gamma} f\right)(x)\right|$$
$$\leq \frac{1}{\Gamma((n + 1)\alpha - k\gamma)} \tag{184}$$
$$\cdot \left(\int_a^x (x - t)^{p((n+1)\alpha - k\gamma - 1)} dt\right)^{\frac{1}{p}} \left(\int_a^x \left|\left(D_a^{(n+1)\alpha} f\right)(t)\right|^q dt\right)^{\frac{1}{q}}$$

$$\leq \frac{1}{\Gamma((n + 1)\alpha - k\gamma)} \frac{(x - a)^{\frac{p((n+1)\alpha - k\gamma - 1) + 1}{p}}}{(p((n + 1)\alpha - k\gamma - 1) + 1)^{\frac{1}{p}}} \tag{185}$$
$$\cdot \left(\int_a^b \left|\left(D_a^{(n+1)\alpha} f\right)(t)\right|^q dt\right)^{\frac{1}{q}}.$$

That is, we have

$$\left|\left(D_a^{k\gamma} f\right)(x)\right|^q$$
$$\leq \frac{1}{(\Gamma((n + 1)\alpha - k\gamma))^q} \tag{186}$$
$$\cdot \frac{(x - a)^{(p((n+1)\alpha - k\gamma - 1) + 1)\frac{q}{p}}}{(p((n + 1)\alpha - k\gamma - 1) + 1)^{\frac{q}{p}}} \left(\int_a^b \left|\left(D_a^{(n+1)\alpha} f\right)(t)\right|^q dt\right).$$

Therefore it holds

$$\int_a^b \left| \left(D_a^{k\gamma} f \right) (x) \right|^q dx$$

$$\leq \frac{1}{(\Gamma ((n+1)\alpha - k\gamma))^q}$$

$$\cdot \frac{(b-a)^{(p((n+1)\alpha - k\gamma - 1) + 1) \frac{q}{p} + 1}}{(p((n+1)\alpha - k\gamma - 1) + 1)^{\frac{q}{p}} \left((p((n+1)\alpha - k\gamma - 1) + 1) \frac{q}{p} + 1 \right)}$$

$$\cdot \left(\int_a^b \left| \left(D_a^{(n+1)\alpha} f \right) (t) \right|^q dt \right)$$

$$= \frac{1}{(\Gamma ((n+1)\alpha - k\gamma))^q} \hspace{5cm} (187)$$

$$\cdot \frac{(b-a)^{q((n+1)\alpha - k\gamma)}}{(p((n+1)\alpha - k\gamma - 1) + 1)^{\frac{q}{p}} q ((n+1)\alpha - k\gamma)}$$

$$\cdot \left(\int_a^b \left| \left(D_a^{(n+1)\alpha} f \right) (t) \right|^q dt \right),$$

proving the claim.

Next we present a very general right fractional Poincaré type inequality:

Theorem 28 *Let $0 < \alpha < 1$, $f : [a, b] \to \mathbb{R}$ such that $f' \in L_\infty ([a, b])$. Assume that $D_{b-}^{\bar{k}\alpha} f \in C ([a, b])$, $\bar{k} = 0, 1, \ldots, n + 1$; $n \in \mathbb{N}$. Suppose that $\left(D_{b-}^{i\alpha} f \right) (b) = 0$, for $i = 0, 2, 3, \ldots, n$. Let $\gamma > 0$ with $\lceil \gamma \rceil = m$, and $p, q > 1 : \frac{1}{p} + \frac{1}{q} = 1$. We further assume that $(k \in \mathbb{N})$*

$$1 > \alpha > \max \left(\frac{m + (k-1)\gamma}{n+1}, \frac{k\gamma q + 1}{(n+1)q} \right).$$

Then

$$\left\| D_{b-}^{k\gamma} f \right\|_q$$

$$\leq \frac{1}{\Gamma ((n+1)\alpha - k\gamma)} \hspace{5cm} (188)$$

$$\cdot \frac{(b-a)^{(n+1)\alpha - k\gamma}}{(p((n+1)\alpha - k\gamma - 1) + 1)^{\frac{1}{p}} q^{\frac{1}{q}} ((n+1)\alpha - k\gamma)^{\frac{1}{q}}} \left\| D_{b-}^{(n+1)\alpha} f \right\|_q.$$

Proof Similar to Theorem 27, using (115). It is omitted.

We continue with a Sobolev type left fractional inequality:

Theorem 29 *Let all here as in Theorem 27. Assume that $r \geq 1$. Then*

$$
\left\| D_a^{k\gamma} f \right\|_r
$$

$$
\leq \frac{1}{\Gamma\left((n+1)\alpha - k\gamma\right)} \tag{189}
$$

$$
\cdot \frac{(b-a)^{(n+1)\alpha - k\gamma - \frac{1}{q} + \frac{1}{r}} \left\| D_a^{(n+1)\alpha} f \right\|_q}{\left(p\left((n+1)\alpha - k\gamma - 1\right) + 1\right)^{\frac{1}{p}} \left[r \left((n+1)\alpha - k\gamma - \frac{1}{q}\right) + 1 \right]^{\frac{1}{r}}}.
$$

Proof As in the proof of Theorem 27, we obtain

$$
\left| \left(D_a^{k\gamma} f \right)(x) \right| \tag{190}
$$

$$
\leq \frac{1}{\Gamma\left((n+1)\alpha - k\gamma\right)}
$$

$$
\cdot \frac{(x-a)^{\frac{p((n+1)\alpha - k\gamma - 1)+1}{p}}}{\left(p\left((n+1)\alpha - k\gamma - 1\right) + 1\right)^{\frac{1}{p}}} \left(\int_a^b \left| \left(D_a^{(n+1)\alpha} f \right)(t) \right|^q dt \right)^{\frac{1}{q}},
$$

$\forall\, x \in [a,b]$. Hence by $r \geq 1$, we derive

$$
\left| \left(D_a^{k\gamma} f \right)(x) \right|^r \tag{191}
$$

$$
\leq \frac{1}{\left(\Gamma\left((n+1)\alpha - k\gamma\right)\right)^r}
$$

$$
\cdot \frac{(x-a)^{\frac{r(p((n+1)\alpha - k\gamma - 1)+1)}{p}}}{\left(p\left((n+1)\alpha - k\gamma - 1\right) + 1\right)^{\frac{r}{p}}} \left\| D_a^{(n+1)\alpha} f \right\|_q^r,
$$

$\forall\, x \in [a,b]$. Consequently, it holds

$$
\int_a^b \left| \left(D_a^{k\gamma} f \right)(x) \right|^r dx \tag{192}
$$

$$
\leq \frac{1}{\left(\Gamma\left((n+1)\alpha - k\gamma\right)\right)^r}
$$

$$
\cdot \frac{(b-a)^{r((n+1)\alpha - k\gamma - 1) + \frac{r}{p} + 1} \left\| D_a^{(n+1)\alpha} f \right\|_q^r}{\left(p\left((n+1)\alpha - k\gamma - 1\right) + 1\right)^{\frac{r}{p}} \left(r\left((n+1)\alpha - k\gamma - 1\right) + \frac{r}{p} + 1 \right)}.
$$

Finally, we get

$$\left\| D_a^{k\gamma} f \right\|_r \tag{193}$$

$$\leq \frac{1}{\Gamma\left((n+1)\alpha - k\gamma\right)}$$

$$\cdot \frac{(b-a)^{(n+1)\alpha - k\gamma - 1 + \frac{1}{p} + \frac{1}{r}} \left\| D_a^{(n+1)\alpha} f \right\|_q}{(p\left((n+1)\alpha - k\gamma - 1\right) + 1)^{\frac{1}{p}} \left[r\left((n+1)\alpha - k\gamma - 1\right) + \frac{r}{p} + 1 \right]^{\frac{1}{r}}}$$

$$= \frac{1}{\Gamma\left((n+1)\alpha - k\gamma\right)} \tag{194}$$

$$\cdot \frac{(b-a)^{(n+1)\alpha - k\gamma - \frac{1}{q} + \frac{1}{r}} \left\| D_a^{(n+1)\alpha} f \right\|_q}{(p\left((n+1)\alpha - k\gamma - 1\right) + 1)^{\frac{1}{p}} \left[r\left((n+1)\alpha - k\gamma - \frac{1}{q}\right) + 1 \right]^{\frac{1}{r}}},$$

proving the claim.

We continue with a Sobolev type right fractional inequality:

Theorem 30 *Let all here as in Theorem 28. Assume that* $r \geq 1$. *Then*

$$\left\| D_{b-}^{k\gamma} f \right\|_r \tag{195}$$

$$\leq \frac{1}{\Gamma\left((n+1)\alpha - k\gamma\right)}$$

$$\cdot \frac{(b-a)^{(n+1)\alpha - k\gamma - \frac{1}{q} + \frac{1}{r}} \left\| D_{b-}^{(n+1)\alpha} f \right\|_q}{(p\left((n+1)\alpha - k\gamma - 1\right) + 1)^{\frac{1}{p}} \left[r\left((n+1)\alpha - k\gamma - \frac{1}{q}\right) + 1 \right]^{\frac{1}{r}}}.$$

Proof As in the proof of Theorem 29, using (115). It is omitted.

We give the following fractional Grüss type inequality:

Theorem 31 *Let* $0 < \alpha < 1$, $f, h : [a, b] \to \mathbb{R}$ *such that* $f', h' \in L_\infty\left([a, b]\right)$. *We assume that* $D_y^{k\alpha} f$, $D_{y-}^{k\alpha} f$, $D_y^{k\alpha} h$, $D_{y-}^{k\alpha} h \in C\left([a, b]\right)$, $\forall \, y \in [a, b]$; $k = 0, 1, 2, \ldots,$ $n+1$; $n \in \mathbb{N}$. *Then*

1.

$$\Delta_{(n+1)\alpha}\left(f, h\right)$$

$$:= \frac{1}{b-a} \int_a^b f(x) h(x) \, dx - \frac{1}{(b-a)^2} \left(\int_a^b f(x) \, dx \right) \left(\int_a^b h(x) \, dx \right)$$

$$+ \frac{1}{2(b-a)^2} \sum_{l=2}^n \frac{1}{\Gamma(i\alpha+2)} \left[\int_a^b \left\{ (y-a)^{i\alpha+1} \left[h(y) \left(D_{y-}^{i\alpha} f \right)(y) + f(y) \left(D_{y-}^{i\alpha} h \right)(y) \right] \right. \right.$$

$$+ (b-y)^{i\alpha+1} \left[h(y) \left(D_y^{i\alpha} f \right)(y) + f(y) \left(D_y^{i\alpha} h \right)(y) \right] \right\} dy \right] \tag{196}$$

$$= \frac{1}{2(b-a)} \int_a^b [f(y) R_1(h,y) + h(y) R_1(f,y)] \, dy =: K_{(n+1)\alpha}(f,h).$$

Above $R_1(f,y)$ is the same as $R_1(y)$ in (20), (21); similarly for $R_1(h,y)$ now it is $R_1(y)$ for h.

2. it holds

$$\left| \Delta_{(n+1)\alpha}(f,h) \right| \leq \frac{(b-a)^{(n+1)\alpha}}{2\Gamma((n+1)\alpha+2)} \tag{197}$$

$$\cdot \left[\|f\|_\infty \left(\sup_{y \in [a,b]} \left(\left\| D_y^{(n+1)\alpha} h \right\|_\infty + \left\| D_{y-}^{(n+1)\alpha} h \right\|_\infty \right) \right) \right.$$

$$\left. + \|h\|_\infty \left(\sup_{y \in [a,b]} \left(\left\| D_y^{(n+1)\alpha} f \right\|_\infty + \left\| D_{y-}^{(n+1)\alpha} f \right\|_\infty \right) \right) \right].$$

3. when $\frac{1}{n+1} \leq \alpha < 1$, we obtain

$$\left| \Delta_{(n+1)\alpha}(f,h) \right|$$
$$\leq \frac{(b-a)^{(n+1)\alpha-1}}{2\Gamma((n+1)\alpha)} \tag{198}$$

$$\cdot \left[\|f\|_\infty \left(\sup_{y \in [a,b]} \left(\left\| D_y^{(n+1)\alpha} h \right\|_{1,[a,b]} + \left\| D_{y-}^{(n+1)\alpha} h \right\|_{1,[a,b]} \right) \right) \right.$$

$$\left. + \|h\|_\infty \left(\sup_{y \in [a,b]} \left(\left\| D_y^{(n+1)\alpha} f \right\|_{1,[a,b]} + \left\| D_{y-}^{(n+1)\alpha} f \right\|_{1,[a,b]} \right) \right) \right].$$

4. when $p, q > 1 : \frac{1}{p} + \frac{1}{q} = 1$, and $\frac{1}{(n+1)q} < \alpha < 1$, we obtain

$$\left| \Delta_{(n+1)\alpha}(f,h) \right| \tag{199}$$

$$\leq \frac{(b-a)^{(n+1)\alpha-\frac{1}{q}}}{2\Gamma((n+1)\alpha)\left((n+1)\alpha+\frac{1}{p}\right)(p((n+1)\alpha-1)+1)^{\frac{1}{p}}}$$

$$\cdot \left[\|f\|_\infty \left(\sup_{y \in [a,b]} \left(\left\| D_y^{(n+1)\alpha} h \right\|_{q,[a,b]} + \left\| D_{y-}^{(n+1)\alpha} h \right\|_{q,[a,b]} \right) \right) \right.$$

$$\left. + \|h\|_\infty \left(\sup_{y \in [a,b]} \left(\left\| D_y^{(n+1)\alpha} f \right\|_{q,[a,b]} + \left\| D_{y-}^{(n+1)\alpha} f \right\|_{q,[a,b]} \right) \right) \right].$$

Proof Notice $\Delta(f,y) = R_1(f,y)$, $\Delta(h,y) = R_1(h,y)$, see (171).

1. We use (20), and we denote $R_1(y)$ as $R_1(f, y)$, see (21), to associate it to f, similarly for $R_1(h, y)$. So we have

$$h(y) f(y)$$

$$= \frac{h(y)}{b-a} \int_a^b f(x)\,dx \tag{200}$$

$$- \sum_{i=2}^n \frac{h(y)}{(b-a)\Gamma(i\alpha+2)} \left[\left(D_{y-}^{i\alpha} f\right)(y)(y-a)^{i\alpha+1} + \left(D_y^{i\alpha} f\right)(y)(b-y)^{i\alpha+1} \right]$$

$$+ h(y) R_1(f, y),$$

$\forall\, y \in [a, b]$. We also can write:

$$f(y) h(y)$$

$$= \frac{f(y)}{b-a} \int_a^b h(x)\,dx \tag{201}$$

$$- \sum_{i=2}^n \frac{f(y)}{(b-a)\Gamma(i\alpha+2)} \left[\left(D_{y-}^{i\alpha} h\right)(y)(y-a)^{i\alpha+1} + \left(D_y^{i\alpha} h\right)(y)(b-y)^{i\alpha+1} \right]$$

$$+ f(y) R_1(h, y),$$

$\forall\, y \in [a, b]$, where $R_1(h, y)$ coresponds to (21) written for h. Then integrating (200) we find

$$\int_a^b f(y) h(y)$$

$$= \frac{\left(\int_a^b f(x)\,dx \right)\left(\int_a^b h(x)\,dx \right)}{b-a} \tag{202}$$

$$- \sum_{i=2}^n \frac{1}{(b-a)\,\Gamma(i\alpha+2)} \int_a^b h(y)$$

$$\cdot \left[\left(D_{y-}^{i\alpha} f\right)(y)(y-a)^{i\alpha+1} + \left(D_y^{i\alpha} f\right)(y)(b-y)^{i\alpha+1} \right] dy$$

$$+ \int_a^b h(y) R_1(f, y)\,dy.$$

And integrating (201) we derive

$$
\int_a^b f\left(y\right) h\left(y\right) dy
$$

$$
= \frac{\left(\int_a^b f\left(x\right) dx\right)\left(\int_a^b h\left(x\right) dx\right)}{b-a} \tag{203}
$$

$$
- \sum_{i=2}^n \frac{1}{(b-a)\,\Gamma\left(i\alpha+2\right)} \int_a^b f\left(y\right)
$$

$$
\cdot \left[\left(D_{y-}^{i\alpha}h\right)\left(y\right)\left(y-a\right)^{i\alpha+1} + \left(D_{y}^{i\alpha}h\right)\left(y\right)\left(b-y\right)^{i\alpha+1}\right] dy
$$

$$
+ \int_a^b f\left(y\right) R_1\left(h, y\right) dy.
$$

Adding (202) and (203), we get:

$$
2\int_a^b f\left(x\right) h\left(x\right) dx
$$

$$
= \frac{2\left(\int_a^b f\left(x\right) dx\right)\left(\int_a^b h\left(x\right) dx\right)}{b-a} \tag{204}
$$

$$
- \sum_{i=2}^n \frac{1}{(b-a)\,\Gamma\left(i\alpha+2\right)}
$$

$$
\cdot \left[\int_a^b \left\{h\left(y\right)\left[\left(D_{y-}^{i\alpha}f\right)\left(y\right)\left(y-a\right)^{i\alpha+1} + \left(D_{y}^{i\alpha}f\right)\left(y\right)\left(b-y\right)^{i\alpha+1}\right]\right.\right.
$$

$$
\left.\left. + f\left(y\right)\left[\left(D_{y-}^{i\alpha}h\right)\left(y\right)\left(y-a\right)^{i\alpha+1} + \left(D_{y}^{i\alpha}h\right)\left(y\right)\left(b-y\right)^{i\alpha+1}\right]\right\} dy\right]
$$

$$
+ \int_a^b \left[f\left(y\right) R_1\left(h, y\right) + h\left(y\right) R_1\left(f, y\right)\right] dy.
$$

Next we divide the last (204) by $2\left(b-a\right)$, and rewrite it properly to obtain:

$$\frac{1}{b-a}\int_a^b f(x)\,h(x)\,dx$$

$$= \frac{\left(\int_a^b f(x)\,dx\right)\left(\int_a^b h(x)\,dx\right)}{(b-a)^2} \tag{205}$$

$$-\frac{1}{2(b-a)^2}\sum_{i=2}^n \frac{1}{\Gamma(i\alpha+2)}\left[\int_a^b \left\{(y-a)^{i\alpha+1}\left[h(y)\left(D_{y-}^{i\alpha}f\right)(y)+f(y)\left(D_{y-}^{i\alpha}h\right)(y)\right]\right.\right.$$

$$\left.\left.+(b-y)^{i\alpha+1}\left[h(y)\left(D_y^{i\alpha}f\right)(y)+f(y)\left(D_y^{i\alpha}h\right)(y)\right]\right\}dy\right]$$

$$+\frac{1}{2(b-a)}\int_a^b [f(y)\,R_1(h,y)+h(y)\,R_1(f,y)]\,dy,$$

proving (196).
2. It holds

$$\left|\Delta_{(n+1)\alpha}(f,h)\right|$$

$$= \left|K_{(n+1)\alpha}(f,h)\right|$$

$$= \frac{1}{2(b-a)}\left|\int_a^b [f(y)\,R_1(h,y)+h(y)\,R_1(f,y)]\,dy\right|$$

$$\le \frac{1}{2(b-a)}\int_a^b [|f(y)||R_1(h,y)|+|h(y)||R_1(f,y)|]\,dy \tag{206}$$

$$\overset{(172)}{\le} \frac{1}{2\Gamma((n+1)\alpha+2)(b-a)^2}$$

$$\cdot \int_a^b \left[|f(y)|\left((b-y)^{(n+1)\alpha+1}\left\|D_y^{(n+1)\alpha}h\right\|_\infty+(y-a)^{(n+1)\alpha+1}\left\|D_{y-}^{(n+1)\alpha}h\right\|_\infty\right)\right.$$

$$\left.+|h(y)|\left((b-y)^{(n+1)\alpha+1}\left\|D_y^{(n+1)\alpha}f\right\|_\infty+(y-a)^{(n+1)\alpha+1}\left\|D_{y-}^{(n+1)\alpha}f\right\|_\infty\right)\right]dy$$

$$\le \frac{(b-a)^{(n+1)\alpha}}{2\Gamma((n+1)\alpha+2)}\left[\|f\|_\infty \sup_{y\in[a,b]}\left(\left\|D_y^{(n+1)\alpha}h\right\|_\infty+\left\|D_{y-}^{(n+1)\alpha}h\right\|_\infty\right)\right.$$

$$\left.+\|h\|_\infty \sup_{y\in[a,b]}\left(\left\|D_y^{(n+1)\alpha}f\right\|_\infty+\left\|D_{y-}^{(n+1)\alpha}f\right\|_\infty\right)\right], \tag{207}$$

proving (197).

3. It holds, when $\frac{1}{n+1} \leq \alpha < 1$, that

$$\left| \Delta_{(n+1)\alpha} (f, h) \right| \tag{208}$$

$$\leq \frac{1}{2(b-a)} \int_a^b \left[|f(y)| |R_1(h, y)| + |h(y)| |R_1(f, y)| \right] dy$$

$$\overset{(29.173)}{\leq} \frac{1}{2(b-a)^2 \Gamma((n+1)\alpha)}$$

$$\cdot \int_a^b \left[|f(y)| \left[(b-y)^{(n+1)\alpha} \left\| D_y^{(n+1)\alpha} h \right\|_{1,[y,b]} + (y-a)^{(n+1)\alpha} \left\| D_{y-}^{(n+1)\alpha} h \right\|_{1,[a,y]} \right] \right.$$

$$\left. + |h(y)| \left[(b-y)^{(n+1)\alpha} \left\| D_y^{(n+1)\alpha} f \right\|_{1,[y,b]} + (y-a)^{(n+1)\alpha} \left\| D_{y-}^{(n+1)\alpha} f \right\|_{1,[a,y]} \right] \right] dy$$

$$\leq \frac{(b-a)^{(n+1)\alpha-1}}{2\Gamma((n+1)\alpha)} \left[\|f\|_\infty \sup_{y\in[a,b]} \left(\left\| D_y^{(n+1)\alpha} h \right\|_{1,[a,b]} + \left\| D_{y-}^{(n+1)\alpha} h \right\|_{1,[a,b]} \right) \right.$$

$$\left. + \|h\|_\infty \sup_{y\in[a,b]} \left(\left\| D_y^{(n+1)\alpha} f \right\|_{1,[a,b]} + \left\| D_{y-}^{(n+1)\alpha} f \right\|_{1,[a,b]} \right) \right], \tag{209}$$

proving (198).

4. It holds, when $\frac{1}{(n+1)q} < \alpha < 1$, that

$$\left| \Delta_{(n+1)\alpha} (f, h) \right|$$

$$\leq \frac{1}{2(b-a)} \int_a^b \left[|f(y)| |R_1(h, y)| + |h(y)| |R_1(f, y)| \right] dy \tag{210}$$

$$\overset{(174)}{\leq} \frac{1}{2\Gamma((n+1)\alpha)(b-a)^2 \left((n+1)\alpha+\frac{1}{p} \right)(p((n+1)\alpha-1)+1)^{\frac{1}{p}}} \tag{211}$$

$$\cdot \int_a^b \left[|f(y)| \left((b-y)^{(n+1)\alpha+\frac{1}{p}} \left\| D_y^{(n+1)\alpha} h \right\|_{q,[y,b]} + (y-a)^{(n+1)\alpha+\frac{1}{p}} \left\| D_{y-}^{(n+1)\alpha} h \right\|_{q,[a,y]} \right) \right.$$

$$\left. + |h(y)| \left((b-y)^{(n+1)\alpha+\frac{1}{p}} \left\| D_y^{(n+1)\alpha} f \right\|_{q,[y,b]} + (y-a)^{(n+1)\alpha+\frac{1}{p}} \left\| D_{y-}^{(n+1)\alpha} f \right\|_{q,[a,y]} \right) \right] dy$$

$$\leq \frac{(b-a)^{(n+1)\alpha+\frac{1}{p}-1}}{2\Gamma((n+1)\alpha) \left((n+1)\alpha+\frac{1}{p} \right)(p((n+1)\alpha-1)+1)^{\frac{1}{p}}} \tag{212}$$

$$\cdot \left[\|f\|_\infty \sup_{y\in[a,b]} \left(\left\| D_y^{(n+1)\alpha} h \right\|_{q,[a,b]} + \left\| D_{y-}^{(n+1)\alpha} h \right\|_{q,[a,b]} \right) \right.$$

$$\left. + \|h\|_\infty \sup_{y\in[a,b]} \left(\left\| D_y^{(n+1)\alpha} f \right\|_{q,[a,b]} + \left\| D_{y-}^{(n+1)\alpha} f \right\|_{q,[a,b]} \right) \right],$$

proving (199).

We use also that a norm is a continuous function. The theorem is proved.

References

1. Anastassiou, G.A.: Quantitative Approximations. Chapman & Hall/CRC, Boca Raton, New York (2001)
2. Anastassiou, G.: On right fractional calculus. Chaos, Solitons Fractals **42**, 365–376 (2009)
3. Anastassiou, G.: Fractional Differentiation Inequalities. Springer, New York (2009)
4. Anastassiou, G.: Fractional representation formulae and right fractional inequalities. Math. Comput. Model. **54**(11–12), 3098–3115 (2011)
5. Anastassiou, G.: Opial type inequalities for functions and their ordinary and balanced fractional derivatives. J. Comput. Anal. Appl. **14**(5), 862–879 (2012)
6. Anastassiou, G.: Advanced fractional Taylor's formulae. J. Comput. Anal. Appl. (2015)
7. Anastassiou, G., Argyros, I.: A convergence analysis for a certain family of extended iterative methods: Part II. Applications to fractional calculus. Ann. Univ. Sci. Bp. Sect. Comp. (2015)
8. Diethelm, K.: The Analysis of Fractional Differential Equations, 1st edn., vol. 2004. Lecture Notes in Mathematics, New York, Heidelberg (2010)
9. Odibat, Z.M., Shawagleh, N.J.: Generalized Taylor's formula. Appl. Math. Comput. **186**, 286–293 (2007)

Erratum to: Analyzing Multivariate Cross-Sectional Poisson Count Using a Quasi-Likelihood Approach: The Case of Trivariate Poisson

Naushad Mamode Khan, Yuvraj Sunecher and Vandna Jowaheer

Erratum to:
G.A. Anastassiou and O. Duman (eds.), *Intelligent*
Mathematics II:Applied Mathematics and Approximation
Theory, **Advances in Intelligent Systems and Computing**
441, DOI 10.1007/978-3-319-30322-2_27

The book was inadvertently published without including a change in one of the equations in page number 410, the variables h(y1, y2, y3) should be written as −h(y1, y2, y3). The erratum book and the chapter have been updated.

The updated original online version for this chapter can be found at
10.1007/978-3-319-30322-2_27

N. Mamode Khan · V. Jowaheer
University of Mauritius, Le Reduit, Moka, Mauritius
e-mail: n.mamodekhan@uom.ac.mu

V. Jowaheer
e-mail: vandnaj@uom.ac.mu

Y. Sunecher (✉)
University of Technology, La Tour Koenig, Port Louis, Mauritius
e-mail: yuvisun@yahoo.com

© Springer International Publishing Switzerland 2016
G.A. Anastassiou and O. Duman (eds.), *Intelligent Mathematics II:*
Applied Mathematics and Approximation Theory, Advances in Intelligent Systems and
Computing 441, DOI 10.1007/978-3-319-30322-2_30

Printed in the United States
By Bookmasters